Cardiac and Vascular Biology

The book series gives an overview on all aspects of state-of-the-art research on the cardiovascular system in health and disease. Basic research aspects of medically relevant topics are covered and the latest advances and methods covering diverse disciplines as epigenetics, genetics, mechanobiology, platelet research or stem cell biology are featured. The book series is intended for researchers, experts and graduates, both basic and clinically oriented, that look for a carefully selected collection of high quality review articles on their respective field of expertise.

More information about this series at http://www.springer.com/series/13128

Dierk Thomas • Carol Ann Remme
Editors

Channelopathies in Heart Disease

Springer

Editors
Dierk Thomas
Department of Cardiology and HCR
(Heidelberg Center for Heart Rhythm
Disorders)
University Hospital Heidelberg
Heidelberg, Germany

DZHK (German Centre for
Cardiovascular Research)
partner site Heidelberg/Mannheim
Heidelberg, Germany

Carol Ann Remme
Dept. of Clinical & Experimental Cardiology
Academic Medical Center, University of
Amsterdam
Amsterdam, The Netherlands

ISSN 2509-7830 ISSN 2509-7849 (electronic)
Cardiac and Vascular Biology
ISBN 978-3-030-08548-3 ISBN 978-3-319-77812-9 (eBook)
https://doi.org/10.1007/978-3-319-77812-9

Printed on acid-free paper

This Springer imprint is published by the registered company Springer Nature Switzerland AG
The registered company address is: Gewerbestrasse 11, 6330 Cham, Switzerland

Contents

Channelopathies in Heart Disease

Introduction and Book Overview

Carol Ann Remme and Dierk Thomas

Abstract

The book "Channelopathies in Heart Disease" provides a translational overview of current state-of-the art research on ion channel (dys)function, cardiac channelopathies, and inherited arrhythmia syndromes. The latest insight on the structure and function of cardiac ion channels and the pro-arrhythmic consequences of their dysfunction is presented. Clinical and genetic characteristics of various inherited channelopathies and arrhythmia syndromes are discussed, in addition to new technologies available to this translational research field.

1.1 Channelopathies in Heart Disease: Background and Book Overview

The majority of cardiac arrhythmias occur in the setting of common (acquired) cardiovascular pathologies associated with structural cardiac abnormalities and/or metabolic dysregulation. In a subset of patients, however, cardiac arrhythmias are the consequence of an inherited arrhythmia syndrome. Mutations in genes encoding

C. A. Remme (✉)
Dept. of Clinical & Experimental Cardiology, Academic Medical Center, University of Amsterdam, Amsterdam, The Netherlands
e-mail: c.a.remme@amc.uva.nl

D. Thomas
Department of Cardiology and HCR (Heidelberg Center for Heart Rhythm Disorders), University Hospital Heidelberg, Heidelberg, Germany

DZHK (German Centre for Cardiovascular Research), partner site Heidelberg/Mannheim, Heidelberg, Germany
e-mail: dierk.thomas@med.uni-heidelberg.de

D. Thomas, C. A. Remme (eds.), *Channelopathies in Heart Disease*, Cardiac and Vascular Biology, https://doi.org/10.1007/978-3-319-77812-9_1

ion channels, transporters, interacting proteins, or regulatory pathways may lead to potentially life-threatening arrhythmias in relatively young and otherwise healthy individuals. During the last two decades, significant progress has been made in the identification of genetic defects underlying inherited channelopathies, which has provided some benefit through elucidation of gene-specific arrhythmia triggers and treatment. However, for many arrhythmia syndromes, clinical management is still hindered by insufficient knowledge of the functional consequences of the mutation in question, the pro-arrhythmic mechanisms involved, and hence the most optimal treatment strategy. In this book, we present the latest insight on the structure and function of cardiac ion channels and the pro-arrhythmic consequences of their dysfunction. Clinical and genetic characteristics of various inherited channelopathies and arrhythmia syndromes are discussed, in addition to new technologies available to this translational research field.

1.2 (Dys)Function of Cardiac Ion Channels

Cardiac electrical activity is a summation of sequential action potentials throughout the heart. The action potential (AP) is the consequence of an orchestrated interplay between various ion channels. Each AP is initiated by a large, rapid influx of sodium (Na^+) through Na^+ channels (I_{Na}), resulting in fast depolarization of the cell membrane and the AP upstroke (Fig. 1.1). Following this phase 0 of the AP, there is a brief repolarizing phase (phase 1), resulting from efflux of potassium (K^+) caused by activation of the transient outward potassium current (I_{to1}). Next, inward flow of calcium (Ca^{2+}) through L-type calcium channels ($I_{Ca,L}$) leads to the plateau phase (phase 2). Finally, the membrane repolarizes to its original state due to activation of the rapid and slow delayed rectifier K^+ channels (conducting the I_{Kr} and I_{Ks} currents, respectively) in phase 3 of the

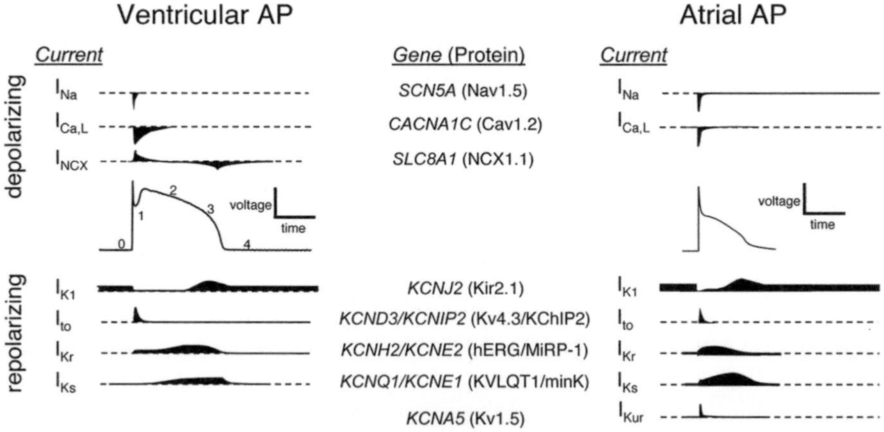

Fig. 1.1 Overview: Primary ionic currents underlying ventricular and atrial action potentials (AP) in the heart (reproduced with permission from Shah et al. 2005)

AP. Adult ventricular and atrial cardiomyocytes (but not nodal cells) also exhibit phase 4 in which the resting membrane potential remains constant due to the presence of the rectifying K^+ current I_{K1}. Due to their close interrelationship, alterations in a particular ion channel (affecting one phase of the AP) will also impact on the function of other ion currents and AP phases. In Chaps. 2–4 of this book, the molecular composition, structure, and function of Na^+, K^+, and Ca^{2+} channels are described. In addition, their regulation, their role in cardiac electrophysiology, and the conditions and consequences associated with their dysfunction are discussed in these chapters. Furthermore, in Chap. 5 the funny current (I_f) is presented, which is the primary contributor of the diastolic pacemaker phase responsible for the automaticity of the conduction system. The chapter will detail the biophysical and modulatory properties of I_f and its potential as a target for pharmacological modulation of heart rate. Ion channel function is tightly interrelated with intracellular ion homeostasis, with Ca^{2+} influx through $I_{Ca,L}$ eliciting the intracellular Ca^{2+} transient which underlies myocyte contraction. The subsequent decline of Ca^{2+} (required for diastolic relaxation) occurs through reuptake into the sarcoplasmic reticulum and extrusion of Ca^{2+} via the Na^+-Ca^{2+} exchanger. Chapter 6 reviews the interrelation between intracellular K^+, Na^+, and Ca^{2+} and discusses the consequences of alterations in their homeostasis for cardiac electrophysiology and arrhythmogenesis.

1.3 Cardiac Channelopathies: Clinical and Genetic Findings

Genetic defects in ion channels and their interacting proteins are associated with various types of clinical arrhythmia syndromes, with mutations in specific ion channels related to different clinical symptoms. Moreover, different types or locations of mutations may be associated with different phenotypes, severity of disease, and treatment strategy and efficacy. During the last two decades, significant progress has been made in genetic studies, which has facilitated the identification of family members and patients at risk. Research on ion channel mutations have also provided essential molecular and biophysical insight into (normal) ion channel function. Furthermore, in some cases elucidation of gene-specific arrhythmia triggers and pharmacology has guided patient treatment and management. However, reduced penetrance and variable disease severity and expressivity among mutation carriers remain problematic. In Chaps. 7–12, the most prevalent inherited cardiac channelopathies and arrhythmia syndromes are reviewed, i.e., long QT syndrome, Brugada syndrome, sinus node disease, Lev-Lenègre syndrome, progressive cardiac conduction disease, catecholaminergic polymorphic ventricular tachycardia, idiopathic ventricular fibrillation, early repolarization syndrome, and atrial fibrillation. For each syndrome, current knowledge on their genetic basis, clinical presentation, diagnosis, risk stratification, and therapy is discussed. Chapter 13 presents an overview of genetic testing and familial clinical screening in patients with (suspected) inherited arrhythmia syndrome in addition to a proposed screening hierarchy according to phenotype. Furthermore, the impact of new sequencing technologies is discussed.

1.4 Research into Cardiac Channelopathies: New Avenues

In recent years, a number of new technologies and innovations have provided essential novel insight into ion channel function and the mechanisms involved in cardiac channelopathies. On the molecular level, novel imaging techniques are significantly improving our knowledge on cardiac nanoscale architecture, ion channel distribution, and function. The methodologies and advantages of one of these techniques, super-resolution fluorescence microscopy, are discussed in Chap. 14, in addition to its application and future potential in cardiac research in general and ion channel (dys)function in particular. Functionally, studies employing transgenic mouse models have demonstrated that heterologous expression systems, traditionally employed for investigating the consequences of ion channel mutations, are not necessarily representative of the situation within the cardiomyocyte environment. As discussed in Chap. 15, transgenic mice allow the investigation of mutation effects in different regions and cell types of the heart, as well as investigation of disease progression. More recently, cardiomyocytes derived from human-induced pluripotent stem cells (hiPSC) have been employed to study mutations in a more physiological environment. These hiPSC, which are reviewed in Chap. 16, appear suitable for the investigation of patient- and disease-specific pharmacology and may provide a tool for studying the role of genetic background.

1.5 Concluding Remark

The current book provides a translational overview of current state-of-the art research on ion channel (dys)function, cardiac channelopathies, and inherited arrhythmia syndromes, bringing together clinical, genetic, and basic science experts.

Compliance with Ethical Standards

Sources of Funding This work was funded by a Priority Medicines Rare Diseases and Orphan Drugs grant (PM-Rare, 113303006 to C.A.R.) from The Netherlands Organization for Health Research and Development (ZonMw) and an Innovational Research Incentives Scheme Vidi grant from ZonMw (grant no. 91714371 to C.A.R.), by the German Cardiac Society and the Hengstberger Foundation (Klaus-Georg and Sigrid Hengstberger Scholarship to D.T.), the German Heart Foundation/German Foundation of Heart Research (F/08/14 to D.T.), the Joachim Siebeneicher Foundation (to D.T.), and the Baden-Wuerttemberg Ministry of Science, Research and Art (Sonderlinie Medizin to D.T.).

Conflict of Interest C.A.R. has previously received research grants from Gilead Sciences. D.T. reports receiving lecture fees/honoraria from Bayer Vital, Boehringer Ingelheim, Bristol-Myers Squibb, Daiichi Sankyo, Medtronic, Pfizer Pharma, Sanofi-Aventis, St. Jude Medical, and ZOLL CMS and research grant support from Daiichi Sankyo. D.T. filed a patent application for the use of K_{2P} potassium channels for altering cardiac electrophysiology.

Ethical Approval This article does not contain any studies with human participants or animals performed by any of the authors.

Reference

Shah M, Akar FG, Tomaselli GF. Molecular basis of arrhythmias. Circulation. 2005;112:2517–29.

Part I

(Dys)Function of Cardiac Ion Channels

Cardiac Sodium Channel (Dys)Function and Inherited Arrhythmia Syndromes

2

Carol Ann Remme

Abstract

Normal cardiac sodium channel function is essential for ensuring excitability of myocardial cells and proper conduction of the electrical impulse within the heart. Cardiac sodium channel dysfunction is associated with an increased risk of arrhythmias and sudden cardiac death. Over the last 20 years, (combined) genetic, electrophysiological, and molecular studies have provided insight into the (dys) function and (dys)regulation of the cardiac sodium channel under physiological circumstances and in the setting of *SCN5A* mutations identified in patients with inherited arrhythmia syndromes. Although our understanding of these sodium channelopathies has increased substantially, important issues remain incompletely understood. It has become increasingly clear that sodium channel distribution, function, and regulation are more complicated than traditionally assumed. Moreover, recent evidence suggests that the sodium channel may play additional, as of yet unrecognized, roles in cardiomyocyte function, which in turn may ultimately also impact on arrhythmogenesis. In this chapter, an overview is provided of the structure and function of the cardiac sodium channel and the clinical and biophysical characteristics of inherited sodium channel dysfunction. In addition, more recent insights into the electrophysiological and molecular aspects of sodium channel dysregulation and dysfunction in the setting of *SCN5A* mutations are discussed.

C. A. Remme (✉)
Dept. of Clinical & Experimental Cardiology, Academic Medical Center, University of Amsterdam, Amsterdam, The Netherlands
e-mail: c.a.remme@amc.uva.nl

D. Thomas, C. A. Remme (eds.), *Channelopathies in Heart Disease*, Cardiac and Vascular Biology, https://doi.org/10.1007/978-3-319-77812-9_2

2.1 Introduction

Influx of sodium ions through voltage-gated sodium channels in cardiomyocytes leads to depolarization of the membrane, thereby initiating the cardiac action potential and triggering the initiation and propagation of action potentials throughout the myocardium. Normal cardiac sodium channel function is therefore essential for ensuring excitability of myocardial cells and proper conduction of the electrical impulse within the heart. The importance of sodium channels for normal cardiac electrical activity is emphasized by the occurrence of potentially lethal arrhythmias in the setting of inherited and acquired sodium channel disease. During common pathological conditions such as myocardial ischemia and heart failure, altered sodium channel function causes conduction disturbances and ventricular arrhythmias. In addition, sodium channel dysfunction caused by mutations in the *SCN5A* gene, encoding the major sodium channel in the heart, is associated with a number of arrhythmia syndromes, including long QT syndrome type 3 (LQT3), cardiac conduction disease, and Brugada syndrome. Over the last 20 years, (combined) genetic, electrophysiological, and molecular studies have provided insight into the (dys)function and (dys)regulation of the cardiac sodium channel under physiological circumstances and in the setting of *SCN5A* mutations identified in patients with inherited arrhythmia syndromes. Although our understanding of these sodium channelopathies has increased substantially, important issues remain incompletely understood. It has become increasingly clear that sodium channel distribution, function, and regulation are more complicated than traditionally assumed. Moreover, recent evidence suggests that the sodium channel may play additional, as of yet unrecognized, roles in cardiomyocyte function, which in turn may ultimately also impact on arrhythmogenesis. Increased knowledge of these issues will be essential for further improvement of diagnosis, risk stratification, and treatment in patients with sodium channelopathies. In this chapter, an overview is provided of the structure and function of the cardiac sodium channel, the clinical and biophysical characteristics of inherited sodium channel dysfunction, and the (limited) therapeutic options for the treatment of cardiac sodium channel disease. In addition, more recent insights into the electrophysiological and molecular aspects of sodium channel dysregulation and dysfunction in the setting of *SCN5A* mutations are discussed.

2.2 Cardiac Sodium Channels: Structure, Function, and Distribution in the Heart

2.2.1 The Cardiac Sodium Channel Gene *SCN5A*

The sodium channel family comprises a total of nine genes (*SCN1A-SCN5A*, *SCN7A-SCN11A*), of which *SCN5A* is the main sodium channel in the heart. The *SCN5A* gene located on human chromosome 3p22 consists of 28 exons and encodes Nav1.5, the pore-forming alpha subunit of the cardiac sodium channel. At least three different *SCN5A* isoforms are known to be expressed in the human heart, of which

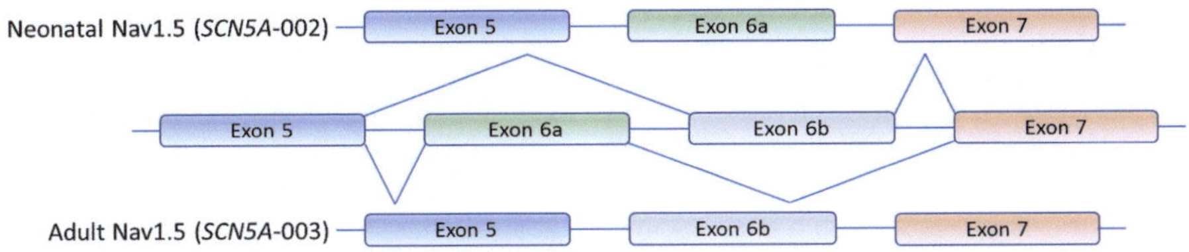

Fig. 2.1 Illustrative scheme of adult and neonatal splice variants of *SCN5A*

SCN5A-003 is the most abundant transcript in adult tissue (reviewed in Veerman et al. 2015) (Fig. 2.1). In contrast, during embryonic development, the "neonatal" *SCN5A*-001 transcript is most abundant. *SCN5A*-001 is replaced by *SCN5A*-003 after birth but may be upregulated during pathophysiological conditions (see Sect. 2.3.1) (Chioni et al. 2005; Schroeter et al. 2010). These two isoforms differ in exon 6 (exon 6b in *SCN5A*-003 and exon 6a in *SCN5A*-001), which results in a difference of seven amino acids between the respective proteins (Onkal et al. 2008; Schroeter et al. 2010). As a result, distinct differences in electrophysiological properties exist between both isoforms, in particular kinetics (Onkal et al. 2008). In addition, the *SCN5A*-014 isoform is also expressed in the human adult heart albeit less abundantly than *SCN5A*-003 (Makielski et al. 2003; Schroeter et al. 2010). This isoform results in an additional glutamine at position 1077 at the boundary of exon 18 (1077Q) and encodes Nav1.5 channels that exhibit a smaller sodium current as compared to *SCN5A*-003 (see Sect. 2.3.1) (Makielski et al. 2003).

2.2.2 Expression and Distribution of *SCN5A*/Nav1.5 in the Heart

SCN5A is highly expressed in the heart, but its expression has also been found in other tissue types, including smooth muscle cells of the intestines (Holm et al. 2002). Within the heart, *SCN5A*/Nav1.5 expression is high in atrial and ventricular myocardium. Nav1.5 protein expression is low to absent in the sinoatrial and atrioventricular nodes but abundant in the His bundle, bundle branches, and Purkinje fibers (Fig. 2.2a,b) (Yoo et al. 2006; Remme et al. 2009a). More specifically, Nav1.5 expression is found in the periphery of the sinoatrial node but not in its central part (Lei et al. 2004). Furthermore, a transmural gradient is observed in the left and right ventricles, with lower *SCN5A* and Nav1.5 abundance in the subepicardium as compared to the subendocardium (Fig. 2.2c) (Remme et al. 2009a). Moreover, evidence is emerging that Nav1.5 is differentially distributed within distinct subcellular microdomains of the cardiomyocyte (discussed in more detail in Sect. 2.8.1).

2.2.3 Sodium Channel Structure and Function

Functional sodium channels are formed by the association of the transmembrane pore-forming α-subunit protein Nav1.5 (220 kDa) with a small (30–40 kDa)

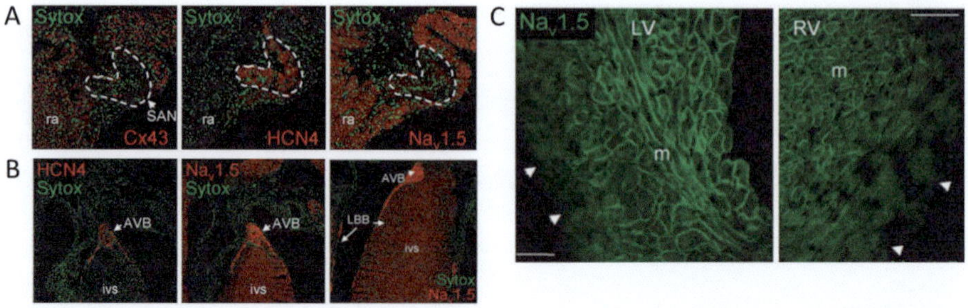

Fig. 2.2 (**a**) Nav1.5 expression is absent from the central sinoatrial node (SAN). (**b**) Nav1.5 is highly expressed in the atrioventricular/His bundle and bundle branches. (**c**) Nav1.5 displays a transmural distribution pattern, with low expression in the subepicardium (indicated by arrowheads). Reproduced from Remme et al. (2009a) with permission

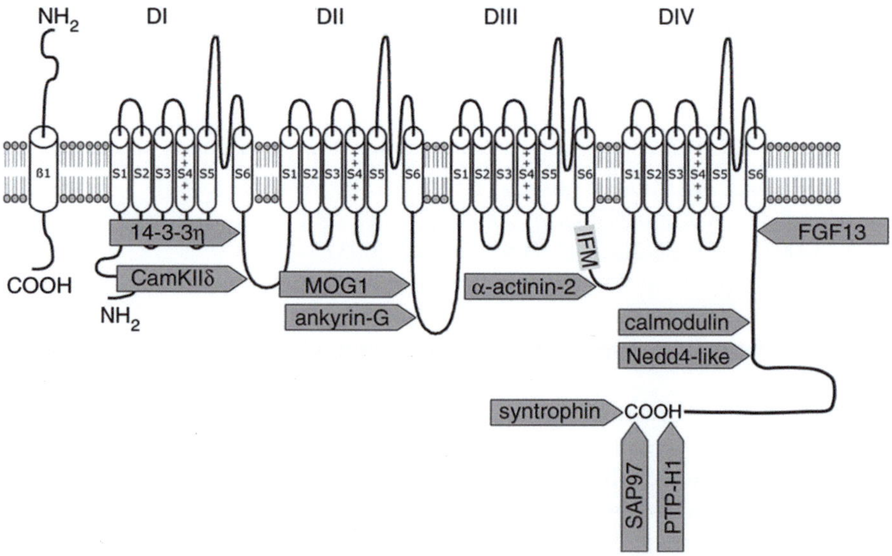

Fig. 2.3 The topology and structure of Nav1.5 and the ancillary modulatory β-subunit. Locations are indicated where interacting proteins bind to the various regions of the channel. Redrawn from Shy et al. (2013), with permission

ancillary modulatory β-subunit and several regulatory proteins. The α-subunit protein Nav1.5 is made up of a cytoplasmic N terminus, four internally homologous domains (DI–DIV) interconnected by cytoplasmic linkers, and a cytoplasmic C-terminal domain (Fig. 2.3). The protein furthermore contains a number of sites relevant for interaction with other proteins, as well as potential phosphorylation sites (see Sects. 2.3.2 and 2.3.3). The DI–DIV domains each consist of six transmembrane α-helical segments (S1–S6), which in turn are interconnected by extracellular and cytoplasmic loops. The four domains fold around an ion-conducting pore, which is lined by the extracellular loops (P-loops) between S5 and S6 segments; the P-loops are considered to contain the channels' selectivity filter for sodium ions (see Kass 2006). The fourth transmembrane segment, S4, is highly charged and acts as the voltage sensor responsible for increased channel permeability (channel activation) during membrane depolarization (Balser 2001). As the membrane depolarizes, outward movement of the positively charged S4 segments results in the opening of the channel pore. As a result, sodium channels become activated within a time span of 1 ms, allowing for sodium influx to commence along the electrochemical gradient. As a result, the membrane is further depolarized (reflecting phase 0 of the cardiac action potential), ultimately enabling activation of L-type calcium channels, calcium influx, and myocardial contraction. Simultaneously, the processes of fast and slow inactivation are initiated leading to closure of the channel pore. Channels remain in closed state until the cell membrane is repolarized, allowing them to recover from inactivation and becoming available for activation again. Fast inactivation occurs within approximately 10 ms and involves an inactivation gate formed by a cluster of three hydrophobic amino acids (isoleucine, phenylalanine, and methionine or IFM motif; see Fig. 2.3) in the intracellular DIII–DIV linker, together with two docking sites located on the intracellular linkers between S4 and S5 of DIII and DIV (West et al. 1992; Kass 2006). The carboxy (COOH-) terminal domain is also involved in the inactivation process and acts through stabilization of the closed gate, thus minimizing channel reopening (Motoike et al. 2004). During prolonged periods of depolarization, the sodium channel enters the state of intermediate and slow inactivation, which require much longer recovery times than fast inactivation (~50 ms and >5 s, respectively). Slow inactivation is an important determinant of sodium channel availability and likely involves conformational changes in the S5–S6 P-loops, although the DIII S4–S5 linker may also be of importance (Kass 2006; Casini et al. 2007). Activation and inactivation properties of sodium channels are tightly regulated during physiological conditions but may be altered in the setting of genetic and acquired sodium channel dysfunction. These alterations in sodium channel function may have profound consequences for cardiac electrophysiology and arrhythmogenesis (see further below in Sect. 2.4).

2.3 Regulation of Cardiac Sodium Channel Expression and Function

2.3.1 (Post-)Transcriptional Regulation

On the transcriptional level, *SCN5A* mRNA expression is regulated by transcription factors that bind to various enhancer and repressor sequences close to or within the promoter region of the gene (Arnolds et al. 2012; van den Boogaard et al. 2012). Of these, Foxo1 and NF-kappaB are involved in gene regulation upon oxidative stress: following production of reactive oxygen species, these transcription factors are translocated to the nucleus and consequently inhibit transcription of *SCN5A* by directly binding the promoter region. The transcription factor *TBX5*, which plays an important role during cardiac development, stimulates *SCN5A* expression in the adult cardiac conduction system through its binding to enhancer elements (Arnolds et al. 2012; Van den Boogaard et al. 2012). Furthermore, a circadian expression pattern of cardiac *Scn5a* mRNA under control of the molecular clock transcription factor *Bmal1* has been demonstrated in mice (Schroder et al. 2013). Posttranscriptionally, alternative splicing of the *SCN5A* gene produces transcript variants with different functional characteristics (see Sect. 2.2.1), which may be of significant functional relevance for disease expressivity in sodium channelopathy. For instance, these variants have been reported to produce differential effects on reduced sodium channel membrane expression in the setting of the Brugada syndrome mutation *SCN5A*-G1406R associated with defective trafficking (Tan et al. 2006). Moreover, as noted in Sect. 2.2.1., *SCN5A* splicing is also developmentally regulated, with the neonatal splice isoform being downregulated after birth (Chioni et al. 2005; Schroeter et al. 2010). Since the sodium current (I_{Na}) generated by the neonatal isoform exhibits alterations in kinetics, including a slower rate of activation and inactivation and a depolarized shift in voltage dependence of activation (Onkal et al. 2008), functional consequences of *SCN5A* mutations may differ significantly in the setting of this isoform. Indeed, this was observed for a mutation identified in a fetus with severe LQTS, where functional analysis demonstrated more severe biophysical defects in the presence of the neonatal isoform, including an increased late I_{Na} (Murphy et al. 2012). Furthermore, during pathophysiological conditions such as heart failure, increased expression of abnormal C-terminal *SCN5A* splicing variants may result in the formation of truncated, nonfunctional sodium channels, thereby predisposing to cardiac conduction slowing and arrhythmogenesis (Shang et al. 2007). Finally, posttranscriptional regulation of *SCN5A* by microRNAs has been demonstrated, in particular miR-219 (Daimi et al. 2015).

2.3.2 Trafficking and Posttranslational Regulation

Like most ion channels, intracellular trafficking of sodium channels involves a complex process starting with assembly in the endoplasmic reticulum (ER), followed by transport to the Golgi apparatus and targeting to and incorporation into the membrane (reviewed by Balse et al. 2012). Once assembled and properly folded, sodium channels are transported to various sarcolemma membrane locations via microtubules in close association with the actin cytoskeleton (Balse et al. 2012). Alterations of the functional properties of microtubules and/or cytoskeletal components have been shown to modulate trafficking of ion channels, including Nav1.5. For instance, the anticancer drug Taxol, which changes the properties of the cytoskeletal component tubulin, affects I_{Na} density and gating in vitro and is associated with increased occurrence of cardiac conduction disorders and arrhythmias in vivo (Rowinsky et al. 1991; Casini et al. 2010). Other posttranslational modifications of Nav1.5 include phosphorylation, glycosylation, S-nitrosylation, ubiquitination, and methylation (reviewed by Rook et al. 2012; Marionneau and Abriel 2015). Conserved amino acid motifs for N-glycosylation are found in the extracellular domain of Nav1.5, and glycosylation may affect sodium channel gating (Zhang et al. 1999; Rook et al. 2012). Ubiquitylation is a well-described process responsible for regulating the number of plasma membrane proteins at the cell surface and is mediated by the binding of Nedd4-2 to the PY motif in the C-terminal domain of Nav1.5 (see Abriel and Staub 2005). Ubiquitylated membrane proteins are internalized and then either subjected to proteasomal or lysosomal degradation or recycled (Abriel and Staub 2005). Phosphorylation of Nav1.5 by PKA, PKC, and calcium-/calmodulin-dependent protein kinase II (CamKII) has been shown to have multiple regulatory effects on sodium channel trafficking and current magnitude, and CamKII may also be relevant for modulating late I_{Na} magnitude during pathophysiological conditions (Marionneau and Abriel 2015; Herren et al. 2013). Finally, sodium channel density and kinetics are furthermore regulated by intracellular calcium levels (Casini et al. 2009), extracellular protons and pH (Jones et al. 2011), reactive oxygen species (Liu et al. 2005), temperature (Amin et al. 2005), and stretch (Beyder et al. 2010).

2.3.3 Regulation of Cardiac Sodium Channels by Interacting Proteins

Sodium channels are not isolated units within the myocyte membrane but are functional components of a macromolecular complex through which they associate with proteins that participate in cell adhesion, signal transduction, and cytoskeleton anchoring (see Abriel 2010). These Nav1.5 interacting proteins in turn regulate sodium channel expression, trafficking, and/or function (Table 2.1). For many of these regulatory proteins, the interaction site within Nav1.5 has been established (Fig. 2.3). The accessory β-subunits (β1–β4 encoded by the *SCN1B-SCN4B* genes) likely bind to the extracellular connecting loops between S5 and S6, allowing them to modulate sodium channel density and kinetics (Malhotra et al. 2001; Ko et al. 2005; Meadows and Isom 2005; Medeiros-Domingo et al. 2007). In the intracellular C-terminal part of Nav1.5, several protein–protein interaction sites have been identified, including the IQ motif, the PY motif, and the PDZ domain-binding motif. Proteins such as PTPH1, SAP97, and syntrophins interact with the PDZ domain-binding motif (Jespersen et al. 2006; Petitprez et al. 2011; Gavillet et al. 2006). The ubiquitin-protein ligase, Nedd4-2, binds to the PY motif, thereby regulating Nav1.5 turnover by ubiquitylating the channel (van Bemmelen et al. 2004). The IQ motif constitutes a binding site for calmodulin (CaM), a ubiquitous calcium-binding protein which is thought to confer sensitivity to intracellular calcium levels on Nav1.5 (see Shy et al. 2013). Fibroblast growth factor homologous factor (FHF) family members can also interact with the C-terminus of Nav1.5, with fibroblast growth factor homologous factor 13 (FGF13) regulating sodium channels and conduction velocity in murine hearts (Wang et al. 2011). The cytosolic adaptor protein 14-3-3η interacts with the DI–DII linker region and modulates steady-state inactivation of the channel (Allouis et al. 2006). The DI–DII linker region furthermore contains multiple PKA and PKC phosphorylation sites in addition to an

Table 2.1 Clinical cardiac phenotypes of the cardiac sodium channel and its interacting proteins

Sodium channel		
Gene	Protein	Clinical cardiac phenotypes
SCN5A	Nav1.5	Long QT syndrome type 3 (LQT3), Brugada syndrome, cardiac conduction disease, sick sinus disease, overlap syndrome, atrial standstill, dilated cardiomyopathy, atrial fibrillation, arrhythmogenic right ventricular cardiomyopathy (ARVC)
Interacting proteins		
Gene	Protein	Clinical cardiac phenotypes
SCN1B	β1	Brugada syndrome, conduction disease, atrial fibrillation
SCN2B	β2	Atrial fibrillation
SCN3B	β3	Brugada syndrome, conduction disease, atrial fibrillation, idiopathic ventricular fibrillation
SCN4B	β4	Long QT syndrome type 10 (LQT10)
SNTA	α-1-syntrophin	Long QT syndrome type 12 (LQT12)
MOG1	MOG1	Brugada syndrome
PTPH1	Protein tyrosine phosphatase H1	–
NEDD4L	Nedd4-2/Nedd4-like	–
CALM	Calmodulin	–
SAP97	SAP97	–
YWHAH	14-3-3η	–
FGF13	Fibroblast growth factor 13 (FGF13)	–
CAMK2D	CamkIId	–
ANK3	Ankyrin-G	–
ACTN2	α-Actinin-2	–
CAV3	Caveolin-3	Long QT syndrome type 9 (LQT9)
GPD1L	Glycerol-3-Phosphate Dehydrogenase 1 Like	Brugada syndrome
PKP2	Plakophilin-2	Arrhythmogenic right ventricular cardiomyopathy (ARVC)
DSG2	Desmoglein-2	Arrhythmogenic right ventricular cardiomyopathy (ARVC)
TCAP	Telethonin	–
ZASP	Z-band alternatively spliced PDZ motif protein	–
SAP97	Synapse-associated protein 97	–
CXADR	Coxsackie and adenovirus receptor (CAR)	–
CASK	Calcium-/calmodulin-dependent serine protein kinase	–

interaction site for CAMKII (see Sect. 2.3.2). Ankyrin-G and MOG1 interact with the II–III linker segment of Nav1.5, and both proteins regulate cell surface expression of sodium channels (Lemaillet et al. 2003; Mohler et al. 2004; Kattygnarath et al. 2011). A member of the superfamily of F-actin cross-linking proteins, α-actinin-2, interacts with the cytoplasmic loop connecting domains III and IV, thereby increasing I_{Na} density without affecting gating properties (Ziane et al. 2010). Other proteins that may directly or indirectly interact with Nav1.5 and functionally regulate the channel include caveolin-3 (Vatta et al. 2006), glycerol-3-phosphate dehydrogenase-like protein (GPD1L) (London et al. 2007), plakophilin-2 (Sato et al. 2009), desmoglein-2 (Rizzo et al. 2012), telethonin (Mazzone et al. 2008), Z-band alternatively spliced PDZ motif protein (ZASP) (Li et al. 2010), the MAGUK proteins SAP97 and calcium-/calmodulin-dependent serine protein kinase (CASK) (Eichel et al. 2016; Petitprez et al. 2011), and coxsackie and adenovirus receptor (CAR) (Marsman et al. 2014) (Table 2.1). The functional relevance of the sodium channel macromolecular complex is underscored by the fact that mutations in genes encoding Nav1.5 interacting proteins are associated with arrhythmia syndromes (see also Sect. 2.5.8) (Kyle and Makielski 2014).

2.4 Biophysical Effects of Sodium Channel Dysfunction and Pro-arrhythmic Consequences

From a biophysical point of view, distinct alterations in gating and other functional properties of the sodium channel may lead to multiple rhythm disturbances (Viswanathan and Balser 2004; Remme et al. 2008). These alterations may be grossly divided into those leading to a loss of sodium channel function and consequent depolarization disturbances and those causing an enhanced late I_{Na} or window current leading to repolarization disturbances. Their electrophysiological and homeostatic consequences may set the stage for potentially life-threatening ventricular arrhythmias (see Fig. 2.4).

2.4.1 Causes and Consequences of Reduced Sodium Channel Availability

Reduced sodium channel availability may occur in the setting of both inherited and acquired pathological conditions, potentially leading to conduction disturbances and ventricular arrhythmias. During myocardial ischemia, local metabolic changes within the myocardium lead to inactivation of the I_{Na} and consequent repression of cardiac excitability and slowing of conduction (Janse and Wit 1989; Fozzard and Makielski 1985). Reduced excitability of ischemic tissue may protect against ventricular arrhythmias, but conduction slowing in the infarcted heart may set the stage for ventricular mechanisms based on reentrant mechanisms (Janse and Wit 1989). Similarly, loss of sodium channel function is also observed during heart failure, leading to conduction slowing and ventricular arrhythmias. The reduced sodium

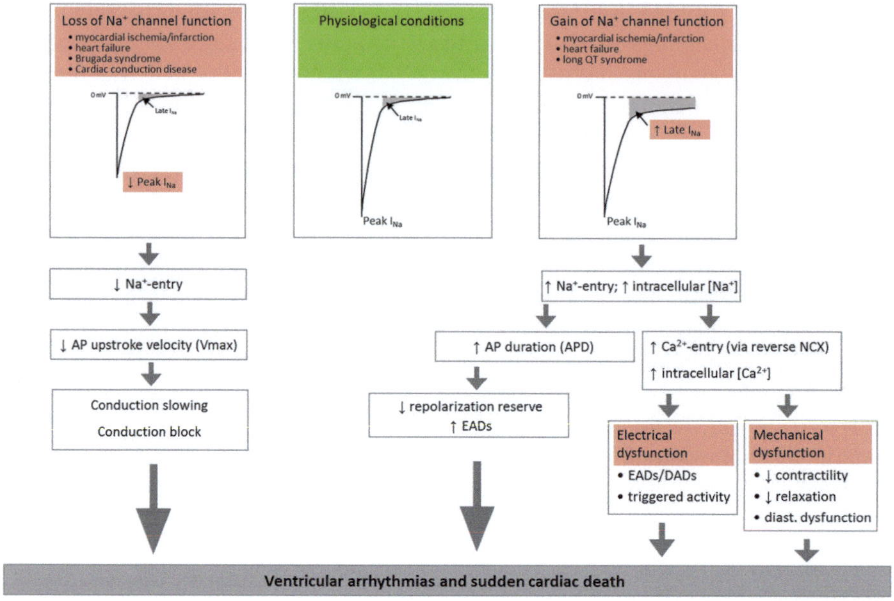

Fig. 2.4 Schematic representation of the arrhythmogenic consequences of reduced peak I_{Na} (left) and increased late I_{Na} (right). EADs, early after depolarizations; DADs, delayed after depolarizations; AP, action potential; diast., diastolic; NCX, sodium–calcium exchanger. Reproduced from Remme and Wilde (2014), with permission

channel availability in failing hearts may be due to alterations in sodium channel mRNA transcript levels and posttranslational regulation (Ufret-Vincenty et al. 2001; Shang et al. 2007). In inherited arrhythmia syndromes, I_{Na} may be reduced as a result of different mechanisms (Fig. 2.5a–c) (Tan et al. 2003; Viswanathan and Balser 2004; Kapplinger et al. 2010). Loss-of-function mutations in *SCN5A* can lead to a decreased number of functional channels on the membrane due to misfolding of the channel and/or altered trafficking. Alternatively, sodium channels may be present on the membrane albeit less functional secondary to reduced conductivity or a shift in the voltage dependence of (in)activation. Moreover, sodium channel trafficking and/or kinetics may be similarly affected consequent to (genetic) alterations in proteins interacting with Nav1.5 (see also Sect. 2.5.8).

2.4.2 Causes and Consequences of Increased Late I_{Na} and/or Window Current

The initial large inward I_{Na} (peak I_{Na}) is for the most part rapidly inactivated, but a small fraction of the current (designated the persistent or late I_{Na}) persists throughout the duration of the action potential (AP) plateau phase. This late I_{Na} is typically small during physiological conditions but may be enhanced in the setting of acquired

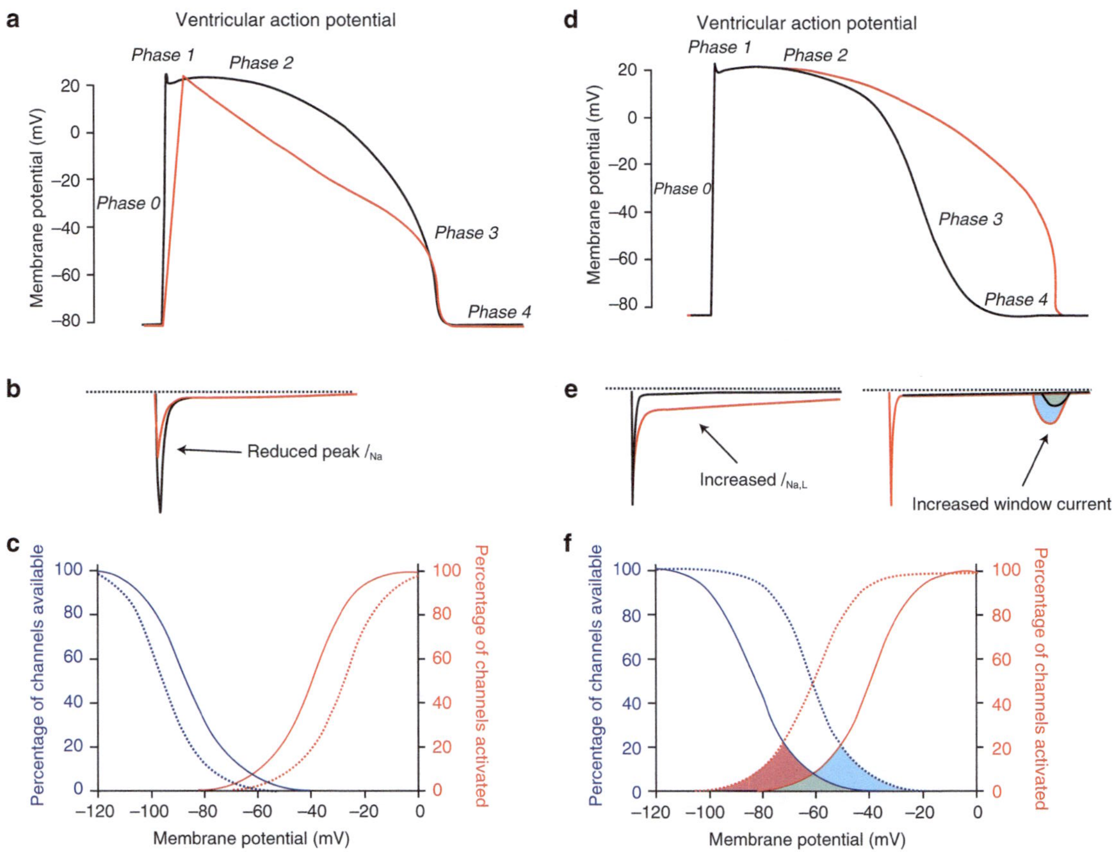

Fig. 2.5 Biophysical alterations underlying loss (**a–c**) and gain (**d–f**) of sodium channel function. (**a**, **b**) Decreased peak current lowers the upstroke velocity of the action potential (red trace). (**c**) Shifts in voltage dependence of (in)activation (dashed lines) that result in loss of function. (**d**) Gain of sodium channel function leads to an increased net influx of Na+-ions and consequent action potential prolongation (red trace). (**e**) Increased late current and increased window current. (**f**) Shifts in voltage dependence of (in)activation (dashed lines) that lead to an increased window current, as illustrated by the red and blue areas under the curve. Reproduced from Veerman et al. (2015), with permission

(ischemia, hypertrophy, and heart failure) and inherited (LQT3) disease (see Remme and Bezzina 2010). Independent of the underlying cause, delayed repolarization and action potential prolongation occur, and early after depolarizations may subsequently trigger torsades de pointes arrhythmias and sudden death. Moreover, enhanced late I_{Na} may alter intracellular sodium and calcium homeostasis, further predisposing to arrhythmias as well as mechanical disturbances such as diastolic dysfunction (see Remme and Wilde 2013). It has been shown that during heart failure, an increased fraction of sodium channels fail to enter the inactivated state resulting in an increased late I_{Na} (Valdivia et al. 2005). The underlying mechanisms are not completely clear, but posttranslational modulation of sodium channels by calcium-dependent pathways (including CamKII) is thought to be involved (Wagner et al. 2006). In the setting of inherited disorders, gain-of-function *SCN5A* mutations may disrupt fast inactivation, thereby allowing for sodium channels to reopen (Bennett et al. 1995) (Fig. 2.5d–f). *SCN5A* mutations may also lead to incomplete or slowed inactivation (resulting in channel openings of longer duration) or a shift in voltage dependence of inactivation with a consequent increased voltage range of incomplete current inactivation (resulting in an increase in window current). These alterations all lead to a persistent inward current (or late I_{Na}) and consequent increased sodium influx during the action potential plateau phase. Alternatively, faster recovery from inactivation (causing increased sodium channel availability) or increased peak I_{Na} density may occur (Wedekind et al. 2001; Rivolta et al. 2001; Clancy et al. 2003). Finally, enhanced late I_{Na} has also been observed secondary to mutations in proteins interacting with Nav1.5 (see also Sect. 2.5.8).

2.5 Inherited Arrhythmia Syndromes Associated with *SCN5A* Mutations

Mutations in *SCN5A* have been implicated in multiple inherited arrhythmia syndromes (Table 2.1). Although caused by mutations in the same ion channel, these clinical syndromes each display distinct phenotypical characteristics, as is discussed below. Moreover, mutations in proteins interacting with Nav1.5 have been associated with different biophysical and clinical arrhythmogenic phenotypes.

2.5.1 Long QT Syndrome Type 3 (LQT3)

Mutations in *SCN5A* can lead to long QT syndrome type 3 (LQT3), which is characterized by prolonged QT intervals on the ECG and increased risk for sudden death due to ventricular tachyarrhythmias, in particular torsades de pointes (see also Chap. 7). LQT3 patients are often bradycardic and display ventricular arrhythmias predominantly during rest or sleep (at slow heart rates) (Schwartz et al. 2001; Schwartz 2006). Compared to other LQTS subtypes, LQT3 patients are particularly at risk for sudden death, and cardiac arrest (rather than syncope) is often the first clinical event (Zareba et al. 2001; Schwartz et al. 2001). Biophysical alterations typically observed secondary to *SCN5A* mutations associated with LQT3 predominantly disrupt fast inactivation of the I_{Na}, allowing for sodium channels to reopen, resulting in a persistent (late) inward current during the action potential plateau phase (Bennett et al. 1995). *SCN5A* mutations less commonly cause LQT3 through slowed inactivation, faster recovery from inactivation, increased window current, and increased peak I_{Na} density (Wedekind et al. 2001; Rivolta et al. 2001; Clancy et al. 2003). At present, standard therapy includes beta-blockers, ICD treatment in patients at high risk of arrhythmias, and pacemaker implantation in selected cases. Additional strategies aimed at inhibiting late I_{Na} are currently under investigation (see Sect. 2.7) (Wilde and Remme 2017).

2.5.2 Brugada Syndrome

Brugada syndrome (BrS) is characterized by an increased risk for ventricular arrhythmias and sudden death occurring mostly during rest or sleep, with an increased incidence in males than females (Antzelevitch 2006). The typical Brugada ECG pattern of ST-segment elevation in the right-precordial leads may be variably present and can be unmasked or increased after administration of Class 1A or 1C antiarrhythmic sodium channel blocking drugs (ajmaline, flecainide) or during exercise (see Meregalli et al. 2005) (see also Chap. 8). Cardiac conduction disease is often also observed in BrS patients, evidenced by prolonged PQ and QRS duration on the ECG (Wilde and Brugada 2011). In approximately 20% of BrS patients, *SCN5A* mutations are identified. The latter are typically "loss-of-function" mutations leading to reduced sodium channel availability, either through decreased trafficking and membrane surface channel expression or through altered channel gating properties including disruption of activation, accelerated inactivation, and impaired recovery from inactivation (Tan et al. 2003; Viswanathan and Balser 2004; Kapplinger et al. 2010). The consequent decreased action potential upstroke velocity underlies the prolongation of PR and QRS intervals often observed in BrS patients, but the right-precordial ST-segment elevation and its relation to arrhythmogenesis are less well understood. Here, two major hypotheses have been proposed, i.e., the repolarization disorder hypothesis involving increased (transmural) heterogeneity in action potential duration versus the depolarization disorder hypothesis encompassing preferential conduction slowing in the right ventricle and/or right ventricular outflow tract (for more details, see also Chap. 8) (Hoogendijk et al. 2010). Structural discontinuities in the subepicardium of the right ventricle may further contribute to the typical BrS ECG pattern as well as the onset of ventricular arrhythmias (Hoogendijk et al. 2010). Treatment options for BrS include implantation of an ICD together with measures aimed at preventing known arrhythmia-provoking factors (for instance, fever) (Antzelevitch et al. 2005). In addition, quinidine (which also blocks Ito) may be used as an adjunct to device therapy to decrease the incidence of ventricular arrhythmias (Belhassen and Viskin 2004).

2.5.3 Progressive Cardiac Conduction Defect (PCCD) and Sick Sinus Syndrome

PCCD, also called Lenègre or Lev disease, is characterized by progressive conduction slowing through the His-Purkinje system, with a right and/or left bundle branch block and QRS widening, leading to complete AV block, syncope, and sudden death (see also Chap. 9). In some cases, inherited PCCD is associated with mutations in *SCN5A* leading to reduced sodium channel availability (Schott et al. 1999; Wolf and Berul 2006; Kyndt et al. 2001). Loss-of-function mutations in *SCN5A* have also been associated with inherited sick sinus syndrome (Benson et al. 2003; Smits et al. 2005). Here, decreased sodium channel availability may cause bradycardia by slowing or blocking conduction from the central sinoatrial region to the surrounding atrial tissue, in addition to a (modest) reduction in automaticity of sinoatrial pacemaking tissue (see Lei et al. 2008). Furthermore, sinus bradycardia and/or arrest may also occur in LQT3 patients with *SCN5A* gain-of-function mutations, where an increased late I_{Na} may cause action potential prolongation (Veldkamp et al. 2003).

2.5.4 Atrial Fibrillation

Atrial fibrillation (AF) is commonly observed in mostly elderly patients with underlying structural cardiac abnormalities, but it may also occur as a hereditary disease in young patients with structurally normal hearts (see also Chap. 12). Both loss-of-function and gain-of-function mutations in *SCN5A* have been described in this familial form, which are thought to induce AF through decreased atrial conduction velocity and increased atrial action potential duration and excitability, respectively (Darbar et al. 2008; Ellinor et al. 2008; Makiyama et al. 2008; Li et al. 2009). In addition, structural remodeling of the atria secondary to sodium channel dysfunction may play a role. Mutations in accessory β-subunits leading to I_{Na} reduction have also been associated with AF (Watanabe et al. 2009; Olesen et al. 2011).

2.5.5 Dilated Cardiomyopathy

Familial forms of dilated cardiomyopathy (DCM) are mostly associated with mutations in cytoskeletal proteins. In rare cases, familial forms of DCM are reported in patients with *SCN5A* mutations (Bezzina et al. 2003; McNair et al. 2004), often presenting in combination with atrial arrhythmias and/or fibrillation (Olson et al. 2005). Interestingly, biophysical properties consistent with both loss and gain of sodium channel function have been observed in *SCN5A* mutations associated with DCM (McNair et al. 2004; Ge et al. 2008; Nguyen et al. 2008; Bezzina and Remme 2008). The mechanisms underlying DCM secondary to *SCN5A* mutations remain unclear and may involve a complex interplay of altered sodium channel current, (preexistent) myocardial structural abnormalities, and occurrence of long-standing (atrial) arrhythmias. In addition, DCM-related mutations (such as *SCN5A*-R219H) may cause a proton leak current, suggesting that intracellular acidification may contribute to the DCM (and arrhythmias) observed in mutation carriers (Gosselin-Badaroudine et al. 2012).

2.5.6 Sodium Channel Overlap Syndrome

In some instances, one single *SCN5A* mutation can result in multiple disease phenotypes even within one affected family, referred to as "sodium channel overlap syndrome." In 1999, our group identified the first such mutation, *SCN5A*-1795insD, in a large Dutch family presenting with extensive variability in type and severity of symptoms, including ECG features of sinus node dysfunction, bradycardia, conduction disease, BrS (ST-segment elevation), and LQT3 (QT-interval prolongation), in addition to nocturnal sudden death (Bezzina et al. 1999; van den Berg et al. 2001). Other *SCN5A* mutations with similar clinical overlap of LQT3, BrS, and conduction disease have since then been reported, including delK1500, D1114N, delF1617, E1784K, and L1786Q (Splawski et al. 2000; Priori et al. 2000; Chen et al. 2005; Makita et al. 2008; Kanters et al. 2014). In addition, combined clinical features of LQT3 and conduction disease have been described in addition to familial DCM combined with conduction disease, atrial arrhythmias, and/or fibrillation (McNair et al. 2004, 2011; Olson et al. 2005; Ge et al. 2008; Nguyen et al. 2008). The simultaneous presence of LQT3 (i.e., gain of function) and BrS or conduction disease (i.e., loss of function) due to one single *SCN5A* mutation may seem unlikely given the apparent opposing biophysical alterations underlying both clinical entities. Indeed, patch-clamp studies of the *SCN5A*-1795insD mutation in two separate expression systems initially revealed opposing and inconclusive effects on I_{Na} density and kinetics (Bezzina et al. 1999; Veldkamp et al. 2000). To assess the biophysical properties of this mutation in native myocytes, we generated transgenic mice carrying the heterozygous *Scn5a*-1798insD$^{/+}$ mutation, equivalent to human *SCN5A*-1795insD (Remme et al. 2006). *Scn5a*-1798insD$^{/+}$ mice displayed similar overlap clinical phenotypes as human 1795insD carriers, including bradycardia; PR, QRS, and QTc prolongation; and right ventricular conduction slowing (a feature of BrS) (Fig. 2.6a). Patch-clamp analysis of action potential characteristics in *Scn5a*-1798insD$^{/+}$ isolated myocytes demonstrated prolongation of cardiac repolarization, predominantly at slow rates and decreased sodium channel availability, especially at high frequencies (Veldkamp et al. 2000; Clancy and Rudy 2002). The mutation caused a drastic reduction in peak I_{Na} density, a delayed time course of fast inactivation, and a small persistent I_{Na}, explaining the observed multiple phenotypes (Fig. 2.6b–c) (Remme et al. 2006). Similarly, both *SCN5A*-E1784K and *SCN5A*-delK1500 (mutations associated with LQT3, BrS, and conduction disease) enhance channel inactivation and reduce peak current density but also significantly increase late I_{Na} magnitude (Grant et al. 2002; Makita et al. 2008). The delF1617 mutation (observed in LQT3 and BrS) displays reduced peak I_{Na} density and impaired recovery from inactivation (loss of function) versus delayed inactivation and increased late I_{Na} (gain of function) (Benson et al. 2003; Chen et al. 2005).

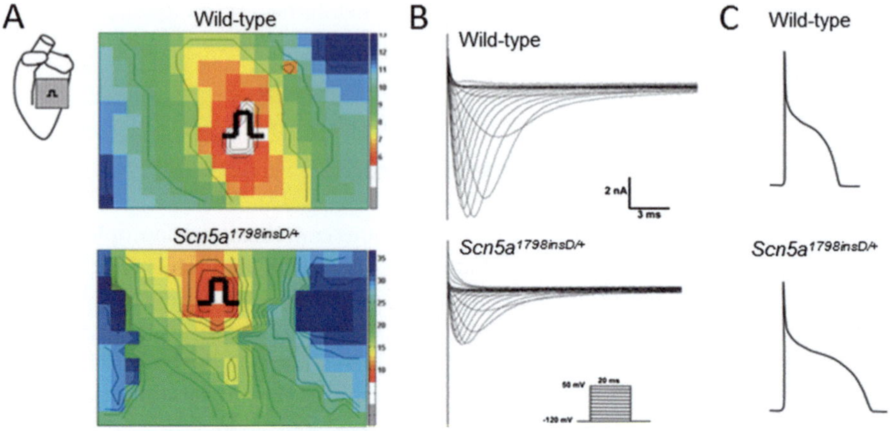

Fig. 2.6 The *Scn5a*-1798insD$^{/+}$ overlap syndrome mouse model shows (**a**) right ventricular conduction slowing, (**b**) reduced peak I_{Na}, and (**c**) enhanced late I_{Na}. Reproduced in part from Remme et al. (2006), with permission

SCN5A mutations associated with both Brugada syndrome and conduction disease (and/or sick sinus syndrome) invariably lead to decreased peak I_{Na} density when studied in heterologous expression systems (Smits et al. 2005; Kyndt et al. 2001; Rossenbacker et al. 2004; Makiyama et al. 2005; Cordeiro et al. 2006). However, for other overlap syndromes, a clear parallel between the mixed clinical phenotype of a certain *SCN5A* mutation and its biophysical properties is not observed (see Wilde and Remme 2017). It must be noted, however, that heterologous expression systems may be limited in their capability of correctly assessing biophysical consequences of a particular *SCN5A* mutation (in particular one associated with a mixed clinical phenotype).

2.5.7 Arrhythmogenic Right Ventricular Cardiomyopathy

Arrhythmogenic right ventricular cardiomyopathy (ARVC) is an inherited disease associated with cardiomyopathic changes (predominantly but not exclusively affecting the right ventricle), ventricular arrhythmias, and sudden death. In most ARVC patients, mutations in desmosomal proteins are identified, including plakoglobin, plakophilin-2, and desmoglein-2 (Calkins et al. 2017). Desmosomal proteins have been shown to interact with Nav1.5, and reduced I_{Na} reduction is a common feature in ARVC disease models (Rizzo et al. 2012; Cerrone et al. 2014; Sato et al. 2009). Recently, rare variants in *SCN5A* were found in ARVC patients that did not carry mutations in ARVC-related desmosomal proteins (Yu et al. 2014; Te Riele et al. 2017). Studies in human iPS-derived cardiomyocytes showed a reduction in I_{Na} and a potential detrimental effect on cell adhesion secondary to one of these

ARVC-related *SCN5A* mutations (Te Riele et al. 2017). Interestingly, ARVC and BrS are both considered to affect predominantly the right ventricle and are both associated with reduced I_{Na}; hence, an increasing overlap between these entities has been suggested (Agullo-Pascual et al. 2014).

2.5.8 Mutations in Genes Encoding Nav1.5-Interacting Proteins

As discussed in Sect. 2.3.3, sodium channels interact with other proteins within the macromolecular complex. The functional relevance of such interactions is evidenced by the fact that mutations in these modulatory proteins are associated with sodium channel dysfunction and arrhythmia (Table 2.1). Mutations in the sodium channel accessory β-subunits have been identified in patients with BrS and/or conduction disease (β1 and β3), atrial fibrillation (β1, β2, and β3), LQTS (β4), and idiopathic ventricular fibrillation (IVF; β3) (Watanabe et al. 2008, 2009; Hu et al. 2009; Medeiros-Domingo et al. 2007; Valdivia et al. 2010; Olesen et al. 2011, 2012). As mentioned above, mutations in desmosomal proteins plakophilin-2 and desmoglein-2 (which interact with Nav1.5) are associated with ARVC, in addition to reduced I_{Na} and conduction abnormalities. Furthermore, the sodium channel-interacting proteins caveolin-3 and alpha1-syntrophin have been implicated in LQTS (Vatta et al. 2006; Wu et al. 2008), whereas mutations in GPD1L and MOG1 have been associated with BrS (London et al. 2007; Kattygnarath et al. 2011; Abriel 2010).

2.6 Variable Expressivity in Cardiac Sodium Channelopathy

2.6.1 Disease Variability in *SCN5A* Mutation Carriers

Significant variability in disease expression has been documented in patients harboring mutations in *SCN5A* (Probst et al. 2009), which may in part depend on the underlying biophysical defect. For instance, truncating loss-of-function mutations in *SCN5A* have been shown to be associated with more severe conduction disease than missense mutations (Meregalli et al. 2009). However, variability in disease expressivity and severity is also observed among family members carrying the same mutation. For example, some *SCN5A*-1795insD mutation carriers display pronounced ECG abnormalities, whereas other family members carrying the same mutation appear unaffected (Bezzina et al. 1999) (Fig. 2.7a). Furthermore, among those presenting with clear ECG alterations, some mutation carriers suffer from arrhythmias and/or sudden death, while others remain symptom-free throughout life. Interestingly, certain *SCN5A*-1795insD mutation carriers display predominantly signs of BrS and/or conduction disease (i.e., a loss-of-function phenotype), whereas other affected family members show mostly QT prolongation (gain-of-function phenotype) (Postema et al. 2009). Similar variability in type and severity of clinical symptoms has been demonstrated in carriers of the overlap mutation *SCN5A*-E1784K (Makita 2009). Thus, both mutation-specific and individual-specific factors

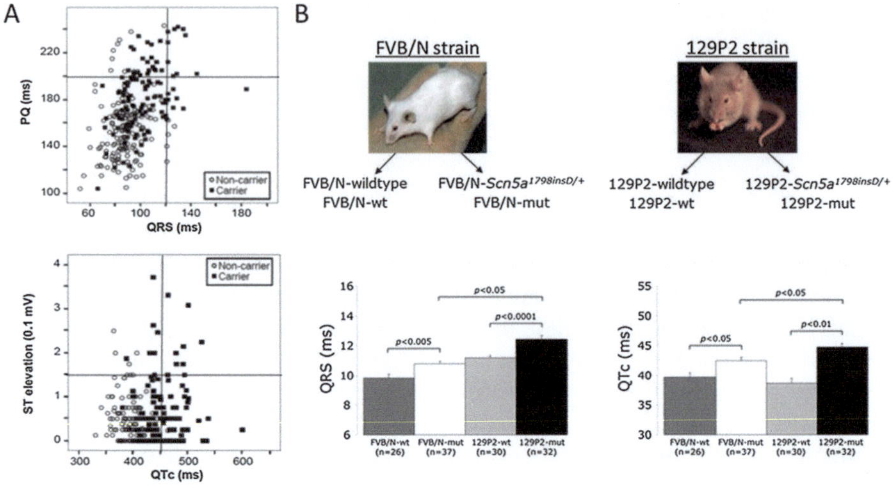

Fig. 2.7 (**a**) Extensive disease variability in *SCN5A*-1795insD mutation carriers; (**b**) severity of conduction (QRS) and repolarization (QTc) disorder differs significantly between two separate inbred mouse strains carrying the *Scn5a*-1798insD$^{/+}$ mutation. Reproduced in part from Remme et al. (2009a, b), with permission

appear to play an important role in determining disease expressivity and severity in sodium channelopathy.

2.6.2 Factors Modulating Disease Expression

Gender has been suggested as a modifier of disease severity in sodium channelopathy. Indeed, this is exemplified by the preponderance of Brugada syndrome in males. Moreover, gender may also affect the expressivity of *SCN5A* mutations, as shown by the observation that in a family carrying the *SCN5A*-G1406R mutation, females displayed predominantly with conduction defects, whereas males presented mostly with clinical features pertaining to the Brugada syndrome (Kyndt et al. 2001). Comorbidities are expected to modulate disease severity, but these remain largely unexplored. Age may constitute another clinical determinant of both disease severity and expressivity. In patients with progressive cardiac conduction disease, age-related fibrosis is thought to play a major role in exacerbating cardiac conduction with advancing age (Probst et al. 2003). Importantly, transgenic mice haplo-insufficient for the cardiac sodium channel *Scn5a* display similar age-dependent structural abnormalities (Royer et al. 2005). Some *SCN5A* mutations have been shown to induce symptoms of BrS specifically during episodes of fever, with channel gating properties affected by increasing temperature (Mok et al. 2003; Keller et al. 2005). Furthermore, membrane expression of certain trafficking-deficient loss-of-function *SCN5A* mutations may be improved by drugs such as the sodium channel blocker mexiletine and the potassium channel blocker cisapride (Valdivia et al. 2004; Liu et al. 2005).

2.6.3 Genetic Modifiers of Sodium Channelopathy

The variable disease expressivity and/or severity often observed among patients (and families) carrying the same sodium channel mutation suggest a potential role for genetic modifiers (Kyndt et al. 2001; van den Berg et al. 2001; Kolder et al. 2012). Identification of disease modifiers is however hindered by the substantial genetic heterogeneity across patients, as different mutations may be associated with different effects and thus also contribute to interindividual variability (Shimizu et al. 2009). We have previously investigated the effects of genetic factors on disease severity by comparing the phenotype in two distinct strains of mice carrying the *Scn5a*-1798insD$^{/+}$ mutation equivalent to *SCN5A*-1795insD in humans. We found that phenotype severity was more pronounced in 129P2 mice as compared to FVBN/J mice and subsequently identified potential modifiers of conduction disease severity, providing the first conclusive evidence that genetic modifiers may determine disease expressivity in cardiac sodium channelopathy (Fig. 2.7b) (Remme et al. 2009b; Scicluna et al. 2011; Lodder et al. 2012). Genetic variation between individuals is derived from the presence of single nucleotide polymorphisms (SNPs) which are frequently observed in the general population. H558R is the most common amino acid-changing polymorphism in *SCN5A* and has a reported prevalence of 9–36%, with a specific distribution among different ethnic populations (Ackerman et al. 2004). A number of studies have demonstrated that an interaction between this polymorphism and *SCN5A* mutations may exert relevant (albeit varying) effects on the functional consequences of mutant sodium channels, including gating properties, plasma membrane targeting, and current density (Viswanathan et al. 2003; Ye et al. 2003; Poelzing et al. 2006; Gui et al. 2010; Marangoni et al. 2011; Shinlapawittayatorn et al. 2011). Clinically, the H558R variant also impacts on ECG parameters and symptoms in BrS patients (Lizotte et al. 2009). A combination of certain polymorphisms (haplotype) within the regulatory (promoter) region of *SCN5A* (commonly observed in Asians) has also been shown to modulate ECG conduction parameters in BrS patients (Bezzina et al. 2006). Further genetic variation is introduced by the presence and relative expression of two main *SCN5A* alternatively spliced variants in cardiac tissue of each individual. The most abundant variant (65% of all *SCN5A* transcripts) is 2015-amino acids long and lacks a glutamine at position 1077 (*SCN5A*-Q1077del), in comparison with the less prominent 2016-amino acid variant *SCN5A*-Q1077 (35% of all transcripts) (Makielski et al. 2003). Phenotype severity of the BrS mutation *SCN5A*-G1406R was found to be dependent on the background splice variant in which it was expressed, with G1406R in combination with the Q1077 variant displaying more severe biophysical alterations (Tan et al. 2006). Similarly, the presence of Q1077del had significant effect on I_{Na} density and kinetics of mutant channels associated with DCM (Cheng et al. 2010).

2.7 Targeting Cardiac Sodium Channels

In sodium channel disease secondary to loss-of-function *SCN5A* mutations, therapeutic options are limited. No genotype-specific treatment strategies are available for Brugada syndrome or conduction disease (Shimizu et al. 2005). Although studies have shown that certain sodium channel blockers can actually restore membrane expression of trafficking defective sodium channels, this does not constitute an attractive therapeutic approach for clinical practice due to the potential deleterious effects of I_{Na} reduction. In the case of gain-of-function *SCN5A* mutations, pharmacological inhibition of the enhanced late I_{Na} is an attractive therapeutic target. While most agents with general sodium channel blocking effects also reduce late I_{Na}, the inhibiting effects of these agents on peak I_{Na} may set the stage for conduction slowing and ventricular arrhythmias, in particular during conditions where sodium channel function is already compromised (i.e., myocardial ischemia) (Echt et al. 1991; Lu et al. 2010). Furthermore, these "classical" sodium channel inhibitors may also impact on other ion channels, in particular potassium channels, potentially limiting their clinical applicability (Wang et al. 1996; Paul et al. 2002; Shimizu and Antzelevitch 1997). Overall, treatment efficacy is variable and mutation-specific (and likely also individual-specific) and as yet impossible to predict based on in vitro findings obtained in heterologous expression systems (see Wilde and Remme 2017). The most specific late I_{Na} inhibitor currently available is ranolazine, a piperazine derivative which is used for the treatment of chronic angina pectoris. Ranolazine may be up to 30- to 40-fold more potent in inhibiting late I_{Na} compared to peak I_{Na}, and small studies have shown potential clinical benefit in LQT3 patients (Moss et al. 2008; Chorin et al. 2016; van den Berg et al. 2014). However, its potency varies between species, cell type, and conditions (see Antzelevitch et al. 2011), and as with other nonselective sodium channel blockers, ranolazine also inhibits the repolarizing potassium current I_{Kr} (Antzelevitch et al. 2004; Schram et al. 2004). Recently, a novel highly selective inhibitor of late I_{Na}, GS-458967, has been identified (Belardinelli et al. 2013), which showed beneficial (anti-arrhythmic) effects in *Scn5a*-1798insD$^{/+}$ cardiomyocytes and human iPS-derived cardiomyocytes from a patient carrying the *SCN5A*-1795insD mutation (Portero et al. 2017). In addition, the analogue compound Eleclazine was recently reported to significantly shorten QTc in LQT3 patients (Zareba et al. 2016), further strengthening the potential clinical applicability of this therapeutic approach.

2.8 Expanding Horizons: Unraveling the Complexity of Cardiac Sodium Channel Function

2.8.1 Distinct Pools of Sodium Channels Within Subcellular Cardiomyocyte Microdomains

As mentioned in Sect. 2.2.2, cardiac sodium channels show inhomogeneous expression within the cardiac conduction system and across the ventricular wall. Additionally, distinct subcellular pools of Nav1.5-based sodium channels within the cardiomyocyte have been demonstrated, in particular at the intercalated disc and lateral membrane regions (Fig. 2.8a) (Verkerk et al. 2007; Lin et al. 2011; Marsman et al. 2014; Shy et al. 2014; Rivaud et al. 2017). Differences in peak I_{Na} amplitude and kinetics between channels located at these two sites have been observed (Fig. 2.8a), although species-specific differences likely exist (Verkerk et al. 2007; Lin et al. 2011; Rivaud et al. 2017). In addition, at the lateral membrane, sodium channels are located both at the cell crest and at the t-tubules, albeit at a lower density in the latter (Bhargava et al. 2013; Rivaud et al. 2017). Within these domains, Nav1.5 channels are not homogeneously distributed but rather grouped in clusters of various sizes and densities (Bhargava et al. 2013; Rivaud et al. 2017). Furthermore, different interacting proteins associate with Nav1.5 at these distinct subcellular domains and are thought to contribute to the differences in channel density and/or kinetics observed between these areas (Fig. 2.8b) (Petitprez et al. 2011; Shy et al.

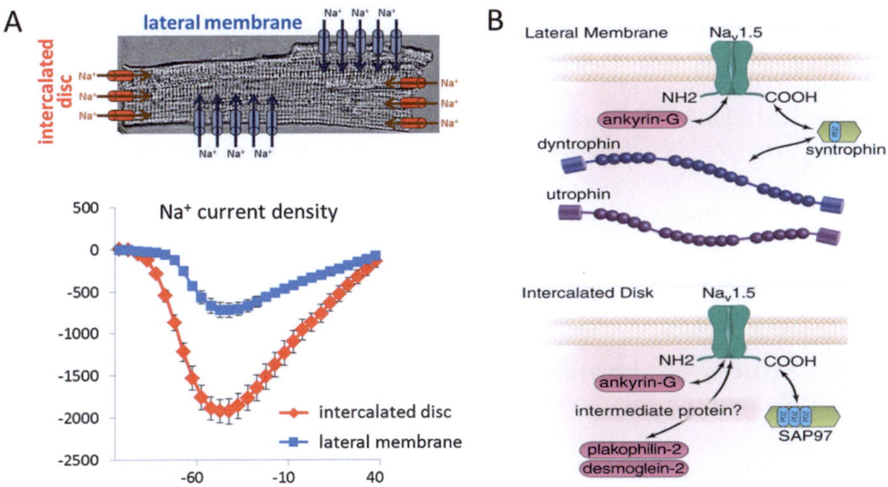

Fig. 2.8 (**a**) Macropatch measurements in mouse ventricular myocytes show significant differences in peak I_{Na} magnitude between the intercalated disc region and the lateral membrane. (**b**) Nav1.5 resides in distinct macromolecular complexes at two different subcellular domains of the cardiac cell: (top) at the lateral membrane where it interacts with the dystrophin–syntrophin complex and (bottom) at the intercalated discs with the MAGUK protein SAP97 (reproduced from Shy et al. (2013), with permission)

2013). Sodium channels located at the lateral membrane are associated with the syntrophin–dystrophin complex, and dystrophin-deficient mdx mice display reduced Nav1.5 expression levels predominantly at the lateral membrane (Petitprez et al. 2011). Similarly, mice lacking the last three amino acids of Nav1.5 essential for the interaction with the syntrophin–dystrophin complex show I_{Na} reduction exclusively at the lateral membrane (Shy et al. 2014). More recently, the MAGUK protein CASK (calcium-/calmodulin-dependent serine protein kinase), which localizes at the lateral membrane in association with dystrophin, was shown to interact with Nav1.5 and control its surface expression specifically at the lateral membrane (Eichel et al. 2016).

Nav1.5 channels in the intercalated disc region (which is devoid of syntrophin) interact with synapse-associated protein 97 (SAP97), plakophilin-2, desmoglein-2, and coxsackie and adenovirus receptor (CAR) (Petitprez et al. 2011; Sato et al. 2009; Rizzo et al. 2012; Marsman et al. 2014) (see also Sect. 2.3.3). Milstein and colleagues showed that SAP97 allows Nav1.5 to interact with the $K_{ir}2.1$ potassium channel and demonstrated a reciprocal modulation between these two channels (Milstein et al. 2012). Intriguingly, ablation of SAP97 in the mouse heart was found to mainly affect potassium channels but did not alter sodium channel function (Gillet et al. 2015). In mouse models of ARVC, loss of plakophilin-2 and a mutation in desmoglein-2 have both been shown to reduce I_{Na} even prior to the development of gross structural abnormalities (Cerrone et al. 2012; Rizzo et al. 2012). Moreover, mice with reduced expression of the coxsackie and adenovirus receptor (CAR, a cell adhesion protein enriched at the intercalated disc) showed preferential I_{Na} reduction in the intercalated disc region associated with increased arrhythmia inducibility during myocardial ischemia (Marsman et al. 2014). Overall, these findings underline the functional relevance of the sodium channel macromolecular complex and their differential regulation at various subcellular regions within the cardiomyocyte (Mohler and Hund 2011). However, little is known about the specific roles for these separate pools of channels or their functional relevance during (patho)physiological conditions. Moreover, the consequences of subcellular diversity in sodium channel composition and function for disease severity and expressivity of *SCN5A* mutations associated with sodium channelopathy and overlap syndrome are as yet unclear but form a challenging topic for future investigations.

2.8.2 Structural Abnormalities Secondary to *SCN5A* Mutations: Relevance for Arrhythmias

Cardiac sodium channelopathies were originally considered pure electrical entities occurring in the absence of structural heart disease, but it is now increasingly recognized that sodium channelopathy can also be associated with the development of cardiac fibrosis, dilatation, and hypertrophy. Indeed, (progressive) cardiac structural abnormalities have been observed in patients with *SCN5A* mutations (Bezzina et al. 2003; Coronel et al. 2005; Frustaci et al. 2005) and in *Scn5a* mouse models (Royer et al. 2005; Zhang et al. 2011; Jeevaratnam et al. 2012). Although it is as yet unclear how a mutation in a cardiac ion channel may ultimately lead to structural changes in the myocardium, a number of mechanisms have been proposed. The sodium channel interacts with many cytoskeletal proteins and components of the extracellular matrix, and one may hypothesize that sodium channel dysfunction may destabilize cytoskeletal integrity. Indeed, studies in mouse embryos and zebrafish demonstrated that a lack or dysfunction of *Scn5a* causes abnormal cardiac structure during development (Papadatos et al. 2002; Nuyens et al. 2001; Chopra et al. 2010). A role for Nav1.5 in cell adhesion has also been proposed, with decreased sodium channel function leading to impaired adhesion and consequent cardiomyopathy (see Sect. 2.5.7) (Te Riele et al. 2017). Moreover, electrical activity-dependent stimulation of pro-fibrotic factors of the transforming growth factor-β (TGFβ) pathway may occur in the setting of sodium channel dysfunction (Leask 2007). Finally, increased late I_{Na} associated with gain-of-function *Scn5a* mutations can disrupt intracellular calcium homeostasis, not only providing a pro-arrhythmic substrate but also predisposing to the development of fibrosis and/or hypertrophy (Lindegger et al. 2009; Remme et al. 2010). As a result, pharmacological late I_{Na} inhibition may prove beneficial by preventing more long-term detrimental effects of intracellular calcium overload (see Remme and Wilde 2013). Taken together, these observations are intriguing since they imply that sodium channels not only determine electrophysiological characteristics of the myocardium but also exert as yet unknown regulatory effects on myocardial structure and function. While different types of *SCN5A* mutations (gain- versus loss-of-function) may lead to structural abnormalities through various mechanisms, the end result is similar, i.e., an increased propensity for cardiac arrhythmias.

2.8.3 Functional Relevance of "Neuronal" Sodium Channels in the Heart

It is now increasingly recognized that other sodium channel isoforms besides *SCN5A*/Nav1.5 are also expressed in the heart and/or cardiomyocytes. These isoforms (*SCN1A*/Nav1.1, *SCN2A*/Nav1.2, *SCN3A*/Nav1.3, *SCN8A*/Nav1.6, *SCN9A*/Nav1.7, and *SCN10A*/Nav1.8) are often designated "neuronal" sodium channels due to their enrichment in central and peripheral neuronal tissues. The majority of these channels display different current kinetics and increased TTX sensitivity compared to Nav1.5 and have been shown to be differentially regulated during pressure overload (Brette and Orchard 2006; Xi et al. 2009). TTX-sensitive, neuronal-type sodium channels comprise approximately 10% of total I_{Na} in cardiomyocytes (Brette and Orchard 2006; Haufe et al. 2005) and are mostly found at the lateral membrane region within t-tubuli (Verkerk et al. 2007; Lin et al. 2011). The functional relevance of these neuronal channels in cardiomyocytes is not yet fully understood, but they are thought to be involved in t-tubular transmission of the action potential to the myocyte interior, intracellular calcium homeostasis, and excitation–contraction coupling (Maier et al. 2002; Noujaim et al. 2012; Torres et al. 2010; Westenbroek et al. 2013). Furthermore, TTX-sensitive sodium channels may underlie, at least in part, the increased late I_{Na} observed during pathophysiological conditions (Xi et al. 2009; Mishra et al. 2015). A recent study furthermore demonstrated that Nav1.6 upregulation was associated with pro-arrhythmic intracellular sodium–calcium dysregulation and that pharmacological Nav1.6 inhibition had anti-arrhythmic effects (Radwański et al. 2016). Clinically, mutations in neuronal sodium channels are associated with epilepsy, but cardiac arrhythmias have only been infrequently reported in disease models (Frasier et al. 2016; Kalume et al. 2013). Hence, the relevance of these channels in inherited arrhythmia syndromes is still largely unclear.

2.9 Conclusions and Future Directions

The last two decades has seen a tremendous increase in our knowledge on inherited sodium channel (dys)function and its consequences for cardiac electrophysiology and arrhythmogenesis, in part driven by the advancement of available technologies. Traditionally, heterologous expression systems have been used to study the biophysical properties of *SCN5A* mutations and correlate them to the clinical disease phenotype. However, results obtained in such cell lines may not always be representative of the situation in the cardiomyocyte environment, as evidenced by findings from a number of transgenic mouse studies (Remme et al. 2006; Watanabe et al. 2011). Additionally, cardiomyocytes derived from human-induced pluripotent stem cells (hiPSC) have proven useful for studying the consequences of mutations in a more physiological environment (see also Chap. 15) (Casini et al. 2017). We and others have shown that hiPSC-derived cardiomyocytes from human *SCN5A* mutation carriers and Scn5a transgenic mice recapitulate the disease phenotype (Davis et al. 2012; Malan et al. 2011; Terrenoire et al. 2013). Moreover, they appear suitable for the investigation of patient- and disease-specific pharmacology and provide a tool for studying the role of genetic background (Portero et al. 2017; Casini et al. 2017). Nevertheless, in vivo, ex vivo, and in vitro studies in transgenic mouse models will remain important for the assessment of the differential effects of sodium channel defects in different regions and cell types of the heart, as well as investigation of disease progression and development of structural abnormalities with age (see also Chap. 16). Ongoing and future studies, aimed at identifying novel components and functions of the sodium channel macromolecular complex, will benefit greatly from new molecular, imaging, and patch-clamp techniques which have recently become available to the field (see also Chap. 14). Such studies are expected to provide new targets for genetic screening and/or future development of therapeutic strategies for sodium channel (dys)function. Continued identification of mutations in known and novel genes in patients with inherited arrhythmia syndromes remains crucial (Bezzina et al. 2013; Veerman et al. 2017). Here, combining genetic and population studies such as GWAS with molecular and functional studies in animals and cell systems will help identify and prioritize novel genes and pathways potentially relevant for cardiac electrophysiology (Scicluna et al. 2011; Kolder et al. 2012; Lodder et al. 2012). In addition, identifying genetic variants modulating cardiac conduction and/or repolarization will ultimately facilitate risk stratification in patients. As such, combined clinical, genetic, and translational studies should ultimately improve diagnosis, treatment, and outcome in patients with cardiac sodium channelopathy.

Compliance with Ethical Standards

Sources of Funding This work was funded by a Priority Medicines Rare Diseases and Orphan Drugs grant (PM-Rare, 113303006) from The Netherlands Organization for Health Research and Development (ZonMw) and an Innovational Research Incentives Scheme Vidi grant from ZonMw (grant no. 91714371).

Conflict of Interest Carol Ann Remme has previously received research grants from Gilead Sciences.

Ethical Approval This article does not contain any studies with human participants or animals performed by any of the authors.

References

Abriel H. Cardiac sodium channel Na(v)1.5 and interacting proteins: physiology and pathophysiology. J Mol Cell Cardiol. 2010;48:2–11.

Abriel H, Staub O. Ubiquitylation of ion channels. Physiology (Bethesda). 2005;20:398–407.

Ackerman MJ, Splawski I, Makielski JC, et al. Spectrum and prevalence of cardiac sodium channel variants among black, white, Asian, and Hispanic individuals: implications for arrhythmogenic susceptibility and Brugada/long QT syndrome genetic testing. Heart Rhythm. 2004;15:600–7.

Agullo-Pascual E, Cerrone M, Delmar M. Arrhythmogenic cardiomyopathy and Brugada syndrome: diseases of the connexome. FEBS Lett. 2014;588(8):1322–30.

Allouis M, Le Bouffant F, Wilders R, Péroz D, Schott JJ, Noireaud J, Le Marec H, Mérot J, Escande D, Baró I. 14-3-3 is a regulator of the cardiac voltage-gated sodium channel Nav1.5. Circ Res. 2006;98:1538–46.

Amin AS, Verkerk AO, Bhuiyan ZA, Wilde AA, Tan HL. Novel Brugada syndrome-causing mutation in ion-conducting pore of cardiac Na^+ channel does not affect ion selectivity properties. Acta Physiol Scand. 2005;185:291–301.

Antzelevitch C. Brugada syndrome. Pacing Clin Electrophysiol. 2006;29:1130–59.

Antzelevitch C, Belardinelli L, Zygmunt AC, Burashnikov A, Di Diego JM, Fish JM, Cordeiro JM, Thomas G. Electrophysiological effects of ranolazine, a novel antianginal agent with antiarrhythmic properties. Circulation. 2004;110:904–10.

Antzelevitch C, Brugada P, Borggrefe M, Brugada J, Brugada R, Corrado D, Gussak I, Le Marec H, Nademanee K, Perez Riera AR, Shimizu W, Schulze-Bahr E, Tan H, Wilde A. Brugada syndrome: report of the second consensus conference. Heart Rhythm. 2005;2:429–40.

Antzelevitch C, Burashnikov A, Sicouri S, Belardinelli L. Electrophysiologic basis for the antiarrhythmic actions of ranolazine. Heart Rhythm. 2011;8(8):1281–90.

Arnolds DE, Liu F, Fahrenbach JP, Kim GH, Schillinger KJ, Smemo S, McNally EM, Nobrega MA, Patel VV, Moskowitz IP. TBX5 drives Scn5a expression to regulate cardiac conduction system function. J Clin Invest. 2012;122:2509–18.

Balse E, Steele DF, Abriel H, Coulombe A, Fedida D, Hatem SN. Dynamic of ion channel expression at the plasma membrane of cardiomyocytes. Physiol Rev. 2012;92(3):1317–58.

Balser JR. The cardiac sodium channel: gating function and molecular pharmacology. J Mol Cell Cardiol. 2001;33:599–613.

Belardinelli L, Liu G, Smith-Maxwell C, Wang WQ, El-Bizri N, Hirakawa R, Karpinski S, Li CH, Hu L, Li XJ, Crumb W, Wu L, Koltun D, Zablocki J, Yao L, Dhalla AK, Rajamani S, Shryock JC. A novel, potent, and selective inhibitor of cardiac late sodium current suppresses experimental arrhythmias. J Pharmacol Exp Ther. 2013;344(1):23–32.

Belhassen B, Viskin S. Pharmacologic approach to therapy of Brugada syndrome: quinidine as an alternative to ICD therapy. In: Antzelevitch C, Brugada P, Brugada J, Brugada R, editors. The Brugada syndrome: from bench to bedside. Oxford: Blackwell Futura; 2004. p. 202–11.

Bennett PB, Yazawa K, Makita N, George AL. Molecular mechanism for an inherited cardiac arrhythmia. Nature. 1995;376:683–5.

Benson DW, Wang DW, Dyment M, Knilans TK, Fish FA, Strieper MJ, Rhodes TH, George AL Jr. Congenital sick sinus syndrome caused by recessive mutations in the cardiac sodium channel gene (SCN5A). J Clin Invest. 2003;112:1019–28.

Beyder A, Rae JL, Bernard C, Strege PR, Sachs F, Farrugia G. Mechanosensitivity of Nav1.5, a voltage-sensitive sodium channel. J Physiol. 2010;588(Pt 24):4969–85.

Bezzina CR, Remme CA. Dilated cardiomyopathy due to sodium channel dysfunction: what is the connection? Circ Arrhythm Electrophysiol. 2008;1:80–2.

Bezzina CR, Veldkamp MW, van den Berg MP, Postma AV, Rook MB, Viersma JW, van Langen IM, Tan-Sindhunata G, Bink-Boelkens MT, van Der Hout AH, Mannens MM, Wilde AA. A single Na^+ channel mutation causing both long-QT and Brugada syndromes. Circ Res. 1999;85:1206–13.

Bezzina CR, Rook MB, Groenewegen WA, Herfst LJ, van der Wal AC, Lam J, Jongsma HJ, Wilde AAM, Mannens MMAM. Compound heterozygosity for mutations (W156X and R225W) in SCN5A associated with severe cardiac conduction disturbances and degenerative changes in the conduction system. Circ Res. 2003;92:159–68.

Bezzina CR, Shimizu W, Yang P, et al. A common sodium channel promoter haplotype in Asian subjects underlies variability in cardiac conduction. Circulation. 2006;113:338–44.

Bezzina CR, Barc J, Mizusawa Y, Remme CA, Gourraud JB, Simonet F, Verkerk AO, Schwartz PJ, Crotti L, Dagradi F, Guicheney P, Fressart V, Leenhardt A, Antzelevitch C, Bartkowiak S, Schulze-Bahr E, Zumhagen S, Behr ER, Bastiaenen R, Tfelt-Hansen J, Olesen MS, Kääb S, Beckmann BM, Weeke P, Watanabe H, Endo N, Minamino T, Horie M, Ohno S, Hasegawa T, Makita N, Nogami A, Shimizu W, Aiba T, Froguel P, Balkau B, Lantieri O, Wiese C, Weber D, Wolswinkel R, Coronel R, Boukens BJ, Charpentier E, Chatel S, Despres A, Gros F, Kyndt F, Lecointe S, Lindenbaum P, Portero V, Violleau J, Gessler M, Tan HL, Roden D, Christoffels VM, Le Marec H, Wilde AA, Probst V, Schott JJ, Dina C, Redon R. Common variants at the SCN5A/SCN10A and HEY2 loci predispose to Brugada syndrome, a rare disease with high risk of sudden cardiac death. Nat Genet. 2013;45(9):1044–9.

Bhargava A, Lin X, Novak P, Mehta K, Korchev Y, Delmar M, Gorelik J. Super-resolution scanning patch clamp reveals clustering of functional ion channels in adult ventricular myocyte. Circ Res. 2013;112(8):1112–20.

Brette F, Orchard CH. Density and sub-cellular distribution of cardiac and neuronal sodium channel isoforms in rat ventricular myocytes. Biochem Biophys Res Commun. 2006;348(3):1163–6.

Calkins H, Corrado D, Marcus F. Risk stratification in arrhythmogenic right ventricular cardiomyopathy. Circulation. 2017;136(21):2068–82.

Casini S, Tan HL, Bhuiyan ZA, Bezzina CR, Barnett P, Cerbai E, Mugelli A, Wilde AA, Veldkamp MW. Characterization of a novel SCN5A mutation associated with Brugada syndrome reveals involvement of DIIIS4-S5 linker in slow inactivation. Cardiovasc Res. 2007;76:418–29.

Casini S, Verkerk AO, van Borren MM, van Ginneken AC, Veldkamp MW, de Bakker JM, Tan HL. Intracellular calcium modulation of voltage-gated sodium channels in ventricular myocytes. Cardiovasc Res. 2009;81:72–81.

Casini S, Tan HL, Demirayak I, Remme CA, Amin AS, Scicluna BP, Chatyan H, Ruijter JM, Bezzina CR, van Ginneken AC, Veldkamp MW. Tubulin polymerization modifies cardiac sodium channel expression and gating. Cardiovasc Res. 2010;85:691–700.

Casini S, Verkerk AO, Remme CA. Human iPSC-derived cardiomyocytes for investigation of disease mechanisms and therapeutic strategies in inherited arrhythmia syndromes: strengths and limitations. Cardiovasc Drugs Ther. 2017;31(3):325–44. https://doi.org/10.1007/s10557-017-6735-0.

Cerrone M, Noorman M, Lin X, Chkourko H, Liang FX, van der Nagel R, Hund T, Birchmeier W, Mohler P, van Veen TA, van Rijen HV, Delmar M. Sodium current deficit and arrhythmogenesis in a murine model of plakophilin-2 haploinsufficiency. Cardiovasc Res. 2012;95(4):460–8.

Cerrone M, Lin X, Zhang M, Agullo-Pascual E, Pfenniger A, Chkourko Gusky H, Novelli V, Kim C, Tirasawadichai T, Judge DP, Rothenberg E, Chen HS, Napolitano C, Priori SG, Delmar M. Missense mutations in plakophilin-2 cause sodium current deficit and associate with a Brugada syndrome phenotype. Circulation. 2014;129(10):1092–103.

Chen T, Inoue M, Sheets MF. Reduced voltage dependence of inactivation in the SCN5A sodium channel mutation delF1617. Am J Physiol Heart Circ Physiol. 2005;288:H2666–76.

Cheng J, Morales A, Siegfried JD, Li D, Norton N, Song J, Gonzalez-Quintana J, Makielski JC, Hershberger RE. SCN5A rare variants in familial dilated cardiomyopathy decrease peak sodium current depending on the common polymorphism H558R and common splice variant Q1077del. Clin Transl Sci. 2010;3(6):287–94.

Chioni AM, Fraser SP, Pani F, Foran P, Wilkin GP, Diss JK, Djamgoz MB. A novel polyclonal antibody specific for the Na(v)1.5 voltage-gated Na(+) channel 'neonatal' splice form. J Neurosci Methods. 2005;147:88–98.

Chopra SS, Stroud DM, Watanabe H, Bennett JS, Burns CG, Wells KS, Yang T, Zhong TP, Roden DM. Voltage-gated sodium channels are required for heart development in zebrafish. Circ Res. 2010;106:1342–50.

Chorin E, Hu D, Antzelevitch C, Hochstadt A, Belardinelli L, Zeltser D, Barajas-Martinez H, Rozovski U, Rosso R, Adler A, Benhorin J, Viskin S. Ranolazine for congenital long-QT syndrome type III: experimental and long-term clinical data. Circ Arrhythm Electrophysiol. 2016;9(10). pii: e004370.

Clancy CE, Rudy Y. Na^+ channel mutation that causes both Brugada and long-QT syndrome phenotypes. A simulation study of mechanism. Circulation. 2002;105:1208–13.

Clancy CE, Tateyama M, Liu H, Wehrens XH, Kass RS. Non-equilibrium gating in cardiac Na^+ channels: an original mechanism of arrhythmia. Circulation. 2003;107:2233–7.

Cordeiro JM, Barajas-Martinez H, Hong K, et al. Compound heterozygous mutations P336L and I1660V in the human cardiac sodium channel associated with the Brugada syndrome. Circulation. 2006;114:2026–33.

Coronel R, Casini S, Koopmann TT, Wilms-Schopman FJG, Verkerk AO, de Groot JR, Bhuiyan Z, Bezzina CR, Veldkamp MW, Linnenbank AC, van der Wal AC, Tan HL, Brugada P, Wilde AA, de Bakker JM. Right ventricular fibrosis and conduction delay in a patient with clinical signs of Brugada syndrome: a combined electrophysiologic, genetic, histopathologic and computational study. Circulation. 2005;112:2769–77.

Daimi H, Lozano-Velasco E, Haj Khelil A, Chibani JB, Barana A, Amorós I, González de la Fuente M, Caballero R, Aranega A, Franco D. Regulation of SCN5A by microRNAs: miR-219 modulates SCN5A transcript expression and the effects of flecainide intoxication in mice. Heart Rhythm. 2015;12:1333–42.

Darbar D, Kannankeril PJ, Donahue BS, Kucera G, Stubblefield T, Haines JL, George AL Jr, Roden DM. Cardiac sodium channel (SCN5A)variants associated with atrial fibrillation. Circulation. 2008;117:1927–35.

Davis RP, Casini S, van den Berg CW, Hoekstra M, Remme CA, Dambrot C, Salvatori D, Oostwaard DW, Wilde AA, Bezzina CR, Verkerk AO, Freund C, Mummery CL. Cardiomyocytes derived from pluripotent stem cells recapitulate electrophysiological characteristics of an overlap syndrome of cardiac sodium channel disease. Circulation. 2012;125:3079–91.

Echt DS, Liebson PR, Mitchell LB, et al. Mortality and morbidity in patients receiving encainide, flecainide, or placebo. The Cardiac Arrhythmia Suppression Trial. N Engl J Med. 1991;324 (12):781–8.

Eichel CA, Beuriot A, Chevalier MY, Rougier JS, Louault F, Dilanian G, Amour J, Coulombe A, Abriel H, Hatem SN, Balse E. Lateral membrane-specific MAGUK CASK down-regulates NaV1.5 channel in cardiac myocytes. Circ Res. 2016;119(4):544–56.

Ellinor PT, Nam EG, Shea MA, Milan DJ, Ruskin JN, Macrae CA. Cardiac sodium channel mutation in atrial fibrillation. Heart Rhythm. 2008;5:99–105.

Fozzard HA, Makielski JC. The electrophysiology of acute myocardial ischemia. Annu Rev Med. 1985;36:275.

Frasier CR, Wagnon JL, Bao YO, McVeigh LG, Lopez-Santiago LF, Meisler MH, Isom LL. Cardiac arrhythmia in a mouse model of sodium channel SCN8A epileptic encephalopathy. Proc Natl Acad Sci U S A. 2016. pii: 201612746.

Frustaci A, Priori SG, Pieroni M, Chimenti C, Napolitano C, Rivolta I, Sanna T, Bellocci F, Russo MA. Cardiac histological substrate in patients with clinical phenotype of Brugada syndrome. Circulation. 2005;112:3680–7.

Gavillet B, Rougier JS, Domenighetti AA, Behar R, Boixel C, Ruchat P, Lehr HA, Pedrazzini T, Abriel H. Cardiac sodium channel Nav1.5 is regulated by a multiprotein complex composed of syntrophins and dystrophin. Circ Res. 2006;99:407–14.

Ge J, Sun A, Paajanen V, Wang S, Su C, Yang Z, Li Y, Wang S, Jia J, Wang K, Zou Y, Gao L, Wang K, Fan Z. Molecular and clinical characterization of a novel SCN5A mutation associated with atrioventricular block and dilated cardiomyopathy. Circ Arrhythm Electrophysiol. 2008;1:83–92.

Gillet L, Rougier JS, Shy D, Sonntag S, Mougenot N, Essers M, Shmerling D, Balse E, Hatem SN, Abriel H. Cardiac-specific ablation of synapse-associated protein SAP97 in mice decreases potassium currents but not sodium current. Heart Rhythm. 2015;12(1):181–92.

Gosselin-Badaroudine P, Keller DI, Huang H, Pouliot V, Chatelier A, Osswald S, Brink M, Chahine M. A proton leak current through the cardiac sodium channel is linked to mixed arrhythmia and the dilated cardiomyopathy phenotype. PLoS One. 2012;7:e38331.

Grant AO, Carboni MP, Neplioueva V, Starmer CF, Memmi M, Napolitano C, Priori S. Long QT syndrome, Brugada syndrome, and conduction system disease are linked to a single sodium channel mutation. J Clin Invest. 2002;110:1201–9.

Gui J, Wang T, Trump D, Zimmer T, Lei M. Mutation-specific effects of polymorphism H558R in SCN5A-related sick sinus syndrome. J Cardiovasc Electrophysiol. 2010;21(5):564–73.

Haufe V, Camacho JA, Dumaine R, Günther B, Bollensdorff C, von Banchet GS, Benndorf K, Zimmer T. Expression pattern of neuronal and skeletal muscle voltage-gated Na+ channels in the developing mouse heart. J Physiol. 2005;564(Pt 3):683–96.

Herren AW, Bers D, Grandi E. Post-translational modifications of the cardiac Na channel: contribution of CaMKII-dependent phosphorylation to acquired arrhythmias. Am J Physiol Heart Circ Physiol. 2013;305(4):H431–45.

Holm AN, Rich A, Miller SM, Strege P, Ou Y, Gibbons S, Sarr MG, Szurszewski JH, Rae JL, Farrugia G. Gastroenterology. 2002;122(1):178–87.

Hoogendijk MG, Potse M, Linnenbank AC, Verkerk AO, den Ruijter HM, van Amersfoorth SC, Klaver EC, Beekman L, Bezzina CR, Postema PG, Tan HL, Reimer AG, van der Wal AC, Ten Harkel AD, Dalinghaus M, Vinet A, Wilde AA, de Bakker JM, Coronel R. Mechanism of right precordial ST-segment elevation in structural heart disease: excitation failure by current-to-load mismatch. Heart Rhythm. 2010;7:238–48.

Hu D, Barajas-Martinez H, Burashnikov E, Springer M, Wu Y, Varro A, Pfeiffer R, Koopmann TT, Cordeiro JM, Guerchicoff A, Pollevick GD, Antzelevitch C. A mutation in the {beta}3 subunit of the cardiac sodium channel associated with Brugada ECG phenotype. Circ Cardiovasc Genet. 2009;2:270–8.

Janse MJ, Wit AL. Electrophysiological mechanisms of ventricular arrhythmias resulting from myocardial ischemia and infarction. Physiol Rev. 1989;69:1049.

Jeevaratnam K, Rewbury R, Zhang Y, Guzadhur L, Grace AA, Lei M, Huang CL. Frequency distribution analysis of activation times and regional fibrosis in murine Scn5a+/- hearts: the effects of ageing and sex. Mech Ageing Dev. 2012;133:591–9.

Jespersen T, Gavillet B, van Bemmelen MX, Cordonier S, Thomas MA, Staub O, Abriel H. Cardiac sodium channel Na(v)1.5 interacts with and is regulated by the protein tyrosine phosphatase PTPH1. Biochem Biophys Res Commun. 2006;348:1455–62.

Jones DK, Peters CH, Tolhurst SA, Claydon TW, Ruben PC. Extracellular proton modulation of the cardiac voltage-gated sodium channel, Nav1.5. Biophys J. 2011;101:2147–56.

Kalume F, Westenbroek RE, Cheah CS, Yu FH, Oakley JC, Scheuer T, Catterall WA. Sudden unexpected death in a mouse model of Dravet syndrome. J Clin Invest. 2013;123(4):1798–808.

Kanters JK, Yuan L, Hedley PL, Stoevring B, Jons C, Bloch Thomsen PE, Grunnet M, Christiansen M, Jespersen T. Flecainide provocation reveals concealed brugada syndrome in a long QT syndrome family with a novel L1786Q mutation in SCN5A. Circ J. 2014;78 (5):1136–43.

Kapplinger JD, Tester DJ, Alders M, Benito B, Berthet M, Brugada J, Brugada P, Fressart V, Guerchicoff A, Harris-Kerr C, Kamakura S, Kyndt F, Koopmann TT, Miyamoto Y, Pfeiffer R, Pollevick GD, Probst V, Zumhagen S, Vatta M, Towbin JA, Shimizu W, Schulze-Bahr E, Antzelevitch C, Salisbury BA, Guicheney P, Wilde AA, Brugada R, Schott JJ, Ackerman MJ. An international compendium of mutations in the SCN5A-encoded cardiac sodium channel in patients referred for Brugada syndrome genetic testing. Heart Rhythm. 2010;7:33–46.

Kass RS. Sodium channel inactivation in heart: a novel role of the carboxyterminal domain. J Cardiovasc Electrophysiol. 2006;17(Suppl 1):S21–5.

Kattygnarath D, Maugenre S, Neyroud N, Balse E, Ichai C, Denjoy I, Dilanian G, Martins RP, Fressart V, Berthet M, Schott JJ, Leenhardt A, Probst V, Le Marec H, Hainque B, Coulombe A, Hatem SN, Guicheney P. MOG1: a new susceptibility gene for Brugada syndrome. Circ Cardiovasc Genet. 2011;4:261–8.

Keller DI, Rougier JS, Kucera JP, et al. Brugada syndrome and fever: genetic and molecular characterization of patients carrying SCN5A mutations. Cardiovasc Res. 2005;67:510–9.

Ko SH, Lenkowski PW, Lee HC, Mounsey JP, Patel MK. Modulation of Na(v)1.5 by beta1- and beta3-subunit co-expression in mammalian cells. Pflugers Arch. 2005;449:403–12.

Kolder IC, Tanck MW, Bezzina CR. Common genetic variation modulating cardiac ECG parameters and susceptibility to sudden cardiac death. J Mol Cell Cardiol. 2012;52:620–9.

Kyle JW, Makielski JC. Diseases caused by mutations in Nav1.5 interacting proteins. Card Electrophysiol Clin. 2014;6(4):797–809.

Kyndt F, Probst V, Potet F, Demolombe S, Chevallier JC, Baró I, Moisan JP, Boisseau P, Schott JJ, Escande D, Le Marec H. Novel SCN5A mutation leading either to isolated cardiac conduction defect or Brugada syndrome in a large French family. Circulation. 2001;104:3081–6.

Leask A. TGFβ, cardiac fibroblasts, and the fibrotic response. Cardiovasc Res. 2007;74:207–12.

Lei M, Jones SA, Liu J, Lancaster MK, Fung SS, Dobrzynski H, Camelliti P, Maier SK, Noble D, Boyett MR. Requirement of neuronal- and cardiac-type sodium channels for murine sinoatrial node pacemaking. J Physiol. 2004;559(Pt 3):835–48.

Lei M, Huang CL, Zhang Y. Genetic Na+ channelopathies and sinus node dysfunction. Prog Biophys Mol Biol. 2008;98:171–8.

Lemaillet G, Walker B, Lambert S. Identification of a conserved ankyrin-binding motif in the family of sodium channel alpha subunits. J Biol Chem. 2003;278:27333–9.

Li Q, Huang H, Liu G, Lam K, Rutberg J, Green MS, Birnie DH, Lemery R, Chahine M, Gollob MH. Gain-of-function mutation of Nav1.5 in atrial fibrillation enhances cellular excitability and lowers the threshold for action potential firing. Biochem Biophys Res Commun. 2009;380:132–7.

Li Z, Ai T, Samani K, Xi Y, Tzeng HP, Xie M, Wu S, Ge S, Taylor MD, Dong JW, Cheng J, Ackerman MJ, Kimura A, Sinagra G, Brunelli L, Faulkner G, Vatta M. A ZASP missense mutation, S196L, leads to cytoskeletal and electrical abnormalities in a mouse model of cardiomyopathy. Circ Arrhythm Electrophysiol. 2010;3:646–56.

Lin X, Liu N, Lu J, Zhang J, Anumonwo JM, Isom LL, Fishman GI, Delmar M. Subcellular heterogeneity of sodium current properties in adult cardiac ventricular myocytes. Heart Rhythm. 2011;8:1923–30.

Lindegger N, Hagen BM, Marks AR, Lederer WJ, Kass RS. Diastolic transient inward current in long QT syndrome type 3 is caused by Ca2+ overload and inhibited by ranolazine. J Mol Cell Cardiol. 2009;47:326–34.

Liu K, Yang T, Viswanathan PC, Roden DM. New mechanism contributing to drug-induced arrhythmia: rescue of a misprocessed LQT3 mutant. Circulation. 2005;112:3239–46.

Lizotte E, Junttila MJ, Dube MP, Hong K, Benito B, DE Zutter M, Henkens S, Sarkozy A, Huikuri HV, Towbin J, Vatta M, Brugada P, Brugada J, Brugada R. Genetic modulation of brugada syndrome by a common polymorphism. J Cardiovasc Electrophysiol. 2009;20(10):1137–41.

Lodder EM, Scicluna BP, Milano A, Sun AY, Tang H, Remme CA, Moerland PD, Tanck MW, Pitt GS, Marchuk DA, Bezzina CR. Dissection of a quantitative trait locus for PR interval duration identifies Tnni3k as a novel modulator of cardiac conduction. PLoS Genet. 2012;8:e1003113.

London B, Michalec M, Mehdi H, Zhu X, Kerchner L, Sanyal S, Viswanathan PC, Pfahnl AE, Shang LL, Madhusudanan M, Baty CJ, Lagana S, Aleong R, Gutmann R, Ackerman MJ, McNamara DM, Weiss R, Dudley SC Jr. Mutation in glycerol-3-phosphate dehydrogenase 1 like gene (GPD1-L) decreases cardiac Na+ current and causes inherited arrhythmias. Circulation. 2007;116:2260–8.

Lu HR, Rohrbacher J, Vlaminckx E, Van Ammel K, Yan GX, Gallacher DJ. Predicting drug-induced slowing of conduction and pro-arrhythmia: identifying the 'bad' sodium current blockers. Br J Pharmacol. 2010;160(1):60–76.

Maier SK, Westenbroek RE, Schenkman KA, Feigl EO, Scheuer T, Catterall WA. An unexpected role for brain-type sodium channels in coupling of cell surface depolarization to contraction in the heart. Proc Natl Acad Sci U S A. 2002;99(6):4073–8.

Makielski JC, Ye B, Valdivia CR, Pagel MD, Pu J, Tester DJ, Ackerman MJ. A ubiquitous splice variant and a common polymorphism affect heterologous expression of recombinant human SCN5A heart sodium channels. Circ Res. 2003;93:821–8.

Makita N. Phenotypic overlap of cardiac sodium channelopathies: individual-specific or mutation-specific? Circ J. 2009;73(5):810–7.

Makita N, Behr E, Shimizu W, Horie M, Sunami A, Crotti L, Schulze-Bahr E, Fukuhara S, Mochizuki N, Makiyama T, Itoh H, Christiansen M, McKeown P, Miyamoto K, Kamakura S, Tsutsui H, Schwartz PJ, George AL Jr, Roden DM. The E1784K mutation in SCN5A is associated with mixed clinical phenotype of type 3 long QT syndrome. J Clin Invest. 2008;118(6):2219–29.

Makiyama T, Akao M, Tsuji K, et al. High risk for bradyarrhythmic complications in patients with Brugada syndrome caused by SCN5A gene mutations. J Am Coll Cardiol. 2005;46:2100–6.

Makiyama T, Akao M, Shizuta S, Doi T, Nishiyama K, Oka Y, Ohno S, Nishio Y, Tsuji K, Itoh H, Kimura T, Kita T, Horie M. A novel SCN5A gain-of-function mutation M1875T associated with familial atrial fibrillation. J Am Coll Cardiol. 2008;52:1326–34.

Malan D, Friedrichs S, Fleischmann BK, Sasse P. Cardiomyocytes obtained from induced pluripotent stem cells with long-QT syndrome 3 recapitulate typical disease-specific features in vitro. Circ Res. 2011;109:841–7.

Malhotra DJ, Chen C, Rivolta I, Abriel H, Malhotra R, Mattei LN, Brosius FC, Kass RS, Isom LL. Characterization of sodium channel alpha and beta-subunits in rat and mouse cardiac myocytes. Circulation. 2001;103:1303–10.

Marangoni S, Di Resta C, Rocchetti M, Barile L, Rizzetto R, Summa A, Severi S, Sommariva E, Pappone C, Ferrari M, Benedetti S, Zaza A. A Brugada syndrome mutation (p.S216L) and its modulation by p.H558R polymorphism: standard and dynamic characterization. Cardiovasc Res. 2011;91(4):606–16.

Marionneau C, Abriel H. Regulation of the cardiac Na+ channel NaV1.5 by post-translational modifications. J Mol Cell Cardiol. 2015;82:36–47.

Marsman RF, Bezzina CR, Freiberg F, Verkerk AO, Adriaens ME, Podliesna S, Chen C, Purfürst B, Spallek B, Koopmann TT, Baczko I, Dos Remedios CG, George AL Jr, Bishopric NH, Lodder EM, de Bakker JM, Fischer R, Coronel R, Wilde AA, Gotthardt M, Remme CA. Coxsackie and adenovirus receptor is a modifier of cardiac conduction and arrhythmia vulnerability in the setting of myocardial ischemia. J Am Coll Cardiol. 2014;63(6):549–59.

Mazzone A, Strege PR, Tester DJ, Bernard CE, Faulkner G, De Giorgio R, Makielski JC, Stanghellini V, Gibbons SJ, Ackerman MJ, Farrugia G. A mutation in telethonin alters Nav1.5 function. J Biol Chem. 2008;283:16537–44.

McNair WP, Ku L, Taylor MR, Fain PR, Dao D, Wolfel E, Mestroni L. SCN5A mutation associated with dilated cardiomyopathy, conduction disorder, and arrhythmia. Circulation. 2004;110:2163–7.

McNair WP, Sinagra G, Taylor MR, Di Lenarda A, Ferguson DA, Salcedo EE, Slavov D, Zhu X, Caldwell JH, Mestroni L. Familial Cardiomyopathy Registry Research Group. SCN5A mutations associate with arrhythmic dilated cardiomyopathy and commonly localize to the voltage-sensing mechanism. J Am Coll Cardiol. 2011;57:2160–8.

Meadows LS, Isom LL. Sodium channels as macromolecular complexes: implications for inherited arrhythmia syndromes. Cardiovasc Res. 2005;67:448–58.

Medeiros-Domingo A, Kaku T, Tester DJ, Iturralde-Torres P, Itty A, Ye B, Valdivia C, Ueda K, Canizales-Quinteros S, Tusié-Luna MT, Makielski JC, Ackerman MJ. SCN4B-encoded sodium channel b4 subunit in congenital long-QT syndrome. Circulation. 2007;116:134–42.

Meregalli PG, Wilde AA, Tan HL. Pathophysiological mechanisms of Brugada syndrome: depolarization disorder, repolarization disorder, or more? Cardiovasc Res. 2005;67:367–78.

Meregalli PG, Tan HL, Probst V, Koopmann TT, Tanck MW, Bhuiyan ZA, Sacher F, Kyndt F, Schott JJ, Albuisson J, Mabo P, Bezzina CR, Le Marec H, Wilde AA. Type of SCN5A mutation determines clinical severity and degree of conduction slowing in loss-of-function sodium channelopathies. Heart Rhythm. 2009;6:341–8.

Milstein ML, Musa H, Balbuena DP, Anumonwo JM, Auerbach DS, Furspan PB, Hou L, Hu B, Schumacher SM, Vaidyanathan R, Martens JR, Jalife J. Dynamic reciprocity of sodium and potassium channel expression in a macromolecular complex controls cardiac excitability and arrhythmia. Proc Natl Acad Sci U S A. 2012;109:E2134–43.

Mishra S, Reznikov V, Maltsev VA, Undrovinas NA, Sabbah HN, Undrovinas A. Contribution of sodium channel neuronal isoform Nav1.1 to late sodium current in ventricular myocytes from failing hearts. J Physiol. 2015;593(6):1409–27.

Mohler PJ, Hund TJ. Membrane-select regulation of cardiac Na(v) channel isoforms. Heart Rhythm. 2011;8:1931–2.

Mohler PJ, Rivolta I, Napolitano C, Lemaillet G, Lambert S, Priori SG, Bennett V. Nav1.5 E1053K mutation causing Brugada syndrome blocks binding to ankyrin-G and expression of Nav1.5 on the surface of cardiomyocytes. Proc Natl Acad Sci U S A. 2004;101:17533–8.

Mok NS, Priori SG, Napolitano C, et al. A newly characterized SCN5A mutation underlying Brugada syndrome unmasked by hyperthermia. J Cardiovasc Electrophysiol. 2003;14:407–11.

Moss AJ, Zareba W, Schwarz KQ, Rosero S, McNitt S, Robinson JL. Ranolazine shortens repolarization in patients with sustained inward sodium current due to type-3 long-QT syndrome. J Cardiovasc Electrophysiol. 2008;19:1289–93.

Motoike HK, Liu H, Glaaser IW, Yang AS, Tateyama M, Kass RS. The Na+ channel inactivation gate is a molecular complex: a novel role of the COOH-terminal domain. J Gen Physiol. 2004;123:155–65.

Murphy LL, Moon-Grady AJ, Cuneo BF, Wakai RT, Yu S, Kunic JD, Benson DW, George AL Jr. Developmentally regulated SCN5A splice variant potentiates dysfunction of a novel mutation associated with severe fetal arrhythmia. Heart Rhythm. 2012;9:590–7.

Nguyen TP, Wang DW, Rhodes TH, George AL Jr. Divergent biophysical defects caused by mutant sodium channels in dilated cardiomyopathy with arrhythmia. Circ Res. 2008;102:364–71.

Noujaim SF, Kaur K, Milstein M, Jones JM, Furspan P, Jiang D, Auerbach DS, Herron T, Meisler MH, Jalife J. A null mutation of the neuronal sodium channel NaV1.6 disrupts action potential propagation and excitation-contraction coupling in the mouse heart. FASEB J. 2012;26(1):63–72.

Nuyens D, Stengl M, Dugarmaa S, Rossenbacker T, Compernolle V, Rudy Y, Smits JF, Flameng W, Clancy CE, Moons L, Vos MA, Dewerchin M, Benndorf K, Collen D, Carmeliet E, Carmeliet P. Abrupt rate accelerations or premature beats cause life-threatening arrhythmias in mice with long-QT3 syndrome. Nat Med. 2001;7:1021–7.

Olesen MS, Jespersen T, Nielsen JB, Liang B, Møller DV, Hedley P, Christiansen M, Varró A, Olesen SP, Haunsø S, Schmitt N, Svendsen JH. Mutations in sodium channel β-subunit SCN3B are associated with early-onset lone atrial fibrillation. Cardiovasc Res. 2011;89:786–93.

Olesen MS, Holst AG, Svendsen JH, Haunso S, Tfelt-Hansen J. SCN1Bb R214Q found in 3 patients: 1 with Brugada syndrome and 2 with lone atrial fibrillation. Heart Rhythm. 2012;9:770–3.

Olson TM, Michels VV, Ballew JD, Reyna SP, Karst ML, Herron KJ, Horton SC, Rodeheffer RJ, Anderson JL. Sodium channel mutations and susceptibility to heart failure and atrial fibrillation. JAMA. 2005;293:447–54.

Onkal R, Mattis JH, Fraser SP, Diss JK, Shao D, Okuse K, Djamgoz MB. Alternative splicing of Nav1.5: an electrophysiological comparison of 'neonatal' and 'adult' isoforms and critical involvement of a lysine residue. J Cell Physiol. 2008;216(3):716–26.

Papadatos GA, Wallerstein PM, Head CE, Ratcliff R, Brady PA, Benndorf K, Saumarez RC, Trezise AE, Huang CL, Vandenberg JI, Colledge WH, Grace AA. Slowed conduction and ventricular tachycardia after targeted disruption of the cardiac sodium channel gene Scn5a. Proc Natl Acad Sci U S A. 2002;99:6210–5.

Paul AA, Witchel HJ, Hancox JC. Inhibition of the current of heterologously expressed HERG potassium channels by flecainide and comparison with quinidine, propafenone and lignocaine. Br J Pharmacol. 2002;136(5):717–29.

Petitprez S, Zmoos AF, Ogrodnik J, Balse E, Raad N, El-Haou S, Albesa M, Bittihn P, Luther S, Lehnart SE, Hatem SN, Coulombe A, Abriel H. SAP97 and dystrophin macromolecular complexes determine two pools of cardiac sodium channels Nav1.5 in cardiomyocytes. Circ Res. 2011;108:294–304.

Poelzing S, Forleo C, Samodell M, et al. SCN5A polymorphism restores trafficking of a Brugada syndrome mutation on a separate gene. Circulation. 2006;114:368–76.

Portero V, Casini S, Hoekstra M, Verkerk AO, Mengarelli I, Belardinelli L, Rajamani S, Wilde AAM, Bezzina CR, Veldkamp MW, Remme CA. Anti-arrhythmic potential of the late sodium current inhibitor GS-458967 in murine Scn5a-1798insD+/- and human SCN5A-1795insD+/- iPSC-derived cardiomyocytes. Cardiovasc Res. 2017;113(7):829–38.

Postema PG, Van den Berg M, Van Tintelen JP, Van den Heuvel F, Grundeken M, Hofman N, Van der Roest WP, Nannenberg EA, Krapels IP, Bezzina CR, Wilde A. Founder mutations in the Netherlands: SCN5a 1795insD, the first described arrhythmia overlap syndrome and one of the largest and best characterised families worldwide. Neth Hear J. 2009;17(11):422–8.

Priori SG, Napolitano C, Gasparini M, et al. Clinical and genetic heterogeneity of right bundle branch block and ST-segment elevation syndrome: a prospective evaluation of 52 families. Circulation. 2000;102:2509–15.

Probst V, Kyndt F, Potet F, Trochu JN, Mialet G, Demolombe S, Schott JJ, Baró I, Escande D, Le Marec H. Haploinsufficiency in combination with aging causes SCN5A-linked hereditary Lenègre disease. J Am Coll Cardiol. 2003;41:643–52.

Probst V, Wilde AA, Barc J, Sacher F, Babuty D, Mabo P, Mansourati J, Le SS, Kyndt F, Le CC, Guicheney P, Gouas L, Albuisson J, Meregalli PG, Le MH, Tan HL, Schott JJ. SCN5A mutations and the role of genetic background in the pathophysiology of Brugada syndrome. Circ Cardiovasc Genet. 2009;2:552–7.

Radwański PB, Ho HT, Veeraraghavan R, Brunello L, Liu B, Belevych AE, Unudurthi SD, Makara MA, Priori SG, Volpe P, Armoundas AA, Dillmann WH, Knollmann BC, Mohler PJ, Hund TJ, Györke S. Neuronal Na+ channels are integral components of pro-arrhythmic Na+/Ca2+ signaling nanodomain that promotes cardiac arrhythmias during β-adrenergic stimulation. JACC Basic Transl Sci. 2016;1(4):251–66.

Remme CA, Bezzina CR. Sodium channel (dys)function and cardiac arrhythmias. Cardiovasc Ther. 2010;28:287–94.

Remme CA, Wilde AA. Late sodium current inhibition in acquired and inherited ventricular (dys) function and arrhythmias. Cardiovasc Drugs Ther. 2013;27:91–101.

Remme CA, Wilde AAM. Targeting sodium channels in cardiac arrhythmia. Curr Opin Pharmacol. 2014;15:53–60.

Remme CA, Verkerk AO, Nuyens D, van Ginneken AC, van Brunschot S, Belterman CN, Wilders R, van Roon MA, Tan HL, Wilde AA, Carmeliet P, de Bakker JM, Veldkamp MW, Bezzina CR. Overlap syndrome of cardiac sodium channel disease in mice carrying the equivalent mutation of human SCN5A-1795insD. Circulation. 2006;114:2584–94.

Remme CA, Wilde AA, Bezzina CR. Cardiac sodium channel overlap syndromes: different faces of SCN5A mutations. Trends Cardiovasc Med. 2008;18:78–87.

Remme CA, Verkerk AO, Hoogaars WM, Aanhaanen WT, Scicluna BP, Annink C, van den Hoff MJ, Wilde AA, van Veen TA, Veldkamp MW, de Bakker JM, Christoffels VM, Bezzina CR. The cardiac sodium channel displays differential distribution in the conduction system and transmural heterogeneity in the murine ventricular myocardium. Basic Res Cardiol. 2009a;104:511–22.

Remme CA, Scicluna BP, Verkerk AO, Amin AS, van Brunschot S, Beekman L, Deneer VH, Chevalier C, Oyama F, Miyazaki H, Nukina N, Wilders R, Escande D, Houlgatte R, Wilde AA, Tan HL, Veldkamp MW, de Bakker JM, Bezzina CR. Genetically determined differences in sodium current characteristics modulate conduction disease severity in mice with cardiac sodium channelopathy. Circ Res. 2009b;104:1283–92.

Remme CA, Baartscheer A, Verkerk AO, van Rijen HV, Zeng D, Belardinelli L, Wilde AA, de Bakker JMT, Bezzina CR. Late sodium current block by ranolazine atttenuates intracellular Na^+ and Ca^{2+} dysregulation in myocytes from *Scn5a*-1798insD$^{/+}$ mice. Heart Rhythm. 2010;7:S160.

Rivaud MR, Augullo-Pascal E, Lin X, Leo-Macias A, Zhang M, Rothenberg E, Bezzina CR, Delmar M, Remme CA. Sodium channel remodeling in subcellular microdomains of murine failing cardiomyocytes. J Am Heart Assoc. 2017;6(12). pii: e007622.

Rivolta I, Abriel H, Tateyama M, Liu H, Memmi M, Vardas P, Napolitano C, Priori SG, Kass RS. Inherited Brugada and long QT-3 syndrome mutations of a single residue of the cardiac sodium channel confer distinct channel and clinical phenotypes. J Biol Chem. 2001;276:30623–30.

Rizzo S, Lodder EM, Verkerk AO, Wolswinkel R, Beekman L, Pilichou K, Basso C, Remme CA, Thiene G, Bezzina CR. Intercalated disc abnormalities, reduced Na+ current density and conduction slowing in desmoglein-2 mutant mice prior to cardiomyopathic changes. Cardiovasc Res. 2012;95:409–18.

Rook MB, Evers MM, Vos MA, Bierhuizen MF. Biology of cardiac sodium channel Nav1.5 expression. Cardiovasc Res. 2012;93:12–23.

Rossenbacker T, Carroll SJ, Liu H, Kuipéri C, de Ravel TJ, Devriendt K, Carmeliet P, Kass RS, Heidbüchel H. Novel pore mutation in SCN5A manifests as a spectrum of phenotypes ranging from atrial flutter, conduction disease, and Brugada syndrome to sudden cardiac death. Heart Rhythm. 2004;1:610–5.

Rowinsky EK, McGuire WP, Guarnieri T, Fisherman JS, Christian MC, Donehower RC. Cardiac disturbances during the administration of taxol. J Clin Oncol. 1991;9:1704–12.

Royer A, van Veen TA, Le Bouter S, Marionneau C, Griol-Charhbili V, Léoni AL, Steenman M, van Rijen HV, Demolombe S, Goddard CA, Richer C, Escoubet B, Jarry-Guichard T, Colledge WH, Gros D, de Bakker JM, Grace AA, Escande D, Charpentier F. Mouse model of SCN5A-linked hereditary Lenègre's disease: age-related conduction slowing and myocardial fibrosis. Circulation. 2005;111:1738–46.

Sato PY, Musa H, Coombs W, Guerrero-Serna G, Patino GA, Taffet SM, Isom LL, Delmar M. Loss of plakophilin-2 expression leads to decreased sodium current and slower conduction velocity in cultured cardiac myocytes. Circ Res. 2009;105:523–6.

Schott JJ, Alshinawi C, Kyndt F, Probst V, Hoorntje TM, Hulsbeek M, Wilde AA, Escande D, Mannens MM, Le Marec H. Cardiac conduction defects associate with mutations in SCN5A. Nat Genet. 1999;23:20–1.

Schram G, Zhang L, Derakhchan K, Ehrlich JR, Belardinelli L, Nattel S. Ranolazine: ion-channel-blocking actions and in vivo electrophysiological effects. Br J Pharmacol. 2004;142(8):1300–8.

Schroder EA, Lefta M, Zhang X, Bartos DC, Feng HZ, Zhao Y, Patwardhan A, Jin JP, Esser KA, Delisle BP. The cardiomyocyte molecular clock, regulation of Scn5a and arrhythmia susceptibility. Am J Phys Cell Phys. 2013;304(10):C954–65. Jan 30. [Epub ahead of print]

Schroeter A, Walzik S, Blechschmidt S, Haufe V, Benndorf K, Zimmer T. Structure and function of splice variants of the cardiac voltage-gated sodium channel Na(v)1.5. J Mol Cell Cardiol. 2010;49:16–24.

Schwartz PJ. The congenital long QT syndromes from genotype to phenotype: clinical implications. J Intern Med. 2006;259:39–47.

Schwartz PJ, Priori SG, Spazzolini C, Moss AJ, Vincent GM, Napolitano C, Denjoy I, Guicheney P, Breithardt G, Keating MT, Towbin JA, Beggs AH, Brink P, Wilde AA, Toivonen L, Zareba W, Robinson JL, Timothy KW, Corfield V, Wattanasirichaigoon D, Corbett C, Haverkamp W, Schulze-Bahr E, Lehmann MH, Schwartz K, Coumel P, Bloise R. Genotype-phenotype correlation in the long-QT syndrome: gene-specific triggers for life-threatening arrhythmias. Circulation. 2001;103:89–95.

Scicluna BP, Tanck MW, Remme CA, Beekman L, Coronel R, Wilde AA, Bezzina CR. Quantitative trait loci for electrocardiographic parameters and arrhythmia in the mouse. J Mol Cell Cardiol. 2011;50:380–9.

Shang LL, Pfahnl AE, Sanyal S, Jiao Z, Allen J, Banach K, Fahrenbach J, Weiss D, Taylor WR, Zafari AM, Dudley SC Jr. Human heart failure is associated with abnormal C-terminal splicing variants in the cardiac sodium channel. Circ Res. 2007;101:1146–54.

Shimizu W, Antzelevitch C. Sodium channel block with mexiletine is effective in reducing dispersion of repolarization and preventing torsade des pointes in LQT2 and LQT3 models of the long-QT syndrome. Circulation. 1997;96(6):2038–47.

Shimizu W, Aiba T, Antzelevitch C. Specific therapy based on the genotype and cellular mechanism in inherited cardiac arrhythmias. Long QT syndrome and Brugada syndrome. Curr Pharmaceut Design. 2005;11:1561–72.

Shimizu W, Moss AJ, Wilde AA, et al. Genotype-phenotype aspects of type 2 long QT syndrome. J Am Coll Cardiol. 2009;54(22):2052–62.

Shinlapawittayatorn K, Du XX, Liu H, Ficker E, Kaufman ES, Deschênes I. A common SCN5A polymorphism modulates the biophysical defects of SCN5A mutations. Heart Rhythm. 2011;8 (3):455–62.

Shy D, Gillet L, Abriel H. Cardiac sodium channel NaV1.5 distribution in myocytes via interacting proteins: the multiple pool model. Biochim Biophys Acta. 2013;1833:886–94.

Shy D, Gillet L, Ogrodnik J, Albesa M, Verkerk AO, Wolswinkel R, Rougier JS, Barc J, Essers MC, Syam N, Marsman RF, van Mil AM, Rotman S, Redon R, Bezzina CR, Remme CA, Abriel H. PDZ domain-binding motif regulates cardiomyocyte compartment-specific NaV1.5 channel expression and function. Circulation. 2014;130(2):147–60.

Smits JPP, Koopmann TT, Wilders R, Veldkamp MW, Opthof T, Bhuiyan ZA, Mannens MM, Balser JR, Tan HL, Bezzina CR, Wilde AA. A mutation in the human cardiac sodium channel (E161K) contributes to sick sinus syndrome, conduction disease and Brugada syndrome in two families. J Mol Cell Cardiol. 2005;38:969–81.

Splawski I, Shen J, Timothy KW, et al. Spectrum of mutations in long-QT syndrome genes. KVLQT1, HERG, SCN5A, KCNE1, and KCNE2. Circulation. 2000;102:1178–85.

Tan HL, Bezzina CR, Smits JP, Verkerk AO, Wilde AA. Genetic control of sodium channel function. Cardiovasc Res. 2003;57:961–73.

Tan BH, Valdivia CR, Song C, Makielski JC. Partial expression defect for the SCN5A missense mutation G1406R depends on splice variant background Q1077 and rescue by mexiletine. Am J Physiol Heart Circ Physiol. 2006;291:H1822–8.

Te Riele AS, Agullo-Pascual E, James CA, Leo-Macias A, Cerrone M, Zhang M, Lin X, Lin B, Sobreira NL, Amat-Alarcon N, Marsman RF, Murray B, Tichnell C, van der Heijden JF, Dooijes D, van Veen TA, Tandri H, Fowler SJ, Hauer RN, Tomaselli G, van den Berg MP, Taylor MR, Brun F, Sinagra G, Wilde AA, Mestroni L, Bezzina CR, Calkins H, Peter van Tintelen J, Bu L, Delmar M, Judge DP. Multilevel analyses of SCN5A mutations in arrhythmogenic right ventricular dysplasia/cardiomyopathy suggest non-canonical mechanisms for disease pathogenesis. Cardiovasc Res. 2017;113(1):102–11.

Terrenoire C, Wang K, Tung KW, Chung WK, Pass RH, Lu JT, Jean JC, Omari A, Sampson KJ, Kotton DN, Keller G, Kass RS. Induced pluripotent stem cells used to reveal drug actions in a long QT syndrome family with complex genetics. J Gen Physiol. 2013;141:61–72.

Torres NS, Larbig R, Rock A, Goldhaber JI, Bridge JH. Na+ currents are required for efficient excitation-contraction coupling in rabbit ventricular myocytes: a possible contribution of neuronal Na+ channels. J Physiol. 2010;588(Pt 21):4249–60.

Ufret-Vincenty CA, Baro DJ, Lederer WJ, Rockman HA, Quinones LE, Santana LF. Role of sodium channel deglycosylation in the genesis of cardiac arrhythmias in heart failure. J Biol Chem. 2001;276:28197–203.

Valdivia CR, Tester DJ, Rok BA, et al. A trafficking defective, Brugada syndrome-causing SCN5A mutation rescued by drugs. Cardiovasc Res. 2004;62:53–62.

Valdivia CR, Chu WW, Pu J, Foell JD, Haworth RA, Wolff MR, Kamp TJ, Makielski JC. Increased late sodium current in myocytes from a canine heart failure model and from failing human heart. J Mol Cell Cardiol. 2005;38:475–83.

Valdivia CR, Medeiros-Domingo A, Ye B, Shen WK, Algiers TJ, Ackerman MJ, Makielski JC. Loss-of-function mutation of the SCN3B-encoded sodium channel {beta}3 subunit associated with a case of idiopathic ventricular fibrillation. Cardiovasc Res. 2010;86:392–400.

Van Bemmelen MX, Rougier JS, Gavillet B, Apothéloz F, Daidié D, Tateyama M, Rivolta I, Thomas MA, Kass RS, Staub O, Abriel H. Cardiac voltage-gated sodium channel Nav1.5 is regulated by Nedd4-2 mediated ubiquitination. Circ Res. 2004;95:284–91.

Van den Berg MP, Wilde AAM, Viersma JW, Brouwer J, Haaksma J, van der Hout AH, Stolte-Dijkstra I, Bezzina TCR, Van Langen IM, Beaufort-Krol GC, Cornel JH 2nd, Crijns HJ. Possible bradycardic mode of death and successful pacemaker treatment in a large family with features of long QT syndrome type 3 and Brugada syndrome. J Cardiovasc Electrophysiol. 2001;12:630–6.

Van den Berg MP, van den Heuvel F, van Tintelen JP, Volders PG, van Gelder IC. Successful treatment of a patient with symptomatic long QT syndrome type 3 using ranolazine combined with a beta-blocker. Int J Cardiol. 2014;171(1):90–2.

Van den Boogaard M, Wong LY, Tessadori F, Bakker ML, Dreizehnter LK, Wakker V, Bezzina CR, 't Hoen PA, Bakkers J, Barnett P, Christoffels VM. Genetic variation in T-box binding element functionally affects SCN5A/SCN10A enhancer. J Clin Invest. 2012;122:2519–30.

Vatta M, Ackerman MJ, Ye B, Makielski JC, Ughanze EE, Taylor EW, Tester DJ, Balijepalli RC, Foell JD, Li Z, Kamp TJ, Towbin JA. Mutant caveolin-3 induces persistent late sodium current and is associated with long-QT syndrome. Circulation. 2006;114:2104–12.

Veerman CC, Wilde AA, Lodder EM. The cardiac sodium channel gene SCN5A and its gene product NaV1.5: role in physiology and pathophysiology. Gene. 2015;573(2):177–87.

Veerman CC, Podliesna S, Tadros R, Lodder EM, Mengarelli I, de Jonge B, Beekman L, Barc J, Wilders R, Wilde AAM, Boukens BJ, Coronel R, Verkerk AO, Remme CA, Bezzina CR. The Brugada syndrome susceptibility gene HEY2 modulates cardiac transmural ion channel patterning and electrical heterogeneity. Circ Res. 2017;121(5):537–48.

Veldkamp MW, Viswanathan PC, Bezzina C, Baartscheer A, Wilde AAM, Balser JR. Two distinct congenital arrhythmias evoked by a multidysfunctional Na^+ channel. Circ Res. 2000;86:e91–7.

Veldkamp MW, Wilders R, Baartscheer A, Zegers JG, Bezzina CR, Wilde AA. Contribution of sodium channel mutations to bradycardia and sinus node dysfunction in LQT3 families. Circ Res. 2003;92:976–83.

Verkerk AO, van Ginneken AC, van Veen TA, Tan HL. Effects of heart failure on brain-type Na+ channels in rabbit ventricular myocytes. Europace. 2007;9:571–7.

Viswanathan PC, Balser JR. Inherited sodium channelopathies: a continuum of channel dysfunction. Trends Cardiovasc Med. 2004;14:28–35.

Viswanathan PC, Benson DW, Balser JR. A common SCN5A polymorphism modulates the biophysical effects of an SCN5A mutation. J Clin Invest. 2003;111:341–6.

Wagner S, Dybkova N, Rasenack EC, Jacobshagen C, Fabritz L, Kirchhof P, Maier SK, Zhang T, Hasenfuss G, Brown JH, Bers DM, Maier LS. Ca/calmodulin-dependent protein kinase II regulates cardiac Na channels. J Clin Invest. 2006;116:3127–8.

Wang DW, Kiyosue T, Sato T, Arita M. Comparison of the effects of class I anti-arrhythmic drugs, cibenzoline, mexiletine and flecainide, on the delayed rectifier K+ current of guinea-pig ventricular myocytes. J Mol Cell Cardiol. 1996;28(5):893–903.

Wang C, Hennessey JA, Kirkton RD, Wang C, Graham V, Puranam RS, Rosenberg PB, Bursac N, Pitt GS. Fibroblast growth factor homologous factor 13 regulates Na+ channels and conduction velocity in murine hearts. Circ Res. 2011;109:775–82.

Watanabe H, Koopmann TT, Le SS, Yang T, Ingram CR, Schott JJ, Demolombe S, Probst V, Anselme F, Escande D, Wiesfeld AC, Pfeufer A, Kääb S, Wichmann HE, Hasdemir C, Aizawa Y, Wilde AA, Roden DM, Bezzina CR. Sodium channel beta1 subunit mutations associated with Brugada syndrome and cardiac conduction disease in humans. J Clin Invest. 2008;118:2260–8.

Watanabe H, Darbar D, Kaiser DW, Jiramongkolchai K, Chopra S, Donahue BS, Kannankeril PJ, Roden DM. Mutations in sodium channel beta1 and beta2 subunits associated with atrial fibrillation. Circ Arrhythm Electrophysiol. 2009;2:268–75.

Watanabe H, Yang T, Stroud DM, Lowe JS, Harris L, Atack TC, Wang DW, Hipkens SB, Leake B, Hall L, Kupershmidt S, Chopra N, Magnuson MA, Tanabe N, Knollmann BC, George AL Jr, Roden DM. Striking in vivo phenotype of a disease-associated human SCN5A mutation producing minimal changes in vitro. Circulation. 2011;124:1001–11.

Wedekind H, Smits JP, Schulze-Bahr E, Arnold R, Veldkamp MW, Bajanowski T, Borggrefe M, Brinkmann B, Warnecke I, Funke H, Bhuiyan ZA, Wilde AA, Breithardt G, Haverkamp W. De novo mutation in the *SCN5A* gene associated with early onset of sudden infant death. Circulation. 2001;104:1158–64.

West JW, Patton DE, Scheuer T, Wang Y, Goldin AL, Catterall WA. A cluster of hydrophobic amino acid residues required for fast Na(+)-channel inactivation. Proc Natl Acad Sci U S A. 1992;89:10910–4.

Westenbroek RE, Bischoff S, Fu Y, Maier SK, Catterall WA, Scheuer T. Localization of sodium channel subtypes in mouse ventricular myocytes using quantitative immunocytochemistry. J Mol Cell Cardiol. 2013;64:69–78.

Wilde AA, Brugada R. Phenotypical manifestations of mutations in the genes encoding subunits of the cardiac sodium channel. Circ Res. 2011;108(7):884–97.

Wilde AA, Remme CA. Therapeutic approaches for Long QT syndrome type 3: an update. Europace. 2017;20:222–4. Apr 11. https://doi.org/10.1093/europace/eux039.

Wolf CM, Berul CI. Inherited conduction system abnormalities--one group of diseases, many genes. J Cardiovasc Electrophysiol. 2006;17:446–55.

Wu G, Ai T, Kim JJ, Mohapatra B, Xi Y, Li Z, Abbasi S, Purevjav E, Samani K, Ackerman MJ, Qi M, Moss AJ, Shimizu W, Towbin JA, Cheng J, Vatta M. Alpha-1-syntrophin mutation and the long QT syndrome: a disease of sodium channel disruption. Circ Arrhythm Electrophysiol. 2008;1:193–201.

Xi Y, Wu G, Yang L, Han K, Du Y, Wang T, Lei X, Bai X, Ma A. Increased late sodium currents are related to transcription of neuronal isoforms in a pressure-overload model. Eur J Heart Fail. 2009;11(8):749–57.

Ye B, Valdivia CR, Ackerman MJ, Makielski JC. A common human SCN5A polymorphism modifies expression of an arrhythmia causing mutation. Physiol Genomics. 2003;12:187–93.

Yoo S, Dobrzynski H, Fedorov VV, Xu SZ, Yamanushi TT, Jones SA, Yamamoto M, Nikolski VP, Efimov IR, Boyett MR. Localization of Na^+ channel isoforms at the atrioventricular junction and atrioventricular node in the rat. Circulation. 2006;114:1360–71.

Yu J, Hu J, Dai X, Cao Q, Xiong Q, Liu X, Liu X, Shen Y, Chen Q, Hua W, Hong K. SCN5A mutation in Chinese patients with arrhythmogenic right ventricular dysplasia. Herz. 2014;39 (2):271–5.

Zareba W, Sattari MN, Rosero S, Couderc JP, Moss AJ. Altered atrial, atrioventricular, and ventricular conduction in patients with the long QT syndrome caused by the DeltaKPQ SCN5A sodium channel gene mutation. Am J Cardiol. 2001;88:1311–4.

Zareba W, Rosero S, McNitt S, Hellawel J, Zheng D, Blair C, et al. Eleclazine: a novel late sodium current inhibitor shortens the QT intervals in LQT3 patients across wide range of heart rates. Heart Rhyhtm. 2016;13:S578.

Zhang Y, Hartmann HA, Satin J. Glycosylation influences voltage-dependent gating of cardiac and skeletal muscle sodium channels. J Membr Biol. 1999;171:195–207.

Zhang T, Yong SL, Drinko JK, Popović ZB, Shryock JC, Belardinelli L, Wang QK. LQTS mutation N1325S in cardiac sodium channel gene SCN5A causes cardiomyocyte apoptosis, cardiac fibrosis and contractile dysfunction in mice. Int J Cardiol. 2011;147:239–45.

Ziane R, Huang H, Moghadaszadeh B, Beggs AH, Levesque G, Chahine M. Cell membrane expression of cardiac sodium channel Na(v)1.5 is modulated by alpha-actinin-2 interaction. Biochemistry. 2010;49:166–78.

Potassium Channels in the Heart

3

Morten B. Thomsen

Abstract

Ionic currents over the plasma membrane through channels are the cornerstone of excitable cells. Human cardiomyocytes are excitable and continuously cycle between a depolarized and a repolarized state every second throughout human life, initiating and coordinating cardiac pump function. Ion channels selective for potassium (K^+) critically participate in cellular repolarization and contribute to stabilizing the diastolic membrane potential, thus shaping the cardiac action potential. Four different subfamilies of potassium channels are present in the heart: small conductance, calcium-activated potassium channels (SK or $K_{Ca}2$), inwardly rectifying potassium channels (K_{ir}), two-pore-domain potassium channels (K_{2P}), and voltage-gated potassium channels (K_V). In the present review, the structure and biophysical function of these cardiac potassium ion channels are reviewed. Moreover, rectification, inactivation, and current dependency on the extracellular potassium concentration are explained.

3.1 Introduction

The largest subfamily of ion channels comprises the potassium (K^+) channels. They modulate cellular excitability and action potential morphology throughout the body controlling diverse functions like endocrine, neuronal, and muscular activity. Depending on the membrane potential and the chemical concentration gradient of potassium, the ions will flow into or out of the cell. In the heart, potassium currents are predominantly outward, thus contributing to cardiac repolarization. There are four main classes of potassium channels, based on structure and function. Channels from all these four classes are expressed in the heart (Table 3.1).

M. B. Thomsen (✉)
Department of Biomedical Sciences, University of Copenhagen, Copenhagen, Denmark
e-mail: mbthom@sund.ku.dk

D. Thomas, C. A. Remme (eds.), *Channelopathies in Heart Disease*, Cardiac and Vascular Biology, https://doi.org/10.1007/978-3-319-77812-9_3

Table 3.1 Overview of the cardiac potassium channels based on the classifications of the International Union of Basic and Clinical Pharmacology (IUPHAR)

Pore-forming protein	Gene	Other names	Current	Recommended reviews
Calcium-activated potassium channels				Wei et al. (2005) and Dong et al. (2016)
$K_{Ca}2.1–2.3$	*KCNN1–3*	SK1–3	I_{KCa}	
Inwardly rectifying potassium channels				Anumonwo and Lopatin (2010) and Kubo et al. (2005)
$K_{ir}2.1–2.3$	*KCNJ2*, *KCNJ12* and *KCNJ4*		I_{K1}	
$K_{ir}3.1$ and 3.4	*KCNJ3* and–*5*	GIRK1 and–4	$I_{K,ach}$	
$K_{ir}6.2$	*KCNJ11*		$I_{K,ATP}$	Zhang et al. (2010)
Two-pore-domain potassium channels				Schmidt et al. (2014), Goldstein et al. (2001, 2005) and Decher et al. (2015)
$K_{2P}2.1$	KCNK2	TREK-1		
$K_{2P}3.1$	*KCNK3*	TASK-1		
Voltage-gated potassium channels				Gutman et al. (2005)
$K_V1.4$	*KCNA4*		$I_{to,slow}$	Patel and Campbell (2005)
$K_V1.5$	*KCNA5*		I_{Kur}	Ravens and Wettwer (2011)
$K_V4.3$	*KCND3*		$I_{to,fast}$	Niwa and Nerbonne (2010)
$K_V7.1$	*KCNQ1*	K_vLQT1	I_{Ks}	Liin et al. (2015)
$K_V11.1$	*KCNH2*	HERG	I_{Kr}	Vandenberg et al. (2012)

The calcium-activated potassium channels (K_{Ca}) open upon a rise in the intracellular calcium. In humans, they are divided into three groups based on the conductance of the single, isolated channel: big, intermediate, and small conductance calcium-activated potassium channels. Only the small conductance potassium currents (SK) are found in the plasma membrane of cardiomyocytes.

The inwardly rectifying potassium channels (K_{ir}) govern three important currents in the heart: the constitutively active, inward rectifier current (I_{K1}), the G-protein activated potassium current modulated by acetylcholine ($I_{K,ACh}$), and the ATP-sensitive potassium current ($I_{K,ATP}$). Common among them is the inward rectification, which is a strong voltage-dependent reduction in channel conductance upon membrane depolarization. The inward current at membrane potentials negative to the reversal potential dominates the current traces, and this is why the family of ion channels has been named inward rectifiers. Notwithstanding, it is the repolarizing, outward current in a narrow window of membrane potentials positive to the reversal potential that is important in adjusting action potential waveform in the heart.

We are only beginning to understand the physiological functions of the cardiac two-pore-domain potassium channels (K_{2P}). Heterologous expression produces instantaneous, non-inactivating potassium currents with little to no voltage dependence. Thus, these currents are often regarded as background or leak currents; however, together with the K_{Ca} cannels, they may be tomorrow's targets in antiarrhythmic drug therapy.

By far the largest body of electrophysiological studies has been performed on the voltage-gated potassium channels (K_V), which encompass at least five species in the heart. These channels activate upon membrane depolarization and with different current kinetics modulate the amplitude and the duration of the cardiac action potential. Two broad classes of voltage-gated potassium channels can be distinguished in the heart: the transient outward currents and delayed, outward rectifying potassium channels. The transient currents activate rapidly and underlie the notch of the human action potential, whereas the delayed currents contribute significantly to the repolarization phase.

Many expression studies have shown the presence of a given potassium channel gene product in cardiac tissue and continued to describe the current in heterologous expression systems. This type of study carries the inherent caveat that the potassium channel may display very different function in the natural environment of the heart. The native current in the cardiomyocyte is not only a function of the α-subunit but also depends on whether the native channels are formed by four identical α-subunits (homotetramers) or several different α-subunits (heterotetramers). Moreover, post-translational modification of the ion channel and molecular interaction partners, in particular, auxiliary or accessory "β"-subunits, significantly affects the current. Hence, Table 3.1 is clearly a simplification: the native current generated by a given α-subunit may have different characteristics in the heart dependent on modifications and molecular partners, which result in a large variation in current appearance based on location within the heart and on potential cardiac remodeling (Nerbonne and Kass 2005). Consequently, this review will include description of auxiliary subunits only to a small degree. Rather, the following will be an overview of the four classes of potassium channels, their structure, and function in relation to the heart. Excellent reviews in this series will cover diseases related to potassium channelopathies, which will thus only be touched upon here when the association to clinical arrhythmia is strong.

3.2 Calcium-Activated Potassium Channels

Small conductance, calcium-activated potassium channels (K_{Ca}2.1–2.3) are located in the plasma membrane of the cardiomyocytes. The channel is composed of homo- or heterotetramers of K_{Ca}2 proteins, each having six transmembrane domains and a pore region (Fig. 3.1; Berkefeld et al. 2010). Moreover, reports of big conductance, calcium-activated potassium channels (K_{Ca}1.1 encoded by *KCNMA1*) in the cardiac mitochondrial membranes are available (Xu et al. 2002), and they may be important during recovery from ischemic episodes (Soltysinska et al. 2014).

Despite sharing the six transmembrane domain topology with the voltage-gated potassium channels, the K_{Ca}2 protein completely lacks a voltage sensor. Consequently,

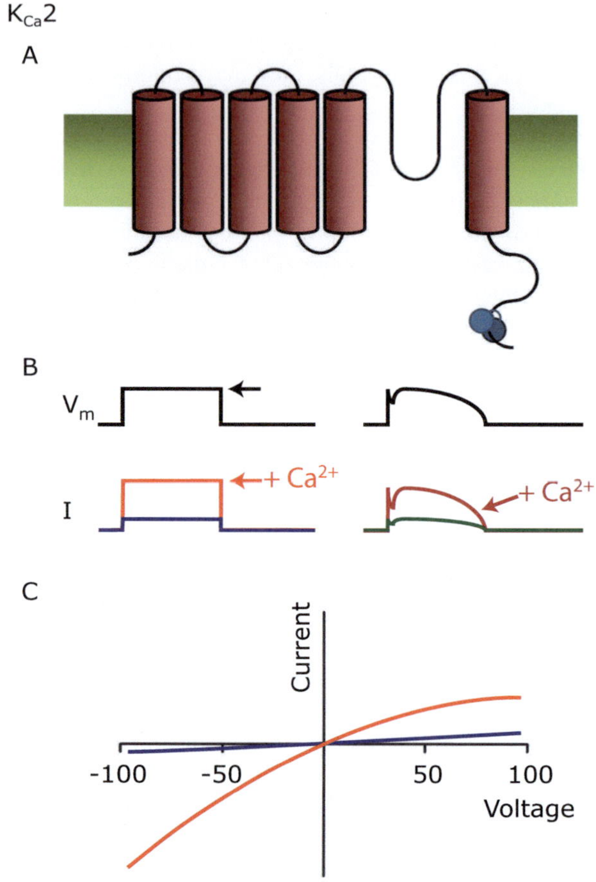

Fig. 3.1 $K_{Ca}2$. A. Membrane topology of $K_{Ca}2$ with six transmembrane domains and one pore-lining segment. The blue dumbbell structure represents the calcium-sensing subunit, calmodulin. Four $K_{Ca}2$ proteins assemble to form a functional channel. Green areas symbolize the plasma membrane; extracellular side is upward. B. Stylized current responses to a square voltage command (left) as used in many experimental settings and during an action potential [right (Berkefeld et al. 2010)]. The current is very small in the absence of calcium (blue and green traces) but becomes significant when the calcium concentration raises above ~0.2 μM. C. Stylized graph showing the relationship between membrane voltage and the current amplitude after activation. In these examples, the graphs have reversal potential (when the current is zero) at 0 mV, indicating that the experiments were performed under conditions where internal and external potassium concentrations are identical (Grunnet et al. 2001)

the small conductance, calcium-activated potassium channels are voltage independent, and channel gating is solely regulated by intracellular calcium. Upon cellular depolarization, the L-type calcium channel activates and increases the calcium concentration in the intracellular, subsarcolemmal microdomain. The nearby $K_{Ca}2$ channel constitutively binds to the calcium-binding protein calmodulin, and the increased calcium concentration opens the $K_{Ca}2$ channel and allows a potassium efflux, which contributes to repolarization of the cardiomyocyte. This repolarization feeds back to the calcium channel that inactivates in a voltage-dependent manner. Accordingly, $K_{Ca}2$ channels limit the calcium transient and thereby modulate the excitation-contraction coupling (Berkefeld et al. 2010). It appears that the calcium channel and the $K_{Ca}2$ channels may even be physically coupled in the cardiomyocytes (Lu et al. 2007; Wang et al. 1999a).

$K_{Ca}2$ channels are predominantly expressed in atrial cardiomyocytes, and it was suggested that ligands for these channels might offer a unique therapeutic opportunity to modify atrial cells without interfering with ventricular myocytes (Tuteja et al. 2005). $K_{Ca}2.2$ knockout in mice prolongs action potentials in atrial cardiomyocytes and increases the vulnerability to atrial arrhythmias (Li et al. 2009), and $K_{Ca}2$ channels were shown to be involved in pro-arrhythmic remodeling in myocardial sleeves in rabbit pulmonary veins, a frequent site for triggering atrial fibrillation (Ozgen et al. 2007). Moreover, $K_{Ca}2.2$ and $K_{Ca}2.3$ are present in human atria, their function is reduced in chronic atrial fibrillation (Skibsbye et al. 2014), and they become upregulated in the ventricles during heart failure (Chang et al. 2013; Bonilla et al. 2014). Notwithstanding, it is still debated whether $K_{Ca}2$ channels significantly affect the morphology of the action potential, as reviewed elsewhere (Ravens and Odening 2017).

3.3 Inwardly Rectifying Potassium Channels

The inwardly rectifying potassium channels are tetramers of α-subunits each with only two transmembrane domains. Four $K_{ir}2.1$ subunits assemble to form the important I_{K1} that contributes to final repolarization of the ventricular action potential and is important in ensuring a stable resting membrane potential. $K_{ir}3.1$ and 3.4 constitute $I_{K,ACh}$, another strong inward rectifier, exclusively found in atrial tissue including the sinus node. The current is activated by acetylcholine, which makes the activity of the channel closely coupled to the activity of the parasympathetic branch of the autonomic nervous system. $K_{ir}6.2$ is the ATP-sensitive potassium channel that shows weaker rectification compared to the other two K_{ir} channels (Anumonwo and Lopatin 2010).

These channels are potassium selective, but do not have a voltage sensor. $K_{ir}2.1$ is open at all times, whereas $K_{ir}3$ opens after stimulation with acetylcholine and $K_{ir}6.2$ is modulated by adenosine mono- and triphosphates (AMP and ATP). The inward rectification is very clear for $K_{ir}2$ and $K_{ir}3$, where large potassium current flows into the cell at membrane potentials negative of the equilibrium potential; however often overlooked is the smaller but physiologically very significant repolarizing current flowing at potentials directly positive to the reversal potential. Cells with large activity of K_{ir} channels are expected to have a resting membrane potential close to the reversal potential of potassium and show very few spontaneous action potentials (Hibino et al. 2010).

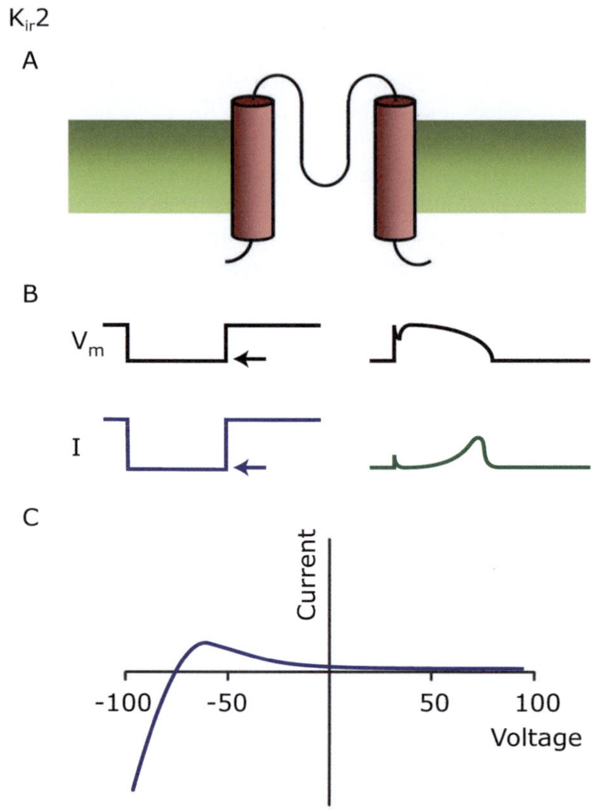

Fig. 3.2 $K_{ir}2$. A. Membrane topology of $K_{ir}2$ with two transmembrane domains and one pore-lining segment. Four $K_{ir}2$ proteins assemble to form a functional channel. Green areas symbolize the plasma membrane; extracellular side is upward. B. Stylized current responses to a square voltage command (left) as used in many experimental settings and during an action potential [right (Anumonwo and Lopatin 2010; Thomsen et al. 2009a)]. Changing the membrane potential, e.g., from 0 mV to −100 mV, generates a large inward current via $K_{ir}2$ (blue trace). During the last phases of the action potential, $K_{ir}2$ contributes to repolarization when the membrane potential is negative. At the resting membrane potential, no net current flows through the channel, because the electrochemical gradients for potassium are in equilibrium. However, the channel is open. C. Stylized graph showing the relationship between membrane voltage and the current amplitude after activation. Block of outward current, predominantly by intracellular spermine, causes inward rectification: a strong voltage-dependent reduction in channel conductance upon membrane depolarization

3.3.1 $K_{ir}2$

The $K_{ir}2$ family consists of $K_{ir}2.1$–2.4, and it appears that $K_{ir}2.1$ is the main component of ventricular I_{K1}, whereas $K_{ir}2.3$ is predominant in the atria (Melnyk et al. 2002). There is no voltage-sensitive gating of the channel. More inward current flows in a voltage-dependent manner, when membrane potentials are below the equilibrium potential for potassium (Fig. 3.2), but this is solely due to increased driving force pulling potassium ions down its electrochemical gradient (Kubo et al. 1993). The channel is only partly selective to potassium ions, and outward current is "contaminated" by magnesium ions and spermine that will block the channel inside the pore (Fakler et al. 1995). This is the mechanism behind the inward rectification—a large voltage-dependent reduction in channel conductance when the membrane becomes more positive: inward current flows freely, but outward current causes channel block (Anumonwo and Lopatin 2010). The physiological consequence is that when the membrane potential is positive to about −20 mV, which is during most of the action potential, there is no I_{K1}. When the cell repolarizes and the membrane potential reaches about −40 mV, $K_{ir}2.1$ is released from spermine/Mg^{2+} block, and I_{K1} contributes to final repolarization.

$K_{ir}2.1$ is open during diastole and keeps a constant resting membrane potential. It follows that I_{K1} is very small during diastole, when the membrane potential is close to the equilibrium potential of potassium (Nichols et al. 1996). The sinus node of the heart has very little expression of $K_{ir}2.1$, which is in part reflected on the unstable diastolic membrane potential in the nodal cells. I_{K1} is suppressed by adrenergic stimulation, especially the inward component (Koumi et al. 1995a, b). Barium ions effectively block the channel reducing both inward and outward current amplitudes (Tamargo et al. 2004).

Congenital loss-of-function mutations in $K_{ir}2.1$ cause Andersen's syndrome. These families with reduced or no $K_{ir}2.1$ have delayed cardiac repolarization, extrasystoles, tachycardia, syncopes, and torsades de pointes arrhythmias, alongside dysmorphic features and periodic paralysis (Plaster et al. 2001). Gain-of-function mutations cause accelerated repolarization manifested as augmented outward current amplitude and short T waves on the electrocardiogram (Priori et al. 2005).

3.3.2 $K_{ir}3.1$ and $K_{ir}3.4$

The channel governing $I_{K,ACh}$ is a heteromeric assembly of $K_{ir}3.1$ and $K_{ir}3.4$ [Fig. 3.3 (Anumonwo and Lopatin 2010)]. The channel complex shows a weaker rectification than $K_{ir}2.1$, and its localization in the heart is almost opposite of $K_{ir}2.1$: $K_{ir}3$ is highly expressed in the atria (Calloe et al. 2013), including the sinoatrial and atrioventricular nodes (Sakmann et al. 1983). Significant $I_{K,ACh}$ is produced only when the parasympathetic branch of the autonomic nervous system is activated and releases the neurotransmitter acetylcholine from postsynaptic nerve terminals in the atrial wall, including the sinoatrial and atrioventricular nodes. Acetylcholine binds to muscarinic receptors (type 2), and this causes dissociation of G proteins at the intracellular side of the plasma membrane. The β-γ-subunits of the G protein activate the $K_{ir}3.1/4$ heteromer via direct biochemical association (Logothetis et al. 1987). This increased $I_{K,ACh}$ contributes to repolarization of the nodal action potential reducing the speed of spontaneous depolarization in phase 4 and making the maximal diastolic potential more negative. This reduces heart rate and increases atrioventricular conduction time. In the non-nodal atrial cardiomyocyte, acetylcholine abbreviates the action potential. The mechanism of rectification is comparable to that of $K_{ir}2$ (Hibino et al. 2010).

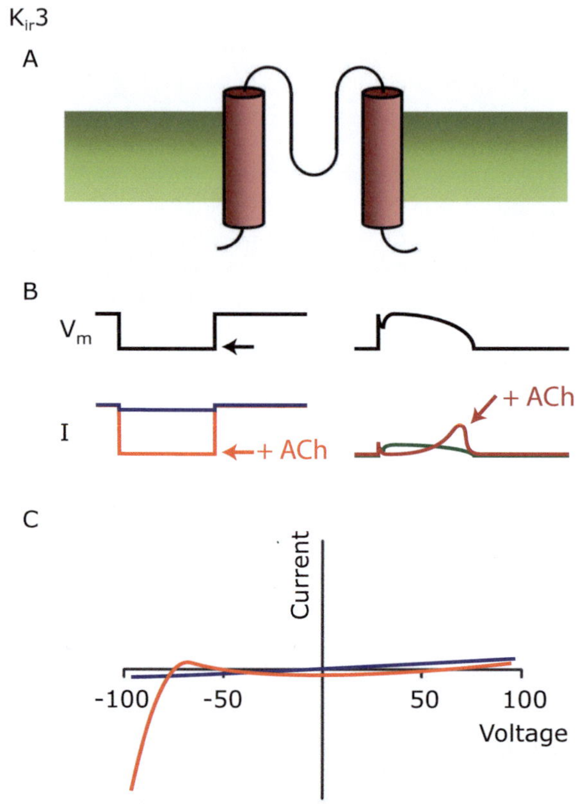

Fig. 3.3 K$_{ir}$3. A. Membrane topology of K$_{ir}$3 with two transmembrane domains and one pore-lining segment. Four K$_{ir}$3 proteins assemble to form a functional channel. Green areas symbolize the plasma membrane; extracellular side is upward. B. Stylized current responses to a square voltage command (left) as used in many experimental settings and during an action potential [right (Calloe et al. 2013)]. In the absence of acetylcholine (ACh), the current is very small; however, it increases upon stimulation of muscarinic receptors. A depolarizing square pulse, e.g., from 0 mV to −100 mV, in the presence of acetylcholine, gives a large inward potassium current through K$_{ir}$3 (orange trace). During an action potential, the acetylcholine-activated repolarizing outward current is more prominent, and this contributes to faster repolarization of the membrane (red trace), which will shorten the action potential. C. Stylized graph showing the relationship between membrane voltage and the current amplitude. In the presence of acetylcholine, the current amplitude is significant and shows strong inward rectification

3.3.3 K$_{ir}$6.2

Assembly of four K$_{ir}$6.2 and four accessory ATP-binding sulfonylurea receptor proteins (SUR2A) makes the ATP-sensitive potassium channel. The resulting current ($I_{K,ATP}$) is inhibited by intracellular ATP and activated by AMP and ADP in a rather complex manner that is presently not completely understood (Zhang et al. 2010); however, it is clear that this current provides a link to cellular metabolism in many cell types, including vascular smooth muscle, skeletal muscle, and the heart. Under normal conditions, the current is negligible, but under ischemic or hypoxic conditions, the current is rather large and contributes significantly to action potential shortening. $K_{ir}6$ shows weaker inward rectification compared to $K_{ir}2$ and $K_{ir}3$. The single channel conductance of $K_{ir}6$ is large, compared to $K_{ir}2$ and K_V channels, and activation of only a few channels is required for markedly augmenting repolarization (Zhang et al. 2010). Most electrophysiological experiments testing the function of $K_{ir}6.2$ use single-channel recordings in conditions where the potassium concentrations are equal inside and outside the membrane patch. Since the channel is not voltage gated, potassium ions flow according to the electrical gradient, and serial application of AMP, ADP, or ATP to open or close the channel, respectively, is used to describe channel kinetics. Classical drugs that increase $I_{K,ATP}$ include pinacidil and cromakalim, whereas glibenclamide is a potent channel antagonist (Tamargo et al. 2004).

3.4 Two-Pore-Domain Potassium Channels

The two-pore-domain potassium channels are dimers of proteins with four transmembrane domains each (Fig. 3.4). The subunits are characterized by having two separate poor-forming loop domains that assemble to form a single transmembrane pore in the dimer (Schmidt et al. 2014). Hence, one K_{2P} subunit can be viewed as two fused K_{ir} subunits. The current flowing through the K_{2P} channel is instantaneous, non-inactivating, and the channel gating is voltage independent. Thus, the channel is in its open state before any change in voltage and is ready to pass potassium ions immediately. At least 15 different K_{2P} channels have been identified in the genome (Ravens and Odening 2016; Yu and Catterall 2004), but it appears that only $K_{2P}2.1$ and $K_{2P}3.1$ have a physiological relevance in the heart (Schmidt et al. 2014). The channels are controlled by multiple physiological factors, including oxygen, pH, arachidonic acid, mechanical stretch, neurotransmitters, G-coupled proteins, and protein kinases (Goldstein et al. 2001). In addition, the channels are sensitive to volatile anesthetics (Tamargo et al. 2004; Goldstein et al. 2005). Initially, the current was termed a “leak” or “background” current; however, it is now clear that this current is highly regulated, and its ubiquitous expression suggests many important physiological functions.

Under physiological conditions with a high intracellular potassium concentration, the K_{2P} channels pass currents only in the outward direction; however, in experimental settings of symmetric potassium concentration inside and outside of the cell, there is a linear current-voltage relationship. This characteristic is called open, outward rectification and is different from the modes of K_{ir} and K_V-channel rectification. The open, outward rectification derives from the unequal potassium concentration across the cell membrane, which makes outward current flow more likely (Goldstein et al. 2001). A background conductance of potassium implies that the channels contribute to stabilizing the resting membrane potential close the equilibrium potential of potassium. Moreover, open channels will shorten the action

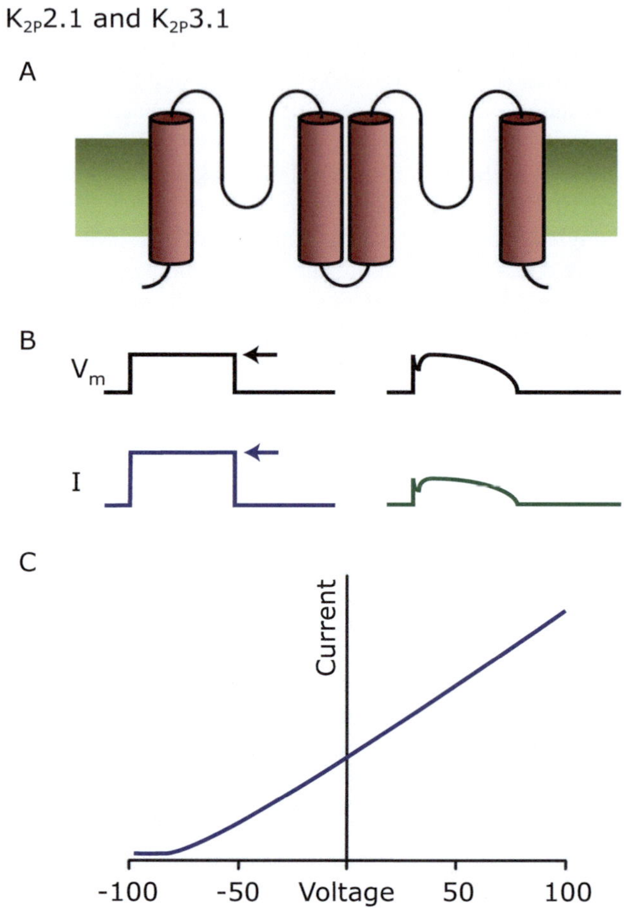

Fig. 3.4 $K_{2P}2.1$ and $K_{2P}3.1$. A. Membrane topology of $K_{2P}2.1$ or $K_{2P}3.1$ with four transmembrane domains and two pore-lining segments. Two $K_{2P}2.1$ and $K_{2P}3.1$ proteins assemble to form a functional channel. Green areas symbolize the plasma membrane; extracellular side is upward. B. Stylized current responses to a square voltage command [left, from Goldstein et al. (2001)] as used in many experimental settings and during an action potential (right). The current flowing through the K_{2P} channel is instantaneous, non-inactivating, and the channel gating is voltage independent. As membrane voltage is changed, the electrochemical gradient for potassium ions is altered, which causes an instantaneous change in the recorded current. C. Stylized graph showing the relationship between membrane voltage and the current amplitude. As the channel does not have any gating (activation, rectification, etc.), the current-voltage relationship is linear. If the internal and external potassium concentrations were identical, the graph would cross the voltage axis at a reversal potential of 0 mV (e.g., Fig. 3.1). Here, the changes in ionic current as a function of membrane voltage is illustrated with more physiological potassium concentrations inside and outside the membrane, and the voltage of no current will be more negative (Goldstein et al. 2001)

potential, potentially suppress afterdepolarizations, and improve the availability of voltage-gated sodium channels (Goldstein et al. 2001).

The K_{2P} channels should not be confused with the "two-pore channels" that are cation-selective ion channels, which appear not to be expressed in the heart (Ishibashi et al. 2000). The protein is made up of two K_V-like six transmembrane domain proteins arranged in series in a single amino acid chain that assembles as dimers in the plasma membrane. Regardless of the name, these channels do not have two separate ion-conducting pores (Ishibashi et al. 2000).

3.4.1 K_{2P}2.1

The K_{2P}2.1 protein is expressed in both human atria and ventricle (Ellinghaus et al. 2005; Gaborit et al. 2007) and was originally named after the "tandem of pore domains in a weak inward rectifying potassium channel (TWIK)" as "TWIK-related channel 1 (TREK-1)." The expression of the *KCNK2* gene encoding K_{2P}2.1 is reduced in humans and animal models with atrial fibrillation, and overexpression of K_{2P}2.1 appears to keep animals in sinus rhythm despite aggressive pacing to induce atrial fibrillation (Lugenbiel et al. 2017). $K_{2P}2.1^{-/-}$ mice are resistant to anesthesia by volatile anesthetics and appear to have increased sensitivity to brain and spinal cord ischemic injuries (Heurteaux et al. 2004). Cardiac-specific deletion of K_{2P}2.1 results in moderate bradycardia and abnormal sinus node excitability during exercise and stress (Unudurthi et al. 2016).

3.4.2 K_{2P}3.1

The K_{2P}3.1 channel appears to be more expressed in the human atria than ventricle (Ellinghaus et al. 2005; Gaborit et al. 2007), which has led to suggestions of using this channel as a target for the treatment of atrial arrhythmias (Schmidt et al. 2014). Formerly named "TWIK-related acid-sensitive potassium channel 1 (TASK-1)," the K_{2P}3.1 channel is blocked by extracellular protons (Lopes et al. 2000). Block of K_{2P}3.1 would theoretically prolong action potential duration and refractoriness primarily in the atria, which provides a safe and reasonably effective manner of converting atrial fibrillation to sinus rhythm. K_{2P}3.1 is expressed in both atria and ventricles in rodents, and genetic deletion of K_{2P}3.1 results in delayed repolarization in mice (Donner et al. 2011; Decher et al. 2011). Atrial fibrillation is associated with an upregulation of K_{2P}3.1 expression, which would contribute to the shortened action potential seen in this disease and thus present us with a potential drug target (Schmidt et al. 2015; Wiedmann et al. 2016). K_{2P}9.1 proteins (encoded by *KCNK9*) form TASK-3 channels, and K_{2P}3.1 and K_{2P}9.1 may form heterodimers with unique electrophysiological properties in vivo (Ravens and Odening 2016; Rinne et al. 2015).

3.5 Voltage-Gated Potassium Channels

The voltage-gated potassium channels (K_V) make up a large and diverse class of ion channels with 12 subgroups identified to date (Gutman et al. 2005). The K_V's are tetramers of α-subunits each with six transmembrane domains and cytosolic C- and N-termini (Snyders 1999). The cardiac action potential is influenced by $K_V1.4$, $K_V1.5$, $K_V4.3$, $K_V7.1$, and $K_V11.1$, whereas the cardiac expression and function of the K_V2 and K_V3 families are presently debated.

The first four transmembrane segments (S1–4) of the K_V form the voltage sensor. Basic residues in the fourth segment hold a number of positive charges, which make the segment move outward upon membrane depolarization. This causes a conformational change of the S1–4 segments that opens the central pore. This process is termed channel activation. Movement of the S4 can be detected electrically as the gating current (Freites et al. 2012), whereas the channel current is flow of potassium through the pore. Key amino acids lining the pore of the channel constitute the selectivity filter, where interactions between potassium ions and water are broken in a process that allows only potassium to be carried through the channel (Kurata and Fedida 2006).

The kinetic properties of the individual K_V channels and the membrane voltage determine the likelihood of having the channel in an activated (open), inactivated (closed), or deactivated (closed) state. The activated channel allows potassium ions to flow through its central pore according to the electrochemical gradient of potassium. After initial activation, some K_V channels inactivate. This is a nonconducting conformational state of the protein that is predominant at prolonged depolarization (Gutman et al. 2005; Snyders 1999), and it is clearly distinct from a deactivated channel (Kurata and Fedida 2006). If a channel has reached a state of inactivation, it takes time and membrane repolarization before the channel can recover and enter a closed or open state. Transitions between channel states are a matter of probability, influenced by membrane voltage and time. For example, 50% of the channels carrying the transient outward currents in murine ventricular cardiomyocytes are inactivated after depolarization to about −30 mV for 2 s (Grubb et al. 2014). The vast majority of the remaining 50% of the channels are open, with only a very small fraction being closed. There are at least two distinct mechanisms of K_V channel inactivation. N-type inactivation is also termed the "ball-and-chain" inactivation, where an N-terminal inactivation domain can access the inner cavity of the pore after channel activation and block current flow. Mostly, the "ball" hangs on the pore-forming subunit, but examples of ball domains on auxiliary subunits are reported [e.g., Morales et al. (1996)]. C-type inactivation is a "collapse" of the extracellular entrance to the channel pore near the selectivity filter, which also abruptly ends the flow of potassium ions (Kurata and Fedida 2006). Some potassium channels have both N- and C-type inactivation. N-type inactivation was described first, involved the NH2 terminus of the channel protein, and hence termed "N-type"; whereas the later-described C-type inactivation was thought to involve a particular sequence near the COOH terminus (Rasmusson et al. 1998). Later, it was clear that C-type inactivation involved many parts of the pore-forming subunit and that it was sensitive to the concentration of extracellular potassium (Lopez-Barneo et al. 1993). Notwithstanding, the term "C-type" is still used to describe this type of channel inactivation.

K_V-channel deactivation occurs when the membrane potential is again repolarized. It can be a rather slow process (e.g., $K_V11.1$) or fast (e.g., $K_V4.2$). The deactivated channel is ready for activation upon membrane depolarization.

The function of some K_V channels is dependent on the extracellular potassium concentration via different mechanisms. This is part of the explanation for action potential shortening and prolongation during hyperkalemia and hypokalemia, respectively (Weiss et al. 2017). Generally, a potassium current will increase when the extracellular potassium concentration is reduced because the driving force for potassium increases. Paradoxically, this is not the case for K_V1, K_V4, K_V11, and $K_{ir}2$. Hypokalemia slows recovery from inactivation of K_V1 and K_V4, so less potassium current is available during the action potential (Weiss et al. 2017; Firek and Giles 1995). Hypokalemia also decreases $K_V11.1$ current when the extracellular potassium concentration decreases (Sanguinetti and Jurkiewicz 1992), because channel inactivation is faster (Yang et al. 1997). Finally, the outward potassium current through $K_{ir}2$ is blocked by magnesium ions and spermine (see above), but extracellular potassium ions destabilize the pore occlusion by magnesium and spermine, which can restore outward potassium current (Nichols et al. 1996). Hyperkalemia depolarizes the resting membrane potential because the reversal potential of potassium becomes less negative. The shortening of the action potential is caused primarily by increased $K_V11.1$ and K_{ir} currents, despite a reduced driving force of potassium (Weiss et al. 2017).

K_V channels give rise to two broad classes of currents: the transient outward potassium currents (I_{to}) and the delayed, rectifying potassium currents (I_K). The channels governing the transient outward currents ($K_V1.4$, $K_V1.5$, $K_V4.3$) activate rapidly and inactivate quite rapidly. They are important early in the cardiac action potential, and due to their rapid activation and subsequent inactivation, they are known as transient currents. The speed of recovery from inactivation after membrane repolarization is quite different among the different I_{to}'s: $K_V4.3$ has a rapid recovery compared to, e.g., $K_V1.4$. This property is often the best way to discriminate between the many K_V currents in native cardiomyocytes that express many different potassium channels (Grubb et al. 2014). The delayed, rectifying potassium channels ($K_V7.1$, $K_V11.1$) have various activation and inactivation kinetics dependent on the channel. The name "inward rectifier" was given at the time when the two current components were not yet distinguished; however, we now know that only $K_V11.1$ shows this characteristic. Via different molecular mechanisms, the largest $K_V7.1$ and $K_V11.1$ currents appear late in the action potential—hence "delayed currents."

3.5.1 $K_V1.4$

The kinetic characteristic of $K_V1.4$ is fast activation and fast inactivation (Fig. 3.5); however, in comparison to K_V4 channels, recovery from inactivation after

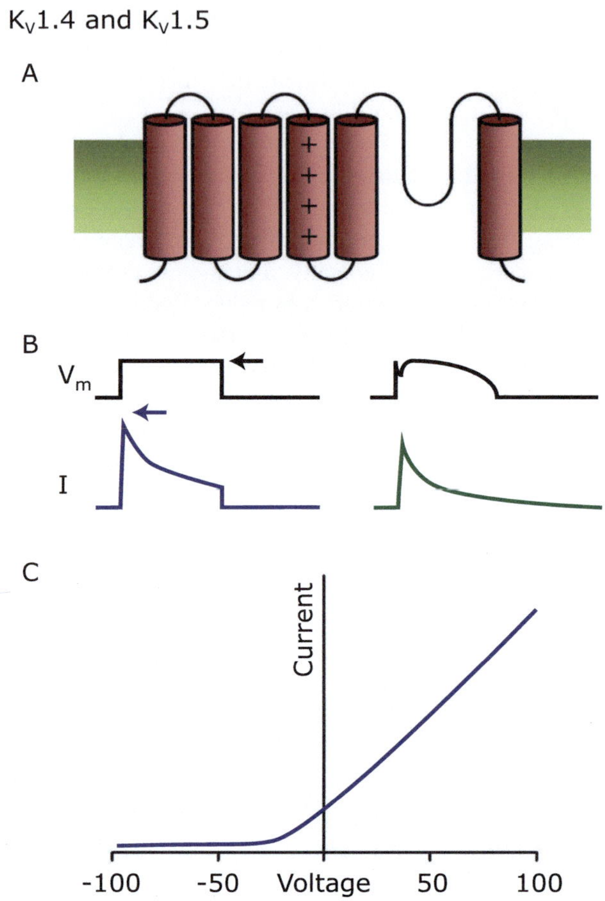

Fig. 3.5 $K_V1.4$ and $K_V1.5$. A. Membrane topology of $K_V1.4$ or $K_V1.5$ with six transmembrane domains and one pore-lining segment. The voltage-sensing domain is indicated by plusses. Four $K_V1.4$ or $K_V1.5$ proteins assemble to form a functional channel. Green areas symbolize the plasma membrane; extracellular side is upward. B. Stylized current responses to a square voltage command (left) as used in many experimental settings and during an action potential (right). The channels will activate upon a depolarizing change in membrane voltage and begin to inactivate, if the membrane is kept depolarized. When the membrane is repolarized, the channel deactivates. The main difference between K_V1 and K_V4 (Fig. 3.6) is the slow recovery from inactivation for K_V1, which is not illustrated on this schematic representation. C. Stylized graph showing the relationship between membrane voltage and the peak current amplitude (arrow in panel **B**). The channels are voltage gated, so they activate (i.e., open) at membrane potentials positive to around ~ -20 mV. At potentials negative to ~ -20 mV, the channels are deactivated (i.e., closed)

depolarization is slow. For this reason, the current generated by $K_V1.4$ is likely $I_{to,slow}$ or simply $I_{to,s}$ (Nerbonne and Kass 2005; London et al. 1998; Guo et al. 1999). Cardiomyocytes disaggregated from human ventricular epicardium show an I_{to} that recovers rapidly from inactivation, whereas myocytes from the endocardial layers have smaller and predominantly a slower recovering I_{to} (Wettwer et al. 1994; Nabauer et al. 1996), suggesting that $K_V1.4$ provides the transient outward potassium current in the endocardium (Niwa and Nerbonne 2010). In support, there is a transmural gradient of $K_V1.4$ mRNA and protein with large expression endocardially (Gaborit et al. 2010), which is in opposite direction of the $I_{to,f}$ gradient which is largest epicardially (see below). $K_V1.4$ does not appear to hold a role in human atrial electrophysiology (Wang et al. 1999b). Currents from septal cardiomyocytes from $K_V1.4^{-/-}$ mice lack the slowly inactivating component of I_{to}, providing further support that $K_V1.4$ underlies $I_{to,s}$ (Guo et al. 1999). Moreover, $K_V1.4$ and $I_{to,s}$ are both upregulated in transgenic mice expressing a dominant negative isoform of that $K_V4.2$, and when the transgene is expressed in $K_V1.4^{-/-}$ mice, both $I_{to,f}$ and $I_{to,s}$ are eliminated (Guo et al. 2000). Based on these experiments, it seems reasonable to suggest that $K_V1.4$ underlies $I_{to,s}$; however there are no specific pharmacological blockers available, limiting the basis of these conclusions to extrapolations from transgenic mice.

3.5.2 $K_V1.5$

The human ultrarapidly activating potassium current I_{Kur} has received considerable interest since it contributes to atrial but not to ventricular repolarization. By blocking this current, atrial-selective pharmacological treatment of atrial fibrillation may be achieved. The murine $I_{K,slow1}$ appears indistinguishable from human I_{Kur}; however, the murine current is not restricted to the atria. It seems reasonable to accept that both $I_{K,slow1}$ and I_{Kur} are governed by $K_V1.5$ (Feng et al. 1997). The first description of the $K_V1.5$ was a current that activated very rapidly at positive membrane potentials and that only inactivated partially (Snyders et al. 1993). In cardiomyocytes at physiological temperatures, I_{Kur} inactivates more completely, but activation is still fast (Fig. 3.5) (Ravens and Wettwer 2011).

Congenital loss-of-$K_V1.5$ function leads to action potential prolongation and triggered activity of the atria and presumably to atrial fibrillation (Olson et al. 2006). From a mechanistic point of view, this seems at odds with a general acceptance of action potential shortening leading to atrial fibrillation (Chen et al. 2003; Wijffels et al. 1995). Moreover, several pharmaceutical companies have pursued development of selective I_{Kur} blockers for treatment of atrial fibrillation, which would be without the potential of ventricular pro-arrhythmia associated with drugs that block currents that are expressed in both atria and ventricles. Nevertheless, atrial fibrillation causes an electrical remodeling of the atria that includes a downregulation of $K_V1.5$ and a different action potential morphology, which could explain the unsuccessful clinical trials using these drugs for treatment for atrial fibrillation (Ravens and Wettwer 2011).

3.5.3 $K_V4.2$ and $K_V4.3$

The pore-forming subunit of $I_{to,f}$ is either $K_V4.2$ or $K_V4.3$, dependent on species. In rodents, both proteins contribute to native $I_{to,f}$ in cardiomyocytes (Niwa and

Nerbonne 2010), whereas $K_V4.3$ seems to be the only isoform in human hearts (Bertaso et al. 2002). Both $K_V4.2$ and $K_V4.3$ interact with several subunits like KChIP2 and DPP6 or DDP10, and these auxiliary proteins are necessary for generating the native current (Grubb et al. 2014, 2015; Thomsen et al. 2009a; Radicke et al. 2005). $K_V4.2$ and $K_V4.3$ are characterized by a fast recovery from inactivation, in addition to their fast activation and fast inactivation (Fig. 3.6). $K_V4.2$ appears to have slightly faster inactivation and recovery from inactivation than $K_V4.3$ (Guo et al. 2002; Liu et al. 2015).

The importance of the K_V4 channel-interacting protein 2 (KChIP2) in generating $I_{to,f}$ is evident from the several independent reports of loss of $I_{to,f}$ after genetic deletion of KChIP2 in mice (Grubb et al. 2014; Thomsen et al. 2009a; Foeger et al. 2013; Kuo et al. 2001). KChIP2 increases $K_V4.2$ and $K_V4.3$ current density by facilitating molecular trafficking of the pore-forming subunit to the plasma membrane, and once there, KChIP2 slows inactivation and accelerates recovery from inactivation (An et al. 2000; Lundby et al. 2010; Grubb et al. 2012). There are four isoforms of KChIPs in the brain (KChIP1–4), but only KChIP2 is expressed in the heart. It is a calcium-sensing protein (An et al. 2000; Grubb et al. 2012); however, calcium binding does not appear to be required for dynamic regulation of the K_V4 channel (An et al. 2000). KChIP2 does not affect other K_V channels, but modulation of $Ca_V1.2$, the cardiac calcium channel, has been shown (Grubb et al. 2015; Foeger et al. 2013; Thomsen et al. 2009b, c). KChIP3 is a transcriptional repressor in the brain, where it translocates from the cytosol to the nucleus upon a raise in cytosolic calcium levels (Ronkainen et al. 2011; Carrion et al. 1999); however, this transcriptional activity is not preserved by KChIP2 in the heart (Winther et al. 2016).

Diaminopeptidyl transferase-like proteins (DPP) are large transmembrane, heavily glucosylated accessory subunits of the K_V4 channels that modulate the voltage dependence of channel kinetics (Radicke et al. 2005). Interestingly, it has been suggested that DPP6 has an important role in generating $I_{to,f}$ in Purkinje cells and that KChIP2 has a smaller role here. Instead, a neuronal calcium sensor was found pivotal for DPP6 modulation in Purkinje cells (Xiao et al. 2013).

The transient outward currents, especially $I_{to,f}$, have an important function in the heart by determining the amplitude of the early phase 1 repolarization. It sets the membrane potential at which the inward calcium current activates and thus modulates the timing and the amplitude of the trigger of calcium-induced calcium release and cardiac contraction (Sah et al. 2003; Cordeiro et al. 2016). In larger mammals, e.g., humans, gradients in phase 1 repolarization across the heart confer synchronization, and augmentation of the cellular calcium release events to coordinate a simultaneous contraction. Loss of early repolarization, as occurs in heart failure, attenuates the sarcolemmal calcium current, desynchronizes cellular calcium release, and contributes to the inefficient pump function of the ventricle. In rodents with shorter action potentials, where $I_{to,f}$ participates in the final repolarization, loss of $I_{to,f}$ leads to action potential prolongation, a longer window for cellular calcium entry and thus a more forceful contraction (Grubb et al. 2015; Sah et al. 2003; Speerschneider et al. 2013). In canines and humans, there is a KChIP2 gradient

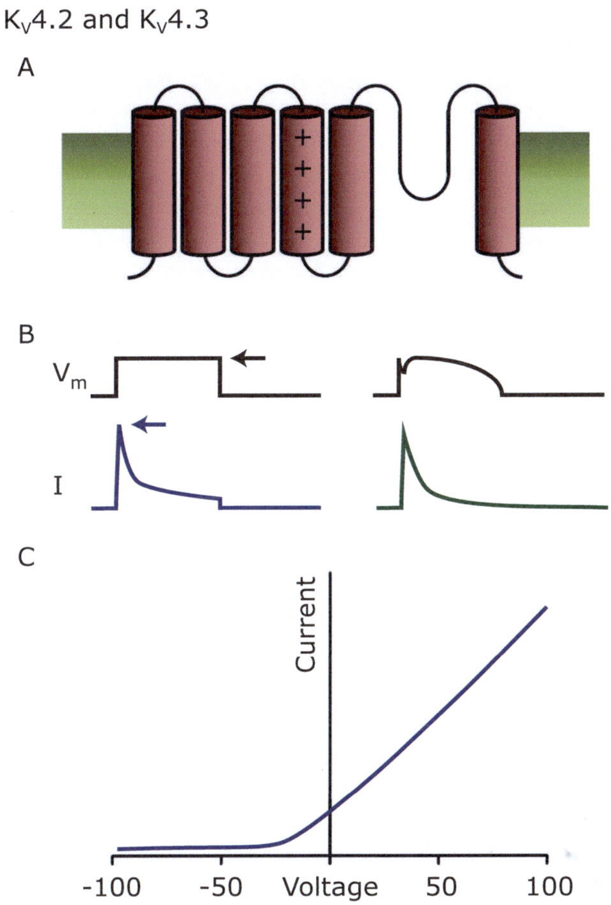

Fig. 3.6 K_V4.2 and K_V4.3. A. Membrane topology of K_V4.2 or K_V4.3 with six transmembrane domains and one pore-lining segment. The voltage-sensing domain is indicated by plusses. Four K_V4 proteins assemble to form a functional channel. Green areas symbolize the plasma membrane; extracellular side is upward. The native subunits, KChIP2 (intercellular) and DPPX (extracellular), are not included in the drawing (Pongs and Schwarz 2010). B. Stylized current responses to a square voltage command (left) as used in many experimental settings and during an action potential [right (Grubb et al. 2014)]. The channels will activate upon a depolarizing change in membrane voltage and begin to inactivate rapidly, if the membrane is kept depolarized. When the membrane is repolarized, the channel deactivates. C. Stylized graph showing the relationship between membrane voltage and the peak current amplitude (arrow in panel **B**). The channels are voltage gated, so they activate (i.e., open) at membrane potentials positive to around ~−20 mV. At potentials negative to ~−20 mV, the channels remain deactivated (i.e., closed)

across the ventricular wall, with largest expression in the epicardial layers, suggesting that KChIP2 underlies the transmural differences in $I_{to,f}$ density (Rosati et al. 2003, 2001; Soltysinska et al. 2009; Calloe et al. 2011). There is only a very small transmural gradient of KChIP2 in mice, whereas a larger gradient of K_V4.2

may explain larger I_{to} in epicardially derived murine cardiomyocytes (Teutsch et al. 2007; Brunet et al. 2004). In the mouse, left-to-right ventricular or apex-base gradients are likely to be more important than transmural gradients (Speerschneider et al. 2017). Nevertheless, different ion channel subunits seem to determine transmural gradients of $I_{to,f}$ in large and small mammals.

Congenital $K_V4.3$ gain of function has been associated with atrial fibrillation, presumably by shortening the atrial action potential and thereby increasing the risk of reentrant circuits (Olesen et al. 2013). In addition, gain-of-function mutations in $K_V4.3$ is linked to Brugada syndrome via mechanisms that are not entirely clear (Giudicessi et al. 2011). 4-Aminopyridine blocks $K_V4.2$ with lower sensitivity [IC50 = 1–5 mM (Fiset et al. 1997; Brouillette et al. 2004)] than it blocks $K_V1.4$ [IC50 = 126 μM (Zhang et al. 1998)] and $K_V1.5$ [IC50 = 50 μM (Bouchard and Fedida 1995)], which can be used as a crude manner of pharmacological current dissection in cardiomyocytes. Heteropoda toxin 2 is a much more selective blocker of $K_V4.2$ (Sanguinetti et al. 1997) and can be used for determining $I_{to,f}$ in native cardiomyocytes (Thomsen et al. 2009a).

3.5.4 $K_V7.1$

In 1990, Sanguinetti and Jurkiewicz were able to separate the two components of the delayed rectifier potassium current (I_K). There is a rapidly activating component, I_{Kr}, that shows prominent inward rectification and a slowly activating component, I_{Ks}, that does not show rectification (Sanguinetti and Jurkiewicz 1990). The pore-forming protein subunit of I_{Ks} was initially named KvLQT1, because loss-of-function mutations in the gene cause long-QT syndrome type 1 (Sanguinetti et al. 1996; Barhanin et al. 1996). Simultaneously, minK was identified as an obligatory accessory subunit of the channel (Sanguinetti et al. 1996; Barhanin et al. 1996). Later, the names for these proteins were changed to $K_V7.1$ and KCNE1, respectively.

$K_V7.1$ alone opens at potentials positive to about −20 mV and activates slowly and shows little inactivation (Volders et al. 1999a, b). When $K_V7.1$ is associated with KCNE1, the current is activated at more positive membrane potentials; it activates and deactivates slower and shows almost no inactivation (Splawski et al. 1997; Seebohm et al. 2003), which resembles the native I_{Ks}.

The slow activation in the absence of noteworthy inactivation causes a continuous buildup of current via $K_V7.1$ during the action potential plateau and makes this current important during phase 3 repolarization (Fig. 3.7). Initially, controversy existed regarding the physiological relevance of $K_V7.1$: on one side were the congenital long-QT syndrome patients with reduced current amplitude, who had delayed repolarization and who were at increased risk of arrhythmias, and on the other, cellular studies of $K_V7.1$, where second-long depolarizations to membrane potentials more positive than the plateau of the action potential were required for buildup of significant current amplitude (Volders et al. 2003). When the selective $K_V7.1$ blockers were developed (e.g., HMR 1556 or chromanol 293B), I_{Ks} block

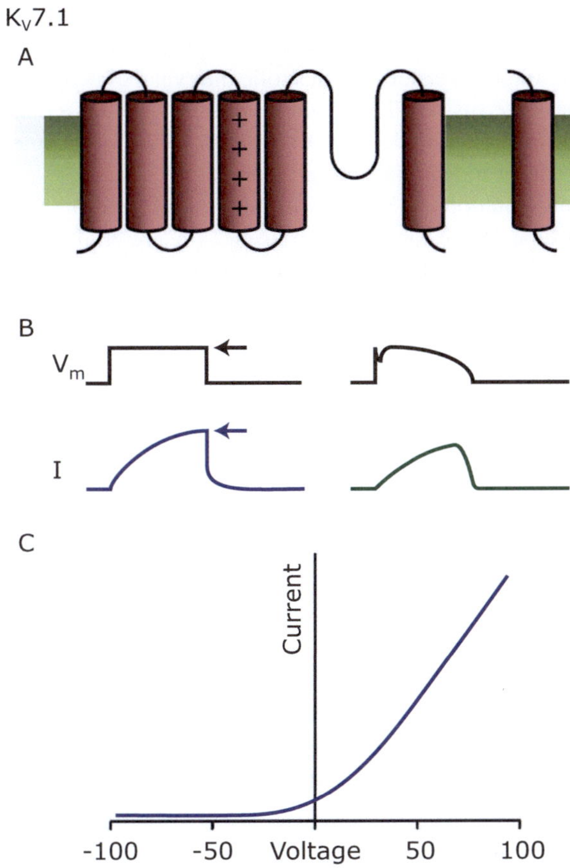

Fig. 3.7 $K_V7.1$. A. Membrane topology of $K_V7.1$ with six transmembrane domains and one pore-lining segment. The voltage-sensing domain is indicated by plusses. Four $K_V7.1$ proteins assemble to form a functional channel. The native subunit, KCNE1, is indicated as a single transmembrane domain protein. Green areas symbolize the plasma membrane; extracellular side is upward. B. Stylized current responses to a square voltage command (left) as used in many experimental settings and during an action potential [right (Volders et al. 2003; Jost et al. 2005)]. The current is slowly activating after a depolarizing step in membrane voltage and reaches maximal activation after several seconds [blue arrow (Sanguinetti et al. 1996)]. C. Stylized graph showing the relationship between membrane voltage and the current amplitude (arrow in panel B). The channels are voltage gated, so they activate (i.e., open) at membrane potentials positive to around ~−20 mV. At potentials negative to ~−20 mV, the channels remain deactivated (i.e., closed)

resulted in negligible action potential prolongation (Jost et al. 2005), in support of a small physiological role for this current. It became clear that β-adrenergic receptor stimulation augments the current and under such circumstances, the physiological role of $K_V7.1$ is clear.

When heart rate increases as a consequence of augmented sympathetic drive, β-adrenergic receptor activation causes a phosphorylation of $K_V7.1$, and this increases the current in the working cardiomyocytes (Volders et al. 2003; Speerschneider and Thomsen 2013; Lundby et al. 2013). This in turn abbreviates the action potential duration, ensuring a diastolic interval that is sufficient for cardiac filling and coronary perfusion. Thus, under physiological circumstances in the presence of a functional autonomic nervous system, $K_V7.1$ contributes significantly to final repolarization of the cardiac action potential.

Loss-of-function mutations in $K_V7.1$ lead to long-QT syndrome type 1, which is the most frequent type of the long-QT syndromes (Wang et al. 1996). The link to physiology is exemplary, as patients with long-QT syndrome type 1 most frequently have arrhythmic events during exercise, where the sympathetic drive increases heart rate but fails to reduce repolarization because $K_V7.1$ is dysfunctional (Schwartz et al. 2001).

3.5.5 $K_V11.1$

$K_V11.1$ is the official name for the channel governing the rapidly activating, delayed, rectifying potassium current, I_{Kr}. The human ether-a-go-go-related gene (*hERG* or *ERG1*), or now *KCNH2*, encodes the pore-forming subunit. $K_V11.1$ is particularly well studied for two main reasons. First, mutations in *KCNH2* underlie 30–40% of the congenital long-QT syndromes (Splawski et al. 2000). Secondly, many cardiac and non-cardiac drugs can block $K_V11.1$, resulting in acquired long-QT syndrome and increased risk of lethal cardiac arrhythmias. Consequently, all new drugs have been tested for $K_V11.1$ block, and development of many potential drugs is discontinued by the pharmaceutical industry due to $K_V11.1$ affinity (Thomsen et al. 2006a).

The $K_V11.1$ channel activates faster than $K_V7.1$, and so it is named the rapidly activating component of I_K (Fig. 3.8). Notwithstanding, the C-type inactivation is much faster than channel activation, and it is voltage dependent to a large degree (Mitcheson and Sanguinetti 1999). This gives rise to a bell-shaped current-voltage relationship and apparent inward rectification, because the inactivation is fast and strong at positive potentials (Perry et al. 2015). Hence, depolarization to positive potentials opens the channel, but this is followed almost immediately by closure due to inactivation.

There are apparent similarities between the current-voltage relationship of I_{K1} and I_{Kr} due to the inward rectification of both channels; however, the mechanisms for rectification are quite different. The inward rectification is a reduction in channel conductance when membrane potential approaches positive values. Inward rectification of I_{Kr} relates to the fast inactivation of the channel (Spector et al. 1996), whereas rectification of I_{K1} is due to block of the channel by intracellular spermine and Mg^{2+} (Fakler et al. 1995).

Toward phase 3 of the action potential, when the depolarizing L-type calcium current is inactivated and $K_V7.1$ has built up some repolarizing current, the negative membrane potential brings the channel out of inactivation, via the open state to deactivation. Because the transition between the open and deactivated state is rather

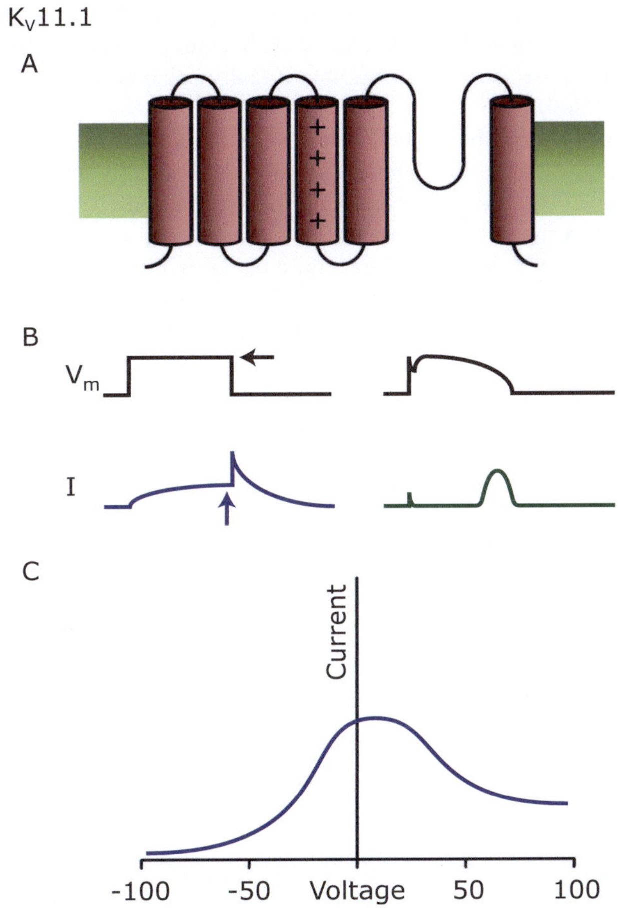

Fig. 3.8 K_V11.1. A. Membrane topology of K_V11.1 with six transmembrane domains and one pore-lining segment. The voltage-sensing domain is indicated by plusses. Four K_V11.1 proteins assemble to form a functional channel. Green areas symbolize the plasma membrane; extracellular side is upward. B. Stylized current responses to a square voltage command [left (Sanguinetti and Jurkiewicz 1990)] as used in many experimental settings and during an action potential [right (Gintant 2000)]. The current activates after a depolarizing step to a test potential (black arrow), and the amplitude of the current can be determined (blue arrow). Upon instantaneous repolarization using a square pulse, the inactivated channels deactivate via an open state that allows a "tail" current flow that can be larger than that at the test potential. During an action potential depolarization, K_V11.1 activates but rapidly inactivates (green). During the gradual repolarization, inactivated K_V11.1 channels recover and allow current flow. C. Stylized graph showing the relationship between membrane voltage and the current amplitude after activation. The voltage-dependent inactivation at positive potentials results in apparent inward rectification: a strong voltage-dependent reduction in channel conductance upon membrane depolarization

slow, the channel is open for a considerable time, and the resulting current is large. Hence, although the channel is activated at the beginning of the action potential, it is toward the end of the action potential that it really becomes important (Perry et al. 2015; Gintant 2000).

As previously stated, the $K_V11.1$ channel has a rich pharmacology. Classical I_{Kr} blockers, like D-sotalol, dofetilide, and E-4031, potently inhibit current flow (Tamargo et al. 2004), thereby prolonging the plateau phase of the action potential and the refractory period. By prolonging the refractory period, the drugs can in theory prevent ventricular and supraventricular arrhythmias (Kober et al. 2000), although treatment is often associated with increased mortality presumably due to ventricular pro-arrhythmia (Waldo et al. 1996). Excessive prolongation of the action potential can induce early afterdepolarizations and large spatial and temporal heterogeneities in refractory periods, a combination that makes the heart particularly vulnerable to torsades de pointes pro-arrhythmia (Thomsen et al. 2004, 2006a, b; Oosterhoff et al. 2010; Winckels et al. 2007). In parallel to the drug-induced long-QT syndrome, a congenital loss-of-function mutation of $K_V11.1$ leads to long-QT syndrome type 2 (Sanguinetti et al. 1996; Barhanin et al. 1996).

3.6 Concluding Remarks

The basic electrophysiological knowledge gathered over the >100 years, since the first recordings of the electrocardiogram and the cardiac action potential, has led to many great successes for clinical electrophysiology. The understanding of basic research provides the basis for clinical therapy but also for drug development and design of potential treatment strategies that can be evaluated clinically. Here, I have briefly characterized the biophysical characteristics of four subfamilies of cardiac potassium channels. Many wide gaps in our knowledge still exist when it comes to the cardiac potassium channels, and it is clear that many important findings are yet to surface.

Acknowledgments The helpful discussions with Dr. Kirstine Calloe are profoundly appreciated.

Compliance with Ethical Standards

Sources of Funding None.

Conflict of Interest The author declares that he has no conflict of interest.

Ethical Approval All animal studies summarized and reviewed in this article were conducted based on international, national, and/or institutional guidelines for the care and use of animals.

References

An WF, Bowlby MR, Betty M, et al. Modulation of A-type potassium channels by a family of calcium sensors. Nature. 2000;403:553–6.

Anumonwo JM, Lopatin AN. Cardiac strong inward rectifier potassium channels. J Mol Cell Cardiol. 2010;48:45–54.

Barhanin J, Lesage F, Guillemare E, Fink M, Lazdunski M, Romey GK. (V)LQT1 and lsK (minK) proteins associate to form the I(Ks) cardiac potassium current. Nature. 1996;384:78–80.

Berkefeld H, Fakler B, Schulte U. Ca2+−activated K+ channels: from protein complexes to function. Physiol Rev. 2010;90:1437–59.

Bertaso F, Sharpe CC, Hendry BM, James AF. Expression of voltage-gated K+ channels in human atrium. Basic Res Cardiol. 2002;97:424–33.

Bonilla IM, Long VP 3rd, Vargas-Pinto P, et al. Calcium-activated potassium current modulates ventricular repolarization in chronic heart failure. PLoS One. 2014;9:e108824.

Bouchard R, Fedida D. Closed- and open-state binding of 4-aminopyridine to the cloned human potassium channel Kv1.5. J Pharmacol Exp Ther. 1995;275:864–76.

Brouillette J, Clark RB, Giles WR, Fiset C. Functional properties of K+ currents in adult mouse ventricular myocytes. J Physiol. 2004;559:777–98.

Brunet S, Aimond F, Li H, et al. Heterogeneous expression of repolarizing, voltage-gated K+ currents in adult mouse ventricles. J Physiol. 2004;559:103–20.

Calloe K, Nof E, Jespersen T, et al. Comparison of the effects of a transient outward potassium channel activator on currents recorded from atrial and ventricular cardiomyocytes. J Cardiovasc Electrophysiol. 2011;22:1057–66.

Calloe K, Goodrow R, Olesen SP, Antzelevitch C, Cordeiro JM. Tissue-specific effects of acetylcholine in the canine heart. Am J Phys Heart Circ Phys. 2013;305:H66–75.

Carrion AM, Link WA, Ledo F, Mellstrom B, Naranjo JR. DREAM is a Ca2+−regulated transcriptional repressor. Nature. 1999;398:80–4.

Chang PC, Turker I, Lopshire JC, et al. Heterogeneous upregulation of apamin-sensitive potassium currents in failing human ventricles. J Am Heart Assoc. 2013;2:e004713.

Chen YH, Xu SJ, Bendahhou S, et al. KCNQ1 gain-of-function mutation in familial atrial fibrillation. Science (New York, NY). 2003;299:251–4.

Cordeiro JM, Calloe K, Aschar-Sobbi R, et al. Physiological roles of the transient outward current Ito in normal and diseased hearts. Front Biosci (Schol Ed). 2016;8:143–59.

Decher N, Wemhoner K, Rinne S, et al. Knock-out of the potassium channel TASK-1 leads to a prolonged QT interval and a disturbed QRS complex. Cell Physiol Biochem. 2011;28:77–86. 19

Decher N, Kiper AK, Rolfes C, Schulze-Bahr E, Rinne S. The role of acid-sensitive two-pore domain potassium channels in cardiac electrophysiology: focus on arrhythmias. Pflugers Arch. 2015;467:1055–67.

Dong DL, Bai YL, Cai BZ. Calcium-activated potassium channels: potential target for cardiovascular diseases. Adv Protein Chem Struct Biol. 2016;104:233–61.

Donner BC, Schullenberg M, Geduldig N, et al. Functional role of TASK-1 in the heart: studies in TASK-1-deficient mice show prolonged cardiac repolarization and reduced heart rate variability. Basic Res Cardiol. 2011;106:75–87.

Ellinghaus P, Scheubel RJ, Dobrev D, et al. Comparing the global mRNA expression profile of human atrial and ventricular myocardium with high-density oligonucleotide arrays. J Thorac Cardiovasc Surg. 2005;129:1383–90.

Fakler B, Brandle U, Glowatzki E, Weidemann S, Zenner HP, Ruppersberg JP. Strong voltage-dependent inward rectification of inward rectifier K+ channels is caused by intracellular spermine. Cell. 1995;80:149–54.

Feng J, Wible B, Li GR, Wang Z, Nattel S. Antisense oligodeoxynucleotides directed against Kv1.5 mRNA specifically inhibit ultrarapid delayed rectifier K+ current in cultured adult human atrial myocytes. Circ Res. 1997;80:572–9.

Firek L, Giles WR. Outward currents underlying repolarization in human atrial myocytes. Cardiovasc Res. 1995;30:31–8.

Fiset C, Clark RB, Shimoni Y, Giles WR. Shal-type channels contribute to the Ca2+-independent transient outward K+ current in rat ventricle. J Physiol. 1997;500(Pt 1):51–64.

Foeger NC, Wang W, Mellor RL, Nerbonne JM. Stabilization of Kv4 protein by the accessory K(+) channel interacting protein 2 (KChIP2) subunit is required for the generation of native myocardial fast transient outward K(+) currents. J Physiol. 2013;591:4149–66.

Freites JA, Schow EV, White SH, Tobias DJ. Microscopic origin of gating current fluctuations in a potassium channel voltage sensor. Biophys J. 2012;102:L44–6.

Gaborit N, Le Bouter S, Szuts V, et al. Regional and tissue specific transcript signatures of ion channel genes in the non-diseased human heart. J Physiol. 2007;582:675–93.

Gaborit N, Varro A, Le Bouter S, et al. Gender-related differences in ion-channel and transporter subunit expression in non-diseased human hearts. J Mol Cell Cardiol. 2010;49:639–46.

Gintant GA. Characterization and functional consequences of delayed rectifier current transient in ventricular repolarization. Am J Phys Heart Circ Phys. 2000;278:H806–17.

Giudicessi JR, Ye D, Tester DJ, et al. Transient outward current (I(to)) gain-of-function mutations in the KCND3-encoded Kv4.3 potassium channel and Brugada syndrome. Heart Rhythm. 2011;8:1024–32.

Goldstein SAN, Bockenhauer D, O'Kelly I, Zilberberg N. Potassium leak channels and the KCNK family of two-p-domain subunits. Nat Rev Neurosci. 2001;2:175–84.

Goldstein SA, Bayliss DA, Kim D, Lesage F, Plant LD, Rajan S. International Union of Pharmacology. LV. Nomenclature and molecular relationships of two-P potassium channels. Pharmacol Rev. 2005;57:527–40.

Grubb S, Calloe K, Thomsen MB. Impact of KChIP2 on cardiac electrophysiology and the progression of heart failure. Front Physiol. 2012;3:118.

Grubb S, Speerschneider T, Occhipinti D, et al. Loss of K+ currents in heart failure is accentuated in KChIP2 deficient mice. J Cardiovasc Electrophysiol. 2014;25:896–904.

Grubb S, Aistrup GL, Koivumaki JT, et al. Preservation of cardiac function by prolonged action potentials in mice deficient of KChIP2. Am J Phys Heart Circ Phys. 2015;309:H481–9.

Grunnet M, Jespersen T, Angelo K, et al. Pharmacological modulation of SK3 channels. Neuropharmacology. 2001;40:879–87.

Guo W, Xu H, London B, Nerbonne JM. Molecular basis of transient outward K+ current diversity in mouse ventricular myocytes. J Physiol. 1999;521(Pt 3):587–99.

Guo W, Li H, London B, Nerbonne JM. Functional consequences of elimination of i(to,f) and i(to, s): early afterdepolarizations, atrioventricular block, and ventricular arrhythmias in mice lacking Kv1.4 and expressing a dominant-negative Kv4 alpha subunit. Circ Res. 2000;87:73–9.

Guo W, Malin SA, Johns DC, Jeromin A, Nerbonne JM. Modulation of Kv4-encoded K(+) currents in the mammalian myocardium by neuronal calcium sensor-1. J Biol Chem. 2002;277:26436–43.

Gutman GA, Chandy KG, Grissmer S, et al. International Union of Pharmacology. LIII. Nomenclature and molecular relationships of voltage-gated potassium channels. Pharmacol Rev. 2005;57:473–508.

Heurteaux C, Guy N, Laigle C, et al. TREK-1, a K+ channel involved in neuroprotection and general anesthesia. EMBO J. 2004;23:2684–95.

Hibino H, Inanobe A, Furutani K, Murakami S, Findlay I, Kurachi Y. Inwardly rectifying potassium channels: their structure, function, and physiological roles. Physiol Rev. 2010;90:291–366.

Ishibashi K, Suzuki M, Imai M. Molecular cloning of a novel form (two-repeat) protein related to voltage-gated sodium and calcium channels. Biochem Biophys Res Commun. 2000;270:370–6.

Jost N, Virag L, Bitay M, et al. Restricting excessive cardiac action potential and QT prolongation: a vital role for IKs in human ventricular muscle. Circulation. 2005;112:1392–9.

Kober L, Bloch Thomsen PE, Moller M, et al. Effect of dofetilide in patients with recent myocardial infarction and left-ventricular dysfunction: a randomised trial. Lancet (London, England). 2000;356:2052–8.

Koumi S, Wasserstrom JA, Ten Eick RE. Beta-adrenergic and cholinergic modulation of inward rectifier K+ channel function and phosphorylation in Guinea-pig ventricle. J Physiol. 1995a;486 (Pt 3):661–78.

Koumi S, Backer CL, Arentzen CE, Sato R. Beta-adrenergic modulation of the inwardly rectifying potassium channel in isolated human ventricular myocytes. Alteration in channel response to beta-18 adrenergic stimulation in failing human hearts. J Clin Invest. 1995b;96:2870–81.

Kubo Y, Baldwin TJ, Jan YN, Jan LY. Primary structure and functional expression of a mouse inward rectifier potassium channel. Nature. 1993;362:127–33.

Kubo Y, Adelman JP, Clapham DE, et al. International Union of Pharmacology. LIV. Nomenclature and molecular relationships of inwardly rectifying potassium channels. Pharmacol Rev. 2005;57:509–26.

Kuo HC, Cheng CF, Clark RB, et al. A defect in the Kv channel-interacting protein 2 (KChIP2) gene leads to a complete loss of I(to) and confers susceptibility to ventricular tachycardia. Cell. 2001;107:801–13.

Kurata HT, Fedida D. A structural interpretation of voltage-gated potassium channel inactivation. Prog Biophys Mol Biol. 2006;92:185–208.

Li N, Timofeyev V, Tuteja D, et al. Ablation of a Ca2+-activated K+ channel (SK2 channel) results in action potential prolongation in atrial myocytes and atrial fibrillation. J Physiol. 2009;587:1087–100.

Liin SI, Barro-Soria R, Larsson HP. The KCNQ1 channel—remarkable flexibility in gating allows for functional versatility. J Physiol. 2015;593:2605–15.

Liu J, Kim KH, Morales MJ, Heximer SP, Hui CC, Backx PH. Kv4.3-encoded fast transient outward current is presented in Kv4.2 knockout mouse Cardiomyocytes. PLoS One. 2015;10: e0133274.

Logothetis DE, Kurachi Y, Galper J, Neer EJ, Clapham DE. The beta gamma subunits of GTP-binding proteins activate the muscarinic K+ channel in heart. Nature. 1987;325:321–6.

London B, Wang DW, Hill JA, Bennett PB. The transient outward current in mice lacking the potassium channel gene Kv1.4. J Physiol. 1998;509(Pt 1):171–82.

Lopes CM, Gallagher PG, Buck ME, Butler MH, Goldstein SA. Proton block and voltage gating are potassium-dependent in the cardiac leak channel Kcnk3. J Biol Chem. 2000;275:16969–78.

Lopez-Barneo J, Hoshi T, Heinemann SH, Aldrich RW. Effects of external cations and mutations in the pore region on C-type inactivation of shaker potassium channels. Recept Channels. 1993;1:61–71.

Lu L, Zhang Q, Timofeyev V, et al. Molecular coupling of a Ca2+-activated K+ channel to L-type Ca2+ channels via alpha-actinin2. Circ Res. 2007;100:112–20.

Lugenbiel P, Wenz F, Syren P, et al. TREK-1 (K2P2.1) K+ channels are suppressed in patients with atrial fibrillation and heart failure and provide therapeutic targets for rhythm control. Basic Res Cardiol. 2017;112:8.

Lundby A, Jespersen T, Schmitt N, et al. Effect of the Ito activator NS5806 on cloned Kv4 channels depends on the accessory protein KChIP2. Br J Pharmacol. 2010;160:2028–44.

Lundby A, Andersen MN, Steffensen AB, et al. In vivo phosphoproteomics analysis reveals the cardiac targets of beta-adrenergic receptor signaling. Sci Signal. 2013;6:rs11.

Melnyk P, Zhang L, Shrier A, Nattel S. Differential distribution of Kir2.1 and Kir2.3 subunits in canine atrium and ventricle. Am J Phys Heart Circ Phys. 2002;283:H1123–33.

Mitcheson JS, Sanguinetti MC. Biophysical properties and molecular basis of cardiac rapid and slow delayed rectifier potassium channels. Cell Physiol Biochem. 1999;9:201–16.

Morales MJ, Wee JO, Wang S, Strauss HC, Rasmusson RL. The N-terminal domain of a K+ channel beta subunit increases the rate of C-type inactivation from the cytoplasmic side of the channel. Proc Natl Acad Sci U S A. 1996;93:15119–23.

Nabauer M, Beuckelmann DJ, Uberfuhr P, Steinbeck G. Regional differences in current density and rate-dependent properties of the transient outward current in subepicardial and subendocardial myocytes of human left ventricle. Circulation. 1996;93:168–77.

Nerbonne JM, Kass RS. Molecular physiology of cardiac repolarization. Physiol Rev. 2005;85:1205–53.

Nichols CG, Makhina EN, Pearson WL, Sha Q, Lopatin AN. Inward rectification and implications for cardiac excitability. Circ Res. 1996;78:1–7.

Niwa N, Nerbonne JM. Molecular determinants of cardiac transient outward potassium current (I (to)) expression and regulation. J Mol Cell Cardiol. 2010;48:12–25.

Olesen MS, Refsgaard L, Holst AG, et al. A novel KCND3 gain-of-function mutation associated with early-onset of persistent lone atrial fibrillation. Cardiovasc Res. 2013;98:488–95.

Olson TM, Alekseev AE, Liu XK, et al. Kv1.5 channelopathy due to KCNA5 loss-of-function mutation causes human atrial fibrillation. Human molecular genetics 2006;15:2185–91.

Oosterhoff P, Thomsen MB, Maas JN, et al. High-rate pacing reduces variability of repolarization and prevents repolarization-dependent arrhythmias in dogs with chronic AV block. J Cardiovasc Electrophysiol. 2010;1384-91(23):21.

Ozgen N, Dun W, Sosunov EA, et al. Early electrical remodeling in rabbit pulmonary vein results from trafficking of intracellular SK2 channels to membrane sites. Cardiovasc Res. 2007;75:758–69.

Patel SP, Campbell DL. Transient outward potassium current, 'Ito', phenotypes in the mammalian left ventricle: underlying molecular, cellular and biophysical mechanisms. J Physiol. 2005;569:7–39.

Perry MD, Ng CA, Mann SA, et al. Getting to the heart of hERG K(+) channel gating. J Physiol. 2015;593:2575–85.

Plaster NM, Tawil R, Tristani-Firouzi M, et al. Mutations in Kir2.1 cause the developmental and episodic electrical phenotypes of Andersen's syndrome. Cell. 2001;105:511–9.

Pongs O, Schwarz JR. Ancillary subunits associated with voltage-dependent K+ channels. Physiol Rev. 2010;90:755–96.

Priori SG, Pandit SV, Rivolta I, et al. A novel form of short QT syndrome (SQT3) is caused by a mutation in the KCNJ2 gene. Circ Res. 2005;96(7):800.

Radicke S, Cotella D, Graf EM, Ravens U, Wettwer E. Expression and function of dipeptidyl-aminopeptidase-like protein 6 as a putative beta-subunit of human cardiac transient outward current encoded by Kv4.3. J Physiol. 2005;565:751–6.

Rasmusson RL, Morales MJ, Wang S, et al. Inactivation of voltage-gated cardiac K+ channels. Circ Res. 1998;82:739–50.

Ravens U, Odening KE. Atrial fibrillation: therapeutic potential of atrial K+ channel blockers. Pharmacol Ther. 2017;176:13–21.

Ravens U, Wettwer E. Ultra-rapid delayed rectifier channels: molecular basis and therapeutic implications. Cardiovasc Res. 2011;89:776–85.

Rinne S, Kiper AK, Schlichthorl G, et al. TASK-1 and TASK-3 may form heterodimers in human atrial cardiomyocytes. J Mol Cell Cardiol. 2015;81:71–80.

Ronkainen JJ, Hanninen SL, Korhonen T, et al. Ca2+-calmodulin-dependent protein kinase II represses cardiac transcription of the L-type calcium channel alpha(1C)-subunit gene (Cacna1c) by DREAM translocation. J Physiol. 2011;2669-86(21):589.

Rosati B, Pan Z, Lypen S, et al. Regulation of KChIP2 potassium channel beta subunit gene expression underlies the gradient of transient outward current in canine and human ventricle. J Physiol. 2001;533:119–25.

Rosati B, Grau F, Rodriguez S, Li H, Nerbonne JM, McKinnon D. Concordant expression of KChIP2 mRNA, protein and transient outward current throughout the canine ventricle. J Physiol. 2003;548:815–22.

Sah R, Ramirez RJ, Oudit GY, et al. Regulation of cardiac excitation-contraction coupling by action potential repolarization: role of the transient outward potassium current (I(to)). J Physiol. 2003;546:5–18.

Sakmann B, Noma A, Trautwein W. Acetylcholine activation of single muscarinic K+ channels in isolated pacemaker cells of the mammalian heart. Nature. 1983;303:250–3.

Sanguinetti MC, Jurkiewicz NK. Two components of cardiac delayed rectifier K+ current. Differential sensitivity to block by class III antiarrhythmic agents. J Gen Physiol. 1990;96:195–215.

Sanguinetti MC, Jurkiewicz NK. Role of external Ca2+ and K+ in gating of cardiac delayed rectifier K+ currents. Pflugers Arch. 1992;420:180–6.

Sanguinetti MC, Curran ME, Zou A, et al. Coassembly of K(V)LQT1 and minK (IsK) proteins to form cardiac I(Ks) potassium channel. Nature. 1996;384:80–3.

Sanguinetti MC, Johnson JH, Hammerland LG, et al. Heteropodatoxins: peptides isolated from spider venom that block Kv4.2 potassium channels. Mol Pharmacol. 1997;51:491–8.

Schmidt C, Wiedmann F, Schweizer PA, Katus HA, Thomas D. Inhibition of cardiac two-pore-domain K+ (K2P) channels—an emerging antiarrhythmic concept. Eur J Pharmacol. 2014;738:250–5.

Schmidt C, Wiedmann F, Voigt N, et al. Upregulation of K(2P)3.1 K+ current causes action potential shortening in patients with chronic atrial fibrillation. Circulation. 2015;132:82–92.

Schwartz PJ, Priori SG, Spazzolini C, et al. Genotype-phenotype correlation in the long-QT syndrome: gene-specific triggers for life-threatening arrhythmias. Circulation. 2001;103:89–95.

Seebohm G, Sanguinetti MC, Pusch M. Tight coupling of rubidium conductance and inactivation in human KCNQ1 potassium channels. J Physiol. 2003;552:369–78.

Skibsbye L, Poulet C, Diness JG, et al. Small-conductance calcium-activated potassium (SK) channels contribute to action potential repolarization in human atria. Cardiovasc Res. 2014;103:156–67.

Snyders DJ. Structure and function of cardiac potassium channels. Cardiovasc Res. 1999;42:377–90.

Snyders DJ, Tamkun MM, Bennett PB. A rapidly activating and slowly inactivating potassium channel cloned from human heart. Functional analysis after stable mammalian cell culture expression. J Gen Physiol. 1993;101:513–43.

Soltysinska E, Olesen SP, Christ T, et al. Transmural expression of ion channels and transporters in human nondiseased and end-stage failing hearts. Pflugers Arch. 2009;459:11–23.

Soltysinska E, Bentzen BH, Barthmes M, et al. KCNMA1 encoded cardiac BK channels afford protection against ischemia-reperfusion injury. PLoS One. 2014;9:e103402.

Spector PS, Curran ME, Zou A, Keating MT, Sanguinetti MC. Fast inactivation causes rectification of the IKr channel. J Gen Physiol. 1996;107:611–9.

Speerschneider T, Thomsen MB. Physiology and analysis of the electrocardiographic T wave in mice. Acta Physiol (Oxf). 2013;209:262–71.

Speerschneider T, Grubb S, Metoska A, Olesen SP, Calloe K, Thomsen MB. Development of heart failure is independent of K+ channel-interacting protein 2 expression. J Physiol. 2013;591:5923–37.

Speerschneider T, Grubb S, Olesen SP, Calloe K, Thomsen MB. Ventricular repolarization time, location of pacing stimulus and current pulse amplitude conspire to determine arrhythmogenicity in mice. Acta Physiol (Oxf). 2017;219:660–8.

Splawski I, Tristani-Firouzi M, Lehmann MH, Sanguinetti MC, Keating MT. Mutations in the hminK gene cause long QT syndrome and suppress IKs function. Nat Genet. 1997;17:338–40.

Splawski I, Shen J, Timothy KW, et al. Spectrum of mutations in long-QT syndrome genes. KVLQT1, HERG, SCN5A, KCNE1, and KCNE2. Circulation. 2000;102:1178–85.

Tamargo J, Caballero R, Gómez R, Valenzuela C, Delpón E. Pharmacology of cardiac potassium channels. Cardiovasc Res. 2004;62:9–33.

Teutsch C, Kondo RP, Dederko DA, Chrast J, Chien KR, Giles WR. Spatial distributions of Kv4 channels and KChip2 isoforms in the murine heart based on laser capture microdissection. Cardiovasc Res. 2007;73:739–49.

Thomsen MB, Verduyn SC, Stengl M, et al. Increased short-term variability of repolarization predicts d-sotalol-induced torsades de pointes in dogs. Circulation. 2004;110:2453–9.

Thomsen MB, Matz J, Volders PG, Vos MA. Assessing the proarrhythmic potential of drugs: current status of models and surrogate parameters of torsades de pointes arrhythmias. Pharmacol Ther. 2006a;112:150–70.

Thomsen MB, Beekman JD, Attevelt NJ, et al. No proarrhythmic properties of the antibiotics Moxifloxacin or azithromycin in anaesthetized dogs with chronic-AV block. Br J Pharmacol. 2006b;149:1039–48.

Thomsen MB, Sosunov EA, Anyukhovsky EP, Ozgen N, Boyden PA, Rosen MR. Deleting the accessory subunit KChIP2 results in loss of I(to,f) and increased I(K,slow) that maintains normal action potential configuration. Heart Rhythm. 2009a;6:370–7.

Thomsen MB, Wang C, Ozgen N, Wang HG, Rosen MR, Pitt GS. Accessory subunit KChIP2 modulates the cardiac L-type calcium current. Circ Res. 2009b;104:1382–9.

Thomsen MB, Foster E, Nguyen KH, Sosunov EA. Transcriptional and electrophysiological consequences of KChIP2-mediated regulation of CaV1.2. Channels (Austin). 2009c;3:308–10.

Tuteja D, Xu D, Timofeyev V, et al. Differential expression of small-conductance Ca2+-activated K + channels SK1, SK2, and SK3 in mouse atrial and ventricular myocytes. Am J Phys Heart Circ Phys. 2005;289:H2714–23.

Unudurthi SD, Wu X, Qian L, et al. Two-pore K+ channel TREK-1 regulates Sinoatrial node membrane excitability. J Am Heart Assoc. 2016;5:e002865.

Vandenberg JI, Perry MD, Perrin MJ, Mann SA, Ke Y, Hill AP. hERG K(+) channels: structure, function, and clinical significance. Physiol Rev. 2012;92:1393–478.

Volders PG, Sipido KR, Carmeliet E, Spatjens RL, Wellens HJ, Vos MA. Repolarizing K+ currents ITO1 and IKs are larger in right than left canine ventricular midmyocardium. Circulation. 1999a;99:206–10.

Volders PG, Sipido KR, Vos MA, et al. Downregulation of delayed rectifier K(+) currents in dogs with chronic complete atrioventricular block and acquired torsades de pointes. Circulation. 1999b;100:2455–61.

Volders PG, Stengl M, van Opstal JM, et al. Probing the contribution of IKs to canine ventricular repolarization: key role for beta-adrenergic receptor stimulation. Circulation. 2003;107:2753–60.

Waldo AL, Camm AJ, deRuyter H, et al. Effect of d-sotalol on mortality in patients with left ventricular dysfunction after recent and remote myocardial infarction. The SWORD investigators. Survival with oral d-Sotalol. Lancet (London, England). 1996;348:7–12.

Wang Q, Curran ME, Splawski I, et al. Positional cloning of a novel potassium channel gene: KVLQT1 mutations cause cardiac arrhythmias. Nat Genet. 1996;12:17–23.

Wang W, Watanabe M, Nakamura T, Kudo Y, Ochi R. Properties and expression of Ca2 +−activated K+ channels in H9c2 cells derived from rat ventricle. Am J Phys. 1999a;276: H1559–66.

Wang Z, Feng J, Shi H, Pond A, Nerbonne JM, Nattel S. Potential molecular basis of different physiological properties of the transient outward K+ current in rabbit and human atrial myocytes. Circ Res. 1999b;84:551–61.

Wei AD, Gutman GA, Aldrich R, Chandy KG, Grissmer S, Wulff H. International Union of Pharmacology. LII. Nomenclature and molecular relationships of calcium-activated potassium channels. Pharmacol Rev. 2005;57:463–72.

Weiss JN, Qu Z, Shivkumar K. Electrophysiology of hypokalemia and hyperkalemia. Circ Arrhythm Electrophysiol. 2017;10(3)

Wettwer E, Amos GJ, Posival H, Ravens U. Transient outward current in human ventricular myocytes of subepicardial and subendocardial origin. Circ Res. 1994;75:473–82.

Wiedmann F, Schmidt C, Lugenbiel P, et al. Therapeutic targeting of two-pore-domain potassium (K(2P)) channels in the cardiovascular system. Clin Sci. 2016;130:643–50.

Wijffels MC, Kirchhof CJ, Dorland R, Allessie MA. Atrial fibrillation begets atrial fibrillation. A study in awake chronically instrumented goats. Circulation. 1995;92:1954–68.

Winckels SK, Thomsen MB, Oosterhoff P, et al. High-septal pacing reduces ventricular electrical remodeling and proarrhythmia in chronic atrioventricular block dogs. J Am Coll Cardiol. 2007;50:906–13.

Winther SV, Tuomainen T, Borup R, Tavi P, Antoons G, Thomsen MB. Potassium Channel interacting protein 2 (KChIP2) is not a transcriptional regulator of cardiac electrical remodeling. Sci Rep. 2016;6:28760.

Xiao L, Koopmann TT, Ordog B, et al. Unique cardiac Purkinje fiber transient outward current beta-subunit composition: a potential molecular link to idiopathic ventricular fibrillation. Circ Res. 2013;112:1310–22.

Xu W, Liu Y, Wang S, et al. Cytoprotective role of Ca2+-activated K+ channels in the cardiac inner mitochondrial membrane. Science. 2002;298:1029–33.

Yang T, Snyders DJ, Roden DM. Rapid inactivation determines the rectification and [K+]o dependence of the rapid component of the delayed rectifier K+ current in cardiac cells. Circ Res. 1997;80:782–9.

Yu FH, Catterall WA. The VGL-Chanome: a protein superfamily specialized for electrical Signaling and ionic homeostasis. Sci STKE. 2004;2004:re15.

Zhang H, Zhu B, Yao JA, Tseng GN. Differential effects of S6 mutations on binding of quinidine and 4-aminopyridine to rat isoform of Kv1.4: common site but different factors in determining blockers' binding affinity. J Pharmacol Exp Ther. 1998;287:332–43.

Zhang H, Flagg TP, Nichols CG. Cardiac sarcolemmal K(ATP) channels: latest twists in a questing tale! J Mol Cell Cardiol. 2010;48:71–5.

Voltage-Gated Calcium Channels and Their Roles in Cardiac Electrophysiology

4

Jordi Heijman, Cristina E. Molina, and Niels Voigt

Abstract

In cardiomyocytes voltage-gated Ca^{2+} channels are major players in cardiac cellular electrophysiology and cellular excitation-contraction coupling. Accordingly, Ca^{2+} channel dysfunction contributes to the development of cardiac arrhythmias and impaired cardiac contractile function. In addition, Ca^{2+} entry through voltage-gated Ca^{2+} channels is an important regulator of gene transcription and cardiac cellular metabolism. In order to fulfil these tasks reliably, Ca^{2+} channels are highly regulated by specific subunit compositions and various signaling pathways. This chapter provides an overview of the role of voltage-gated Ca^{2+} channels in cardiac cellular electrophysiology and summarizes their molecular composition, biophysical properties, and regulatory mechanisms, with a special focus on L-type Ca^{2+} channels.

4.1 Introduction

Sarcolemmal Ca^{2+} influx into cardiomyocytes through Ca^{2+}-permeable ion channels plays a critical role in cellular physiology (Bers 2008). On the one hand, electrogenic Ca^{2+} influx results in depolarization of the cellular membrane potential, thereby

J. Heijman
Department of Cardiology, CARIM School for Cardiovascular Diseases, Maastricht University, Maastricht, The Netherlands
e-mail: jordi.heijman@maastrichtuniversity.nl

C. E. Molina · N. Voigt (✉)
Institute of Pharmacology and Toxicology, University Medical Center Göttingen, Georg-August University Göttingen, Göttingen, Germany

DZHK (German Center for Cardiovascular Research), partner site Göttingen, Göttingen, Germany
e-mail: cristina.molina@med.uni-goettingen.de; niels.voigt@med.uni-goettingen.de

D. Thomas, C. A. Remme (eds.), *Channelopathies in Heart Disease*, Cardiac and Vascular Biology, https://doi.org/10.1007/978-3-319-77812-9_4

contributing to the action potential (AP) plateau in atrial and ventricular cardiomyocytes and to cellular automaticity in the sinoatrial and atrioventricular nodes (Heijman et al. 2016). On the other hand, Ca^{2+} ions are important intracellular second-messenger molecules, involved in excitation-contraction (EC) coupling, regulation of cellular function by phosphorylation, and regulation of gene expression (Fearnley et al. 2011; Harada et al. 2012a; Yue et al. 2011; Kreusser and Backs 2014; Rose et al. 2012; Du et al. 2010).

Voltage-activated Ca^{2+} channels represent the major route of Ca^{2+} entry into cardiomyocytes in response to depolarizations of the cellular membrane potential (Rose and Backx 2014). So far, 10 members of the family of voltage-gated Ca^{2+} channels have been identified in mammals. Based on their biophysical and pharmacological properties, they are further grouped into L-, P/Q-, R-, and T-type Ca^{2+} channels. However, only L-type and T-type Ca^{2+} channels (LTCC and TTCC, respectively) are expressed in cardiomyocytes, whereas the others are most abundant in neurons (Bers 2008; Catterall 2011; Rose and Backx 2014). LTCC are expressed in all cardiomyocytes, whereas TTCC are restricted to myocytes in the sinoatrial (SAN) and atrioventricular (AVN) nodes (Vassort et al. 2006). In addition, TTCC are expressed in atrial myocytes in a species-dependent manner. Smaller species (mouse, guinea pig) show higher TTCC current amplitudes compared to larger species (rabbit, pig), and TTCC appear to be absent in human atrial cardiomyocytes (Catterall 2011; Ono and Iijima 2010).

In this chapter, we provide an overview of the role of Ca^{2+} channels in cellular electrophysiology and intracellular Ca^{2+} handling. We introduce their molecular composition followed by their characteristic biophysical properties and an overview of the most important regulatory mechanisms. We will focus predominantly on the LTCC given its central role in cardiac pathophysiology but will highlight the major differences between LTCC and TTCC.

4.2 The Role of L-Type and T-Type Ca^{2+} Currents in Cardiac Cellular Electrophysiology

Depending on the cell type, Ca^{2+} channels are involved in several fundamental electrophysiological processes in the heart (Bers 2008). In sinoatrial and atrioventricular nodes, depolarizing Ca^{2+} currents through TTCC and $Ca_v1.3$ LTCC have been suggested to contribute to diastolic depolarization and may thereby play an important role in pacemaker activity and regulation of heart rate (Capel and Terrar 2015; Bers 2008; Vassort et al. 2006). On the other hand, $Ca_v1.2$-mediated L-type Ca^{2+} current ($I_{Ca,L}$) is responsible for the AP upstroke in SAN and AVN myocytes. Inhibition of $I_{Ca,L}$ is therefore a commonly used approach to inhibit AV-nodal conduction to reduce ventricular rate in patients with atrial fibrillation (rate control) (Kirchhof et al. 2016).

In the working myocardium, LTCC are a major contributor to the AP "plateau" at which the membrane potential remains at a relatively depolarized level for a few hundred milliseconds prior to being repolarized (Fig. 4.1) (Heijman et al. 2016; Bers

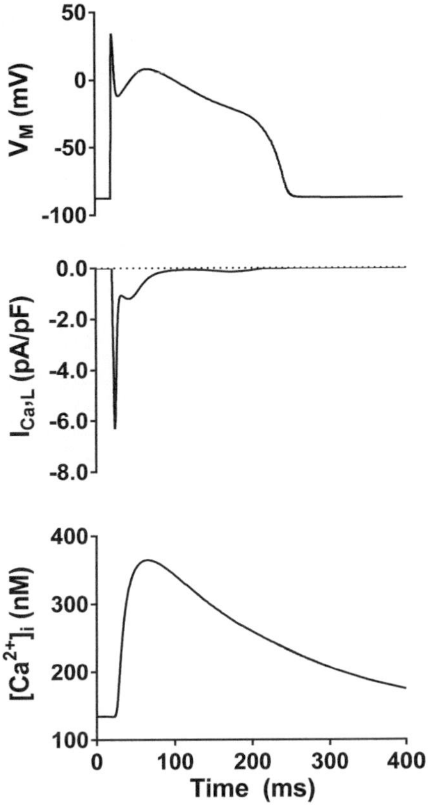

Fig. 4.1 L-type Ca^{2+} current ($I_{Ca,L}$) activity during a ventricular action potential. Membrane potential (V_M), Ca^{2+} transient, and $I_{Ca,L}$ during 1 Hz pacing in the Heijman et al. canine ventricular cardiomyocyte model (Heijman et al. 2011)

2008). This long plateau phase is a hallmark of cardiac cellular electrophysiology. The Ca^{2+} entering the cardiomyocyte through LTCC also rapidly triggers a much greater Ca^{2+} release from the sarcoplasmic reticulum (SR) through Ca^{2+}-release channels known as "ryanodine receptor channels" (RyRs, RyR2=cardiac form), a process termed Ca^{2+}-induced Ca^{2+} release (Fig. 4.2) (Voigt et al. 2012a; Grandi et al. 2011; Bers 2008). During the resulting intracellular Ca^{2+} transient (Voigt et al. 2014; Voigt et al. 2012b), Ca^{2+} binds to troponin-C in the myofilaments and initiates cardiomyocyte contraction, a process referred to as excitation-contraction coupling (EC coupling) (Bers 2002). During diastole Ca^{2+} is removed from the cytosol by Ca^{2+} reuptake into the SR, mediated via the SR Ca^{2+} ATPase ("SERCA," SERCA2a=predominant cardiac form) and by Ca^{2+} extrusion into the extracellular space via forward-mode Na^+/Ca^{2+} exchanger (NCX, NCX1=cardiac form) (Bers 2008; Voigt et al. 2012b; Hohendanner et al. 2013).

In ventricular cardiomyocytes LTCC are located mainly within membrane invaginations, so called t-tubules (Fig. 4.2). This organization allows their close interaction with RyR2 on the SR within dyadic junctions throughout the whole volume of the myocyte. Thus, t-tubules conduct membrane depolarizations to LTCC deep in the cell, activating RyR2 throughout the cellular volume (Song et al. 2005) and ensuring simultaneous uniform SR Ca^{2+} release. In contrast, atrial cardiomyocytes do not have

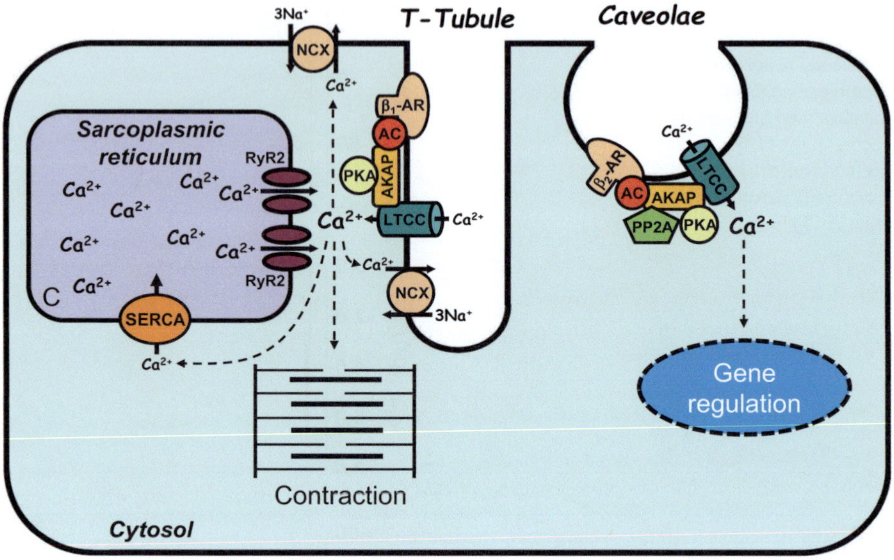

Fig. 4.2 Distinct L-type Ca^{2+} channel (LTCC) macromolecular complexes within t-tubules and caveolae. Ca^{2+} enters ventricular cardiomyocytes through t-tubular LTCC and triggers Ca^{2+} release from the sarcoplasmic reticulum (SR) through Ca^{2+}-release channels known as "ryanodine receptor channels" (RyRs, RyR2=cardiac form). The released Ca^{2+} binds to troponin-C in the myofilaments and initiates cardiomyocyte contraction. During diastole Ca^{2+} is removed from the cytosol by Ca^{2+} reuptake into the SR, mediated via the SR Ca^{2+} ATPase ("SERCA," SERCA2a=predominant cardiac form) and by Ca^{2+} extrusion into the extracellular space via forward-mode Na^+/Ca^{2+} exchanger (NCX, NCX1=cardiac form). LTCC complexes within t-tubules include β_1-adrenergic receptor (β_1-AR), adenylyl cyclase (AC), and protein kinase A (PKA). Additional LTCC subpopulations in caveolae are implicated in signaling to the nucleus to regulate the transcription of genes. Caveolar LTCC complexes are composed of β_2-AR, AC, PKA, and protein phosphatase 2A (PP2A)

an extensive t-tubule network or may have only rudimentary t-tubular structures. As such, LTCC are expressed mainly around the periphery of atrial cardiomyocytes (Bootman et al. 2011; Brandenburg et al. 2016; Richards et al. 2011). Therefore, the close interaction of RyRs and LTCC occurs only in the immediate subsarcolemmal space, and the SR Ca^{2+} release occurs as a centripetal Ca^{2+} wave: RyR2s close to the sarcolemma get activated first and sequentially activate neighboring RyR2s toward the cell-center (Greiser et al. 2009; Wakili et al. 2010). Recently, Brandenburg et al. have provided evidence for voluminous axial tubules that harbor LTCC with extensive junctions to RyR2 in the SR in atrial cardiomyocytes. Although less extensive than the t-tubular system in ventricular cardiomyocytes, these axial tubules may be connected to the cellular surface membrane by the sparse t-tubule network, thereby partially synchronizing intracellular Ca^{2+} release from the SR (Brandenburg et al. 2016). Given the critical role of LTCC in EC coupling, their inhibition can have profound negative inotropic effects.

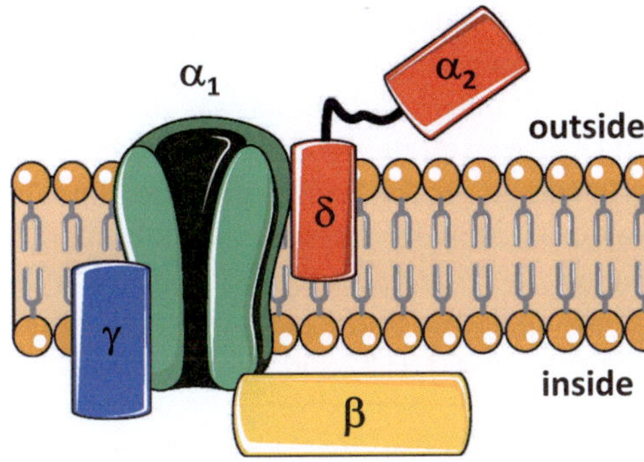

Fig. 4.3 Molecular composition of the L-type Ca^{2+} channel. See text for further details

4.3 Molecular Composition of LTCC and TTCC

4.3.1 L-Type Ca^{2+} Channels

The LTCC consist of a pore forming α_1-subunit and the accessory β-, α_2-δ-, and γ-subunits (Catterall 2000, 2011) (Fig. 4.3). Of the four currently known α_1-subunits ($Ca_v1.1$, *CACNA1S*; $Ca_v1.2$, *CACNA1C*; $Ca_v1.3$, *CACNA1D;* and $Ca_v1.4$, *CACNA1F*), only two are expressed in the heart (Schram et al. 2002). Similar to cardiac voltage-gated Na^+ channels, $Ca_v1.2$ and $Ca_v1.3$ consist of four domains with six transmembrane segments each, containing pore-forming loops and the voltage sensor as well as drug-binding sites. Whereas the $Ca_v1.2$-mediated L-type Ca^{2+} current ($I_{Ca,L}$) represents a major route of Ca^{2+} entry in all cardiomyocytes, $Ca_v1.3$-mediated Ca^{2+} currents are predominantly found in the SAN, conduction system, and atrial cardiomyocytes.

The β-subunits bind to a single site on a loop between domains I and II of the α-subunit (AID, α interaction domain) (Chen et al. 2004). The β_2-isoform (*CACNB2*) is the major cardiac isoform. Co-expression of this β-subunit with the Cav1.2 pore-forming subunit enhances activation and inactivation kinetics, shifts the inactivation curve to more negative potentials (see below), and enhances affinity to dihydropyridines (Mitterdorfer et al. 1994). In addition, Cavβ acts as a chaperone that antagonizes an endoplasmatic reticulum retention signal in the α_1-subunit, thereby supporting the correct folding and membrane targeting of the channel (Arikkath and Campbell 2003). In agreement, recent work has shown that a Cavβ_2-mimetic peptide can increase LTCC trafficking, overcoming dysregulation of LTCC density in obesity (Rusconi et al. 2016).

The α_2-δ-subunits are encoded by the same genes (*CACNAD1–4*) (Jay et al. 1991). They are cleaved after transcription and re-linked with disulfide bonds. Of the four known isoforms, α_2-δ_1 and α_2-δ_3 are expressed in the heart. α_2-δ-subunits have been suggested to facilitate the formation of functional sarcolemmal Ca^{2+} channels but may also alter channel gating, shifting voltage dependence of activation and inactivation to more hyperpolarized potentials (see below) (Davies et al. 2007).

The Ca^{2+} channel γ-subunit is encoded by eight genes with γ_4-, γ_6-, γ_7-, and γ_8-subunits expressed in the heart (*CACNG4*, *CACNG6–8*). The γ-subunits alter activation and inactivation properties of the channel in expression systems (Yang et al. 2011). However, their role in cardiac cellular electrophysiology remains unclear.

In addition to these core subunits, the LTCC contain numerous other associated proteins within its large macromolecular complex. These proteins can regulate channel function (e.g., through phosphorylation, as discussed below) or channel trafficking and targeting [e.g., the protein myoscape, which has recently been shown to interact with LTCC and to influence its surface expression (Eden et al. 2016)].

4.3.2 T-Type Ca^{2+} Channels

TTCC are mediated by the Ca_v3 family of Ca^{2+} channels, consisting of $Ca_v3.1$–$Ca_v3.3$ (*CACNA1G*, *CACNA1H*, and *CACNA1I*, respectively) (Catterall 2011). Interestingly, these α_1 subunits share only <25% amino acid sequence identity with Ca_v1 channels (Catterall 2011), suggesting an early divergence during evolution and potentially providing a structural basis for the differences in electrophysiological properties (discussed below). In the heart, TTCC consist predominantly of Cav3.1 and Cav3.2, whereby the exact balance is species and condition dependent (Perez-Reyes 2003). In contrast to LTCC, the α_2-δ and β subunits do not appear to modulate TTCC gating, although they may affect trafficking of the α_1 subunits (Dubel et al. 2004).

4.4 Voltage-Gated Ca^{2+} Channels: Biophysical Properties

$I_{Ca,L}$ was first described in Purkinje fibers and described as "slow inward current" in order to distinguish this current from the fast sodium current (Reuter 1967, 1979). Shortly afterward the presence of $I_{Ca,L}$ in all types of cardiomyocytes including atrial, SAN, and AVN was proven (Vassort et al. 2006). $I_{Ca,L}$ is named after its characteristic slow voltage-dependent inactivation resulting in "long-lasting" activity of the current, as well as large single-channel conductance and open time. Furthermore, $Ca_v1.2$-carried $I_{Ca,L}$ is characterized by high voltage of activation [$V_{1/2}$ for activation between −10 mV and −15 mV (Mangoni et al. 2006)], marked upregulation in response to the activation of cAMP-dependent phosphorylation pathways, and sensitivity to Ca^{2+} channel antagonists (1,4-dihydropyridines, phenylalkylamines, benzothiazepines) (Bers 2002; Catterall 2011; Rose and Backx 2014; Tang et al. 2016). $Ca_v1.3$-mediated currents activate at more negative membrane potentials ($V_{1/2}$ of activation between −40 mV and −30 mV), and their inactivation curve is also shifted to the left (Mangoni et al. 2006). These biophysical properties are consistent with the notion that $Ca_v1.3$ contributes to the diastolic depolarization in the SAN and thus is an important determinant of the pacemaker activity in the heart.

The second family of Ca^{2+} currents was initially described in dorsal root ganglion neuron in starfish eggs (Hagiwara et al. 1975). In contrast to $I_{Ca,L}$, these Ca^{2+}

currents showed rapid voltage-dependent inactivation and were therefore termed T-type Ca^{2+} currents ($I_{Ca,T}$) for their "transient" openings and small ("tiny") single-channel conductance. In addition, T-type Ca^{2+} currents are activated at much more negative membrane potentials and were insensitive to conventional Ca^{2+} channel inhibitors (Vassort et al. 2006; Catterall 2011). Because of their relatively negative activation potential ($V_{1/2}$ of activation between −60 mV and −50 mV), which is within the diastolic range, TTCC have been suggested to contribute to diastolic depolarization in the SAN and AVN (Mangoni et al. 2006; Capel and Terrar 2015; Vassort et al. 2006). In contrast, because of their small amplitude and their rapid inactivation, the contribution of TTCC to Ca^{2+} influx in ventricular cardiomyocytes is negligible (Catterall 2011; Ono and Iijima 2010).

4.4.1 Activation of Ca^{2+} Channels

In patch-clamp experiments, $I_{Ca,L}$ is activated by a depolarizing step-pulse protocol. Following depolarization from the resting membrane potential, the current amplitude reaches a peak value within 5–7 ms followed by inactivation of the current despite sustained depolarization (Fig. 4.4) (Bers and Perez-Reyes 1999; Voigt et al. 2012b, 2014; Bers 2001; Hadley and Hume 1987). The current reaches its maximal amplitude in response to depolarizations between 0 and +10 mV. Despite maximum activation of the channel, further depolarizations result in smaller $I_{Ca,L}$ amplitudes as the voltage approaches the reversal potential and the driving force (E_m–E_{rev}) for Ca^{2+} currents diminishes (Fig. 4.4). The balance between channel activation and reduced driving force results in the typical bell-shaped $I_{Ca,L}$ current-voltage relationship. In order to illustrate the channel activation independent of the driving force, the conductance of the channel, i.e., the LTCC-mediated current divided by the driving force, is plotted versus the membrane potential. This results in a typical sigmoidal activation curve with $V_{1/2}$ of activation typically seen between −10 and −15 mV (Fig. 4.5) (Mangoni et al. 2006).

4.4.2 Inactivation of Ca^{2+} Channels

Following sustained depolarization, $I_{Ca,L}$ undergoes a typical decay which depends on time, voltage, and intracellular Ca^{2+} (Bers and Perez-Reyes 1999). The contribution of voltage-dependent inactivation and intracellular Ca^{2+}-dependent inactivation (VDI and CDI, respectively) can be determined experimentally by replacing Ca^{2+} as a charge carrier with Ba^{2+} or monovalent ions. In absence of divalent cations, the LTCC-mediated current shows very slow inactivation, which under these conditions is largely due to voltage-dependent inactivation and from which the name of the channel (L-type) was derived. Using Ba^{2+} as a charge carrier results in a faster inactivation of the current, although the Ba^{2+}-mediated current inactivation is still much slower than in the presence of Ca^{2+} as a charge carrier (Bers and Perez-Reyes 1999; Bers 2008; Catterall 2011).

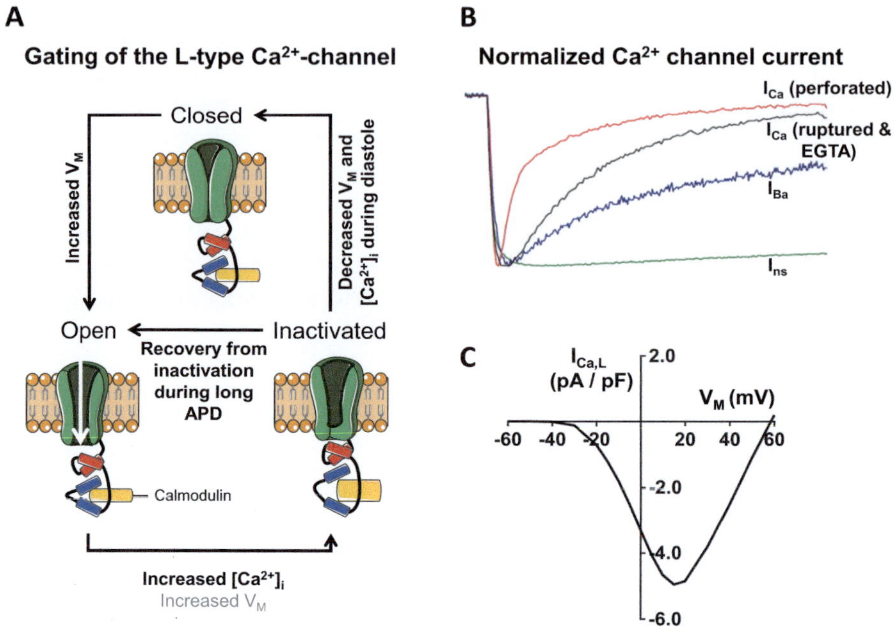

Fig. 4.4 L-type Ca^{2+} current activation and inactivation. (**a**) Schematic illustration of three gating states of the L-type Ca^{2+} channel (LTCC). Following depolarization from the resting membrane potential, the channel is activated within 5–7 ms followed by inactivation of the current despite sustained depolarization. (**b**) LTCC inactivation with Ca^{2+}, Ba^{2+}, or monovalent cations (ns) as the charge carrier. Currents were measured at 0 mV except for I_{ns} at −30 mV to obtain comparable activation, and peak currents were normalized. I_{Ca} with sarcoplasmic reticulum (SR) Ca^{2+} release was recorded using the perforated patch-clamp technique and 2 mM external Ca^{2+}. I_{Ca} with no SR Ca^{2+} release was recorded in the whole-cell configuration with 10 mM EGTA in the pipette. I_{Ba} was recorded in the whole-cell configuration with 2 mM external Ba^{2+} and 10 mM EGTA in the pipette. I_{ns} was measured in divalent cation-free conditions. Reprinted with kind permission from Bers (2001). (**c**) Current-voltage relationship of canine ventricular L-type Ca^{2+} current modeled according to Heijman et al. (2011)

CDI of LTCC represents a feedback control mechanism to prevent extensive Ca^{2+} overload by increasing LTCC inactivation, thereby decreasing Ca^{2+} influx, in the presence of high intracellular Ca^{2+}. Interestingly, buffering intracellular Ca^{2+} transients with ethylene glycol tetra-acetic acid (EGTA) slowed CDI of LTCC-mediated current but did not completely abolish it. These data indicate that both Ca^{2+} entering the cell through LTCC, which is not buffered by the relatively slow buffering properties of EGTA, and the Ca^{2+} released from the SR during systolic Ca^{2+} transients contribute to the LTCC inactivation (Richard et al. 2006). Calmodulin (CaM) attached to the C-terminus of the LTCC α-subunit has been identified as the Ca^{2+} sensor for CDI (Qin et al. 1999; Sanchez-Alonso et al. 2016). In particular, Ca^{2+} binding to CaM leads to a conformational change allowing an intracellular loop of the α-subunit to interact with the channel pore and induce inactivation.

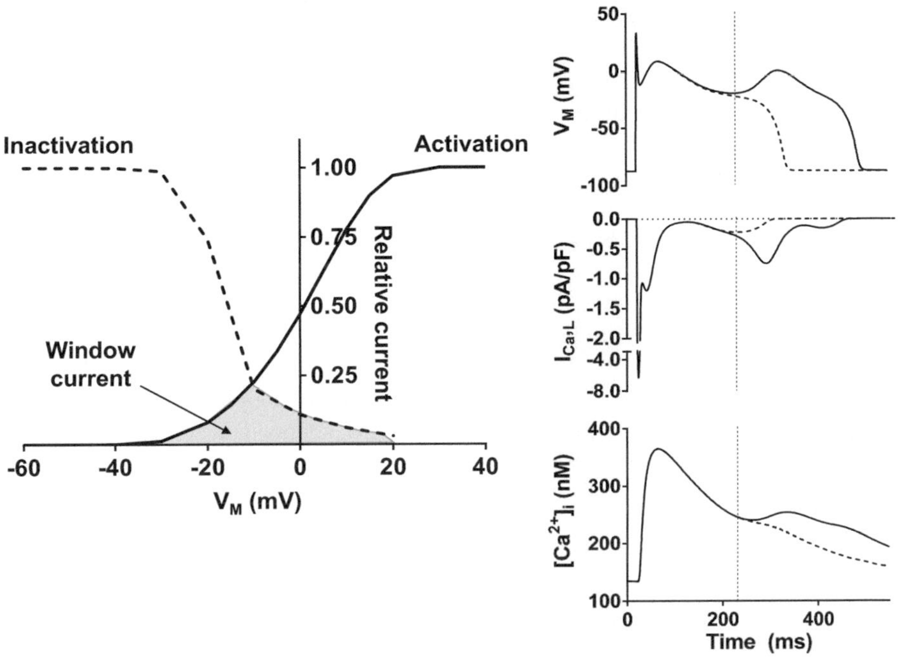

Fig. 4.5 L-type Ca^{2+} current may contribute to the development of early afterdepolarizations. (**a**) Steady-state activation and inactivation curves of L-type Ca^{2+} current modeled according to Heijman et al. (2011). Overlap of both curves (indicated in gray) illustrates the presence of a window current, which may contribute to the generation of early afterdepolarizations (**b**, EADs)

The channel availability during a depolarizing pulse depends on the preceding holding potential and follows a sigmoidal relationship with $V_{1/2}$ of inactivation between −35 mV and −45 mV, depending on the molecular channel composition, which varies from species to species (Fig. 4.5) (Yuan et al. 1996; Benitah et al. 2010). Upon repolarization from a depolarizing voltage pulse, the LTCC recover from voltage-dependent inactivation, thereby increasing the availability of the channel. The time course of recovery from inactivation is generally assessed using a paired pulse protocol with variable inter-pulse duration and in large mammals has a time constant between 20 and 70 ms (Fulop et al. 2004; Heijman et al. 2011; Li et al. 2000). If there is insufficient time for complete recovery between consecutive depolarizing pulses during fast rates, LTCC can accumulate in the inactivated state, resulting in a rate-dependent reduction of $I_{Ca,L}$ (Li et al. 2000).

4.5 Phosphorylation-Dependent Regulation of LTCC

The marked upregulation in response to cAMP-dependent protein phosphorylation pathways is a major characteristic of $I_{Ca,L}$ that contributes significantly to the positive inotropic effect of β-adrenergic receptor (β-AR) stimulation (Catterall

2011; Rose and Backx 2014; Heijman et al. 2011). Moreover, given the central role of Ca^{2+} influx through LTCC in cardiomyocyte function, it is not surprising that LTCC are regulated by multiple local signal transduction cascades. Most of them involve phosphorylation of the LTCC complex by protein kinases and dephosphorylation by protein phosphatases (Hofmann et al. 2014; Treinys and Jurevicius 2008).

4.5.1 β-Adrenergic Signaling and cAMP-Activated Protein Kinase A

It is well known that stimulation of the sympathetic nervous system leads to an increase in heart rate and cardiac contractility as part of the so-called "fight or flight" response (Ripplinger et al. 2016; Reuter 1974). This is mainly mediated by the release of catecholamines, which stimulate G_s-protein-coupled β-AR, leading to increased cAMP levels and activation of cAMP-dependent protein kinase A (PKA). Cardiac LTCC are a major target for PKA-mediated phosphorylation, and it was shown already in the 1970s that PKA-mediated phosphorylation of LTCC results in increased $I_{Ca,L}$ amplitude, prolonging the AP plateau and increasing contractility (Reuter 1974). Within the following years, several possible phosphorylation sites at the Cav1.2 α-subunit (Ser1627, Ser1700, Thr1704, Ser1928) or β-subunit (Ser143, Ser459, Ser165, Ser478, Ser479) have been proposed, mainly based on data obtained in expression systems (Hofmann et al. 2014). However, in vivo experiments with mice in which these sites were muted or truncated revealed mostly unaltered fight-or-flight reactions (Brandmayr et al. 2012). In agreement, recent work has shown that Cav1.2 activity in the brain, but not in the heart, requires phosphorylation of Ser1928, suggesting that there are tissue-specific differences in Cav1.2 regulation (Qian et al. 2017). The exact PKA-phosphorylation sites responsible for the PKA-mediated $I_{Ca,L}$ regulation in the heart are still under debate (Hofmann et al. 2014).

The PP2A catalytic subunit can bind directly to the LTCC α_{1C} subunit, and channel-bound PP2A plays an important role in the inhibition of $I_{Ca,L}$, although other mechanisms of PP2A/LTCC interaction have also been described (Heijman et al. 2017). PKA is targeted to LTCCs by A-kinase anchoring proteins (AKAPs) (Wong and Scott 2004). AKAP15 interacts with the C-terminus of $Ca_v1.2$ and targets PKA near the phosphorylation site Ser1928 (Gray et al. 1997; Hulme et al. 2002). AKAPs are involved in the organization of LTCC in distinct subcellular compartments with specific molecular compositions. Examples of such LTCC subpopulations include those found in t-tubules and those in caveolar plasma membrane domains outside of t-tubules (Fig. 4.2) (Balijepalli et al. 2006; Best and Kamp 2012). This spatial organization allows precise local regulation of $I_{Ca,L}$ (Sanchez-Alonso et al. 2016). LTCC complexes within t-tubules include β_1-AR, adenylyl cyclase, and PKA, whereas caveolar LTCC complexes are composed of β_2-AR, adenylyl cyclases, PKA, PP2A, and caveolin 3 (Balijepalli et al. 2006; Best and Kamp 2012; Rose and Backx 2014). The differential compartmentalization of $Ca_v1.2$ channels results in different regulation, involving stimulation by β_1-AR in the t-tubule or β_2-AR in the caveolae (Nikolaev et al. 2010).

To ensure this compartment-specific regulation of $I_{Ca,L}$ by β_1-AR or β_2-AR, local cAMP compartmentation is required. An important mechanism that limits cAMP diffusion between cellular compartments is the cAMP degradation by PDEs (Mika et al. 2012). There are 11 families of PDEs, 4 of which are classically responsible for cAMP degradation in the heart: PDE1, PDE2, PDE3, and PDE4 (Osadchii et al. 2005). PDE2–4 contribute to the regulation of $I_{Ca,L}$ in atria and ventricle (Mehel et al. 2013; Molina et al. 2012; Rivet-Bastide et al. 1997; Vandecasteele et al. 2001).

4.5.2 Ca^{2+}/Calmodulin-Dependent Protein Kinase (CaMKII)

CaMKII-dependent phosphorylation of the LTCC has been linked to an increase of LTCC activity in response to increased stimulation frequencies, often referred to as "facilitation" (Richard et al. 2006). "Facilitation" reflects a Ca^{2+}-dependent increase of $I_{Ca,L}$ amplitude and slowing of $I_{Ca,L}$ inactivation, which does not occur with other divalent cations such as Ba^{2+} as charge carrier. Facilitation occurs likely as a result of CaMKII-mediated phosphorylation (Lee 1987; Lee et al. 2006). However, the physiological role of this process is not entirely clear. It may counterbalance the direct Ca^{2+}-dependent inactivation described above.

4.5.3 G_q-Protein-Coupled Receptors and Activation of Protein Kinase C

Protein kinase C (PKC) activation is a major step in the signaling cascade activated by G_q-coupled receptors such as α_1- and α_2-AR, endothelin, angiotensin II, and muscarinic receptors. Although there is evidence for PKC-dependent phosphorylation of $Ca_v1.2$ and Cavβ2, the consequences of PKC activation on $I_{Ca,L}$ remain controversial (Kamp and Hell 2000; Shistik et al. 1998). PKC-dependent activation can increase (Alden et al. 2002) or decrease (Yue et al. 2004) $I_{Ca,L}$ or can even have biphasic effects (Weiss et al. 2004). For a detailed discussion, we refer the interested reader to Benitah et al. (2010).

4.5.4 Nitric-Oxide Dependent Signaling and Activation of Protein Kinase G

Results about the effects of protein kinase G (PKG)-mediated LTCC phosphorylation on $I_{Ca,L}$ are discrepant and may depend on cell type and experimental conditions (in vivo vs. in vitro). PKG activation occurs in response to NO-dependent stimulation of guanylate cyclase and the resulting increase of cyclic guanosine monophosphate (cGMP) levels (Tamargo et al. 2010; Hare 2003). The rise in cGMP may influence $I_{Ca,L}$ not only by direct PKG-mediated phosphorylation but also by cGMP-dependent activation (PDE2) or inhibition (PDE3) of PDEs that control cAMP (Vandecasteele et al. 2001). cGMP-dependent signaling may thereby modulate the LTCC response to β-AR stimulation (Abi-Gerges et al. 2001, 2002). Whether cGMP-dependent signaling facilitates or inhibits cAMP-signaling seems to depend on tissue and species (Han et al. 1994, 1996; Wang et al. 2000; Kirstein et al. 1995; Martynyuk et al. 1997).

Furthermore, NO may regulate $I_{Ca,L}$ activity in a cGMP-independent manner. The α_{1C}-subunit of the LTCC (Cav1.2) contains more than 10 cysteine residues that may be nitrosylated and involved in LTCC regulation. Nitrosylation of LTCC has been suggested to inhibit $I_{Ca,L}$ in atrial and ventricular myocytes (Sun et al. 2006; Carnes et al. 2007; Rozmaritsa et al. 2014). Interestingly, S-nitrosylation increases during ischemia reperfusion and AF, thereby contributing to a reduction in $I_{Ca,L}$, SR-Ca^{2+} load, and Ca^{2+}-induced cardiac injury, which represents an important cardioprotective mechanism (Tamargo et al. 2010).

4.5.5 Adenosine Monophosphate-Activated Protein Kinase

Adenosine monophosphate-activated protein kinase (AMPK), a serine/threonine kinase, is activated in response to increased AMP/ATP ratios and represents a sensor of cellular energy status. AMPK has been recently shown to physically associate with $Ca_v1.2$ and regulate $I_{Ca,L}$. This may contribute to adaptation of cellular activity to the energy metabolism (Harada et al. 2012b, 2015).

4.6 The Role of L-Type Ca^{2+} Currents in Cardiac Pathophysiology

Dysregulation of LTCC and TTCC at the transcriptional, translational, and post-translational level plays an important role in a wide variety of diseases, as recently summarized in a number of reviews (Venetucci et al. 2012; Heijman et al. 2014; Adams and Snutch 2007; Bartos et al. 2015). Here, we summarize LTCC dysregulation [since TTCC are absent from working myocardium in humans (Catterall 2011; Ono and Iijima 2010)] in two major cardiovascular pathologies: atrial fibrillation and heart failure.

In patients with long-standing persistent (chronic) atrial fibrillation, $I_{Ca,L}$ is significantly reduced, contributing to the proarrhythmic AP shortening and the reduction in Ca^{2+}-transient amplitude, which plays an important role in the stroke-promoting atrial contractile dysfunction (Voigt et al. 2012b, 2014). The molecular mechanisms underlying the reduced $I_{Ca,L}$ are diverse and can involve reduced expression of Cav1.2, increased microRNA-328-mediated reduction in Cav1.2 protein levels, or altered S-nitrosylation, phosphorylation, or calpain-mediated degradation of LTCC (Heijman et al. 2014). In addition, dysregulation of Cavβ_2 expression (Ling et al. 2017), reducing trafficking of LTCC to the plasma membrane, may contribute. In contrast to chronic atrial fibrillation, $I_{Ca,L}$ is not reduced in patients with paroxysmal atrial fibrillation that were in normal sinus rhythm when the tissue was obtained (Voigt et al. 2014), suggesting that the reduction in $I_{Ca,L}$ is a protective mechanism to reduce Ca^{2+} entry in response to a persistent high activation rate. Consistent with this notion, rapid pacing of atrial cardiomyocytes results in reduction of $I_{Ca,L}$ within hours (Qi et al. 2008).

Data on $I_{Ca,L}$ in heart failure are variable, with some studies reporting a decrease in $I_{Ca,L}$, whereas most others found no differences in $I_{Ca,L}$ between human ventricular cardiomyocytes from failing and non-failing hearts (Richard et al. 1998; Bartos et al. 2015; Nattel et al. 2007). Similarly, the mRNA level of Cav1.2 was either unchanged (Ambrosi et al. 2013) or increased (Soltysinska et al. 2009) in end-stage heart failure patients compared to non-failing controls. Some studies suggest a blunted response of $I_{Ca,L}$ to β-adrenergic stimulation in heart failure, partly due to increased basal phosphorylation (Bartos et al. 2015; Nattel et al. 2007). However, little is known about alterations in the composition of the LTCC macromolecular complex in heart failure.

Activation and inactivation curves of $I_{Ca,L}$ are overlapping resulting in a "window current," a voltage range, which enables a persistent $I_{Ca,L}$ through LTCC that are activated but do not completely inactivate. If AP duration is prolonged, for example, in heart failure, LTCC will spend more time in this voltage range, allowing $I_{Ca,L}$ to recover from inactivation and become conducting again (Fig. 4.5). Under these conditions, the increase in $I_{Ca,L}$ may induce an abnormal membrane depolarization. These so-called early afterdepolarizations may trigger Torsades de Pointes (TdP) arrhythmias at the tissue scale, which are an important cause for sudden cardiac death in patients with heart failure and long QT syndromes (Weiss et al. 2010).

Long QT syndrome type-8 (also known as Timothy syndrome, TS) has been attributed to gain-of-function mutations such as G406R, G402R, and A1473G that lead to the disruption of the LTCC inactivation (Betzenhauser et al. 2015; Napolitano and Antzelevitch 2011). In addition to marked QT prolongation at birth, intrauterine bradycardia and AV conduction block are characteristic hallmarks of TS. The characteristic extra-cardiac phenotype including dysmorphic facial features, syndactyly, and autism often leads to the diagnosis. The occurrence of ventricular tachycardia is the major cause for the limited life expectancy with average survival of 2–3 years. On the other hand, loss-of-function mutations in the α_1, β_2, and $\alpha_2\delta$ subunits of the LTCC have been found to be associated with early repolarization syndrome and Brugada syndrome. The loss of function may thereby result from changes in gating, permeation, or trafficking. A more detailed overview on Ca^{2+} channel mutations and their role in cardiac arrhythmia syndromes is beyond the scope of this chapter and therefore discussed elsewhere in this book and in excellent recent reviews (Betzenhauser et al. 2015; Napolitano and Antzelevitch 2011).

4.7 Conclusions

In cardiomyocytes LTCC-mediated Ca^{2+} currents are the major route for Ca^{2+} entry. Most importantly, $I_{Ca,L}$ is a major player in cardiac cellular electrophysiology and in cellular EC coupling (Bers 2002, 2008). In addition, Ca^{2+} entry through LTCC is an important regulator of gene transcription and cardiac cellular metabolism (Makary et al. 2011; Wakili et al. 2011). In order to fulfil these tasks reliably, LTCC are highly regulated by specific subunit compositions and various signaling pathways (Benitah et al. 2010; Catterall 2000, 2011; Kohlhaas et al. 2017). According to their diverse role in several important physiological mechanisms, it is not surprising that abnormalities in LTCC activity contribute to the pathophysiology of many common heart diseases such as atrial fibrillation, ischemia reperfusion injury, and heart failure (as discussed elsewhere in this book). Furthermore, in channelopathies affecting LTCC, both loss-of-function and gain-of-function mutations lead to rare arrhythmia syndromes, which are discussed more extensively elsewhere in this book.

Compliance with Ethical Standards

Funding N.V. is supported by the DZHK, by the German Research Foundation (DFG VO 1568/3-1, IRTG1816 RP12, SFB1002 TPA13) and by the Else-Kröner-Fresenius Stiftung (EKFS 2016_A20). J.H. is supported by the Netherlands Organization for Scientific Research (NWO/ZonMw Veni 91616057).

Conflict of Interest Jordi Heijman declares that he has no conflict of interest. Cristina E. Molina declares that she has no conflict of interest. Niels Voigt declares that he has no conflict of interest.

Ethical Approval This article does not contain any studies with human participants or animals performed by any of the authors.

References

Abi-Gerges N, Fischmeister R, Mery PF. G protein-mediated inhibitory effect of a nitric oxide donor on the L-type Ca^{2+} current in rat ventricular myocytes. J Physiol. 2001;531(Pt 1):117–30.

Abi-Gerges N, Szabo G, Otero AS, Fischmeister R, Mery PF. NO donors potentiate the beta-adrenergic stimulation of $I_{Ca,L}$ and the muscarinic activation of $I_{K,ACh}$ in rat cardiac myocytes. J Physiol. 2002;540(Pt 2):411–24.

Adams PJ, Snutch TP. Calcium channelopathies: voltage-gated calcium channels. Subcell Biochem. 2007;45:215–51.

Alden KJ, Goldspink PH, Ruch SW, Buttrick PM, Garcia J. Enhancement of L-type Ca^{2+} current from neonatal mouse ventricular myocytes by constitutively active PKC-betaII. Am J Physiol Cell Physiol. 2002;282(4):C768–74. https://doi.org/10.1152/ajpcell.00494.2001.

Ambrosi CM, Yamada KA, Nerbonne JM, Efimov IR. Gender differences in electrophysiological gene expression in failing and non-failing human hearts. PLoS One. 2013;8(1):e54635. https://doi.org/10.1371/journal.pone.0054635.

Arikkath J, Campbell KP. Auxiliary subunits: essential components of the voltage-gated calcium channel complex. Curr Opin Neurobiol. 2003;13(3):298–307.

Balijepalli RC, Foell JD, Hall DD, Hell JW, Kamp TJ. Localization of cardiac L-type Ca^{2+} channels to a caveolar macromolecular signaling complex is required for beta(2)-adrenergic regulation. Proc Natl Acad Sci U S A. 2006;103(19):7500–5. https://doi.org/10.1073/pnas.0503465103.

Bartos DC, Grandi E, Ripplinger CM. Ion channels in the heart. Compr Physiol. 2015;5 (3):1423–64. https://doi.org/10.1002/cphy.c140069.

Benitah JP, Alvarez JL, Gomez AM. L-type Ca^{2+} current in ventricular cardiomyocytes. J Mol Cell Cardiol. 2010;48(1):26–36. https://doi.org/10.1016/j.yjmcc.2009.07.026.

Bers DM. Excitation-contraction coupling and cardiac contractile force. 2nd ed. Dordrecht: Springer Science+Business Media; 2001. https://doi.org/10.1007/978-94-010-0658-3.

Bers DM. Cardiac excitation-contraction coupling. Nature. 2002;415(6868):198–205. https://doi.org/10.1038/415198a.

Bers DM. Calcium cycling and signaling in cardiac myocytes. Annu Rev Physiol. 2008;70:23–49. https://doi.org/10.1146/annurev.physiol.70.113006.100455.

Bers DM, Perez-Reyes E. Ca channels in cardiac myocytes: structure and function in Ca influx and intracellular Ca release. Cardiovasc Res. 1999;42(2):339–60.

Best JM, Kamp TJ. Different subcellular populations of L-type Ca^{2+} channels exhibit unique regulation and functional roles in cardiomyocytes. J Mol Cell Cardiol. 2012;52(2):376–87. https://doi.org/10.1016/j.yjmcc.2011.08.014.

Betzenhauser MJ, Pitt GS, Antzelevitch C. Calcium channel mutations in cardiac arrhythmia syndromes. Curr Mol Pharmacol. 2015;8(2):133–42.

Bootman MD, Smyrnias I, Thul R, Coombes S, Roderick HL. Atrial cardiomyocyte calcium signalling. Biochim Biophys Acta. 2011;1813(5):922–34. https://doi.org/10.1016/j.bbamcr.2011.01.030.

Brandenburg S, Kohl T, Williams GS, Gusev K, Wagner E, Rog-Zielinska EA, Hebisch E, Dura M, Didie M, Gotthardt M, Nikolaev VO, Hasenfuss G, Kohl P, Ward CW, Lederer WJ, Lehnart SE. Axial tubule junctions control rapid calcium signaling in atria. J Clin Invest. 2016;126 (10):3999–4015. https://doi.org/10.1172/JCI88241.

Brandmayr J, Poomvanicha M, Domes K, Ding J, Blaich A, Wegener JW, Moosmang S, Hofmann F. Deletion of the C-terminal phosphorylation sites in the cardiac beta-subunit does not affect the basic beta-adrenergic response of the heart and the Ca(v)1.2 channel. J Biol Chem. 2012;287 (27):22584–92. https://doi.org/10.1074/jbc.M112.366484.

Capel RA, Terrar DA. The importance of Ca^{2+}-dependent mechanisms for the initiation of the heartbeat. Front Physiol. 2015;6:80. https://doi.org/10.3389/fphys.2015.00080.

Carnes CA, Janssen PM, Ruehr ML, Nakayama H, Nakayama T, Haase H, Bauer JA, Chung MK, Fearon IM, Gillinov AM, Hamlin RL, Van Wagoner DR. Atrial glutathione content, calcium current, and contractility. J Biol Chem. 2007;282(38):28063–73. https://doi.org/10.1074/jbc.M704893200.

Catterall WA. Structure and regulation of voltage-gated Ca^{2+} channels. Annu Rev Cell Dev Biol. 2000;16:521–55. https://doi.org/10.1146/annurev.cellbio.16.1.521.

Catterall WA. Voltage-gated calcium channels. Cold Spring Harb Perspect Biol. 2011;3(8): a003947. https://doi.org/10.1101/cshperspect.a003947.

Chen YH, Li MH, Zhang Y, He LL, Yamada Y, Fitzmaurice A, Shen Y, Zhang H, Tong L, Yang J. Structural basis of the alpha1-beta subunit interaction of voltage-gated Ca^{2+} channels. Nature. 2004;429(6992):675–80. https://doi.org/10.1038/nature02641.

Davies A, Hendrich J, Van Minh AT, Wratten J, Douglas L, Dolphin AC. Functional biology of the alpha(2)delta subunits of voltage-gated calcium channels. Trends Pharmacol Sci. 2007;28 (5):220–8. https://doi.org/10.1016/j.tips.2007.03.005.

Du J, Xie J, Zhang Z, Tsujikawa H, Fusco D, Silverman D, Liang B, Yue L. TRPM7-mediated Ca^{2+} signals confer fibrogenesis in human atrial fibrillation. Circ Res. 2010;106(5):992–1003. https://doi.org/10.1161/CIRCRESAHA.109.206771.

Dubel SJ, Altier C, Chaumont S, Lory P, Bourinet E, Nargeot J. Plasma membrane expression of T-type calcium channel alpha(1) subunits is modulated by high voltage-activated auxiliary subunits. J Biol Chem. 2004;279(28):29263–9. https://doi.org/10.1074/jbc.M313450200.

Eden M, Meder B, Volkers M, Poomvanicha M, Domes K, Branchereau M, Marck P, Will R, Bernt A, Rangrez A, Busch M, German Mouse Clinic C, Hrabe de Angelis M, Heymes C, Rottbauer W, Most P, Hofmann F, Frey N. Myoscape controls cardiac calcium cycling and contractility via regulation of L-type calcium channel surface expression. Nat Commun. 2016;7:11317. https://doi.org/10.13155/ncomms11317.

Fearnley CJ, Roderick HL, Bootman MD. Calcium signaling in cardiac myocytes. Cold Spring Harb Perspect Biol. 2011;3(11):a004242. https://doi.org/10.1101/cshperspect.a004242.

Fulop L, Banyasz T, Magyar J, Szentandrassy N, Varro A, Nanasi PP. Reopening of L-type calcium channels in human ventricular myocytes during applied epicardial action potentials. Acta Physiol Scand. 2004;180(1):39–47. https://doi.org/10.1046/j.0001-6772.2003.01223.x.

Grandi E, Pandit SV, Voigt N, Workman AJ, Dobrev D, Jalife J, Bers DM. Human atrial action potential and Ca^{2+} model: sinus rhythm and chronic atrial fibrillation. Circ Res. 2011;109 (9):1055–66. https://doi.org/10.1161/CIRCRESAHA.111.253955.

Gray PC, Tibbs VC, Catterall WA, Murphy BJ. Identification of a 15-kDa cAMP-dependent protein kinase-anchoring protein associated with skeletal muscle L-type calcium channels. J Biol Chem. 1997;272(10):6297–302.

Greiser M, Neuberger HR, Harks E, El-Armouche A, Boknik P, de Haan S, Verheyen F, Verheule S, Schmitz W, Ravens U, Nattel S, Allessie MA, Dobrev D, Schotten U. Distinct contractile and molecular differences between two goat models of atrial dysfunction: AV block-induced atrial dilatation and atrial fibrillation. J Mol Cell Cardiol. 2009;46(3):385–94. https://doi.org/10.1016/j.yjmcc.2008.11.012.

Hadley RW, Hume JR. An intrinsic potential-dependent inactivation mechanism associated with calcium channels in guinea-pig myocytes. J Physiol. 1987;389:205–22.

Hagiwara S, Ozawa S, Sand O. Voltage clamp analysis of two inward current mechanisms in the egg cell membrane of a starfish. J Gen Physiol. 1975;65(5):617–44.

Han X, Shimoni Y, Giles WR. An obligatory role for nitric oxide in autonomic control of mammalian heart rate. J Physiol. 1994;476(2):309–14.

Han X, Kobzik L, Balligand JL, Kelly RA, Smith TW. Nitric oxide synthase (NOS3)-mediated cholinergic modulation of Ca^{2+} current in adult rabbit atrioventricular nodal cells. Circ Res. 1996;78(6):998–1008.

Harada M, Luo X, Qi XY, Tadevosyan A, Maguy A, Ordog B, Ledoux J, Kato T, Naud P, Voigt N, Shi Y, Kamiya K, Murohara T, Kodama I, Tardif JC, Schotten U, Van Wagoner DR, Dobrev D, Nattel S. Transient receptor potential canonical-3 channel-dependent fibroblast regulation in atrial fibrillation. Circulation. 2012a;126(17):2051–64. https://doi.org/10.1161/CIRCULATIONAHA.112.121830.

Harada M, Nattel SN, Nattel S. AMP-activated protein kinase: potential role in cardiac electrophysiology and arrhythmias. Circ Arrhythm Electrophysiol. 2012b;5(4):860–7. https://doi.org/10.1161/CIRCEP.112.972265.

Harada M, Tadevosyan A, Qi X, Xiao J, Liu T, Voigt N, Karck M, Kamler M, Kodama I, Murohara T, Dobrev D, Nattel S. Atrial fibrillation activates AMP-dependent protein kinase and its regulation of cellular calcium handling: potential role in metabolic adaptation and prevention of progression. J Am Coll Cardiol. 2015;66(1):47–58. https://doi.org/10.1016/j.jacc.2015.04.056.

Hare JM. Nitric oxide and excitation-contraction coupling. J Mol Cell Cardiol. 2003;35(7):719–29.

Heijman J, Volders PG, Westra RL, Rudy Y. Local control of beta-adrenergic stimulation: effects on ventricular myocyte electrophysiology and Ca^{2+}-transient. J Mol Cell Cardiol. 2011;50 (5):863–71. https://doi.org/10.1016/j.yjmcc.2011.02.007.

Heijman J, Voigt N, Nattel S, Dobrev D. Cellular and molecular electrophysiology of atrial fibrillation initiation, maintenance, and progression. Circ Res. 2014;114(9):1483–99. https://doi.org/10.1161/CIRCRESAHA.114.302226.

Heijman J, Erfanian Abdoust P, Voigt N, Nattel S, Dobrev D. Computational models of atrial cellular electrophysiology and calcium handling, and their role in atrial fibrillation. J Physiol. 2016;594(3):537–53. https://doi.org/10.1113/JP271404.

Heijman J, Ghezelbash S, Wehrens XH, Dobrev D. Serine/threonine phosphatases in atrial fibrillation. J Mol Cell Cardiol. 2017;103:110–20. https://doi.org/10.1016/j.yjmcc.2016.12.009.

Hofmann F, Flockerzi V, Kahl S, Wegener JW. L-type CaV1.2 calcium channels: from in vitro findings to in vivo function. Physiol Rev. 2014;94(1):303–26. https://doi.org/10.1152/physrev.00016.2013.

Hohendanner F, Ljubojevic S, MacQuaide N, Sacherer M, Sedej S, Biesmans L, Wakula P, Platzer D, Sokolow S, Herchuelz A, Antoons G, Sipido K, Pieske B, Heinzel FR. Intracellular dyssynchrony of diastolic cytosolic [Ca^{2+}] decay in ventricular cardiomyocytes in cardiac remodeling and human heart failure. Circ Res. 2013;113(5):527–38. https://doi.org/10.1161/CIRCRESAHA.113.300895.

Hulme JT, Ahn M, Hauschka SD, Scheuer T, Catterall WA. A novel leucine zipper targets AKAP15 and cyclic AMP-dependent protein kinase to the C terminus of the skeletal muscle Ca^{2+} channel and modulates its function. J Biol Chem. 2002;277(6):4079–87. https://doi.org/10.1074/jbc.M109814200.

Jay SD, Sharp AH, Kahl SD, Vedvick TS, Harpold MM, Campbell KP. Structural characterization of the dihydropyridine-sensitive calcium channel alpha 2-subunit and the associated delta peptides. J Biol Chem. 1991;266(5):3287–93.

Kamp TJ, Hell JW. Regulation of cardiac L-type calcium channels by protein kinase A and protein kinase C. Circ Res. 2000;87(12):1095–102.

Kirchhof P, Benussi S, Kotecha D, Ahlsson A, Atar D, Casadei B, Castella M, Diener HC, Heidbuchel H, Hendriks J, Hindricks G, Manolis AS, Oldgren J, Popescu BA, Schotten U, Van Putte B, Vardas P, Agewall S, Camm J, Baron Esquivias G, Budts W, Carerj S, Casselman F, Coca A, De Caterina R, Deftereos S, Dobrev D, Ferro JM, Filippatos G, Fitzsimons D, Gorenek B, Guenoun M, Hohnloser SH, Kolh P, Lip GY, Manolis A, McMurray J, Ponikowski P, Rosenhek R, Ruschitzka F, Savelieva I, Sharma S, Suwalski P, Tamargo JL, Taylor CJ, Van Gelder IC, Voors AA, Windecker S, Zamorano JL, Zeppenfeld K. 2016 ESC Guidelines for the management of atrial fibrillation developed in collaboration with EACTS. Europace. 2016;18(11):1609–78. https://doi.org/10.1093/europace/euw295.

Kirstein M, Rivet-Bastide M, Hatem S, Benardeau A, Mercadier JJ, Fischmeister R. Nitric oxide regulates the calcium current in isolated human atrial myocytes. J Clin Invest. 1995;95(2):794–802. https://doi.org/10.1172/JCI117729.

Kohlhaas M, Nickel AG, Maack C. Mitochondrial energetics and calcium coupling in the heart. J Physiol. 2017. https://doi.org/10.1113/JP273609.

Kreusser MM, Backs J. Integrated mechanisms of CaMKII-dependent ventricular remodeling. Front Pharmacol. 2014;5:36. https://doi.org/10.3389/fphar.2014.00036.

Lee KS. Potentiation of the calcium-channel currents of internally perfused mammalian heart cells by repetitive depolarization. Proc Natl Acad Sci U S A. 1987;84(11):3941–5.

Lee TS, Karl R, Moosmang S, Lenhardt P, Klugbauer N, Hofmann F, Kleppisch T, Welling A. Calmodulin kinase II is involved in voltage-dependent facilitation of the L-type Cav1.2 calcium channel: identification of the phosphorylation sites. J Biol Chem. 2006;281(35):25560–7. https://doi.org/10.1074/jbc.M508661200.

Li D, Melnyk P, Feng J, Wang Z, Petrecca K, Shrier A, Nattel S. Effects of experimental heart failure on atrial cellular and ionic electrophysiology. Circulation. 2000;101(22):2631–8.

Ling TY, Wang XL, Chai Q, Lu T, Stulak JM, Joyce LD, Daly RC, Greason KL, Wu LQ, Shen WK, Cha YM, Lee HC. Regulation of cardiac CACNB2 by microRNA-499: potential role in atrial fibrillation. BBA Clin. 2017;7:78–84. https://doi.org/10.1016/j.bbacli.2017.02.002.

Makary S, Voigt N, Maguy A, Wakili R, Nishida K, Harada M, Dobrev D, Nattel S. Differential protein kinase C isoform regulation and increased constitutive activity of acetylcholine-regulated potassium channels in atrial remodeling. Circ Res. 2011;109(9):1031–43. https://doi.org/10.1161/CIRCRESAHA.111.253120.

Mangoni ME, Couette B, Marger L, Bourinet E, Striessnig J, Nargeot J. Voltage-dependent calcium channels and cardiac pacemaker activity: from ionic currents to genes. Prog Biophys Mol Biol. 2006;90(1–3):38–63. https://doi.org/10.1016/j.pbiomolbio.2005.05.003.

Martynyuk AE, Kane KA, Cobbe SM, Rankin AC. Role of nitric oxide, cyclic GMP and superoxide in inhibition by adenosine of calcium current in rabbit atrioventricular nodal cells. Cardiovasc Res. 1997;34(2):360–7.

Mehel H, Emons J, Vettel C, Wittkopper K, Seppelt D, Dewenter M, Lutz S, Sossalla S, Maier LS, Lechene P, Leroy J, Lefebvre F, Varin A, Eschenhagen T, Nattel S, Dobrev D, Zimmermann WH, Nikolaev VO, Vandecasteele G, Fischmeister R, El-Armouche A. Phosphodiesterase-2 is up-regulated in human failing hearts and blunts beta-adrenergic responses in cardiomyocytes. J Am Coll Cardiol. 2013;62(17):1596–606. https://doi.org/10.1016/j.jacc.2013.05.057.

Mika D, Leroy J, Vandecasteele G, Fischmeister R. PDEs create local domains of cAMP signaling. J Mol Cell Cardiol. 2012;52(2):323–9. https://doi.org/10.1016/j.yjmcc.2011.08.016.

Mitterdorfer J, Froschmayr M, Grabner M, Striessnig J, Glossmann H. Calcium channels: the beta-subunit increases the affinity of dihydropyridine and Ca^{2+} binding sites of the alpha 1-subunit. FEBS Lett. 1994;352(2):141–5.

Molina CE, Leroy J, Richter W, Xie M, Scheitrum C, Lee IO, Maack C, Rucker-Martin C, Donzeau-Gouge P, Verde I, Llach A, Hove-Madsen L, Conti M, Vandecasteele G, Fischmeister R. Cyclic adenosine monophosphate phosphodiesterase type 4 protects against atrial arrhythmias. J Am Coll Cardiol. 2012;59(24):2182–90. https://doi.org/10.1016/j.jacc.2012.01.060.

Napolitano C, Antzelevitch C. Phenotypical manifestations of mutations in the genes encoding subunits of the cardiac voltage-dependent L-type calcium channel. Circ Res. 2011;108 (5):607–18. https://doi.org/10.1161/CIRCRESAHA.110.224279.

Nattel S, Maguy A, Le Bouter S, Yeh YH. Arrhythmogenic ion-channel remodeling in the heart: heart failure, myocardial infarction, and atrial fibrillation. Physiol Rev. 2007;87(2):425–56. https://doi.org/10.1152/physrev.00014.2006.

Nikolaev VO, Moshkov A, Lyon AR, Miragoli M, Novak P, Paur H, Lohse MJ, Korchev YE, Harding SE, Gorelik J. Beta2-adrenergic receptor redistribution in heart failure changes cAMP compartmentation. Science. 2010;327(5973):1653–7. https://doi.org/10.1126/science.1185988.

Ono K, Iijima T. Cardiac T-type Ca(2+) channels in the heart. J Mol Cell Cardiol. 2010;48 (1):65–70. https://doi.org/10.1016/j.yjmcc.2009.08.021.

Osadchii O, Norton G, Woodiwiss A. Inotropic responses to phosphodiesterase inhibitors in cardiac hypertrophy in rats. Eur J Pharmacol. 2005;514(2–3):201–8. https://doi.org/10.1016/j.ejphar.2005.03.022.

Perez-Reyes E. Molecular physiology of low-voltage-activated t-type calcium channels. Physiol Rev. 2003;83(1):117–61. https://doi.org/10.1152/physrev.00018.2002.

Qi XY, Yeh YH, Xiao L, Burstein B, Maguy A, Chartier D, Villeneuve LR, Brundel BJ, Dobrev D, Nattel S. Cellular signaling underlying atrial tachycardia remodeling of L-type calcium current. Circ Res. 2008;103(8):845–54. https://doi.org/10.1161/CIRCRESAHA.108.175463.

Qian H, Patriarchi T, Price JL, Matt L, Lee B, Nieves-Cintron M, Buonarati OR, Chowdhury D, Nanou E, Nystoriak MA, Catterall WA, Poomvanicha M, Hofmann F, Navedo MF, Hell JW. Phosphorylation of Ser1928 mediates the enhanced activity of the L-type Ca^{2+} channel Cav1.2 by the beta$_2$-adrenergic receptor in neurons. Sci Signal. 2017;10(463). https://doi.org/10.1126/scisignal.aaf9659.

Qin N, Olcese R, Bransby M, Lin T, Birnbaumer L. Ca^{2+}-induced inhibition of the cardiac Ca2+ channel depends on calmodulin. Proc Natl Acad Sci U S A. 1999;96(5):2435–8.

Reuter H. The dependence of slow inward current in Purkinje fibres on the extracellular calcium-concentration. J Physiol. 1967;192(2):479–92.

Reuter H. Localization of beta adrenergic receptors, and effects of noradrenaline and cyclic nucleotides on action potentials, ionic currents and tension in mammalian cardiac muscle. J Physiol. 1974;242 (2):429–51.

Reuter H. Properties of two inward membrane currents in the heart. Annu Rev Physiol. 1979;41:413–24. https://doi.org/10.1146/annurev.ph.41.030179.002213.

Richard S, Leclercq F, Lemaire S, Piot C, Nargeot J. Ca^{2+} currents in compensated hypertrophy and heart failure. Cardiovasc Res. 1998;37(2):300–11.

Richard S, Perrier E, Fauconnier J, Perrier R, Pereira L, Gomez AM, Benitah JP. 'Ca^{2+}-induced Ca^{2+} entry' or how the L-type Ca^{2+} channel remodels its own signalling pathway in cardiac cells. Prog Biophys Mol Biol. 2006;90(1-3):118–35. https://doi.org/10.1016/j.pbiomolbio.2005.05.005.

Richards MA, Clarke JD, Saravanan P, Voigt N, Dobrev D, Eisner DA, Trafford AW, Dibb KM. Transverse tubules are a common feature in large mammalian atrial myocytes including human. Am J Physiol Heart Circ Physiol. 2011;301(5):H1996–2005. https://doi.org/10.1152/ajpheart.00284.2011.

Ripplinger CM, Noujaim SF, Linz D. The nervous heart. Prog Biophys Mol Biol. 2016;120 (1-3):199–209. https://doi.org/10.1016/j.pbiomolbio.2015.12.015.

Rivet-Bastide M, Vandecasteele G, Hatem S, Verde I, Benardeau A, Mercadier JJ, Fischmeister R. cGMP-stimulated cyclic nucleotide phosphodiesterase regulates the basal calcium current in human atrial myocytes. J Clin Invest. 1997;99(11):2710–8. https://doi.org/10.1172/JCI119460.

Rose RA, Backx PH. Calcium channels in the heart. In: Zipes DP, Jalife J, editors. Cardiac electrophysiology – from cell to bedside, vol. 6. New York: Elsevier; 2014. p. 13–22.

Rose RA, Belke DD, Maleckar MM, Giles WR. Ca^{2+} entry through TRP-C channels regulates fibroblast biology in chronic atrial fibrillation. Circulation. 2012;126(17):2039–41. https://doi.org/10.1161/CIRCULATIONAHA.112.138065.

Rozmaritsa N, Christ T, Van Wagoner DR, Haase H, Stasch JP, Matschke K, Ravens U. Attenuated response of L-type calcium current to nitric oxide in atrial fibrillation. Cardiovasc Res. 2014;101 (3):533–42. https://doi.org/10.1093/cvr/cvt334.

Rusconi F, Ceriotti P, Miragoli M, Carullo P, Salvarani N, Rocchetti M, Di Pasquale E, Rossi S, Tessari M, Caprari S, Cazade M, Kunderfranco P, Chemin J, Bang ML, Polticelli F, Zaza A, Faggian G, Condorelli G, Catalucci D. Peptidomimetic targeting of Cavbeta2 overcomes dysregulation of the L-type calcium channel density and recovers cardiac function. Circulation. 2016;134(7):534–46. https://doi.org/10.1161/CIRCULATIONAHA.116.021347.

Sanchez-Alonso JL, Bhargava A, O'Hara T, Glukhov AV, Schobesberger S, Bhogal N, Sikkel MB, Mansfield C, Korchev YE, Lyon AR, Punjabi PP, Nikolaev VO, Trayanova NA, Gorelik J. Microdomain-specific modulation of L-type calcium channels leads to triggered ventricular arrhythmia in heart failure. Circ Res. 2016;119(8):944–55. https://doi.org/10.1161/CIRCRESAHA.116.308698.

Schram G, Pourrier M, Melnyk P, Nattel S. Differential distribution of cardiac ion channel expression as a basis for regional specialization in electrical function. Circ Res. 2002;90 (9):939–50.

Shistik E, Ivanina T, Blumenstein Y, Dascal N. Crucial role of N terminus in function of cardiac L-type Ca^{2+} channel and its modulation by protein kinase C. J Biol Chem. 1998;273 (28):17901–9.

Soltysinska E, Olesen SP, Christ T, Wettwer E, Varro A, Grunnet M, Jespersen T. Transmural expression of ion channels and transporters in human nondiseased and end-stage failing hearts. Pflugers Arch. 2009;459(1):11–23. https://doi.org/10.1007/s00424-009-0718-3.

Song LS, Guatimosim S, Gomez-Viquez L, Sobie EA, Ziman A, Hartmann H, Lederer WJ. Calcium biology of the transverse tubules in heart. Ann N Y Acad Sci. 2005;1047:99–111. https://doi.org/10.1196/annals.1341.009.

Sun J, Picht E, Ginsburg KS, Bers DM, Steenbergen C, Murphy E. Hypercontractile female hearts exhibit increased S-nitrosylation of the L-type Ca^{2+} channel alpha1 subunit and reduced ischemia/reperfusion injury. Circ Res. 2006;98(3):403–11. https://doi.org/10.1161/01.RES.0000202707.79018.0a.

Tamargo J, Caballero R, Gomez R, Delpon E. Cardiac electrophysiological effects of nitric oxide. Cardiovasc Res. 2010;87(4):593–600. https://doi.org/10.1093/cvr/cvq214.

Tang L, Gamal El-Din TM, Swanson TM, Pryde DC, Scheuer T, Zheng N, Catterall WA. Structural basis for inhibition of a voltage-gated Ca^{2+} channel by Ca^{2+} antagonist drugs. Nature. 2016;537 (7618):117–21. https://doi.org/10.1038/nature19102.

Treinys R, Jurevicius J. L-type Ca^{2+} channels in the heart: structure and regulation. Medicina (Kaunas). 2008;44(7):491–9.

Vandecasteele G, Verde I, Rucker-Martin C, Donzeau-Gouge P, Fischmeister R. Cyclic GMP regulation of the L-type Ca^{2+} channel current in human atrial myocytes. J Physiol. 2001;533 (Pt 2):329–40.

Vassort G, Talavera K, Alvarez JL. Role of T-type Ca^{2+} channels in the heart. Cell Calcium. 2006;40(2):205–20. https://doi.org/10.1016/j.ceca.2006.04.025.

Venetucci L, Denegri M, Napolitano C, Priori SG. Inherited calcium channelopathies in the pathophysiology of arrhythmias. Nat Rev Cardiol. 2012;9(10):561–75. https://doi.org/10.1038/nrcardio.2012.93.

Voigt N, Nattel S, Dobrev D. Proarrhythmic atrial calcium cycling in the diseased heart. Adv Exp Med Biol. 2012a;740:1175–91. https://doi.org/10.1007/978-94-007-2888-2_53.

Voigt N, Li N, Wang Q, Wang W, Trafford AW, Abu-Taha I, Sun Q, Wieland T, Ravens U, Nattel S, Wehrens XH, Dobrev D. Enhanced sarcoplasmic reticulum Ca^{2+} leak and increased Na^{+}-Ca^{2+} exchanger function underlie delayed after depolarizations in patients with chronic atrial fibrillation. Circulation. 2012b;125(17):2059–70. https://doi.org/10.1161/CIRCULATIONAHA.111.067306.

Voigt N, Heijman J, Wang Q, Chiang DY, Li N, Karck M, Wehrens XH, Nattel S, Dobrev D. Cellular and molecular mechanisms of atrial arrhythmogenesis in patients with paroxysmal atrial fibrillation. Circulation. 2014;129(2):145–56. https://doi.org/10.1161/CIRCULATIONAHA.113.006641.

Wakili R, Yeh YH, Yan Qi X, Greiser M, Chartier D, Nishida K, Maguy A, Villeneuve LR, Boknik P, Voigt N, Krysiak J, Kaab S, Ravens U, Linke WA, Stienen GJ, Shi Y, Tardif JC, Schotten U, Dobrev D, Nattel S. Multiple potential molecular contributors to atrial hypocontractility caused by atrial tachycardia remodeling in dogs. Circ Arrhythm Electrophysiol. 2010;3(5):530–41. https://doi.org/10.1161/CIRCEP.109.933036.

Wakili R, Voigt N, Kaab S, Dobrev D, Nattel S. Recent advances in the molecular pathophysiology of atrial fibrillation. J Clin Invest. 2011;121(8):2955–68. https://doi.org/10.1172/JCI46315.

Wang Y, Wagner MB, Joyner RW, Kumar R. cGMP-dependent protein kinase mediates stimulation of L-type calcium current by cGMP in rabbit atrial cells. Cardiovasc Res. 2000;48(2):310–22.

Weiss S, Doan T, Bernstein KE, Dascal N. Modulation of cardiac Ca^{2+} channel by Gq-activating neurotransmitters reconstituted in Xenopus oocytes. J Biol Chem. 2004;279(13):12503–10. https://doi.org/10.1074/jbc.M310196200.

Weiss JN, Garfinkel A, Karagueuzian HS, Chen PS, Qu Z. Early afterdepolarizations and cardiac arrhythmias. Heart Rhythm. 2010;7(12):1891–9. https://doi.org/10.1016/j.hrthm.2010.09.017.

Wong W, Scott JD. AKAP signalling complexes: focal points in space and time. Nat Rev Mol Cell Biol. 2004;5(12):959–70. https://doi.org/10.1038/nrm1527.

Yang L, Katchman A, Morrow JP, Doshi D, Marx SO. Cardiac L-type calcium channel (Cav1.2) associates with gamma subunits. FASEB J. 2011;25(3):928–36. https://doi.org/10.1096/fj.10-172353.

Yuan WL, Ginsburg KS, Bers DM. Comparison of sarcolemmal calcium channel current in rabbit and rat ventricular myocytes. J Physiol London. 1996;493(3):733–46.

Yue Y, Qu Y, Boutjdir M. Beta- and alpha-adrenergic cross-signaling for L-type Ca^{2+} current is impaired in transgenic mice with constitutive activation of epsilon PKC. Biochem Biophys Res Commun. 2004;314(3):749–54.

Yue L, Xie J, Nattel S. Molecular determinants of cardiac fibroblast electrical function and therapeutic implications for atrial fibrillation. Cardiovasc Res. 2011;89(4):744–53. https://doi.org/10.1093/cvr/cvq329.

HCN Channels and Cardiac Pacemaking

5

Annalisa Bucchi, Chiara Piantoni, Andrea Barbuti, Dario DiFrancesco, and Mirko Baruscotti

Abstract

Cardiomyocytes located in the central part of the sinoatrial node are responsible for generating the electrical rhythm of the heart since they are endowed with the fastest automaticity of the entire conduction system. The source of this automaticity is the diastolic pacemaker phase which consists of the slow depolarization that links the end of each action potential with the beginning of the next, and the funny current ("I_f") is the primary contributor of this phase. Each f-channel results from the assembly of four single subunits belonging to the family of the hyperpolarization-activated cyclic nucleotide-gated (HCN) channels which includes four isoforms (HCN1–HCN4). The biophysical and modulatory properties of the f/HCN current will be presented together with some of the underlying molecular details which have been partly unraveled by the recent structural definition of the channel obtained by cryo-electron microscopy studies. The chapter will also provide an extensive review of the mutations of the HCN4 channels in humans associated with sinus arrhythmias and left ventricular noncompaction cardiomyopathy. Functional studies based on HCN transgenic and knockout mouse models confirm the importance of the I_f current in sustaining the pacemaker activity since its suppression affects the cardiac performance and autonomic modulation of heart rate. These studies also provide the evidence that cardiac HCN currents are required for proper cardiac development and embryo survival.

Finally, the clinical relevance of HCN channels as targets of drugs aimed to selectively reduce the heart rate will be also discussed.

A. Bucchi · C. Piantoni · A. Barbuti · D. DiFrancesco · M. Baruscotti (✉)
Department of Biosciences, The PaceLab and "Centro Interuniversitario di Medicina Molecolare e Biofisica Applicata", Università degli Studi di Milano, Milano, Italy
e-mail: annalisa.bucchi@unimi.it; chiara.piantoni@unimi.it; andrea.barbuti@unimi.it; dario.difrancesco@unimi.it; mirko.baruscotti@unimi.it

D. Thomas, C. A. Remme (eds.), *Channelopathies in Heart Disease*, Cardiac and Vascular Biology, https://doi.org/10.1007/978-3-319-77812-9_5

5.1 Spontaneous Activity of Sinoatrial Node Cells and the Native "Pacemaker"

The first medical observation that the heart has an intrinsic automaticity which persists also when the heart is removed from the body was made by Claudius Galenus in the second century AD, but only in 1907 the structure responsible for the initiation of the heartbeat, the sinoatrial node (SAN), was identified by Keith and Flack (Silverman et al. 2006; Keith and Flack 1907).

SAN cells are specialized myocytes which generate rhythmic action potentials (APs) that spread along preferential routes (Conduction System tissue) to the entire heart to trigger the orderly sequence of contractions of the heart chambers. Sinoatrial APs differ from those of the atrial and ventricular working myocytes in many aspects, but the most relevant is the lack of a stable resting potential since after SAN cells have reached the maximum diastolic potential (MDP, around −60 mV), the membrane slowly depolarizes up to the threshold (around −40 mV) for the initiation of a new AP; in so doing this phase sets the time interval between consecutive APs and thus the heart rate. This slow depolarization is commonly known as "pacemaker" phase, phase 4, or slow diastolic depolarization (DD, Fig. 5.1a, left). Several of the molecular details of the fascinating puzzle of electrical events that generate the pacemaker activity of SAN cells and allow our heart to beat and sustain our metabolic needs have now been identified (Mangoni and Nargeot 2008); in this chapter we will focus on the role of the pacemaker "I_f" current (Fig. 5.1a, right). The I_f current, discovered in 1979 in rabbit SAN cells (Brown et al. 1979), has the following features: (1) activation in hyperpolarization, (2) mixed Na^+/K^+ permeability, and (3) modulation by the second messenger cAMP; these aspects will be here briefly discussed (for a more accurate treatment, see DiFrancesco et al. 1986; Baruscotti et al. 2005).

1. *Activation in hyperpolarization.* The I_f current has the unusual property of being activated on membrane hyperpolarization rather than on depolarization, and the activation and deactivation kinetics are S-shaped with an initial "delay" followed by the real "gating" process (DiFrancesco and Ferroni 1983; DiFrancesco 1984). The reported values of its voltage dependence are largely variable with activation threshold and half-activation (V½) values in the range of −35/−70 mV and −52/−90 mV, respectively (Baruscotti et al. 2005). Such a large scattering of the data, which is unusual for most other ion channels, may be accounted for by several causes, both biological and methodological. A real biological heterogeneity is indeed caused by the intrinsic differences in the voltage dependence of the I_f current recorded in different areas of the SA node since it activates at progressively more negative voltages when moving from the center to the periphery (Boyett et al. 2000). In addition, the presence of "run-down," a well-known phenomenon which progressively reduces the I_f current and shifts its availability curve during patch-clamp recordings, should also be considered (DiFrancesco et al. 1986). Although a comprehensive molecular understanding of the run-down process is still incomplete, part of it is due to the depletion of f-channel

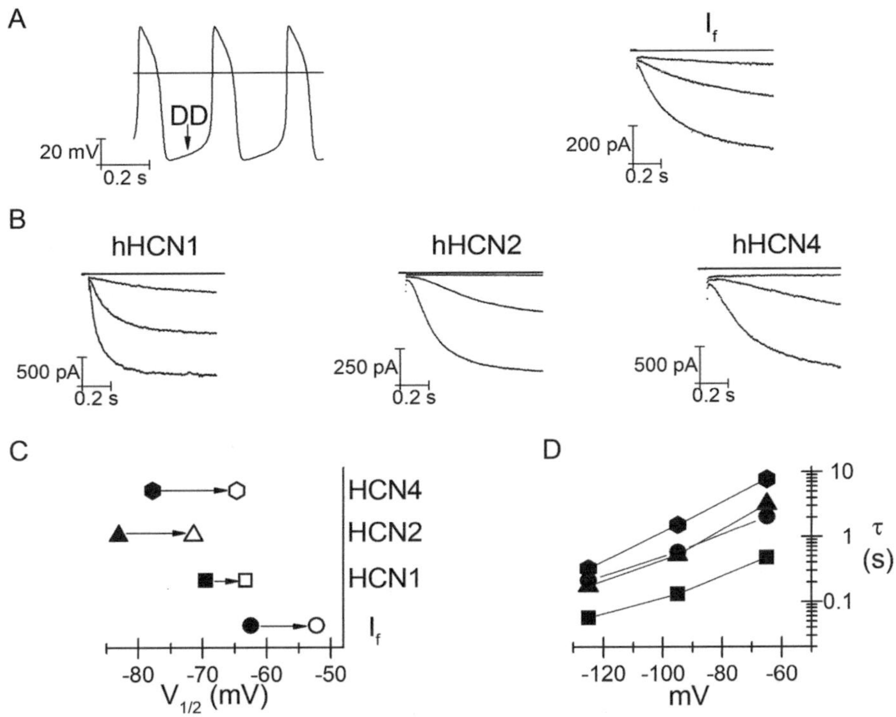

Fig. 5.1 Properties of native sinoatrial pacemaker (f) and homotetrameric HCN currents. (**a**) Sample action potentials recordings (left) and native I_f current traces (right) recorded in rabbit SAN cells. Current traces were acquired during voltage steps at −65 mV, −95 mV, and −125 mV (hp −35 mV). (**b**) Sample hHCN1, hHCN2, and hHCN4 clonal current traces elicited by voltage steps at −65 mV, −95 mV, and −125 mV (hp −35 mV). (**c**) Comparison of half-maximal activation values (V½) of native I_f, hHCN1, hHCN2, and hHCN4 currents in control conditions (filled symbols) and in the presence of saturating concentrations of cAMP (open symbols). Both V½ values obtained in control condition and the cAMP-induced shifts display significant differences (data obtained from Altomare et al. 2003; Stieber et al. 2005; Moroni et al. 2000; Baruscotti et al. 2017). (**d**) Comparison of activation time constants of HCN currents and native I_f. Activation time constants were obtained by fitting current traces in A with a single exponential function after an initial delay

modulators, such as cAMP and phosphoinositide PI(4,5)P2, which occurs during whole-cell patch-clamp recordings (Pian et al. 2006). Finally, the different experimental conditions and recording protocols used by different laboratories may also represent an additional source of variability of the observed differences in the voltage dependence.

2. *Mixed Na^+/K^+ permeability.* f-channels are permeable to both Na^+ and K^+ ions with a reversal potential around −10/−20 mV (DiFrancesco and Ojeda 1980; DiFrancesco 1981b; DiFrancesco et al. 1986); thanks to the recently obtained 3D resolution of the channel structure, the structural molecular elements governing

this mixed selectivity have now been identified (Lee and MacKinnon 2017). This Na^+/K^+ permeability is fundamental for the generation of the diastolic depolarization phase, indeed; although the channel is ~3.7 to fourfold more permeable to K^+ than to Na^+ ions (DiFrancesco 1981b; Frace et al. 1992), the inward and depolarizing Na^+ flux prevails at diastolic voltages (about −40 to −60 mV in the SAN).

3. *Modulation by the second messenger cAMP.* The control of f-channels kinetics by the second messenger cAMP represents an important physiological mechanism used by the neurohormonal system to adapt the cardiac chronotropism to the metabolic demand of the body (Brown et al. 1979; DiFrancesco and Tromba 1987, 1988; DiFrancesco et al. 1989). In SAN cells the stimulation of β-adrenoreceptors (β-ARs) by catecholamines activates the stimulatory G protein (Gαs) and the adenylyl cyclase (AC) leading to the increase in cAMP cell content. cAMP molecules are direct modulators since they can bind to the f-channels and in so doing exert a modulatory action which favors the equilibrium toward the open state which can be quantitatively described as a shift of the activation curve toward more positive voltages (Fig. 5.1c) (DiFrancesco and Tortora 1991; DiFrancesco and Mangoni 1994). This molecular event ultimately results in an increased inward f-current and a steeper diastolic depolarization and therefore a cardiac acceleration (Bucchi et al. 2007). According to Barbuti et al. (2007), SAN cells express both β1- and β2-AR subtypes; however, β2 stimulation determines a more relevant shift of the I_f current and consequently a more pronounced rate acceleration than β1 stimulation. These functional results are nicely paralleled by the evidence that membrane microdomains such as caveolae are rich in β2-ARs and f-channels, while β1 receptors are mainly outside these membrane regions (Barbuti et al. 2007). SAN cells are also abundantly innervated by vagal terminals which release Acetylcholine (ACh). In the presence of a cholinergic stimulus, the muscarinic receptors activate the inhibitory G protein (Gαi) which then inhibits the AC, therefore, ultimately leading to the following set of events: a decrease in cAMP levels, a shift of the activation curve of f-channels toward more negative potentials, and a decline of both the pacemaker current and heart rate.

The I_f current is not only expressed in SAN cells, but it is also present in other cardiac regions such as atrioventricular node (AVN) cells and Purkinje fibers, where it activates at more negative voltages compared to SAN myocytes (Munk et al. 1996; Hancox et al. 1993; DiFrancesco 1981a, b). The presence of both spontaneous activity and of the I_f current has also been reported in cells isolated from the region surrounding the rabbit tricuspid valve and from canine and rabbit pulmonary sleeves that are extensions of the left atrial myocardium into the pulmonary veins (Anumonwo et al. 1990; Chen et al. 2000, 2009; Suenari et al. 2012). While the functional roles of these currents are still unexplored, it is interesting that AV blocks were observed in association with knockout of the cardiac HCN4 channel in a mouse model (Baruscotti et al. 2011); also interesting is that the pulmonary veins are often the target of the ablation procedure in human patients suffering from atrial fibrillation, raising the question whether anomalous

HCN-dependent activity may represent a molecular and functional derangement contributing to AF.

Taken together all these data strengthen the association between the presence of spontaneous activity and the expression of the pacemaker current.

5.2 Clonal HCN Currents

Clonal and native f-channels have a tetrameric composition, with single subunits belonging to the hyperpolarization-activated cyclic nucleotide-gated (HCN) channel family, and in mammalians four isoforms (HCN1–HCN4) have been identified (Ludwig et al. 1998; Santoro et al. 1998; Baruscotti et al. 2010). Each HCN subunit has intracellular N- and C-termini and a central core domain organized in six transmembrane segments (S1–S6), with a positively charged S4 acting as the voltage sensor, and a pore region between S5 and S6 carrying the GYG signature typical of K^+-permeable channels (Fig. 5.2). A cyclic nucleotide-binding domain (CNBD) is localized in the C-terminus and is connected to the S6 segment by the C-linker (see next section for structural details). The primary structures of the four HCN isoforms are 80–90% identical in the transmembrane core and in the CNBD but diverge in the amino- and carboxy-terminal cytoplasmic regions (Viscomi et al. 2001).

1. *Biophysical and modulatory properties:* Heterologous expression and in vivo studies have shown that the different HCN isoforms can assemble both as homotetramers and heterotetramers (with the exception of HCN2-HCN3 heteromers, Much et al. 2003) to yield functional channels with properties similar to native f-channels: mixed Na^+ and K^+ permeability, activation upon hyperpolarization, time-dependent activation and deactivation, and modulation by the second messenger cAMP. Despite these qualitative similarities, heterologous expression of homotetrameric channels has revealed that different isoforms exhibit important quantitative differences in their kinetic aspects and in cAMP modulation. For example, HCN1 channels exhibit the more positive position of the activation curve followed by HCN4, HCN3, and HCN2 (Baruscotti et al. 2010), while the activation kinetics become progressively slower according to the following order: HCN1, HCN2, HCN3, and HCN4 (Fig. 5.1b–d).

   cAMP-induced modulation of f/HCN currents represents a key process by which the autonomic nervous system fine-tunes the slope of the diastolic depolarization of SAN cells and therefore SAN activity and heart rate (DiFrancesco 1993). As it will be discussed more thoroughly in the next section of this chapter, cAMP binding to CNBD stabilizes the open state of the channel gate, and, in kinetic terms, this action results in a positive shift of the activation curve which is larger for HCN2 and HCN4 than for HCN1 (Sartiani et al. 2017; Baruscotti et al. 2005; Chen et al. 2001) (Fig. 5.1c). According to a large part of the literature, the opening of HCN1 channels is only weakly facilitated by the binding of cAMP molecules. However, this interpretation has recently been challenged by some authors who suggest that HCN1 channels are stably associated with cAMP

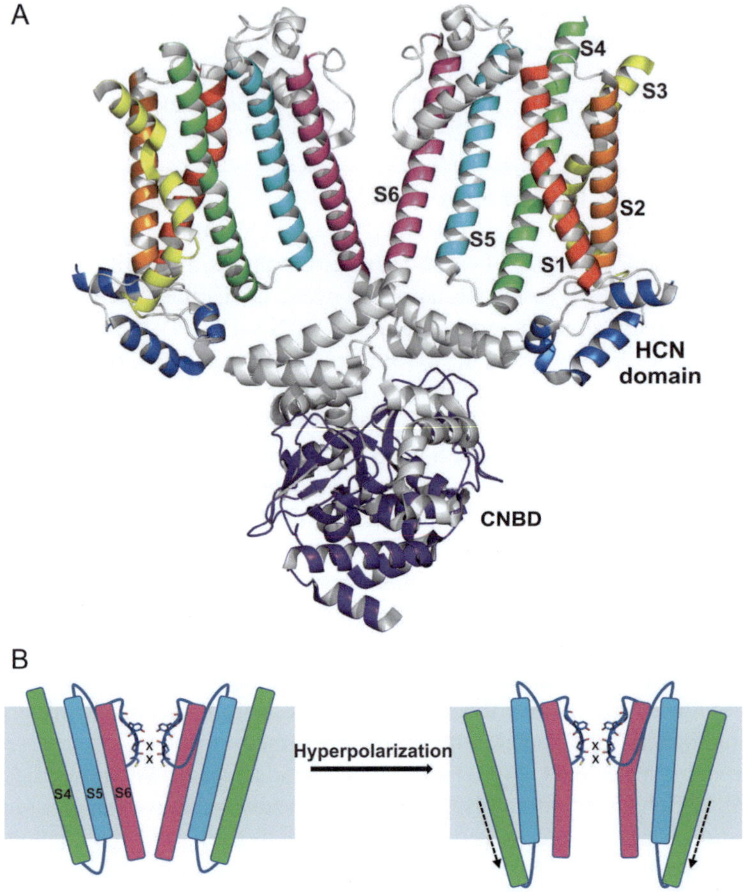

Fig. 5.2 Structural organization of the HCN1 channel. (**a**) Structure of the HCN1 channel (pdb 5u6p) obtained by Lee and MacKinnon (2017). Only two of the four subunits are shown for clarity. The transmembrane segments (S1, S2, S3, S4, S5, and S6), the cyclic nucleotide-binding domains (CNBD), and the HCN domains are indicated. (**b**) Schematic representations of the S4-S5-P-S6 regions of the HCN1 channel in the closed (left) and open form (right), derived from the cryo-EM structures of hHCN1 (Lee and MacKinnon 2017). Only two of the four subunits are shown for clarity. Residues lining the selectivity filter are shown in stick mode, and the only two cation binding sites are represented as x. Upon hyperpolarization, the downward displacement of the S4 helix drives a series of conformational changes in the subunits allowing the S6 helices to open (see text for details)

molecules (Lolicato et al. 2011; Chow et al. 2012); according to this view, any further increase in cAMP concentration would not induce any additional modulation thus resulting in an apparent weak modulation. Interestingly, this hypothesis could also account for the evidence that the activation curve of HCN1 channels is the most positive among those of HCN isoforms. Finally, cAMP stimulation of HCN3 is peculiar and unclear since it has been described as a lack of or a small negative shift (−2.9/−5 mV) (Stieber et al. 2005; Mistrik et al. 2005).

Heterologous expression of various isoforms revealed that heteromeric channels possess kinetic and modulatory properties intermediate between those of the individual components (Ishii et al. 2001; Chen et al. 2001; Ulens and Tytgat 2001; Altomare et al. 2003; Much et al. 2003). Although at present a detailed understanding of the structural and functional mechanisms responsible for differences in the voltage dependence and in the time course of activation is still missing, there is evidence indicating that the C-terminus region contributes to these processes. Indeed, replacement of the HCN4 C-terminus by that of HCN1 caused a strong acceleration of activation and deactivation rates and a decreased response to cAMP (Viscomi et al. 2001).

2. *Tissue distribution in the adult heart and during development.* Molecular investigations have reported the presence of various HCN isoforms in mammalian cardiac conduction tissue; however, their distribution and relative expression are extremely different and differently developmentally regulated. In the human adult healthy heart, HCN1, HCN2, and HCN4 proteins are highly expressed in the SAN, and mRNA transcripts of *Hcn1*, *Hcn3*, and *Hcn4* have been detected in Purkinje fibers (PF) (Gaborit et al. 2007; Chandler et al. 2009; Li et al. 2015). Several studies have identified the presence of HCN4 channels (both transcripts and proteins) in the atrioventricular node (AVN) and pulmonary veins of different species including humans, thus providing a molecular identification for the f-currents that are normally recorded in these cells at physiological voltages (Greener et al. 2009, 2011; Dobrzynski et al. 2003; Li et al. 2014; Ye et al. 2015; Yamamoto et al. 2006).

   Although HCN1 and HCN4 are the main isoforms in the mammalian SAN (Chandler et al. 2009; Li et al. 2015; Brioschi et al. 2009), their expression either alone or in combination failed to reproduce the sinoatrial I_f current. Several molecular mechanisms may account for this discrepancy, and they are collectively referred to as "context dependence" (Qu et al. 2002); the term context dependence intends to highlight the fact that the ultimate functional behavior of the channel also depends on the interacting and accessory proteins (i.e., Mink-related protein (MiRP) 1, caveolin3, KCR1, and SAP97), phosphorylation state, membrane phospholipid (i.e., phosphatidylinositol 4,5-bisphosphate, PIP_2), surrounding membrane composition and fluidity, interaction with cyclic dinucleotides (c-di-GMP and 2′3′-cGAMP), and likely other yet unidentified mechanisms (Baruscotti et al. 2010). Small amounts of HCN2 and HCN4 proteins have also been detected in the adult human atria and ventricles, while HCN1 were noted only in the atria (Chandler et al. 2009; Stillitano et al. 2008; Li et al. 2015).

   Despite the fact that pacemaker currents are functionally irrelevant in healthy atria and ventricles because of their small densities and negative, nonphysiological threshold for activation (Porciatti et al. 1997; Hoppe and Beuckelmann 1998; Hoppe et al. 1998), we are now aware that kinetic alterations and/or overexpression of these channels is often observed in association with cardiac disease such as chronic AF, heart failure (Stillitano et al. 2008, 2013; Cerbai et al. 2001; Hoppe et al. 1998). These pathology-related alterations of HCN expression are likely associated with the arrhythmic profiles often observed in these conditions.

Several studies have indeed elegantly shown that, in addition to generating the pacemaker activity in the adult heart, f-currents are also critical for a correct cardiac development. Indeed in the mouse embryo, significant levels of *Hcn4* mRNA can be detected as early as embryonic day (ED) 7.5 in the cardiac crescent (Garcia-Frigola et al. 2003). As development progresses (ED8), *Hcn4* can be found in ventricular progenitors of the first heart field which drive the peristaltic contraction of the heart tube (Liang et al. 2013), even though these cells will not form the mature sinus node. From ED9.5 the expression of the HCN4 channels will be progressively restricted to the sinus venous, the region that will become the adult SAN. Only few data on the expression of other HCN isoforms in the developing heart are available: Stieber et al. (2003), for example, showed that in global HCN4 knockout mice at ED9.5, the HCN1 and HCN3 isoforms are expressed and may account for the residual I_f current recorded in embryonic cardiomyocytes.

5.3 Structural Hallmarks of HCN Channels

HCN channels have been the focus of intense molecular investigation aiming at the identification of the structural domains associated with specific functional features as mentioned previously. The recent resolution of the cryo-electron microscopy structure of the human HCN1 in the closed state has substantially advanced our knowledge in this field (Lee and MacKinnon 2017). The 3D structure has indeed provided a structural interpretation of the following aspects: (1) the mixed permeability of HCN channels, (2) the activation upon hyperpolarization, and (3) the cAMP-induced facilitation of the close-to-open transition. We will now briefly review the main aspects of this structure-function association.

1. *Mixed Na^+/K^+ permeability of HCN channels.* The ionic selectivity of tetrameric HCN channels is determined by the presence within the hairpin-shaped pore region comprised between the S5 and S6 transmembrane segments of four GYG triplets which form the selectivity filter (SF) of K^+ channels. A puzzling question that has long remained unanswered was why the GYG arrangement of K^+ channels restricts the permeability to K^+ ions ($P_{Na}/P_k > 1000:1$), while HCN channels are instead highly permeable to Na^+ ions [P_K/P_{Na} of 2–5:1 (Ludwig et al. 1999; Santoro et al. 1998; Moroni et al. 2000)]. The cryo-EM structure has elegantly shown that the outer half of the SF of HCN channels is more dilated than that of purely K^+ selective channels, and this forces the SF filter of HCN channels to have only two cation binding sites instead of the four present in pure selective K^+ channels

(Fig. 5.2b). The presence of four binding sites in K^+ channels allows the simultaneous coordination in the SF of two ions with the consequence that since K^+ ions are more stably coordinated than Na^+ ions, the presence in the SF of a K^+ ion opposes the permeation of Na^+ ions limiting the permeability to K^+ ions. On the other hand, the structural organization of the SF of HCN channels limits to one the number of ions that the SF can "host" at the same time, thus allowing a mixed Na^+/K^+ permeability (Lee and MacKinnon 2017).

2. *Activation upon hyperpolarization.* Pacemaker channels are hyperpolarization-activated channels, and this rather unique property is caused by an unusually long S4 segment which, at depolarized potentials, protrudes in the cytoplasm (Fig. 5.2b). The consequence of the anomalous S4 segments is that at depolarized potentials the S4–S5 linkers interact and stabilize the four C-linkers (C-linker disk) in a position which forces the activation gate at the bottom of the S6 helices (HCN1: Val390, Thr 394, and Gln 398) to assume a tight (closed) conformation. During membrane hyperpolarization, the S4 segments are driven further in the cytoplasmic direction, and this displacement ultimately twists and releases the constraint of the S6 segments that are now free to assume a more energetically stable conformation corresponding to the open state of the channel. A novel information provided by cryo-EM investigation is the presence of a particular domain in the N-terminus which is unique to HCN channels and for this reason is called "HCN domain" (Fig. 5.2a). This domain, composed by the 45 amino acids preceding the transmembrane segment S1, and arranged in three α-helices, takes contact with both the S4 helix of the same subunit and the C-linker of the adjacent subunit and acts as an additional element that determines the structural stability of the closed channel.
3. *cAMP-induced facilitation of the close-to-open transition.* Binding of cAMP molecules to the CNBDs induces a structural rearrangement of these domains and the rotation of the C-linker elements; this movement determines a small displacement of the S6 segments and the opening of the channels (Lee and MacKinnon 2017). In other words, cAMP binding contributes to remove channel inhibition. It is interesting to note that functional experiments have previously shown that membrane hyperpolarization and cAMP binding are allosteric partners in regulating the opening processes (DiFrancesco 1999; Altomare et al. 2001) and it is thus likely that the structural movement and final state of the gate induced by voltage are identical to those occurring upon cAMP binding.

5.4 Pathophysiological Aspects of HCN Channel Mutations

The results obtained with in vitro studies on the role of pacemaker channels provided the logical background for genetic studies aiming at the identification of HCN channel mutations associated with inheritable cardiac impulse generation and conduction dysfunctions. This search has been extremely fruitful for the *HCN4* isoform, while no pathological mutations in *HCN1*, *HCN2*, or *HCN3* genes have been so far reported. In the following part, we will discuss the HCN4 disease-causing mutations according to their position within the different structural and functional domains of the channel.

5.4.1 Mutations of the N-Terminus (P257S)

In a recent study, Macri et al. (2014) searched for the presence of HCN4 mutations in patients with early-onset atrial fibrillation and identified seven novel variants (seven in the AF population and three in the control population with no history of AF): p. K189R, p.P257S, p.T822M, p.G885R, p.P945S, p.G1077S, and p.E1193Q. Expression studies of these mutations were carried out in CHO cells, and, with the exception of the P257S located in the N-terminus, no functional differences in current densities and kinetic features were found. When homomerically expressed, the P257S variant did not yield any measurable currents likely because mutant channels were retained in the cytoplasm as revealed by immunocytochemistry. Thus, it was concluded that the P257S mutation disrupts membrane trafficking. Whether in native cardiac cells this trafficking defective mechanism leads to a reduced current expression is a possibility that could explain the clinical features observed in the single patient carrying the heterozygous mutation.

5.4.2 Mutations of the S4 Segment (R393H) and of the S4–S5 Linker (A414G)

5.4.2.1 R393H

The mutation p.R393H (c.1178G>A) has recently been identified by Ishikawa and coworkers in three family members presenting different types of cardiac dysfunctions: bradyarrhythmia (proband), sinus arrest and supraventricular escape beats (brother), and dilated cardiomyopathy and atrial fibrillation (father) (Ishikawa et al. 2017). Heterologous in vitro expression of mutant channels in tsA-201 cells resulted in a marked reduction of the current (~−77% for homomeric R393H/R393H and ~−57% for heteromeric wt/R393H), but trafficking defects were excluded. Half-activation values ($V_{½}$) and time constants of activation did not differ between heteromeric and wt currents, while a significant difference was found in the slope factors. The authors also report that cAMP-induced modulation was maintained in heteromeric channels, but it could not be evaluated in homomeric channels.

5.4.2.2 A414G

Milano et al. identified the mutation p.A414G (c.1241C>G), located in the S4–S5 linker, in three related individuals (father and sons) variously affected by sinus bradycardia and left ventricular noncompaction cardiomyopathy (LVNC), AF, and atrial standstill (Milano et al. 2014). Patch-clamp studies in transiently transfected CHO cells showed that the voltage dependence of heteromeric wt/A414G channels was negatively shifted by 23.9 mV, thus causing a significant reduction of the current density in the pacemaker range of potentials (−50/−60 mV). These data, together with the established notion that the S4–S5 linker is an important structural element that couples voltage-dependent movement of the S4 segment to the gating structures, support the conclusion of a causative association between the p.A414G mutation and the bradyarrhythmic phenotype observed in mutant carriers. The presence of LVNC has also been reported in association with the p.695X and p.883R HCN4 mutations (Schweizer et al. 2014; see also below). Although a clear understanding of this association is still missing, Milano et al. hypothesize that LVNC may be congenital or secondary to sinus bradycardia (Milano et al. 2014). During embryonic heart development, HCN4 channels are expressed in progenitors of the first heart field (that will later contribute to the formation of the working myocardium; Barbuti and Robinson 2015); for this reason dysfunctional embryonic HCN4 currents may be the direct cause of the structural alterations observed in the adult cardiac tissue. Alternatively, LVNC may represent a remodeling process which is caused by the chronic bradycardia.

5.4.3 Mutations of the HCN4 Selectivity Filter (G480R, Y481H, G482R)

A structural and functional hallmark of HCN channels, as well as of all members of the extended K^+ superfamily, is the presence in the selectivity filter of the conserved GYG motif (in HCN4 G480-Y481-G482) whose role has been discussed in one of the previous sections. Genetic investigations have identified inheritable HCN4 channelopathies associated with the following loss-of-function mutations: p. G480R (Nof et al. 2007), p.Y481H (Milano et al. 2014), and p.G482R (Milano et al. 2014; Schweizer et al. 2014; Ishikawa et al. 2017).

Clinical data show that patients carrying either the mutation p.Y481H or p. G482R are commonly affected by sinus bradycardia and LVNC (Milano et al. 2014; Schweizer et al. 2014; Ishikawa et al. 2017). Other phenotypes such as mitral valve prolapse, syncope, AF, and first-degree AV block have also been variously reported. Heterologous expression of Y481H and G482R mutant channels allowed to define these mutations as loss of function, because they both determine a large decrease of the current density (Milano et al. 2014; Schweizer et al. 2014). Different results are reported for the effects of the mutations on the voltage dependence: according to Milano et al. (2014), the mutations p.Y481H and p.G482R induce large negative shifts of the activation curves (V½: wt/wt −68.4 mV, wt/Y481H −112.3 mV, wt/G482R −107.1 mV), while no significant shift was found by Schweizer et al. (2014) (V½: wt/wt −94.6 mV, wt/G482R −91.5 mV). Since the V½ values of wt HCN4 currents reported by the two groups differ by about 26 mV, any attempt to interpret the differences in the voltage dependence of mutant channels is difficult.

The clinical phenotypes associated with the p.G480R mutation are much less disruptive [asymptomatic sinus bradycardia and no structural heart abnormalities (Nof et al. 2007)]. Expression studies in HEK cells showed that homomeric G480R currents were practically absent (~90% reduction), while heterologous expression of heteromeric wt/G480R channels yielded two separate set of cells: one with currents identical to the wt and the other with currents similar to the homomeric G480R expression. Western blot experiments carried out in the heterologous expression system confirmed an extremely faint signal for homomeric p.G480R mutant proteins. It is certainly puzzling that while expression studies indicate a striking phenotypic effect, individuals carrying the p.G480R mutation only display a mild bradycardia. Of particular interest is the observation that among all HCN4 channel mutations, only those affecting the pore residues G481 and Y482 are also associated with LVNC. Although at present only speculative, this may reflect a defective role of HCN4 channels during embryonic development (leading to structural alteration of ventricular tissue) rather than a consequence of the altered pacemaker activity. As illustrated in the previous section, the structure of the selectivity filter of the HCN1 channel has been identified by means of the cryo-electron microscopy (Lee and MacKinnon 2017); under this simplified assumption, the residues G480 and Y481 should form the bottleneck of the selectivity filter, while residue G482 is more externally rotated and should not contribute to ionic coordination (Fig. 5.2; see also the previous section). Under this simplified assumption, the mutation of the residue G482 should have a smaller functional impact on the phenotypic clinical aspects than mutations of the residues G480 and Y481. This obviously is not the case, and the reasons for this dichotomy are unclear.

5.4.4 Mutations of the P Region-S6 Extracellular Loop (A485V, V492F)

5.4.4.1 A485V

Laish-Farkash identified the presence of the p.A485V (c.1454C>T) mutation in three individuals affected by symptomatic sinus bradycardia and normal structural heart conditions (Laish-Farkash et al. 2010). The first patient was hospitalized for cardiac arrest during intense exercise, the second patient for pre-syncopal events, and the third one for paroxysmal atrial fibrillation. Although the individuals were unrelated, they all shared a Moroccan-Jewish heritage, thus suggesting a common origin of the mutation. Bradycardia, dizziness, and pre-syncopal events were largely present in other family members carrying the mutation. Current densities elicited by homomeric and heteromeric A485V expression in oocytes and HEK293 cells were severely reduced, and Western blot experiments in A485V-transfected HEK 203 cells confirmed that mutant proteins were ~50% less than wild-type channels.

5.4.4.2 V492F

The heterozygous mutation p.V492F causing a substitution of the hydrophobic valine with the aromatic phenylalanine has recently been identified in a patient with suspected Brugada syndrome by Biel et al. (2016). While the hHCN4 V492 is highly conserved for the HCN4 isoform, this residue is absent in HCN1 and HCN2 isoforms where it is substituted by leucine (hHCN1) and isoleucine (hHCN2) hydrophobic residues. Expression of mutant heteromeric channels in HEK cells reduced currents which retained cAMP modulation, while homomeric expression resulted in virtually no currents. Whether this mutation is responsible of the Brugada syndrome and why no sinus arrhythmias were observed is an open question.

5.4.5 Mutations of the C-Linker (R524Q, K530N, D553N, 573X)

The C-linker region, which in the hHCN4 channel extends from residue S522 to residue A599, connects the S6 transmembrane domain to the cyclic nucleotide-binding domain. The C-linker is not only a structural bridge, but it is also a functional key domain since in the presence of cAMP, the "C-linker disk," formed by the assembly of the four C-linkers, rotates and in doing so favors the opening of the channel gate. Several mutations have been identified in this region.

5.4.5.1 R524Q

We recently identified the first HCN4 gain-of-function mutation (p.R524Q) in siblings of an Italian family affected by the inappropriate sinus tachycardia (IST, Baruscotti et al. 2017); IST is a rare clinical syndrome, and diseased individuals exhibit a faster than expected cardiac rate both at rest and during moderate physical activity (Baruscotti et al. 2016, 2017; Codvelle 1939; Sheldon et al. 2015; Vedantham and Scheinman 2017). The residue p.R524 is a highly evolutionary conserved arginine which is positioned in the A′ α-helix of the C-linker region of HCN channels. In the homomeric wild-type channel, the four p.R524 residues assemble as a positively charged ring located in a region where multiple interactions between the S4–S5 linker, the S5 helix, and the C-linker are at work to control channel gating. Heterologously expressed homomeric R524Q channels have a ~21-fold higher sensitivity to the second messenger cAMP (dose-response Kd values of 0.08 and 1.67 μM for R524Q and wild-type channels, respectively) but no difference in their intrinsic voltage dependence (measured in inside-out condition in the absence of cAMP). Since cAMP and membrane hyperpolarization exert an allosteric control of pacemaker channel availability, an increase in cAMP sensitivity ultimately results in an increased I_f current flowing during diastole, therefore leading to tachycardia (Fig. 5.3a) (Baruscotti et al. 2017).

Our study did not address the structural reason for the increase in cAMP sensitivity, however it provided additional support to the conclusion that the C-linkers are structural elements central to the modulation of the channel.

The concept of cAMP affinity for HCN channels is not only central to the interpretation of the role of the R524Q mutation, but it could also represent a more general molecular mechanism of IST. Indeed, there is evidence that a large fraction (~50%) of IST patients have β-adrenergic receptor autoantibodies, a condition that causes a larger-than-normal cAMP production (Chiale et al. 2006) and, hence, pacemaker channel activation. Taken together these studies support the concept that nonphysiological cAMP overproduction and increased HCN sensitivity to cAMP are molecular mechanisms responsible for IST.

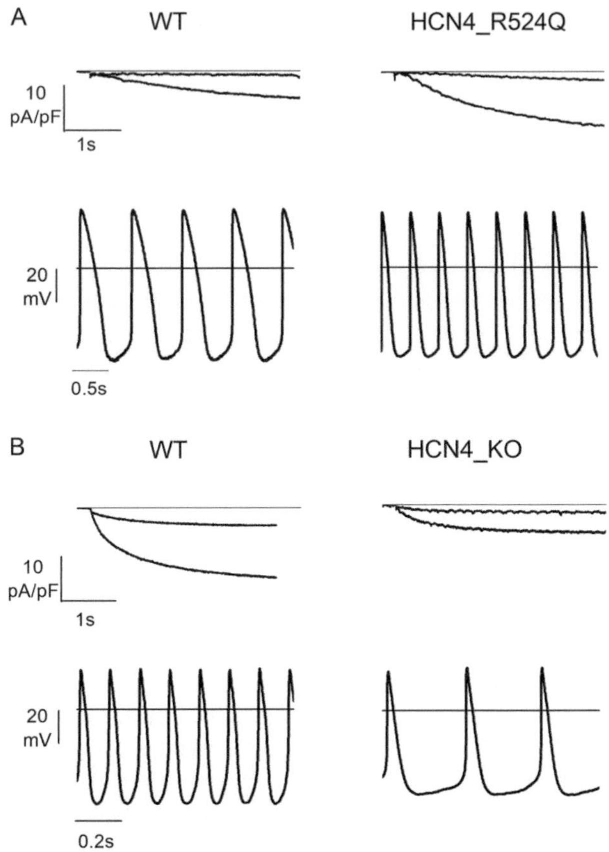

Fig. 5.3 The I_f current density influences the spontaneous rates of rabbit adult SAN myocytes and rat newborn cardiac myocytes. (**a**) The gain-of-function HCN4 R524Q mutation increases both the pacemaker current and the spontaneous activity of rat newborn cardiac myocytes. Sample current traces (steps to −55 and −75 mV, hp = −35 mV; top), and APs (bottom) recorded from rat newborn cardiac myocytes overexpressing either the WT (left) or the R524Q mutated (right) channels. (**b**) The knockout of HCN4 channels decreases both the I_f current and the spontaneous activity of SAN cells. Sample I_f traces (steps to −55 and −75, hp = −35 mV; top) and APs (bottom) recorded from SAN cells isolated from control (left) and HCN4 KO (right)

5.4.5.2 K530N

The heterogeneity of the phenotypic manifestations of mutations of the C-linker region is further demonstrated by the clinical signs associated with the K530N mutation which was identified in members of a family affected by mild asymptomatic sinus bradycardia, age-dependent tachy-brady syndrome, and persistent atrial fibrillation (Duhme et al. 2013).

Similarly to p.R524Q, the p.K530N mutation is located in the A′ helix of the C-linker and replaces a positively charged lysine with an asparagine. Expression studies of mutant channels unexpectedly showed that the K530N mutation had a significant impact on the functional properties of the channel only in the heteromeric condition. For example, V½ of activation were wt, −87.5 mV; K530N/K530N, −88.8 mV; and wt/K530, −101.7 mV, while cAMP-induced V½ shifts were wt, +14.3 mV; K530N/K530N, +17.4 mV; and wt/K530, +21.8 mV. Given the heterozygous condition of the affected carriers, we could speculate that while the negative shift of the voltage dependence could account for the bradycardic condition, the increased response to cAMP stimulation may represent the mechanisms associated with the tachycardic aspect of the tachy-brady syndrome.

Since K530 residues are involved in the subunit-subunit interaction, the authors proposed that in the heteromeric condition, residues N530 likely alter the integrity of this interaction and cause an inhibitory action on the voltage-dependent gating (i.e., favor the closed state). However, when all four K530 residues are mutated (homomeric mutants), proper channel gating is maintained.

5.4.5.3 D553N

The heterozygous mutation p.D553N was originally discovered by Ueda and collaborators (Ueda et al. 2004) in three members of a family affected by severe bradycardia and QT prolongation; recurrent syncope and polymorphic ventricular tachycardia were also observed in the proband. The electrophysiological properties of mutant channels have been analyzed by heterologous expression in COS7 cells: a large reduction of the current density (~−65%−75% and ~−92% for heteromeric and homomeric channels, respectively) and no alterations of the voltage dependence were reported (Ueda et al. 2004; Netter et al. 2012). However, while Ueda et al. (2004) observed a trafficking impairment with cytoplasmic retention of the channels, this was not reported by Netter et al. (2012). Although a thorough investigation was not carried out, Netter et al. (2012) showed that homomeric D553N channels failed to be modulated by high cAMP concentration.

5.4.5.4 573X

The first mutation ever identified in the *HCN4* gene was the single-base 1631delC deletion found in a single index patient. This mutation, located in the C-linker coding region, caused a frameshift and an early stop codon resulting in the truncated protein 573X lacking part of the C-linker, the CNBD, and the C-terminus (Schulze-Bahr et al. 2003). The clinical characterization included the following symptoms: sinus bradycardia, episodes of syncope, intermittent atrial fibrillation, and chronotropic incompetence. Lack of cAMP-dependent modulation was reported both in the homomeric and in the heteromeric channels suggesting a dominant negative action, while immunofluorescence experiments indicated normal trafficking to the plasma membrane. The 573X mutation represents a milestone for HCN4-related channelopathies, and although a conclusive causative association between the genotype and the clinical phenotype of this mutation was impossible because it was based on a single patient, we can now retrospectively state that this association has been clearly supported by all the findings that followed its discovery.

5.4.6 Mutations of the Cyclic Nucleotide-Binding Domain (S672R, 695X) and of the C-Terminus (P883R, G1097W)

The CNBD domain is the hallmark of HCN and CNG (cyclic nucleotide-gated) channels and is also present in Erg/Eag channels, and its structural architecture is similar to that of the catabolite gene activator protein (CAP) of *E. coli* and to that of the cAMP binding site of the protein kinase A (Craven and Zagotta 2006; Biel et al. 2009). In HCN4 channels the CNBD extends from residue D600 to residue D712, and binding of cAMP molecules to this site results in an allosteric facilitation of the voltage-dependent channel opening process. Following the CNBD, there is the proper C-terminus.

5.4.6.1 S672R

The heterozygous loss-of-function mutation p.S672R was identified by our group in bradycardic members of an Italian family. Mean heart rates calculated for mutation carriers was 52.2 bpm ($n = 15$), while those presenting a normal phenotype had a mean value of 73.2 bpm ($n = 12$). Given the large number of genetically related individuals investigated, it was possible to quantitatively assess the co-segregation between the phenotype (bradycardia) and genotype (p.S672R) and the presence of a tight linkage was confirmed by a LOD score value of 5.47. Sequence alignment analysis confirmed that the wild-type S672 residue is highly preserved in the phylogenetic tree from invertebrates to humans, an observation supporting its functional importance. Extensive electrophysiological investigation has shown that despite its localization within the CNBD, this mutation does not influence the affinity of the channel for the second messenger cAMP but rather renders the channel less sensitive to opening secondary to membrane hyperpolarization (V½: −76.1 mV and −84.5 mV for wild-type and homomeric mutant channels, respectively), and this action fully resembles that of the muscarinic modulation of the current.

5.4.6.2 695X

During a screening specifically aimed at the identification of common causative genetic defects leading to familial electromechanical disorders (sinus node disease and noncompaction cardiomyopathy), Schweizer et al. (2014) identified the novel mutations p.695X and p.P883R (in addition to the p.G482R previously described).

The mutation p.695X was found during a candidate gene study in genetically related heterozygous patients affected by sinus bradycardia and noncompaction cardiomyopathy (Schweizer et al. 2010, 2014). Because of the presence of an early stop codon, the truncated 695X protein lacks the CNBD, and expression studies in HEK 293 cells coherently showed that both homo- and heteromeric mutant channels were insensitive to cAMP-induced modulation. In addition, a rightward shift (+7.4 mV) of the half-activation voltage (V½) was observed in homomeric mutant channels.

Despite the lack of cAMP sensitivity of mutant channels, patients carrying the heterozygous 695X mutation exhibit normal chronotropic control and were able to reach normal maximal heart rates. In this respect it is worth noting that patients with the 573X mutation, which also removed the CNBD, were instead affected by chronotropic incompetence; the reason of this difference is unclear.

5.4.6.3 P883R

The missense mutation p.P883R was found in a single unrelated patient with sinus bradycardia, noncompaction cardiomyopathy, and paroxysmal atrial fibrillation; however, the phenotypic features of p.P833R mutant channels were never analyzed (Schweizer et al. 2014).

Patients carrying either the p.695X or the p.P833R exhibit the combined noncompaction cardiomyopathy and sinus bradycardia phenotypes.

5.4.6.4 G1097W

The mutation p.G1097W, located in the terminal part of the C-terminus, was found in a single patient with AV nodal dysfunction (AV block) but normal sinus node activity (Zhou et al. 2014). Upon heterologous expression, mutant current densities were significantly reduced and exhibit a negative shift of their voltage dependence ($V_{½}$: wt, −86.6 mV; homomeric, −98.6 mV; and heteromeric, −94.2 mV). Modulation by cAMP was preserved. Since this mutation was found in a single patient and no sinus node alteration was observed, the genotypic-phenotypic association can only be hypothesized.

5.5 HCN Knockout and Transgenic Mice

The information obtained with in vitro single-cell experiments have provided a wealth of details on the importance of the cardiac pacemaker current; however, these studies lack the integration in the context of the entire organism. For this reason, different transgenic models have been developed to clarify the physiological contribution of HCN channels to cardiac pacemaking and heart development, as well as to improve our knowledge of their pathological relevance in arrhythmias. As discussed in the previous sections, the HCN4 isoform is the most functionally relevant in the mammalian SA Node, and its contribution to cardiac pacemaking and heart development has been studied by means of several HCN4 transgenic/KO mouse lines.

The first HCN4 knockout mice were developed by Stieber et al. (2003) who generated global and cardiac-specific constitutive HCN4 KO models by deleting the exon 4 which encodes for the pore region and the sixth transmembrane segment (TM6). Both global and cardiac-specific homozygous KO mice died between the embryonic days 9.5 and 11.5, but no structural abnormalities were noted in the developing heart. When isolated, the hearts of KO mice retained the intrinsic automaticity albeit they had a slower pace (−37% at day E9.5) than hearts isolated from wild-type embryos. Interestingly, action potential recordings carried out in single cardiomyocytes isolated from embryonic hearts of both models demonstrated the absence of a "mature" pacemaker phenotype which was instead observed in cells isolated from control wild-type embryo hearts. In agreement with these findings, HCN4 KO cardiomyocytes exhibit a 75–90% reduction of the I_f current, and the residual current could not be modulated by cAMP as opposed to a 10.3 mV positive shift of the activation curve observed for the wild-type I_f current. When the chronotropic behaviors of both single myocytes and whole hearts isolated from HCN4 KO mice were challenged with a membrane-permeable cAMP analog (8-Br-cAMP), no significant effects were observed. The finding that in the HCN4 KO constitutive models the I_f current and the spontaneous activities of both isolated hearts and cardiomyocytes are insensitive to the adrenergic mediator cAMP, suggests that the I_f current represents a critical cAMP-sensitive element contributing to the chronotropic control of the heart at this developmental stage.

Although the lack of cAMP modulation might at first appeared as an additional "minor side effect" in comparison to the embryo lethality, its full relevance became clear a few years later when Harzheim et al. (2008) developed a cAMP-insensitive HCN4 knock-in (KI) transgenic model with a similar embryo lethality. The *HCN4* gene of this mouse was engineered to introduce the single amino acid exchange (R669Q) within the CNBD, and this mutation abolished the cAMP-induced modulation. Homozygous R669Q mice normally expressed the HCN4 protein, as verified by immunolabelling and Western blot experiments; however, they too died between embryonic days E11 and E12. Electrophysiological analysis confirmed that maximal f-channel conductance was similar in KI and wild-type cells, but the activation curve was shifted negatively by −13.2 mV in KI cells likely because of the lack of basal cAMP-induced modulation. In agreement with a reduced contribution of the I_f current in KI cells, automaticity of single cells and isolated heart was also reduced by 37% (at E9.5) and 40%, respectively. Experiments with isolated hearts also confirmed that neither adrenergic stimulation nor cAMP increase could enhance the beating rate of homozygous KI hearts.

Taken together the global, the cardiac specific, and the R669Q-KI constitutive KO models allow to conclude that the HCN4 current is necessary for proper cardiac embryo development and survival, and this crucial role appears to be related to the ability of the current to respond to cAMP/adrenergic stimulation. According to Harzheim et al. (2008), the need for a strong chronotropic response to catecholaminergic stimuli is indeed a vital aspect of the developing heart since it counteracts the otherwise life-threatening bradycardia which may occur during transient hypoxic states. However, a further element to consider is the evidence that the integrity of the HCN current is mandatory for the correct cell cycle progression of proliferating stem cells (Lau et al. 2011; Omelyanenko et al. 2016); whether this is an additional mechanism responsible for HCN4 KO embryo lethality is a question of remarkable interest.

Because of the lethality of KO embryos, the role of the pacemaker current during adulthood could not be evaluated by constitutive models. This limitation was soon overcome by Herrmann et al. (2007) and by Hoesl et al. (2008) who developed two inducible models, i.e. models where functional silencing of the *HCN4* gene could be obtained by knockout of the exon 4 (pore-S6 region) in a time-dependent manner. The two models differed in that the knockout was inducible and global in one case (Herrmann et al. 2007), while it was inducible but specifically restricted to all the cells of the organisms expressing the HCN4 channel in the other case (Hoesl et al. 2008). Despite the differences in the way they were engineered, the two models ultimately resulted in identical phenotypes, that is, the generation of non-functional HCN4 proteins in the entire organism. Single-cell studies reported that the I_f current density was indeed similarly reduced (by 75–80%) in both models, and 45–90% of SAN cells lacked spontaneous activity (Herrmann et al. 2007; Hoesl et al. 2008). In vivo recordings in adult mice showed that heart rate, ECG parameters, and chronotropic response to adrenergic stimulation were not different between control and KO mice; however, recurrent sinus pauses and increased response to muscarinic stimulation were observed after KO induction. The presence of sinus pauses (which was also observed in isolated hearts) was maximal at heart rates of 350–450 bpm; of note is the observation that these rates (350–450 bpm) are slightly lower than the intrinsic heart rate of mice (Yaniv et al. 2016; Larson et al. 2013), and therefore sinus pauses are maximal in a condition where the vagal tone prevails. In line with this observation, the authors also report that in KO mice the bradycardic response to muscarinic stimulation was maintained, and pauses' duration was increased; on the contrary when the heart rate was raised by intense activity, therefore by robust sympathetic input, the number of sinus pauses decreased. These observations thus suggest that SAN cells of HCN4 KO mice have an "unstable" and "weak" diastolic pacemaker depolarization as a consequence of the paucity of the I_f current. This instability is clearly exacerbated by the parasympathetic-induced bradycardia, while it is partly resolved by the adrenergic action which increases the f-current. These results, which were partly unexpected giving the embryo lethality of the HCN KO constitutive models, suggest that in the adult mouse, additional mechanisms likely contribute to pacemaker generation and/or that the remaining I_f current after KO induction (20–25%) may be sufficient to drive the diastolic depolarization. Also relevant is the observation that while in the embryo heart the removal of the cAMP-dependent modulation of I_f impairs the chronotropic modulation, this is not the case in the adult where obviously other adrenergic modulatory mechanisms are at work.

As previously noted, the models by Hermann et al. (2007) and Hoesl et al. (2008) allowed to study the role of HCN4 channels in the adult animal; however, a possible confounding element of these models is that the knockout procedure occurred throughout all the cells of the organisms. Our laboratory therefore further improved the study model by creating a mouse line where the HCN4 knockout process was inducible and strictly restricted to cardiac cells (Baruscotti et al. 2011). This mouse thus presented the noticeable advantage of not altering the functional contribution of HCN4 channels in non-cardiac HCN4-expressing tissues (such as some types of neurons). This model exhibited a quite severe arrhythmic phenotype: severe bradycardia with a reduction of heart rate up to 47% (but sinus pauses were not observed), prolongation of the PQ interval, and AV block which progressed to complete AV block and heart arrest. SAN cells isolated from these mice consistently displayed a reduction of both spontaneous rate (up to ~60%) and I_f density (70%) highlighting therefore a solid cause-effect quantitative dependence between the reduction of the I_f current and SAN cells/cardiac rates (Fig. 5.3b).

In addition, despite the permanence of robust β-adrenergic chronotropic responses both in single SAN cells and in freely moving animals, the maximal rates attained were inferior to those elicited in wild-type conditions (cells, −43%; animals, −31%). We thus concluded that presence of an intact I_f current is required for proper cardiac impulse generation and modulation. While the association between HCN4 channels and SAN activity was obviously expected, our model provided the evidence for an active functional presence of HCN4 currents also in the AVN, suggesting an additional role of the HCN4 current associated with impulse conduction.

Alig et al. (2009) also developed a cardiac-specific and inducible HCN4 transgenic mouse model (Tet-Off system) based on the inducible and cardiac-specific expression of the human HCN4 isoform carrying the mutation 573X. This mutation, originally identified in a single patient by Schultze-Bahr, renders the HCN4 channel insensitive to cAMP modulation (Schulze-Bahr et al. 2003). The loss of cAMP modulation was indeed verified in single-cell experiments, since the I_f current recorded in SAN cells isolated from adult 573X mice was characterized by a negative shift of the voltage-dependent activation (~20 mV) and by the finding that isoprenaline could not increase the I_f current. Spontaneous activity of mutant cells was generally highly arrhythmic; however, a regular spontaneous AP rate could be restored in the presence of isoprenaline even though the maximal rate attainable was lower than that observed in wild-type cells. Adult mice expressing the HCN4X channels displayed a significant decrease of sinus rate both at rest and during exercise. Taken together these observations point to the conclusion that basal cAMP cellular content ensures a tonic control of the I_f current at rest and that an additional cAMP-dependent increase of the I_f can be recruited during adrenergic stimuli (i.e., during activity). This study shows that removal of cAMP-modulation during adulthood is not lethal, while embryo lethality was instead observed by Harzeim et al. (2008); despite this difference both models strongly provide the evidence that cAMP modulation of HCN4 channels is a key element to maintain proper basal heart rate as well as to provide a depolarization reserve that can be readily used upon the adrenergic stimulation and cytoplasmic cAMP increase.

The study of HCN4 transgenic models and the pathological role of HCN4 mutations underline the predominant role of this isoform during cardiac embryonic development and in the adult conduction tissue. However, since HCN1 and HCN2 expression have also been found in the human SAN, the cardiac phenotypes of these HCN1 and HCN2 KO models were also evaluated. Both HCN1 and HCN2 global and a cardiac-specific constitutive KO mice completed the embryonic development, indicating that these isoforms are not critical for cardiac development (Ludwig et al. 2003; Fenske et al. 2013; Nolan et al. 2003). Fenske et al. (2013) developed an inducible and global HCN1 KO mouse with sinus bradycardia, recurrent sinus pauses, increased heart rate variability, and lower maximal heart. Single-cell studies confirmed a strong reduction of the I_f current (~40%) and of both basal (−13%) and maximal (−13%) firing rates (Fenske et al. 2013). Somewhat similar results were observed by Ludwig et al. in adult HCN2 KO mice which exhibited sinus dysrhythmia which however disappeared in the presence of adrenergic stimulation (Ludwig et al. 2003). The I_f current measured in HCN2 KO SAN cells was ~30% less than in wild-type cells but was still modulated by cAMP (Ludwig et al. 2003). These data thus indicate that HCN2 channels provide a limited contribution to the adult pacemaker generation, and this contribution is mainly restricted to the basal rhythm. The physiological role of HCN3 isoform in the heart is still unclear; however, the HCN3 constitutive and global KO mouse developed by Fenske et al. (2011) did not show relevant phenotypic cardiac alteration in their chronotropic behavior.

Taken together the results obtained by the use of HCN transgenic mouse lines confirm the relevance of these channels in the control of cardiac rate. However, the reader should be aware that given the obvious differences in cardiac rates between the murine models and the human heart, any attempt to interpret transgenic mouse data in the context of a unifying kinetic scheme of rhythm generation and control in humans may not be fully correct.

An inducible double HCN2/HCN4 knockout model restricted to working cardiomyocytes was developed by Hofmann et al. (2012) as a tool to study the mechanism underlying the arrhythmogenesis associated with cardiac hypertrophy and failure. Several studies had indeed previously demonstrated that the ventricular I_f current increases during these pathological states (Cerbai and Mugelli 2006). The study of Hofmann and colleagues (Hofmann et al. 2012) showed that even though HCN2 and HCN4 are the main isoforms expressed in the healthy ventricle, the isoform that is upregulated during cardiac hypertrophy is HCN1. Interestingly they also reported that when hypertrophy was induced after the knockout of the HCN2 and HCN4 isoforms, the risk of arrhythmogenesis was diminished.

5.6 HCN Channels: A Pharmacological Target for Therapy and Disease Treatment

The relevance of the I_f current in setting the slope of the diastolic depolarization of SAN cells makes it an important pharmacological target for selective modulation of heart rate. In the last few decades, different I_f blockers have been developed and extensively characterized by in vitro and in vivo studies (Bois et al. 1996; Monnet et al. 2001; Bucchi et al. 2002, 2006; Vilaine et al. 2003). These drugs, called "pure heart rate-lowering" agents, include alinidine (ST567), zatebradine (UL-FS49), cilobradine (DK_AH26), ZD-7288, and ivabradine (S16257). Apart from ivabradine, these drugs never reached the market due to the presence of undesired side effects such as block of cardiac K^+ and/or Ca^{2+} channels and block of neuronal HCN channels. Ivabradine is the only member of this family which caused minimal side effects (mild visual symptoms), and for this reason its efficacy has been extensively tested in three clinical trials developed to assess the beneficial effects of a selective reduction of heart rate in patients (1) with chronic heart failure with left ventricle systolic dysfunction [SHIFT trial (Swedberg et al. 2010)], (2) with both stable coronary artery disease and left ventricle systolic dysfunction [BEAUTIFUL trial (Fox et al. 2008)], and (3) with stable CAD without overt heart failure and left ventricle systolic dysfunction [SIGNIFY trial (Fox et al. 2014)]. These trials confirmed the effectiveness of ivabradine in relieving the symptoms of chronic stable *angina pectoris* in patients with coronary artery disease with normal sinus rhythm. The antianginal effect of ivabradine relays on its ability to selectively reduce heart rate; it is this rate reduction that improves oxygen supply to cardiomyocytes due both to an increase in the duration of the diastolic coronary perfusion time and to a reduction in cell oxygen consumption. Indeed, ivabradine reduced hospitalization of patients with stable, symptomatic chronic heart failure with reduced left ventricular ejection fraction ($\leq$35%) and in sinus rhythm with a resting heart rate $\geq$70 bpm. Ivabradine can be used as a single agent when β-blockers are not tolerated or contraindicated or as add-on therapy when adequate heart rate control is not achieved. In contrast to the conventional pharmacological agents used to reduce heart rate (β-adrenergic blockers and calcium channel antagonists), ivabradine, at clinically approved doses, does not affect myocardial contractility, atrioventricular conduction, and hemodynamic parameters (DiFrancesco and Camm 2004; Sulfi and Timmis 2006). Finally, it is also worth noting that several case reports indicate beneficial effects of ivabradine for the treatment of inappropriate sinus tachycardia, postural orthostatic tachycardia syndrome, cardiogenic shock, and in case of uncontrolled heart rate following heart transplantation (Oliphant et al. 2016).

Ivabradine blocks native sinoatrial f-channels by entering from the intracellular side, and the resulting decrease in the pacemaker current causes a decrease of the slope of the diastolic depolarization of the action potential and thus of the heart automaticity (Fig. 5.4a, b) (Bois et al. 1996; Bucchi et al. 2002, 2007; Thollon et al. 1997).

An important feature of ivabradine block is its "use dependence" since the block is stronger when the channels repetitively cycle between the open and closed states, and this results in a block efficiency which is increased at high rates (tachycardia).

Investigation of the ivabradine binding site in HCN channels has highlighted that drug binding occurs in the water-filled cavity lined below the internal portion of the pore. In particular, the major determinant of ivabradine binding to the HCN4 channel is the structural integrity of the floor of the cavity in the closed state, which is formed by the side chains of the S6 residues Y506 and I510 (Bucchi et al. 2013).

Despite the existence of differences in the molecular details of the ivabradine block of HCN1 and HCN4 channels, the K_d values are similar (2.0 and 0.94 μM, for HCN4 and HCN1, respectively; from Bucchi et al. 2006), thus suggesting that

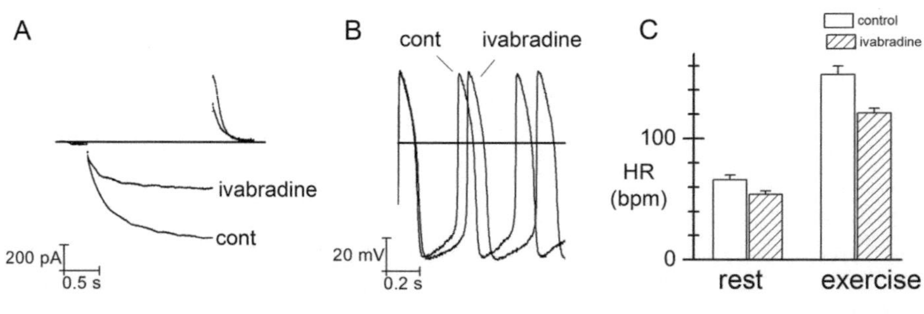

Fig. 5.4 Effect of ivabradine on native I_f, SAN Action potentials, and heart rate (HR). (**a**) Superimposed sample I_f current traces recorded from rabbit SAN cell in control condition and during ivabradine (3 μM) steady-state block. Currents were elicited by voltage steps at −100/+40 mV from a holding potential of −35 mV. Ivabradine reduced the I_f current by about 60%. (**b**) Superimposed sample AP traces recorded from rabbit SAN in control condition and following perfusion with ivabradine (3 μM). Ivabradine reduced the spontaneous AP rate by about 20%. (**c**) Effect of a single oral dose of ivabradine (30 mg) on the heart rate of healthy human volunteers at rest and during exercise (10 min exercise on a bicycle ergometer at increasing workloads). 2 hours after ivabradine intake, the heart rate was reduced by about 10% and 20% at rest and during exercise, respectively. Data shown in panel (**c**) are from Joannides et al. (2006)

ivabradine should not act as an isoform-selective drug. It is therefore this lack of selectivity that accounts for the partial block of HCN1 channels largely expressed in the retina that cause luminous phenomena/phosphenes which are mild side effects affecting vision (Cervetto et al. 2007; Demontis et al. 2009; Oliphant et al. 2016). It is thus important to develop a second-generation of isoform-selective blockers since this would allow to also target neurological conditions associated with dysfunctional neuronal HCN channels.

5.7 Conclusions

Since the discovery of the I_f current and the cloning of *HCN* channel genes a large amount of data have been collected on the properties and the physiopathological functions of this current using different models and approaches (including single cells, animal models, and newly published cryo-electron microscopy data on the structure of the human HCN1 channel). Although the cellular processes contributing, directly or indirectly, to pacemaker activity are many, the above data clearly indicate that the pacemaker current plays a substantial role in the generation and control of the SAN spontaneous activity. Evidence for this role includes (1) I_f activation range overlapping that of diastolic depolarization, (2) I_f-mediated control of cardiac rate by autonomic transmitters, (3) correlation between I_f expression in a given cell type and the presence of spontaneous activity, and (4) use of a specific f-channel blocker (ivabradine) as heart rate-reducing agent. HCN4 channel mutations also indicate a clear association between I_f and impulse generation, since most of the HCN4 dysfunctional mutations reported are associated with brady (loss-of-function) or tachy (gain-of-function) arrhythmias. In addition, experimental data also relate loss-of-function mutations with more complex pathologies such as AF, AV blocks, and structural abnormalities, thus suggesting new and still unexplored roles of HCN4 in the heart.

Compliance with Ethical Standards

Conflict of Interest The authors declare that they have no conflict of interest.

Ethical Approval This article does not contain any studies with human participants or animals performed by any of the authors.

References

Alig J, Marger L, Mesirca P, Ehmke H, Mangoni ME, Isbrandt D. Control of heart rate by cAMP sensitivity of HCN channels. Proc Natl Acad Sci U S A. 2009;106(29):12189–94. https://doi.org/10.1073/pnas.0810332106.

Altomare C, Bucchi A, Camatini E, Baruscotti M, Viscomi C, Moroni A, DiFrancesco D. Integrated allosteric model of voltage gating of HCN channels. J Gen Physiol. 2001;117(6):519–32.

Altomare C, Terragni B, Brioschi C, Milanesi R, Pagliuca C, Viscomi C, Moroni A, Baruscotti M, DiFrancesco D. Heteromeric HCN1-HCN4 channels: a comparison with native pacemaker channels from the rabbit sinoatrial node. J Physiol. 2003;549(Pt 2):347–59. https://doi.org/10.1113/jphysiol.2002.027698.

Anumonwo JM, Delmar M, Jalife J. Electrophysiology of single heart cells from the rabbit tricuspid valve. J Physiol. 1990;425:145–67.

Barbuti A, Robinson RB. Stem cell-derived nodal-like cardiomyocytes as a novel pharmacologic tool: insights from sinoatrial node development and function. Pharmacol Rev. 2015;67:368–88. https://doi.org/10.1124/pr.114.009597.

Barbuti A, Terragni B, Brioschi C, DiFrancesco D. Localization of f-channels to caveolae mediates specific beta2-adrenergic receptor modulation of rate in sinoatrial myocytes. J Mol Cell Cardiol. 2007;42(1):71–8. https://doi.org/10.1016/j.yjmcc.2006.09.018.

Baruscotti M, Bucchi A, DiFrancesco D. Physiology and pharmacology of the cardiac pacemaker ("funny") current. Pharmacol Ther. 2005;107(1):59–79. https://doi.org/10.1016/j.pharmthera.2005.01.005.

Baruscotti M, Barbuti A, Bucchi A. The cardiac pacemaker current. J Mol Cell Cardiol. 2010;48(1):55–64. https://doi.org/10.1016/j.yjmcc.2009.06.019.

Baruscotti M, Bucchi A, Viscomi C, Mandelli G, Consalez G, Gnecchi-Rusconi T, Montano N, Casali KR, Micheloni S, Barbuti A, DiFrancesco D. Deep bradycardia and heart block caused by inducible cardiac-specific knockout of the pacemaker channel gene Hcn4. Proc Natl Acad Sci U S A. 2011;108(4):1705–10. https://doi.org/10.1073/pnas.1010122108.

Baruscotti M, Bianco E, Bucchi A, DiFrancesco D. Current understanding of the pathophysiological mechanisms responsible for inappropriate sinus tachycardia: role of the If "funny" current. J Interv Card Electrophysiol. 2016;46(1):19–28. https://doi.org/10.1007/s10840-015-0097-y.

Baruscotti M, Bucchi A, Milanesi R, Paina M, Barbuti A, Gnecchi-Ruscone T, Bianco E, Vitali-Serdoz L, Cappato R, DiFrancesco D. A gain-of-function mutation in the cardiac pacemaker HCN4 channel increasing cAMP sensitivity is associated with familial inappropriate sinus tachycardia. Eur Heart J. 2017;38(4):280–8. https://doi.org/10.1093/eurheartj/ehv582.

Biel M, Wahl-Schott C, Michalakis S, Zong X. Hyperpolarization-activated cation channels: from genes to function. Physiol Rev. 2009;89(3):847–85. https://doi.org/10.1152/physrev.00029.2008.

Biel S, Aquila M, Hertel B, Berthold A, Neumann T, DiFrancesco D, Moroni A, Thiel G, Kauferstein S. Mutation in S6 domain of HCN4 channel in patient with suspected Brugada syndrome modifies channel function. Pflugers Arch – Eur J Physiol. 2016;468(10):1663–71. https://doi.org/10.1007/s00424-016-1870-1.

Bois P, Bescond J, Renaudon B, Lenfant J. Mode of action of bradycardic agent, S 16257, on ionic currents of rabbit sinoatrial node cells. Br J Pharmacol. 1996;118(4):1051–7.

Boyett MR, Honjo H, Kodama I. The sinoatrial node, a heterogeneous pacemaker structure. Cardiovasc Res. 2000;47(4):658–87.

Brioschi C, Micheloni S, Tellez JO, Pisoni G, Longhi R, Moroni P, Billeter R, Barbuti A, Dobrzynski H, Boyett MR, DiFrancesco D, Baruscotti M. Distribution of the pacemaker HCN4 channel mRNA and protein in the rabbit sinoatrial node. J Mol Cell Cardiol. 2009;47(2):221–7. https://doi.org/10.1016/j.yjmcc.2009.04.009.

Brown HF, DiFrancesco D, Noble SJ. How does adrenaline accelerate the heart? Nature. 1979;280(5719):235–6.

Bucchi A, Baruscotti M, DiFrancesco D. Current-dependent block of rabbit sino-atrial node I(f) channels by ivabradine. J Gen Physiol. 2002;120(1):1–13.

Bucchi A, Tognati A, Milanesi R, Baruscotti M, DiFrancesco D. Properties of ivabradine-induced block of HCN1 and HCN4 pacemaker channels. J Physiol. 2006;572(Pt 2):335–46. https://doi.org/10.1113/jphysiol.2005.100776.

Bucchi A, Baruscotti M, Robinson RB, DiFrancesco D. Modulation of rate by autonomic agonists in SAN cells involves changes in diastolic depolarization and the pacemaker current. J Mol Cell Cardiol. 2007;43(1):39–48. https://doi.org/10.1016/j.yjmcc.2007.04.017.

Bucchi A, Baruscotti M, Nardini M, Barbuti A, Micheloni S, Bolognesi M, DiFrancesco D. Identification of the molecular site of ivabradine binding to HCN4 channels. PLoS One. 2013;8(1):e53132. https://doi.org/10.1371/journal.pone.0053132.

Cerbai E, Mugelli A. I(f) in non-pacemaker cells: role and pharmacological implications. Pharmacol Res. 2006;53(5):416–23. https://doi.org/10.1016/j.phrs.2006.03.015.

Cerbai E, Sartiani L, DePaoli P, Pino R, Maccherini M, Bizzarri F, DiCiolla F, Davoli G, Sani G, Mugelli A. The properties of the pacemaker current I(F)in human ventricular myocytes are modulated by cardiac disease. J Mol Cell Cardiol. 2001;33(3):441–8. https://doi.org/10.1006/jmcc.2000.1316.

Cervetto L, Demontis GC, Gargini C. Cellular mechanisms underlying the pharmacological induction of phosphenes. Br J Pharmacol. 2007;150(4):383–90. https://doi.org/10.1038/sj.bjp.0706998.

Chandler NJ, Greener ID, Tellez JO, Inada S, Musa H, Molenaar P, Difrancesco D, Baruscotti M, Longhi R, Anderson RH, Billeter R, Sharma V, Sigg DC, Boyett MR, Dobrzynski H. Molecular architecture of the human sinus node: insights into the function of the cardiac pacemaker. Circulation. 2009;119(12):1562–75. https://doi.org/10.1161/CIRCULATIONAHA.108.804369.

Chen YJ, Chen SA, Chang MS, Lin CI. Arrhythmogenic activity of cardiac muscle in pulmonary veins of the dog: implication for the genesis of atrial fibrillation. Cardiovasc Res. 2000;48(2):265–73.

Chen S, Wang J, Siegelbaum SA. Properties of hyperpolarization-activated pacemaker current defined by coassembly of HCN1 and HCN2 subunits and basal modulation by cyclic nucleotide. J Gen Physiol. 2001;117(5):491–504.

Chen YC, Pan NH, Cheng CC, Higa S, Chen YJ, Chen SA. Heterogeneous expression of potassium currents and pacemaker currents potentially regulates arrhythmogenesis of pulmonary vein cardiomyocytes. J Cardiovasc Electrophysiol. 2009;20(9):1039–45. https://doi.org/10.1111/j.1540-8167.2009.01480.x.

Chiale PA, Garro HA, Schmidberg J, Sanchez RA, Acunzo RS, Lago M, Levy G, Levin M. Inappropriate sinus tachycardia may be related to an immunologic disorder involving cardiac beta adrenergic receptors. Heart Rhythm. 2006;3(10):1182–6. https://doi.org/10.1016/j.hrthm.2006.06.011.

Chow SS, Van Petegem F, Accili EA. Energetics of cyclic AMP binding to HCN channel C terminus reveal negative cooperativity. J Biol Chem. 2012;287(1):600–6. https://doi.org/10.1074/jbc.M111.269563.

Codvelle MMBH. Permanent sinus tachycardia without high frequency functional disorders. Bulletins et Mémoires de la Société Médicale des Hôpitaux de Paris. 1939;54:1849–52.

Craven KB, Zagotta WN. CNG and HCN channels: two peas, one pod. Annu Rev Physiol. 2006;68:375–401. https://doi.org/10.1146/annurev.physiol.68.040104.134728.

Demontis GC, Gargini C, Paoli TG, Cervetto L. Selective Hcn1 channels inhibition by ivabradine in mouse rod photoreceptors. Invest Ophthalmol Vis Sci. 2009;50(4):1948–55. https://doi.org/10.1167/iovs.08-2659.

DiFrancesco D. A new interpretation of the pace-maker current in calf Purkinje fibres. J Physiol. 1981a;314:359–76.

DiFrancesco D. A study of the ionic nature of the pace-maker current in calf Purkinje fibres. J Physiol. 1981b;314:377–93.

DiFrancesco D. Characterization of the pace-maker current kinetics in calf Purkinje fibres. J Physiol. 1984;348:341–67.

DiFrancesco D. Pacemaker mechanisms in cardiac tissue. Annu Rev Physiol. 1993;55:455–72. https://doi.org/10.1146/annurev.ph.55.030193.002323.

DiFrancesco D. Dual allosteric modulation of pacemaker (f) channels by cAMP and voltage in rabbit SA node. J Physiol. 1999;515(Pt 2):367–76.

DiFrancesco D, Camm JA. Heart rate lowering by specific and selective I(f) current inhibition with ivabradine: a new therapeutic perspective in cardiovascular disease. Drugs. 2004;64(16):1757–65.

DiFrancesco D, Ferroni A. Delayed activation of the cardiac pacemaker current and its dependence on conditioning pre-hyperpolarizations. Pflugers Arch – Eur J Physiol. 1983;396(3):265–7.

DiFrancesco D, Mangoni M. Modulation of single hyperpolarization-activated channels (i(f)) by cAMP in the rabbit sino-atrial node. J Physiol. 1994;474(3):473–82.

DiFrancesco D, Ojeda C. Properties of the current if in the sino-atrial node of the rabbit compared with those of the current iK, in Purkinje fibres. J Physiol. 1980;308:353–67.

DiFrancesco D, Tortora P. Direct activation of cardiac pacemaker channels by intracellular cyclic AMP. Nature. 1991;351(6322):145–7. https://doi.org/10.1038/351145a0.

DiFrancesco D, Tromba C. Acetylcholine inhibits activation of the cardiac hyperpolarizing-activated current, if. Pflugers Arch – Eur J Physiol. 1987;410(1–2):139–42.

DiFrancesco D, Tromba C. Muscarinic control of the hyperpolarization-activated current (If) in rabbit sino-atrial node myocytes. J Physiol. 1988;405:493–510.

DiFrancesco D, Ferroni A, Mazzanti M, Tromba C. Properties of the hyperpolarizing-activated current (if) in cells isolated from the rabbit sino-atrial node. J Physiol. 1986;377:61–88.

DiFrancesco D, Ducouret P, Robinson RB. Muscarinic modulation of cardiac rate at low acetylcholine concentrations. Science. 1989;243(4891):669–71.

Dobrzynski H, Nikolski VP, Sambelashvili AT, Greener ID, Yamamoto M, Boyett MR, Efimov IR. Site of origin and molecular substrate of atrioventricular junctional rhythm in the rabbit heart. Circ Res. 2003;93(11):1102–10. https://doi.org/10.1161/01.RES.0000101913.95604.B9.

Duhme N, Schweizer PA, Thomas D, Becker R, Schroter J, Barends TR, Schlichting I, Draguhn A, Bruehl C, Katus HA, Koenen M. Altered HCN4 channel C-linker interaction is associated with familial tachycardia-bradycardia syndrome and atrial fibrillation. Eur Heart J. 2013;34 (35):2768–75. https://doi.org/10.1093/eurheartj/ehs391.

Fenske S, Mader R, Scharr A, Paparizos C, Cao-Ehlker X, Michalakis S, Shaltiel L, Weidinger M, Stieber J, Feil S, Feil R, Hofmann F, Wahl-Schott C, Biel M. HCN3 contributes to the ventricular action potential waveform in the murine heart. Circ Res. 2011;109(9):1015–23. https://doi.org/10.1161/CIRCRESAHA.111.246173.

Fenske S, Krause SC, Hassan SI, Becirovic E, Auer F, Bernard R, Kupatt C, Lange P, Ziegler T, Wotjak CT, Zhang H, Hammelmann V, Paparizos C, Biel M, Wahl-Schott CA. Sick sinus syndrome in HCN1-deficient mice. Circulation. 2013;128(24):2585–94. https://doi.org/10.1161/CIRCULATIONAHA.113.003712.

Fox K, Ford I, Steg PG, Tendera M, Ferrari R, Investigators B. Ivabradine for patients with stable coronary artery disease and left-ventricular systolic dysfunction (BEAUTIFUL): a randomised, double-blind, placebo-controlled trial. Lancet. 2008;372(9641):807–16. https://doi.org/10.1016/S0140-6736(08)61170-8.

Fox K, Ford I, Steg PG, Tardif JC, Tendera M, Ferrari R, Investigators S. Ivabradine in stable coronary artery disease without clinical heart failure. N Engl J Med. 2014;371(12):1091–9. https://doi.org/10.1056/NEJMoa1406430.

Frace AM, Maruoka F, Noma A. External K+ increases Na+ conductance of the hyperpolarization-activated current in rabbit cardiac pacemaker cells. Pflugers Arch – Eur J Physiol. 1992;421 (2–3):97–9.

Gaborit N, Le Bouter S, Szuts V, Varro A, Escande D, Nattel S, Demolombe S. Regional and tissue specific transcript signatures of ion channel genes in the non-diseased human heart. J Physiol. 2007;582(Pt 2):675–93. https://doi.org/10.1113/jphysiol.2006.126714.

Garcia-Frigola C, Shi Y, Evans SM. Expression of the hyperpolarization-activated cyclic nucleotide-gated cation channel HCN4 during mouse heart development. Gene Expr Patterns. 2003;3(6):777–83.

Greener ID, Tellez JO, Dobrzynski H, Yamamoto M, Graham GM, Billeter R, Boyett MR. Ion channel transcript expression at the rabbit atrioventricular conduction axis. Circ Arrhythm Electrophysiol. 2009;2(3):305–15. https://doi.org/10.1161/CIRCEP.108.803569.

Greener ID, Monfredi O, Inada S, Chandler NJ, Tellez JO, Atkinson A, Taube MA, Billeter R, Anderson RH, Efimov IR, Molenaar P, Sigg DC, Sharma V, Boyett MR, Dobrzynski H. Molecular architecture of the human specialised atrioventricular conduction axis. J Mol Cell Cardiol. 2011;50(4):642–51. https://doi.org/10.1016/j.yjmcc.2010.12.017.

Hancox JC, Levi AJ, Lee CO, Heap P. A method for isolating rabbit atrioventricular node myocytes which retain normal morphology and function. Am J Physiol. 1993;265(2 Pt 2):H755–66.

Harzheim D, Pfeiffer KH, Fabritz L, Kremmer E, Buch T, Waisman A, Kirchhof P, Kaupp UB, Seifert R. Cardiac pacemaker function of HCN4 channels in mice is confined to embryonic development and requires cyclic AMP. EMBO J. 2008;27(4):692–703. https://doi.org/10.1038/emboj.2008.3.

Herrmann S, Stieber J, Stockl G, Hofmann F, Ludwig A. HCN4 provides a 'depolarization reserve' and is not required for heart rate acceleration in mice. EMBO J. 2007;26(21):4423–32. https://doi.org/10.1038/sj.emboj.7601868.

Hoesl E, Stieber J, Herrmann S, Feil S, Tybl E, Hofmann F, Feil R, Ludwig A. Tamoxifen-inducible gene deletion in the cardiac conduction system. J Mol Cell Cardiol. 2008;45(1):62–9. https://doi.org/10.1016/j.yjmcc.2008.04.008.

Hofmann F, Fabritz L, Stieber J, Schmitt J, Kirchhof P, Ludwig A, Herrmann S. Ventricular HCN channels decrease the repolarization reserve in the hypertrophic heart. Cardiovasc Res. 2012;95(3):317–26. https://doi.org/10.1093/cvr/cvs184.

Hoppe UC, Beuckelmann DJ. Modulation of the hyperpolarization-activated inward current (If) by antiarrhythmic agents in isolated human atrial myocytes. Naunyn Schmiedeberg's Arch Pharmacol. 1998;358(6):635–40.

Hoppe UC, Jansen E, Sudkamp M, Beuckelmann DJ. Hyperpolarization-activated inward current in ventricular myocytes from normal and failing human hearts. Circulation. 1998;97(1):55–65.

Ishii TM, Takano M, Ohmori H. Determinants of activation kinetics in mammalian hyperpolarization-activated cation channels. J Physiol. 2001;537(Pt 1):93–100.

Ishikawa T, Ohno S, Murakami T, Yoshida K, Mishima H, Fukuoka T, Kimoto H, Sakamoto R, Ohkusa T, Aiba T, Nogami A, Sumitomo N, Shimizu W, Yoshiura KI, Horigome H, Horie M, Makita N. Sick sinus syndrome with HCN4 mutations shows early onset and frequent association with atrial fibrillation and left ventricular noncompaction. Heart Rhythm. 2017;14(5):717–24. https://doi.org/10.1016/j.hrthm.2017.01.020.

Joannides R, Moore N, Iacob M, Compagnon P, Lerebours G, Menard JF, Thuillez C. Comparative effects of ivabradine, a selective heart rate-lowering agent, and propranolol on systemic and cardiac haemodynamics at rest and during exercise. Br J Clin Pharmacol. 2006;61(2):127–37. https://doi.org/10.1111/j.1365-2125.2005.02544.x.

Keith A, Flack M. The form and nature of the muscular connections between the primary divisions of the vertebrate heart. J Anat Physiol. 1907;41(Pt 3):172–89.

Laish-Farkash A, Glikson M, Brass D, Marek-Yagel D, Pras E, Dascal N, Antzelevitch C, Nof E, Reznik H, Eldar M, Luria D. A novel mutation in the HCN4 gene causes symptomatic sinus bradycardia in Moroccan Jews. J Cardiovasc Electrophysiol. 2010;21(12):1365–72. https://doi.org/10.1111/j.1540-8167.2010.01844.x.

Larson ED, St Clair JR, Sumner WA, Bannister RA, Proenza C. Depressed pacemaker activity of sinoatrial node myocytes contributes to the age-dependent decline in maximum heart rate. Proc Natl Acad Sci U S A. 2013;110(44):18011–6. https://doi.org/10.1073/pnas.1308477110.

Lau YT, Wong CK, Luo J, Leung LH, Tsang PF, Bian ZX, Tsang SY. Effects of hyperpolarization-activated cyclic nucleotide-gated (HCN) channel blockers on the proliferation and cell cycle progression of embryonic stem cells. Pflugers Arch – Eur J Physiol. 2011;461(1):191–202. https://doi.org/10.1007/s00424-010-0899-9.

Lee CH, MacKinnon R. Structures of the human HCN1 hyperpolarization-activated channel. Cell. 2017;168(1–2):111–20. e111. https://doi.org/10.1016/j.cell.2016.12.023.

Li YD, Hong YF, Zhang Y, Zhou XH, Ji YT, Li HL, Hu GJ, Li JX, Sun L, Zhang JH, Xin Q, Yusufuaji Y, Xiong J, Tang BP. Association between reversal in the expression of hyperpolarization-activated cyclic nucleotide-gated (HCN) channel and age-related atrial fibrillation. Med Sci Monit. 2014;20:2292–7. https://doi.org/10.12659/MSM.892505.

Li N, Csepe TA, Hansen BJ, Dobrzynski H, Higgins RS, Kilic A, Mohler PJ, Janssen PM, Rosen MR, Biesiadecki BJ, Fedorov VV. Molecular mapping of sinoatrial node HCN channel expression in the human heart. Circ Arrhythm Electrophysiol. 2015;8(5):1219–27. https://doi.org/10.1161/CIRCEP.115.003070.

Liang X, Wang G, Lin L, Lowe J, Zhang Q, Bu L, Chen Y, Chen J, Sun Y, Evans SM. HCN4 dynamically marks the first heart field and conduction system precursors. Circ Res. 2013;113 (4):399–407. https://doi.org/10.1161/CIRCRESAHA.113.301588.

Lolicato M, Nardini M, Gazzarrini S, Moller S, Bertinetti D, Herberg FW, Bolognesi M, Martin H, Fasolini M, Bertrand JA, Arrigoni C, Thiel G, Moroni A. Tetramerization dynamics of C-terminal domain underlies isoform-specific cAMP gating in hyperpolarization-activated cyclic nucleotide-gated channels. J Biol Chem. 2011;286(52):44811–20. https://doi.org/10.1074/jbc.M111.297606.

Ludwig A, Zong X, Jeglitsch M, Hofmann F, Biel M. A family of hyperpolarization-activated mammalian cation channels. Nature. 1998;393(6685):587–91. https://doi.org/10.1038/31255.

Ludwig A, Zong X, Stieber J, Hullin R, Hofmann F, Biel M. Two pacemaker channels from human heart with profoundly different activation kinetics. EMBO J. 1999;18(9):2323–9. https://doi.org/10.1093/emboj/18.9.2323.

Ludwig A, Budde T, Stieber J, Moosmang S, Wahl C, Holthoff K, Langebartels A, Wotjak C, Munsch T, Zong X, Feil S, Feil R, Lancel M, Chien KR, Konnerth A, Pape HC, Biel M, Hofmann F. Absence epilepsy and sinus dysrhythmia in mice lacking the pacemaker channel HCN2. EMBO J. 2003;22(2):216–24. https://doi.org/10.1093/emboj/cdg032.

Macri V, Mahida SN, Zhang ML, Sinner MF, Dolmatova EV, Tucker NR, McLellan M, Shea MA, Milan DJ, Lunetta KL, Benjamin EJ, Ellinor PT. A novel trafficking-defective HCN4 mutation is associated with early-onset atrial fibrillation. Heart Rhythm. 2014;11(6):1055–62. https://doi.org/10.1016/j.hrthm.2014.03.002.

Mangoni ME, Nargeot J. Genesis and regulation of the heart automaticity. Physiol Rev. 2008;88 (3):919–82. https://doi.org/10.1152/physrev.00018.2007.

Milano A, Vermeer AM, Lodder EM, Barc J, Verkerk AO, Postma AV, van der Bilt IA, Baars MJ, van Haelst PL, Caliskan K, Hoedemaekers YM, Le Scouarnec S, Redon R, Pinto YM, Christiaans I, Wilde AA, Bezzina CR. HCN4 mutations in multiple families with bradycardia and left ventricular noncompaction cardiomyopathy. J Am Coll Cardiol. 2014;64(8):745–56. https://doi.org/10.1016/j.jacc.2014.05.045.

Mistrik P, Mader R, Michalakis S, Weidinger M, Pfeifer A, Biel M. The murine HCN3 gene encodes a hyperpolarization-activated cation channel with slow kinetics and unique response to cyclic nucleotides. J Biol Chem. 2005;280(29):27056–61. https://doi.org/10.1074/jbc.M502696200.

Monnet X, Ghaleh B, Colin P, de Curzon OP, Giudicelli JF, Berdeaux A. Effects of heart rate reduction with ivabradine on exercise-induced myocardial ischemia and stunning. J Pharmacol Exp Ther. 2001;299(3):1133–9.

Moroni A, Barbuti A, Altomare C, Viscomi C, Morgan J, Baruscotti M, DiFrancesco D. Kinetic and ionic properties of the human HCN2 pacemaker channel. Pflugers Arch – Eur J Physiol. 2000;439(5):618–26.

Much B, Wahl-Schott C, Zong X, Schneider A, Baumann L, Moosmang S, Ludwig A, Biel M. Role of subunit heteromerization and N-linked glycosylation in the formation of functional hyperpolarization-activated cyclic nucleotide-gated channels. J Biol Chem. 2003;278 (44):43781–6. https://doi.org/10.1074/jbc.M306958200.

Munk AA, Adjemian RA, Zhao J, Ogbaghebriel A, Shrier A. Electrophysiological properties of morphologically distinct cells isolated from the rabbit atrioventricular node. J Physiol. 1996;493 (Pt 3):801–18.

Netter MF, Zuzarte M, Schlichthorl G, Klocker N, Decher N. The HCN4 channel mutation D553N associated with bradycardia has a C-linker mediated gating defect. Cell Physiol Biochem. 2012;30(5):1227–40. https://doi.org/10.1159/000343314.

Nof E, Luria D, Brass D, Marek D, Lahat H, Reznik-Wolf H, Pras E, Dascal N, Eldar M, Glikson M. Point mutation in the HCN4 cardiac ion channel pore affecting synthesis, trafficking, and functional expression is associated with familial asymptomatic sinus bradycardia. Circulation. 2007;116(5):463–70. https://doi.org/10.1161/CIRCULATIONAHA.107.706887.

Nolan MF, Malleret G, Lee KH, Gibbs E, Dudman JT, Santoro B, Yin D, Thompson RF, Siegelbaum SA, Kandel ER, Morozov A. The hyperpolarization-activated HCN1 channel is important for motor learning and neuronal integration by cerebellar Purkinje cells. Cell. 2003;115(5):551–64.

Oliphant CS, Owens RE, Bolorunduro OB, Jha SK. Ivabradine: a review of labeled and off-label uses. Am J Cardiovasc Drugs. 2016;16(5):337–47. https://doi.org/10.1007/s40256-016-0178-z.

Omelyanenko A, Sekyrova P, Andang M. ZD7288, a blocker of the HCN channel family, increases doubling time of mouse embryonic stem cells and modulates differentiation outcomes in a context-dependent manner. SpringerPlus. 2016;5:41. https://doi.org/10.1186/s40064-016-1678-7.

Pian P, Bucchi A, Robinson RB, Siegelbaum SA. Regulation of gating and rundown of HCN hyperpolarization-activated channels by exogenous and endogenous PIP2. J Gen Physiol. 2006;128(5):593–604. https://doi.org/10.1085/jgp.200609648.

Porciatti F, Pelzmann B, Cerbai E, Schaffer P, Pino R, Bernhart E, Koidl B, Mugelli A. The pacemaker current I(f) in single human atrial myocytes and the effect of beta-adrenoceptor and A1-adenosine receptor stimulation. Br J Pharmacol. 1997;122(5):963–9. https://doi.org/10.1038/sj.bjp.0701473.

Qu J, Altomare C, Bucchi A, DiFrancesco D, Robinson RB. Functional comparison of HCN isoforms expressed in ventricular and HEK 293 cells. Pflugers Arch – Eur J Physiol. 2002;444(5):597–601. https://doi.org/10.1007/s00424-002-0860-7.

Santoro B, Liu DT, Yao H, Bartsch D, Kandel ER, Siegelbaum SA, Tibbs GR. Identification of a gene encoding a hyperpolarization-activated pacemaker channel of brain. Cell. 1998;93(5):717–29.

Sartiani L, Mannaioni G, Masi A, Novella Romanelli M, Cerbai E. The hyperpolarization-activated cyclic nucleotide-gated channels: from biophysics to pharmacology of a unique family of ion channels. Pharmacol Rev. 2017;69(4):354–95. https://doi.org/10.1124/pr.117.014035.

Schulze-Bahr E, Neu A, Friederich P, Kaupp UB, Breithardt G, Pongs O, Isbrandt D. Pacemaker channel dysfunction in a patient with sinus node disease. J Clin Invest. 2003;111(10):1537–45. https://doi.org/10.1172/JCI16387.

Schweizer PA, Duhme N, Thomas D, Becker R, Zehelein J, Draguhn A, Bruehl C, Katus HA, Koenen M. cAMP sensitivity of HCN pacemaker channels determines basal heart rate but is not critical for autonomic rate control. Circ Arrhythm Electrophysiol. 2010;3(5):542–52. https://doi.org/10.1161/CIRCEP.110.949768.

Schweizer PA, Schroter J, Greiner S, Haas J, Yampolsky P, Mereles D, Buss SJ, Seyler C, Bruehl C, Draguhn A, Koenen M, Meder B, Katus HA, Thomas D. The symptom complex of familial sinus node dysfunction and myocardial noncompaction is associated with mutations in the HCN4 channel. J Am Coll Cardiol. 2014;64(8):757–67. https://doi.org/10.1016/j.jacc.2014.06.1155.

Sheldon RS, Grubb BP II, Olshansky B, Shen WK, Calkins H, Brignole M, Raj SR, Krahn AD, Morillo CA, Stewart JM, Sutton R, Sandroni P, Friday KJ, Hachul DT, Cohen MI, Lau DH, Mayuga KA, Moak JP, Sandhu RK, Kanjwal K. 2015 Heart Rhythm Society expert consensus statement on the diagnosis and treatment of postural tachycardia syndrome, inappropriate sinus tachycardia, and vasovagal syncope. Heart Rhythm. 2015;12(6):e41–63. https://doi.org/10.1016/j.hrthm.2015.03.029.

Silverman ME, Grove D, Upshaw CB Jr. Why does the heart beat? The discovery of the electrical system of the heart. Circulation. 2006;113(23):2775–81. https://doi.org/10.1161/CIRCULATIONAHA.106.616771.

Stieber J, Herrmann S, Feil S, Loster J, Feil R, Biel M, Hofmann F, Ludwig A. The hyperpolarization-activated channel HCN4 is required for the generation of pacemaker action potentials in the embryonic heart. Proc Natl Acad Sci U S A. 2003;100(25):15235–40. https://doi.org/10.1073/pnas.2434235100.

Stieber J, Stockl G, Herrmann S, Hassfurth B, Hofmann F. Functional expression of the human HCN3 channel. J Biol Chem. 2005;280(41):34635–43. https://doi.org/10.1074/jbc.M502508200.

Stillitano F, Lonardo G, Zicha S, Varro A, Cerbai E, Mugelli A, Nattel S. Molecular basis of funny current (If) in normal and failing human heart. J Mol Cell Cardiol. 2008;45(2):289–99. https://doi.org/10.1016/j.yjmcc.2008.04.013.

Stillitano F, Lonardo G, Giunti G, Del Lungo M, Coppini R, Spinelli V, Sartiani L, Poggesi C, Mugelli A, Cerbai E. Chronic atrial fibrillation alters the functional properties of If in the human atrium. J Cardiovasc Electrophysiol. 2013;24(12):1391–400. https://doi.org/10.1111/jce.12212.

Suenari K, Cheng CC, Chen YC, Lin YK, Nakano Y, Kihara Y, Chen SA, Chen YJ. Effects of ivabradine on the pulmonary vein electrical activity and modulation of pacemaker currents and calcium homeostasis. J Cardiovasc Electrophysiol. 2012;23(2):200–6. https://doi.org/10.1111/j.1540-8167.2011.02173.x.

Sulfi S, Timmis AD. Ivabradine – the first selective sinus node I(f) channel inhibitor in the treatment of stable angina. Int J Clin Pract. 2006;60(2):222–8. https://doi.org/10.1111/j.1742-1241.2006.00817.x.

Swedberg K, Komajda M, Bohm M, Borer JS, Ford I, Dubost-Brama A, Lerebours G, Tavazzi L, Investigators S. Ivabradine and outcomes in chronic heart failure (SHIFT): a randomised placebo-controlled study. Lancet. 2010;376(9744):875–85. https://doi.org/10.1016/S0140-6736(10)61198-1.

Thollon C, Bidouard JP, Cambarrat C, Lesage L, Reure H, Delescluse I, Vian J, Peglion JL, Vilaine JP. Stereospecific in vitro and in vivo effects of the new sinus node inhibitor (+)-S 16257. Eur J Pharmacol. 1997;339(1):43–51.

Ueda K, Nakamura K, Hayashi T, Inagaki N, Takahashi M, Arimura T, Morita H, Higashiuesato Y, Hirano Y, Yasunami M, Takishita S, Yamashina A, Ohe T, Sunamori M, Hiraoka M, Kimura A. Functional characterization of a trafficking-defective HCN4 mutation, D553N, associated with cardiac arrhythmia. J Biol Chem. 2004;279(26):27194–8. https://doi.org/10.1074/jbc.M311953200.

Ulens C, Tytgat J. Functional heteromerization of HCN1 and HCN2 pacemaker channels. J Biol Chem. 2001;276(9):6069–72. https://doi.org/10.1074/jbc.C000738200.

Vedantham V, Scheinman MM. Familial inappropriate sinus tachycardia: a new chapter in the story of HCN4 channelopathies. Eur Heart J. 2017;38(4):289–91. https://doi.org/10.1093/eurheartj/ehv635.

Vilaine JP, Bidouard JP, Lesage L, Reure H, Peglion JL. Anti-ischemic effects of ivabradine, a selective heart rate-reducing agent, in exercise-induced myocardial ischemia in pigs. J Cardiovasc Pharmacol. 2003;42(5):688–96.

Viscomi C, Altomare C, Bucchi A, Camatini E, Baruscotti M, Moroni A, DiFrancesco D. C terminus-mediated control of voltage and cAMP gating of hyperpolarization-activated cyclic nucleotide-gated channels. J Biol Chem. 2001;276(32):29930–4. https://doi.org/10.1074/jbc.M103971200.

Yamamoto M, Dobrzynski H, Tellez J, Niwa R, Billeter R, Honjo H, Kodama I, Boyett MR. Extended atrial conduction system characterised by the expression of the HCN4 channel and connexin45. Cardiovasc Res. 2006;72(2):271–81. https://doi.org/10.1016/j.cardiores.2006.07.026.

Yaniv Y, Ahmet I, Tsutsui K, Behar J, Moen JM, Okamoto Y, Guiriba TR, Liu J, Bychkov R, Lakatta EG. Deterioration of autonomic neuronal receptor signaling and mechanisms intrinsic to heart pacemaker cells contribute to age-associated alterations in heart rate variability in vivo. Aging Cell. 2016;15(4):716–24. https://doi.org/10.1111/acel.12483.

Ye W, Song Y, Huang Z, Zhang Y, Chen Y. Genetic regulation of sinoatrial node development and pacemaker program in the venous pole. J Cardiovasc Dev Dis. 2015;2(4):282–98. https://doi.org/10.3390/jcdd2040282.

Zhou J, Ding WG, Makiyama T, Miyamoto A, Matsumoto Y, Kimura H, Tarutani Y, Zhao J, Wu J, Zang WJ, Matsuura H, Horie M. A novel HCN4 mutation, G1097W, is associated with atrioventricular block. Circ J. 2014;78(4):938–42.

Dysregulation of Ionic Homeostasis: Relevance for Cardiac Arrhythmias

6

Claire Hopton, Luigi Venetucci, and Miriam Lettieri

Abstract

The action potential is formed by the interaction of various sarcolemmal ionic currents. These currents are produced by the flow of ions which is driven by ionic electrochemical gradient across the sarcolemma and is mediated by ion channels. Alterations in ion channel function or trans-sarcolemmal ionic electrochemical gradients lead to alteration in ionic current and action potential which can cause arrhythmias. Extracellular and intracellular concentration of some ions can modulate the gating of ion channels. Therefore, the maintenance of the correct intra- and extracellular ionic concentrations and trans-sarcolemmal ionic gradients (ionic homeostasis) is essential for the electrical function of the heart. Alterations in ionic homeostasis can lead to profound alterations in cardiac electrophysiology and arrhythmias. In this chapter will review how dysregulation of ionic homeostasis can lead to arrhythmias with a particular emphasis on channelopathies.

During each cardiac cycle electrical activation originates in the sinus node and propagates rapidly, first through the atria and then through the ventricles. At the level of each cardiac myocyte, this electrical activation is constituted by the onset of an action potential (AP). The AP initiates and modulates cardiac contraction through a process called excitation-contraction coupling. The AP is produced by sequential activation of various sarcolemmal ion channels that allow the flow of ions (ionic

C. Hopton · M. Lettieri
Division of Cardiovascular Sciences, University of Manchester, Manchester, UK
e-mail: claire.hopton@manchester.ac.uk; miriam.lettieri@postgrad.manchester.ac.uk

L. Venetucci (✉)
Division of Cardiovascular Sciences, University of Manchester, Manchester Heart Centre, Manchester Foundation Trust, Manchester, UK
e-mail: luigi.venetucci@manchester.ac.uk

D. Thomas, C. A. Remme (eds.), *Channelopathies in Heart Disease*, Cardiac and Vascular Biology, https://doi.org/10.1007/978-3-319-77812-9_6

currents), mainly sodium (Na^{+}), potassium (K^{+}), and calcium (Ca^{2+}), along their electrochemical gradient and change membrane potential. Dysfunction of various ion channels can cause alterations in the size, duration or shape of the AP which can all predispose to arrhythmias. To allow the flow of ions, it is also important that the correct electrochemical gradient between the extracellular space and the intracellular space is constantly maintained. This is mainly determined by concentration gradients. Alterations in these gradients will affect ionic currents and therefore will affect AP characteristics. In addition, intracellular and extracellular concentrations of certain ions (Na^{+}, K^{+}, and Ca^{2+}) directly modulate the gating of various ion channels, and this can also produce profound alterations in AP characteristics and cardiac electrophysiology. In this chapter we will review how the alterations in the homeostasis of Na^{+}, K^{+}, and Ca^{2+} alter cardiac electrophysiology and can lead to arrhythmias with a particular emphasis on channelopathies.

6.1 Calcium

Calcium (Ca^{2+}) exerts important action both in the intracellular and the extracellular space.

6.1.1 Intracellular Ca^{2+}

In cardiac myocytes Ca^{2+} plays a pivotal role in the excitation-contraction coupling process (Bers 2002). As mentioned above this is the process that translates the electrical activation (AP) into mechanical activation (cardiac contraction). This process is schematically illustrated in the next section (Fig. 6.1).

6.1.1.1 Excitation-Contraction Coupling

This process starts in each cardiac myocyte with the onset of an AP. The AP triggers a rapid and transient change in cytosolic Ca^{2+} levels which is also known as systolic Ca transient. Ca^{2+} binds to the myofilaments activating them and initiating contraction. The process that mediates the onset of the Ca^{2+} transient is called Ca^{2+}-induced Ca^{2+} release (CICR) because the influx of Ca^{2+} through the sarcolemmal L-type channels (which are activated by the AP) triggers the release of a much bigger quantity of Ca^{2+} from the main intracellular Ca^{2+} store, the sarcoplasmic reticulum (SR). The SR channel that modulates this release is called the ryanodine receptor (RyR). Following contraction myofilament inactivation and relaxation is achieved by a rapid decline of Ca^{2+} back to precontraction levels. There are three main mechanisms that modulate this decline in Ca^{2+}: (1) termination of SR Ca^{2+} release by inactivation of RyR, (2) reuptake of Ca^{2+} in the SR by the SR Ca^{2+} ATPase (SERCA), and (3) extrusion of Ca^{2+} through the Na^{+}/Ca^{2+} exchanger (NCX) that couples the influx of three Na^{+} ions with the efflux of one Ca^{2+} ion (resulting in a net inward current). At this stage it is important to remember that to maintain Ca^{2+}

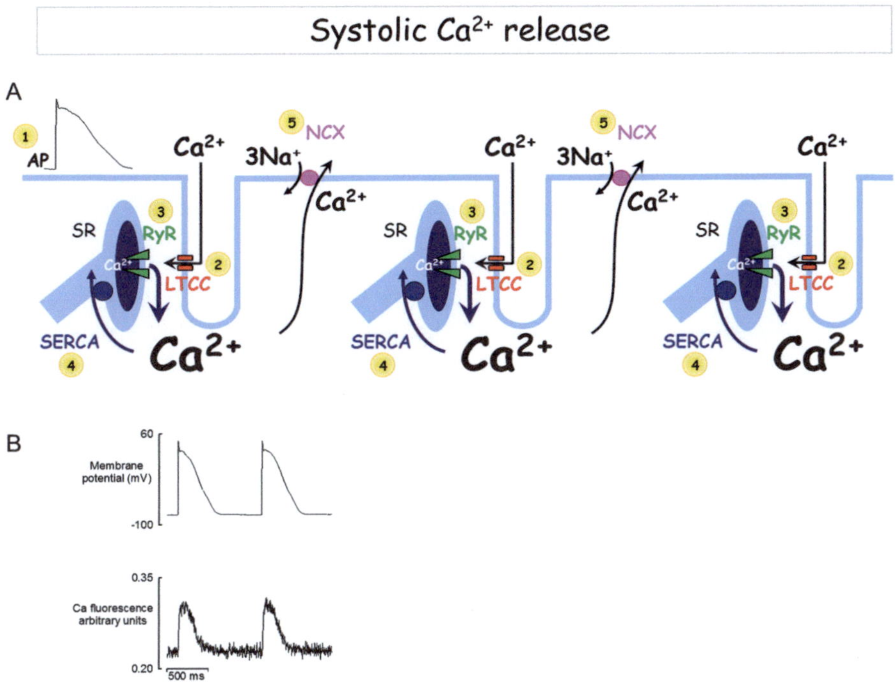

Fig. 6.1 (**a**) Schematic representation of mechanism responsible for the onset of Ca^{2+} transient in cardiac myocytes. The arrival of an AP (1) opens L-type Ca^{2+} channels (LTCC, in red) on T-tubules (2). Influx of a small amount of Ca^{2+} through LTCC activates ryanodine receptor (RyR, in red) on the SR and promotes synchronous release of Ca^{2+} (3). This Ca^{2+} activates the myofilaments and initiates contraction. Ca^{2+} decay and relaxation are produced by the activity of sarcoplasmic reticulum Ca^{2+} ATPase (SERCA, in blue) (4) and Na^{+}/Ca^{2+} exchanger (NCX, in magenta) (5). (**b**) Example of AP (top) and Ca^{2+} transient (bottom) from a ventricular cardiac myocyte

homeostasis at steady state, Ca^{2+} influx through the L-type Ca^{2+} channels has to equal Ca^{2+} efflux via NCX.

Structural Organization and Modulation of Ca^{2+} Release The amplitude of the Ca^{2+} transient determines the level of activation of the myofilaments and therefore is one of the main regulators of cardiac contractility. To understand how Ca^{2+} transient amplitude is modulated, we need to review the ultrastructural organization of the Ca^{2+} release machinery. To facilitate Ca^{2+} release, RyRs and L-type Ca^{2+} channels are organized into Ca^{2+} release units (CRUs) (Franzini-Armstrong et al. 1999). These CRUs are formed by close apposition of the SR to the sarcolemma. The majority of CRUs form along T-tubules (deep invagination of the sarcolemma) with a minority forming along the external membrane. This close apposition provides co-localization of L-type channels (situated on the sarcolemma) with RyR (situated on the SR membrane). Activation of a single CRU produces a Ca^{2+} spark. This is a localized, transient increase in Ca^{2+} concentration. The AP synchronizes the activation of CRUs, and the Ca^{2+} transient is formed by the summation of multiple Ca^{2+} sparks. Therefore, the amplitude the Ca^{2+} transient is determined by the number of CRUs activated (number of sparks) and the amount of Ca^{2+} released by each unit (spark amplitude). The number of units activated is mainly determined by the amplitude of the L-type Ca^{2+} current while the amount of Ca^{2+} released by each is controlled mainly by SR Ca^{2+} content. Cardiac myocytes have a complex system to regulate SR Ca^{2+} content. In the next section we will briefly review what controls SR Ca^{2+} content and how SR content modulates Ca^{2+} release.

Regulation of SR Ca^{2+} Content The amount of Ca^{2+} stored in the SR depends on the balance between Ca^{2+} released from the SR through RyR and Ca^{2+} reuptake via SERCA. Therefore, stimulation of SERCA via phosphorylation of its accessory protein phospholamban (PLN) will increase SR Ca^{2+} content (Hussain and Orchard 1997). On the contrary, stimulation of the ryanodine receptor (via calcium/calmodulin-dependent protein kinase II [CAMKII] phosphorylation or drugs) will produce a reduction in SR Ca^{2+} content (Trafford et al. 2000). SR Ca^{2+} content is also controlled by trans-sarcolemmal Ca^{2+} fluxes. Increases in Ca^{2+} influx through Ca^{2+} current and/or decrease in Ca^{2+} efflux via NCX will lead to an increase in the amount of Ca^{2+} available for reuptake into the SR and therefore an increase in SR Ca^{2+} content. It is important to remember that the Ca^{2+} transient modulates both Ca^{2+} influx by determining Ca^{2+} dependent inactivation of the L-type Ca^{2+} current and Ca^{2+} efflux by determining the size of the NCX-mediated current that controls efflux (Trafford et al. 1997). This ability is essential for the cell to maintain Ca^{2+} balance and prevent excessive Ca^{2+} accumulation. For example, if we reduce Ca^{2+} efflux via NCX (e.g., by increasing intracellular Na^{+}), this will in turn lead to an increase in SR Ca^{2+} content and increased Ca^{2+} transient amplitude which in turn will decrease Ca^{2+} influx via L-type Ca^{2+} current (Ca^{2+}-dependent inactivation) and increase Ca^{2+} efflux via NCX so that a new balance is reached where influx equals efflux and there is no further accumulation of Ca^{2+}.

How Does SR Ca^{2+} Modulate Transient Amplitude? Various studies have demonstrated that the relationship between SR Ca^{2+} content and Ca^{2+} transient amplitude is not linear but follows an exponential cubic pattern (Eisner 2014). This clearly suggests that SR Ca^{2+} content does not play a simple passive role by providing the concentration gradient for Ca^{2+} release but exerts an active modulatory role on Ca^{2+} release. A large body of evidence demonstrates that the RyR gating is modulated by SR Ca^{2+} concentration so that RyR opening increases with increases in SR Ca^{2+} concentration. Originally it was believed that calsequestrin, an accessory protein that binds to the SR on its luminal side, acted as the main Ca^{2+} sensor for RyR (Györke and Terentyev 2008). However, more recent evidence clearly demonstrates that in the absence of calsequestrin, the RyR is still able to sense Ca^{2+}. In addition, an elegant paper has identified the Ca^{2+}-sensing domain in RyR (Chen et al. 2014).

Physiological Modulators of the Ca^{2+} Transient In vivo there are two main physiological modulators of Ca^{2+} transient: beta adrenergic stimulation and heart rate.

Beta adrenergic stimulation Beta adrenergic stimulation is the main physiological modulator of Ca^{2+} transient amplitude and cardiac contractility. Normally adrenergic stimulation occurs during exercise and emotional stress. It increases heart rate and cardiac contractility. The effects on heart rate are due to direct stimulation of the sinus node, while the effects on cardiac contractility are due both to direct stimulation of myocytes and to the increase in heart rate. Direct stimulation of myocytes increases Ca^{2+} transient amplitude and contractility by three main mechanisms (Hussain and Orchard 1997; Curran et al. 2007):

1. It increases Ca^{2+} current amplitude via protein kinase A (PKA) phosphorylation of the L-type Ca^{2+} channel.
2. It stimulates SR Ca^{2+} reuptake activity via SERCA and increases SR Ca^{2+} content. This is achieved by PKA phosphorylation of PLN, an accessory protein of SERCA that modulates its activity.
3. It promotes CAMKII-dependent phosphorylation of RyR that increases its open probability.

Heart rate Increasing heart rate is associated with increased amplitude of the Ca^{2+} transient and increased contractility. This phenomenon is called positive force-frequency relationship, and it is observed mainly in larger species. Three main mechanisms mediate this phenomenon (Endoh 2004):

1. Increased intracellular Na^+ that reduces NCX-mediated Ca^{2+} efflux and increases SR Ca^{2+} content.
2. Frequency-dependent phosphorylation of PLN that increases SERCA activity and SR Ca^{2+} content. This is mediated by CAMKII.
3. Frequency-dependent phosphorylation of RyR that increases its opening, again mediated by CAMKII.

Pharmacological Modulation of Ca^{2+} Cycling Ca^{2+} transient and contractility can also be modulated pharmacologically. Digoxin is a widely available drug used in the treatment of heart failure and atrial fibrillation. It exerts powerful inotropic action by increasing Ca^{2+} transient amplitude. This effect is due to inhibition of the cardiac Na^+ pump which increases intracellular Na^+ levels. Increasing cytosolic Na^+ levels decreases Ca^{2+} efflux through the NCX (see section on Na^+ homeostasis) and increases SR Ca^{2+} content. Recent evidence also suggests that digoxin promotes CAMKII-dependent phosphorylation of RyR and increases its open probability.

6.1.1.2 Ca^{2+} Cycling and Arrhythmias

In various disease states, the SR can also release Ca^{2+} independently from an AP. This process is called a Ca^{2+} wave or spontaneous Ca^{2+} release. It is not synchronous like systolic Ca^{2+} release, it starts in one point of the cells and propagates along the cells as a wave of Ca^{2+}-induced Ca^{2+} release (Venetucci et al. 2008). As this wave of Ca^{2+} release propagates along the cell, it activates NCX which generates an inward current and produces either an early afterdepolarization

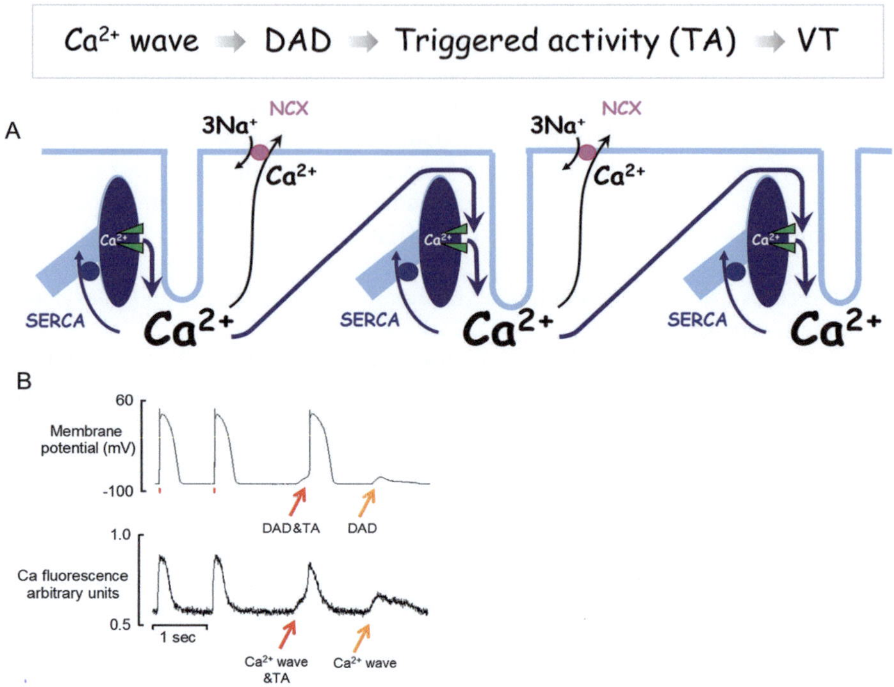

Fig. 6.2 (**a**) Schematic representation of mechanism responsible for the onset of Ca^{2+} waves and delayed afterdepolarizations (DADs). Under conditions that cause disproportionate increase in SR Ca^{2+} content, Ca^{2+} release can start spontaneously (independently of an AP) in any part of the cell and propagate along the cell as a wave of Ca^{2+} release. As this wave propagates along the cell, it activates NCX which generates an inward current which produces a DAD. The DAD can activate cellular Na+ channels and generate an AP. (**b**) Examples of Ca^{2+} wave and DAD followed by triggered activity (red arrow) and Ca^{2+} wave and DAD without triggered activity (orange arrow)

(EAD if it occurs during the AP) or a delayed afterdepolarization (DAD if it occurs after completion of the AP) (Fig. 6.2). Several studies have demonstrated that Ca^{2+} waves occur when SR Ca^{2+} concentration reaches a critical level also known as SR threshold. This threshold at which spontaneous Ca^{2+} release occurs can be altered by modulating the sensitivity of the RyR to activation by Ca^{2+} (both on the luminal and cytosolic sides) (Venetucci et al. 2007). Ca^{2+} waves are clinically relevant for arrhythmogenesis in a number of inherited arrhythmia syndromes.

Catecholaminergic Polymorphic Ventricular Tachycardia (CPVT) This is a genetic arrhythmia syndrome characterized by the onset of polymorphic ventricular tachycardia (VT) during exertion and emotional stress, a situation in which there is adrenergic stimulation of the heart (van der Werf and Wilde 2013). Polymorphic VT can degenerate into ventricular fibrillation (VF) and cause sudden death. The great majority of CPVT cases are caused by mutations of the gene that encodes for

RyR (*RYR2*) (Priori et al. 2001). A small proportion of cases have been linked with mutations in genes which encode for accessory proteins of RyR, such as calsequestrin, triadin, and calmodulin. Mutations in *RYR2* typically increase the RyR Ca^{2+} sensitivity and decrease the threshold for Ca^{2+} waves (Venetucci et al. 2012), this facilitates the onset of Ca^{2+} and DADs. Ca^{2+} waves DADs and arrhythmias occur mainly after adrenergic stimulation. Following adrenergic stimulation SR content will reach the threshold for Ca^{2+} waves. This occurs because adrenergic stimulation exerts dual action: (1) It increases SR Ca^{2+} content via stimulation of SERCA. (2) It decreases the threshold for Ca^{2+}waves by promoting CAMKII-mediated phosphorylation of RyR.

Long QT Syndrome (LQTS) The classical arrhythmia associated with long QT syndrome is torsades de pointes (TdP). The main arrhythmia mechanism responsible for TdP is re-entry. Often this re-entry is started by ventricular ectopic beats which are mainly caused by triggered activity due either to EADs or DADs. Studies performed on transgenic animal models of LQTS clearly show that in some forms of LQT there are significant alterations of intracellular Ca^{2+} handling that lead to the onset of both EADs and DADs. Terentyev et al. (2014) studied Ca^{2+} handling in a transgenic rabbit model of LQT-2 (caused by mutations of the K^+ channel gene *KCNH2*). They demonstrated that the transgenic hearts had high levels of RyR phosphorylation. This increases channel activity and predisposes to the development of Ca^{2+} waves, EADs, DADs, and arrhythmias. Work on a mouse model of LQT-3 (Fredj et al. 2006) (caused by mutations of the Na^+ channel gene SCN5A) showed that cells isolated from the transgenic heart had a high incidence of DADs and triggered activity especially at slow heart rates. This is likely to be secondary to higher levels of intracellular Na^+ consequent to enhanced late sodium current associated with the mutation that leads to reduced Ca^{2+} extrusion via NCX and Ca^{2+} overload. It is therefore possible that preventing the onset of DADs and EADs could prove to be a valuable therapeutic strategy. Beta blockers are the main treatment for both CPVT and LQTS; however there are patients who still develop arrhythmias despite full beta blockade. The discovery that abnormalities in Ca^{2+} handling play an important role in the genesis of arrhythmias in these conditions has generated a large interest in developing new strategies to target Ca^{2+} handling. In particular there has been great interest in RyR as a new therapeutic target, and several compounds have been developed. Recent studies have also suggested that some clinically available drugs such as flecainide (Watanabe et al. 2009) and carvedilol (Zhou et al. 2011) are RyR inhibitors. Flecainide is a powerful Na^+ channel blocker which has been shown to be effective in preventing arrhythmias in patients with CPVT, especially when added to beta blockers (van der Werf et al. 2011). There is still controversy on whether RyR blockade mediates some of these antiarrhythmic properties. Carvedilol is a beta blocker routinely used in the treatment of heart failure. It exists as a racemic mixture (having equal amounts of the R and S enantiomers), and it has been suggested that the R enantiomer is a RyR blocker but does not have any beta-blocking properties (Zhang et al. 2015). Studies on a mouse model of CPVT show that the R enantiomer is very effective in preventing

Table 6.1 Causes of hypocalcemia and hypercalcemia

Common causes of hypocalcemia
Vitamin D deficiency
Altered vitamin D metabolism (disease affecting the liver or kidney medications)
Pseudohypoparathyroidism
Hypomagnesemia, hypermagnesemia
Common causes of hypercalcemia
Hyperparathyroidism
Multiple myeloma
Cancer
Excessive vitamin D intake

arrhythmias following infusions with adrenaline and caffeine (caffeine is an activator of RyR used to induce arrhythmias in mice).

6.1.2 Disorders of Extracellular Ca^{2+}

Both hypocalcemia and hypercalcemia are frequently encountered in clinical practice. Approximately 50% of circulating Ca^{2+} is bound to albumin so Ca^{2+} levels must always be corrected for albumin levels.

6.1.2.1 Hypocalcemia

Causes of hypocalcemia are listed in Table 6.1. The main clinical manifestation of hypocalcemia is neuromuscular irritability that causes symptoms such as tetany, paresthesia, and numbness. The main ECG manifestation is QT prolongation with prominent prolongation of the ST segment. At the level of the cardiac action potential, hypocalcemia predominantly causes prolongation of the plateau (phase 2). A modelling study (Grandi et al. 2009) has suggested that this is mainly due to slowed inactivation of the L-type Ca^{2+} current because of reduction in Ca^{2+}-dependent inactivation. In the literature there are several reports of QT prolongation and TdP triggered by hypocalcemia. As hypocalcemia can prolong the QT interval, patients with LQTS should take particular care to avoid medications or conditions that can lower extracellular calcium (such as diuretics, aminoglycosides, and vitamin D deficiency).

6.1.2.2 Hypercalcemia

Hypercalcemia is observed mainly in the setting of hyperparathyroidism and malignancies (see Table 6.1). It can cause AP and QT shortening; however this does not tend to cause arrhythmias. The AP and QT shortening have been attributed to faster inactivation of Ca^{2+} current during the plateau of the AP and also reduction in the inward depolarizing NCX current. It is also important to remember that raising extracellular Ca^{2+} stabilizes resting membrane potential and increases the extent of depolarization needed to initiate an AP.

6.2 Sodium

Na^+ is the main extracellular cation and plays an essential role in the regulation of extracellular volume and plasma osmolarity. The concentration of Na^+ inside cardiac myocytes is significantly lower than in the extracellular space; however intracellular Na^+ plays an essential modulatory role in various processes (Murphy and Eisner 2006). Alterations in intracellular Na^+ concentration have been linked with the onset of arrhythmias while alterations in extracellular Na^+ concentrations are rarely associated with arrhythmias. In this section we will mainly focus on the modulation of intracellular Na^+ concentration and arrhythmias.

6.2.1 Intracellular Na^+ Concentrations

Intracellular Na^+ concentration is a key regulator of transmembrane flux of various ions and metabolites. In addition, recent evidence suggests that intracellular Na^+ levels modulate mitochondrial function. Cardiac myocytes have a complex system to regulate intracellular Na^+ concentration. To illustrate the mechanisms that control intracellular Na^+ concentration, it is useful to separately discuss Na^+ influx and Na^+ efflux pathways.

6.2.1.1 Na^+ Influx Pathways

In cardiac myocytes there are three main routes of Na^+ influx: voltage-dependent Na^+ channels, NCX, and Na/H exchanger.

Na^+ Channels The activation of voltage-dependent Na^+ channels is responsible for the upstroke of the AP. Most of the current is mediated by the cardiac isoform Nav1.5 which is encoded by the *SCN5A* gene. The gating of the channels is characterized by rapid succession of activation and inactivation (Remme 2013) so that the current almost completely inactivates within 5 ms of activation. During the plateau of the AP, a very small non-inactivating Na^+ current can still be present. This is called late Na^+ current and is mainly present in Purkinje cells and ventricular myocytes. The late Na^+ current modulates AP duration. Mutations of *SCN5A* that increase late Na^+ current are responsible for LQT-3. Na^+ channels represent the main route of Na^+ influx. When the heart rate increases, Na^+ influx through these channels increases, and this results in an increase in intracellular Na^+ concentration. Increasing late Na^+ current (which can be caused by ischemia, LQT-3 and heart failure) also produces greater Na^+ influx and higher intracellular Na^+ levels. Cardiac myocytes also express some neuronal Na^+ channels, and recent evidence suggests that these neuronal channels are mainly localized at the level of the CRUs where they play an important role in modulating the concentration of Na^+ and therefore Ca^{2+} in near RyRs (Radwański et al. 2015).

NCX As mentioned above, NCX represents the main route of Ca^{2+} efflux from the cells. It uses the energy provided by Na^+ influx to pump Ca^{2+} out of the cell. Three

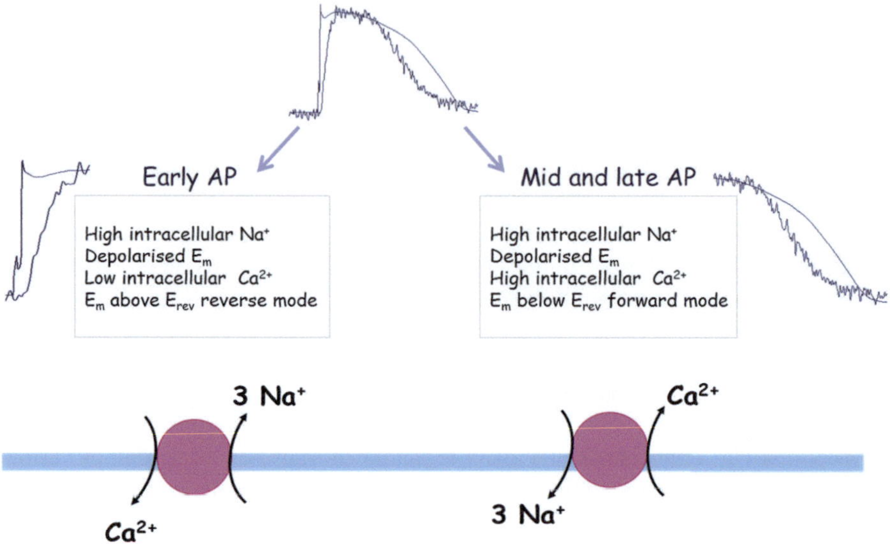

Fig. 6.3 Schematic representation of NCX function. NCX reversal potential changes with changes in intracellular Na^+ and Ca^{2+} concentration within myocytes. This together with the changes in membrane potential affects the direction of NCX function. In early AP (left), after upstroke of AP and before the onset of the Ca^{2+}, Na^+ levels are high and Ca^{2+} levels are low. These changes move reversal potential to more negative values; therefore the reversal potential is below membrane potential and the NCX function is in reverse mode (Na^+ out, Ca^{2+} in). During mid- and late AP (right), the increase in intracellular Ca^{2+} increases reversal potential which now is above membrane potential and NCX is in forward mode (Na^+ in Ca^{2+} out)

Na^+ (three positive charges) ions are exchanged for one Ca^{2+} ion (two positive charges); therefore NCX generates a current. The NCX can work in two directions: forward mode (Na^+ influx, Ca^{2+} efflux resulting in an inward current) and reverse mode (Na^+ efflux, Ca^{2+} influx resulting in an outward current). The direction of flux is determined by transmembrane Na^+ and Ca^{2+} concentration gradients and membrane potential (Fig. 6.3) (Matsuoka and Hilgemann 1992). In simple terms, forward mode will occur when membrane potential is below reversal potential, while reverse mode will occur when membrane potential is above reversal potential. An increase in intracellular Na^+ (which decreases transmembrane Na^+ gradient) will shift reversal potential toward more negative potentials, while increases in intracellular Ca^{2+} will shift reversal potential toward more positive potentials. Under normal conditions NCX works mainly in the forward mode. Pathological conditions that lead to increase in intracellular Na^+ and decrease in Ca^{2+} transient amplitude can lead to an increase in the reverse mode. One of the main abnormalities detected in many models of heart failure is a reduction in SERCA function and an increase in NCX expression. The net result of these changes is that myocytes become more reliant on NCX for Ca^{2+} removal. This promotes Na^+ influx and can contribute to the Na^+ overload that is typically observed in heart failure.

Na/H Exchanger This exchanger provides one of the main H^+ efflux mechanisms from cells; therefore it is involved in the modulation of intracellular pH. It uses the energy provided by the influx of Na^+ ions to pump H^+ out of the cell. The exchanger is electroneutral because the ratio between Na^+ ion influx and H efflux is one to one. The exchanger activity is stimulated by intracellular acidosis. The exchanger is very active during ischemia and reperfusion and in heart failure and hypertrophy. In these situations there is substantial Na^+ influx, and this contributes to the Na^+ overload observed in these conditions.

6.2.1.2 Na^+ Efflux Mechanisms

The main route of Na^+ extrusion from the cell is provided by the Na-K pump. This utilizes the energy provided by hydrolysis of one ATP molecule to pump three Na^+ ions outside the cell in exchange of two K^+ ions inside the cell. Therefore, the pump is electrogenic and generates an outward current. The pump is modulated by the accessory protein phospholemman (PLM) (Shattock 2009) that at baseline conditions reduces pump activity by decreasing the affinity to Na^+ ions. Adrenergic stimulation promotes phosphorylation of PLM, and this removes the inhibitory action of PLM on the pump and increases pump activity. Pump activity is also modulated by extracellular levels of K^+ (Eisner and Lederer 1979). The Na^+ pump is the main target of digoxin which has been used for the treatment of heart failure and atrial fibrillation. Several studies have detected downregulation of Na^+ pump in animal models of heart failure, and this contributes to Na^+ overload observed in heart failure.

6.2.2 Intracellular Na^+ Homeostasis and Channelopathies

Long QT As mentioned above LQT-3 is caused by mutations in *SCN5A,* the gene that encodes for the cardiac Na^+ channel. These mutations disrupt channel inactivation and produce persistent late Na^+ current that causes significant AP prolongation. Late Na^+ current (see Fig. 6.4 left) and AP prolongation tend to be more pronounced at slow heart rates, and this explains why these patients typically develop arrhythmias during sleep and at rest when the heart rate tends to be low. Recent evidence suggests that increased intracellular Na^+ levels predispose to the Ca^{2+} overload by modulating NCX function (Lindegger et al. 2009). This Ca^{2+} overload predisposes to the formation of Ca^{2+} waves that can cause both EADs and DADs which in turn can act as a trigger for arrhythmias.

CPVT Recent evidence suggests that modulation of intracellular Na^+ levels is a new potential antiarrhythmic strategy for CPVT (Sikkel et al. 2013). This originated from the observation that flecainide, a powerful Na^+ channel blocker, exerts significant antiarrhythmic actions in CPVT. It was originally suggested that this is due to RyR-blocking properties of flecainide. However, elegant work from the McLeod Group (Sikkel et al. 2013) clearly suggests that part of the action of flecainide is due to reduction of intracellular Na^+ levels which reduces RyR opening and prevents the

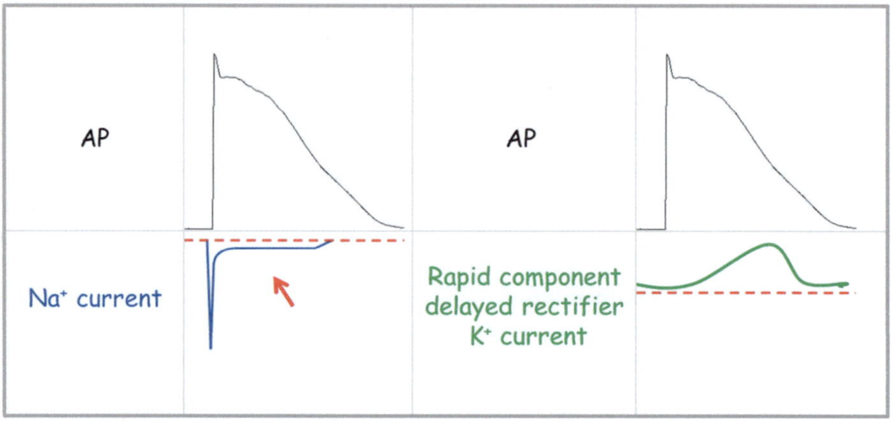

Fig. 6.4 Schematic representations of Na^+ current (in blue) with late component (red arrow) and delayed rectifier K^+ current (in green). *AP* action potential

onset of Ca^{2+} waves and DADs. Recent evidence from the Gyorke (Radwański et al. 2015, 2016) group has also suggested that Na^+ concentration in CRUs plays an important role in the modulation of RyR gating by modulating Ca^{2+} concentration via NCX. Reduction in Na^+ levels near RyR in CRUs also reduces Ca^{2+} concentration, and this reduces RyR opening and prevents Ca^{2+} waves. One important factor involved in the modulation of Na^+ levels in CRUs is the activity of neuronal Na^+ channels. Inhibition of this channel, either with low dose TTX or riluzole, prevents Ca^{2+} waves and DADs in cellular models and also arrhythmias in mice (Radwański et al. 2015).

6.3 Potassium

K^+ is mainly an intracellular cation. K^+ electrochemical gradient is responsible for setting the resting membrane potential because during diastole there is selective membrane conductance to K^+ ions through the inward rectifier channel. Extracellular K^+ concentration modulates the function of various ion channels, and changes in extracellular K^+ concentration produce profound alterations in AP characteristics. Alterations in extracellular K^+ concentration are the most common electrolyte disorder encountered in clinical practice and can lead to life-threatening arrhythmias. The normal extracellular K^+ concentration is between 3.5 and 5 mmol/l. In the presence of extracellular potassium concentration below 3.5 mmol/l, there is hypokalemia, while concentrations above 5 mmol/l constitute hyperkalemia. We will discuss separately the effects of hypokalemia and hyperkalemia on the cardiac AP and arrhythmias.

Table 6.2 Causes of hypokalemia and hyperkalemia

Common causes of hypokalemia
Renal loss (renal tubular acidosis, diuretics, hyperaldosteronism and magnesium deficiency)
Gastrointestinal loss (diarrhea and vomiting)
Intracellular shift (alkalosis and insulin)
Common causes of hyperkalemia
Increased intake
Reduced renal excretion (renal failure, hypoaldosteronism, ACE inhibitors and K-sparing diuretics)
Extracellular shift (acidosis and cell necrosis)

6.3.1 Hypokalemia

It is estimated that in the USA, 20% of patients admitted to a hospital are affected by hypokalemia. Causes of hypokalemia (see Table 6.2) can be divided into four broad categories: (1) reduced intake, (2) intracellular shift, (3) renal loss, and (4) extrarenal loss. Renal loss, especially due to diuretics (loop diuretics and thiazides), is the most common cause.

6.3.1.1 Electrophysiological Effects

The electrophysiological effects of hypokalemia are very pronounced and can be attributed to three main actions: (1) hyperpolarization of resting membrane potential, (2) inhibition of rapid delayed rectifier K^+ channel, and (3) inhibition of the sodium-potassium pump.

Hyperpolarization of Resting Membrane Potential The potassium electrochemical gradient is responsible for setting the membrane potential. Hypokalemia will increase this gradient and therefore move the equilibrium potential toward a more hyperpolarized level. However, it is interesting to note that the conductance of the inward rectifier channel, which is responsible for diastolic membrane potential (El-Sherif and Turitto 2011), is influenced by extracellular K^+ levels and channel conductance decreases when potassium levels decrease. Because of this mechanism, at K^+ levels below 3 mmol/l, the membrane potential hyperpolarization is less pronounced than what it would be predicted to be on the basis of the change in K^+ concentration. The important consequence of this hyperpolarization shift is an increased difference between resting membrane potential and activation potential for sodium channels (in the working myocardium) and calcium channels (in sinus and AV nodes) that result in decreased cardiac excitability.

Reduction of Rapid Delayed Rectifier Current (IKr) The rapid delayed rectifier current is one of the main cardiac repolarizing currents. As shown in Fig. 6.4 (right), it is inactive at the peak of the AP (phase 1), gradually activates during the plateau (phase 2), and reaches its peak amplitude during the final phase of repolarization (phase 3). Reduction in this current results in AP prolongation. Mutations of the gene (*KCNH2*) that encodes for the channel responsible for IKr (hERG channel) cause the LQT-2, while most of the antiarrhythmic drugs used in clinical practice have effects on IKr and produce AP and QT prolongation (sotalol, amiodarone, dofetilide, and flecainide). One of the key characteristics of IKr is that the amplitude of the current depends on extracellular K^+ concentration. Low extracellular K^+ level reduces the current amplitude (Sanguinetti et al. 1995). These changes are opposite to what it would be expected from the increase in the electrochemical gradient that hypokalemia produces. It has been suggested that extracellular K^+ ions interact with the channel pore and modulate collapse-of-pore type inactivation (Smith et al. 1996). The main consequence of this effect on IKr current is that hypokalemia can produce AP prolongation and QT prolongation on the ECG. This effect tends to be much more pronounced in the presence of QT-causing mutations or in conjunction with QT-prolonging drugs (Hancox et al. 2008).

Inhibition of Na-K Pump The Na-K pump maintains the correct concentrations of Na^+ and K^+ inside the cell. Using ATP energy it pumps two K^+ ions inside the cell and three Na^+ ions outside the cell (against their concentration gradients). Extracellular K^+ concentration is a key regulator of the pump activity (Aronsen et al. 2015). Lowering K^+ reduces pump activity. One of the key consequences of a reduction in the pump activity is an increase in intracellular Na^+. As we have seen in the previous section, Na^+ is a key modulator of the sarcolemmal NCX, which is the main system of Ca^{2+} removal from the cell. This will result in decreased Ca^{2+} efflux from the cell and an increase in the amount of Ca^{2+} stored in the SR and Ca^{2+} overload. This overload predisposes to Ca^{2+} waves which can cause EADs and DADs. Therefore, one of the consequences of hypokalemia is increased triggered activity.

6.3.1.2 Hypokalemia and Inherited Arrhythmia Syndromes

Hypokalemia and Long QT As we have seen in the previous section, hypokalemia can produce AP and QT prolongation. Therefore, it is no surprise that hypokalemia is particularly dangerous in patients with long QT syndrome, and it is one of the factors that can precipitate arrhythmias in these patients. This is very relevant in patients with LQT-2 which is caused by mutation of the gene that encodes for the hERG channel (*KCNH2*). The proarrhythmic effect of hypokalemia in these patients is not solely due to the interaction with the hERG channel. There is growing evidence that the modulation of the Na-K pump also plays an important role (Pezhouman et al. 2015). As we have mentioned above, inhibition of the Na^+ pump leads to Ca^{2+} overload and triggered activity. This triggered activity can initiate TdP. This is one of the mechanisms responsible for the initiation of arrhythmias especially in patients with LQT-2.

Hypokalemia and CPVT As we mentioned above, one of the key determinants of the onset of arrhythmias in CPVT is an increase in SR Ca^{2+} content that predisposes to Ca^{2+} waves, DADs, and triggered activity. Hypokalemia can precipitate arrhythmias by inhibition of the Na-K pump and stimulation of the reverse mode of the NCX, both of which cause Ca^{2+} overload which can trigger afterdepolarizations.

6.3.2 Hyperkalemia

Hyperkalemia is the second most common electrolyte disorder encountered in clinical practice. Potassium is mainly excreted through the kidney; therefore hyperkalemia is most commonly caused by reduced renal excretion. This can be due to renal failure, drugs (ACE inhibitors, spironolactone, and NSAIDs), and hypoaldosteronism. Hyperkalemia can also be caused by a shift of K^+ from the intracellular to the extracellular space. This can be caused by acidosis, insulin deficiency, or trauma with severe muscle necrosis (Table 6.2).

6.3.2.1 Electrophysiological Effects

Similar to hypokalemia, the electrophysiological effects of hyperkalemia are due to modulation of the resting membrane potential and modulation of hERG. At mild levels of hyperkalemia, the predominant effect is the increase in IKr current (despite decrease in driving force) which tends to shorten the AP duration and on the ECG causing tall-peaked T waves in the precordial leads. With more severe levels of hyperkalemia, the effect on the resting membrane potential becomes predominant. Increasing extracellular potassium lowers the potassium electrochemical gradient, and this leads to a shift in resting membrane potentials toward more depolarized values (Diercks et al. 2004). This shift causes partial inactivation of the cardiac Na^+ channels which will result in slower upstroke velocity of the AP. This is one of the main determinants of conduction velocity in the heart, and therefore hyperkalemia will result in impaired conduction in the heart with slowing of AV and intraventricular conduction. Interestingly, different areas of the heart have different sensitivities to hyperkalemia. The most sensitive chambers are the atria. In the presence of moderate hyperkalemia, the main ECG signs are prolongation of PR interval and flattening of P waves which reflect slowing of atrial conduction. As hyperkalemia worsens, the ventricles are affected and QRS widening occurs. In the presence of very severe hyperkalemia, the depolarizing shift of resting membrane potential causes total inactivation of Na^+ channels and the heart becomes non-excitable. This principle is utilized during cardioplegia for heart surgery. In this situation the heart is stopped by applying a solution containing high K^+ concentration.

6.3.2.2 Hyperkalemia and Brugada Syndrome

Brugada syndrome is a channelopathy characterized by a typical ECG pattern recorded in the right precordial leads and the onset of polymorphic ventricular tachycardia. The ECG pattern can be intermittent and can be elicited pharmacologically by the use of Na^+ channel blockers. In the literature there are several reports of the Brugada pattern caused by hyperkalemia (Postema et al. 2011). This is likely to be related to the effect of hyperkalemia on resting membrane potential and Na^+ channel inactivation. It is therefore of paramount importance to avoid hyperkalemia in patients with Brugada syndrome.

6.4 Conclusions

Ionic homeostasis is essential for the electrical and mechanical function of the heart. Complex systems control intracellular and extracellular ionic concentrations and trans-sarcolemmal concentration gradients. Dysregulation of ionic homeostasis can have profound effects on the electrical properties of the heart and result in arrhythmias. This is very important for channelopathies because dysregulation of ionic homeostasis can exacerbate the effects of the ion channel dysfunction that causes the channelopathy. Often normalization of ionic homeostasis can be an effective therapeutic option and should always be attempted.

Compliance with Ethical Standards

Funding Claire Hopton is funded by a BHF Clinical Research Training Fellowship. Miriam Lettieri is funded by a BHF Studentship.

Conflict of Interest Claire Hopton declares that she has no conflict of interest. Luigi Venetucci declares that he has no conflict of interest. Miriam Lettieri declares that she has no conflict of interest.

Ethical Approval This article does not contain any studies with human participants or animals performed by any of the authors.

References

Aronsen JM, Skogestad J, Lewalle A, et al. Hypokalaemia induces Ca(2+) overload and Ca(2+) waves in ventricular myocytes by reducing Na(+),K(+)-ATPase α2 activity. J Physiol. 2015;593 (6):1509–21.

Bers DM. Cardiac excitation-contraction coupling. Nature. 2002;415(6868):198–205.

Chen W, Wang R, Chen B, et al. The ryanodine receptor store-sensing gate controls Ca2+ waves and Ca2+-triggered arrhythmias. Nat Med. 2014;20(2):184–92.

Curran J, Hinton MJ, Ríos E, et al. Beta-adrenergic enhancement of sarcoplasmic reticulum calcium leak in cardiac myocytes is mediated by calcium/calmodulin-dependent protein kinase. Circ Res. 2007;100(3):391–8.

Diercks DB, Shumaik GM, Harrigan RA, et al. Electrocardiographic manifestations: electrolyte abnormalities. J Emerg Med. 2004;27(2):153–60.

Eisner D. Calcium in the heart: from physiology to disease. Exp Physiol. 2014;99(10):1273–82.

Eisner DA, Lederer WJ. Inotropic and arrhythmogenic effects of potassium-depleted solutions on mammalian cardiac muscle. J Physiol. 1979;294:255–77.

El-Sherif N, Turitto G. Electrolyte disorders and arrhythmogenesis. Cardiol J. 2011;18(3):233–45.

Endoh M. Force-frequency relationship in intact mammalian ventricular myocardium: physiological and pathophysiological relevance. Eur J Pharmacol. 2004;500(1-3):73–86.

Franzini-Armstrong C, Protasi F, Ramesh V. Shape, size, and distribution of Ca(2+) release units and couplons in skeletal and cardiac muscles. Biophys J. 1999;77:1528–39.

Fredj S, Lindegger N, Sampson KJ, et al. Altered Na+ channels promote pause-induced spontaneous diastolic activity in long QT syndrome type 3 myocytes. Circ Res. 2006;99(11):1225–32.

Grandi E, Pasqualini FS, Pes C, et al. Theoretical investigation of action potential duration dependence on extracellular Ca2+ in human cardiomyocytes. J Mol Cell Cardiol. 2009;46 (3):332–42.

Györke S, Terentyev D. Modulation of ryanodine receptor by luminal calcium and accessory proteins in health and cardiac disease. Cardiovasc Res. 2008;77(2):245–55.

Hancox JC, McPate MJ, El Harchi A, Zhang YH. The hERG potassium channel and hERG screening for drug-induced torsades de pointes. Pharmacol Ther. 2008;119:118–32.

Hussain M, Orchard CH. Sarcoplasmic reticulum Ca2+ content, L-type Ca2+ current and the Ca2+ transient in rat myocytes during β-adrenergic stimulation. J Physiol. 1997;505:385–402.

Lindegger N, Hagen BM, Marks AR, et al. Diastolic transient inward current in long QT syndrome type 3 is caused by Ca2+ overload and inhibited by ranolazine. J Mol Cell Cardiol. 2009;47 (2):326–34.

Matsuoka S, Hilgemann DW. Steady-state and dynamic properties of cardiac sodium-calcium exchange. Ion and voltage dependencies of the transport cycle. J Gen Physiol. 1992;100 (6):963–1001.

Murphy E, Eisner DA. Regulation of intracellular and mitochondrial sodium in health and disease. J Mol Cell Cardiol. 2006;41(5):782–4.

Pezhouman A, Singh N, Song Z. Molecular basis of hypokalemia-induced ventricular fibrillation. Circulation. 2015;132(16):1528–37.

Postema PG, Vlaar AP, DeVries JH, Tan HL. Familial Brugada syndrome uncovered by hyperkalaemic diabetic ketoacidosis. Europace. 2011;13(10):1509–10.

Priori SG, Napolitano C, Tiso N, et al. Mutations in the cardiac ryanodine receptor gene (hRyR2) underlie catecholaminergic polymorphic ventricular tachycardia. Circulation. 2001;103 (2):196–200.

Radwański PB, Brunello L, Veeraraghavan R, et al. Neuronal Na+ channel blockade suppresses arrhythmogenic diastolic Ca2+ release. Cardiovasc Res. 2015;106(1):143–52.

Radwański PB, Ho HT, Veeraraghavan R, et al. Neuronal Na+ channels are integral components of pro-arrhythmic Na+/Ca2+ signaling nanodomain that promotes cardiac arrhythmias during β-adrenergic stimulation. JACC Basic Transl Sci. 2016;1(4):251–66.

Remme CA. Cardiac sodium channelopathy associated with SCN5A mutations: electrophysiological, molecular and genetic aspects. J Physiol. 2013;591(17):4099–116.

Sanguinetti MC, Jiang C, Curran ME, Keating MT. A mechanistic link between an inherited and an acquired cardiac arrhythmia: HERG encodes the IKr potassium channel. Cell. 1995;81 (2):299–307.

Shattock MJ. Phospholemman: its role in normal cardiac physiology and potential as a druggable target in disease. Curr Opin Pharmacol. 2009;9(2):160–6.

Sikkel MB, Collins TP, Rowlands C, et al. Flecainide reduces Ca(2+) spark and wave frequency via inhibition of the sarcolemmal sodium current. Cardiovasc Res. 2013;98(2):286–96.

Smith PL, Baukrowitz T, Yellen G. The inward rectification mechanism of the HERG cardiac potassium channel. Nature. 1996;379:833–6.

Terentyev D, Rees CM, Li W, et al. Hyperphosphorylation of RyRs underlies triggered activity in transgenic rabbit model of LQT2 syndrome. Circ Res. 2014;115(11):919–28.

Trafford AW, Díaz ME, Negretti N, Eisner DA. Enhanced Ca2+ current and decreased Ca2+ efflux restore sarcoplasmic reticulum Ca2+ content after depletion. Circ Res. 1997;81(4):477–84.

Trafford AW, Díaz ME, Sibbring GC. Eisner DA Modulation of CICR has no maintained effect on systolic Ca2+: simultaneous measurements of sarcoplasmic reticulum and sarcolemmal Ca2+ fluxes in rat ventricular myocytes. J Physiol. 2000;522(Pt 2):259–70.

van der Werf C, Wilde AA. Catecholaminergic polymorphic ventricular tachycardia: from bench to bedside. Heart. 2013;99(7):497–504.

van der Werf C, Kannankeril PJ, Sacher F, et al. Flecainide therapy reduces exercise-induced ventricular arrhythmias in patients with catecholaminergic polymorphic ventricular tachycardia. J Am Coll Cardiol. 2011;57(22):2244–54.

Venetucci LA, Trafford AW, Eisner DA. Increasing ryanodine receptor open probability alone does not produce arrhythmogenic calcium waves: threshold sarcoplasmic reticulum calcium content is required. Circ Res. 2007;100(1):105–11.

Venetucci LA, Trafford AW, O'Neill SC, Eisner DA. The sarcoplasmic reticulum and arrhythmogenic calcium release. Cardiovasc Res. 2008;77(2):285–92.

Venetucci L, Denegri M, Napolitano C, Priori SG. Inherited calcium channelopathies in the pathophysiology of arrhythmias. Nat Rev Cardiol. 2012;9(10):561–75.

Watanabe H, Chopra N, Laver D, et al. Flecainide prevents catecholaminergic polymorphic ventricular tachycardia in mice and humans. Nat Med. 2009;15(4):380–3.

Zhang J, Zhou Q, Smith CD, et al. Non-β-blocking R-carvedilol enantiomer suppresses Ca2+ waves and stress-induced ventricular tachyarrhythmia without lowering heart rate or blood pressure. Biochem J. 2015;470(2):233–42.

Zhou Q, Xiao J, Jiang D, et al. Carvedilol and its new analogs suppress arrhythmogenic store overload-induced Ca2+ release. Nat Med. 2011;17(8):1003–9.

Part II

Cardiac Channelopathies: Clinical and Genetic Findings

Long and Short QT Syndromes

7

Lia Crotti, Maria-Christina Kotta, and Silvia Castelletti

Abstract

The long and short QT syndromes are genetically transmitted arrhythmogenic diseases characterized by an abnormal QTc on the basal ECG and by an increased risk of life-threatening arrhythmias. While in the long QT syndrome well-established diagnostic criteria are available as well as effective treatments, in the short QT syndrome, much less is known in terms of diagnosis, risk stratification and pharmacological treatment. In this chapter we discuss for each syndrome current knowledge on their genetic basis, clinical presentation, diagnosis, risk stratification and therapy. Furthermore, multisystem disorders associated with a prolongation of the QT, such as the Jervell and Lange-Nielsen syndrome, the Timothy syndrome, the ankyrin-B syndrome and the Andersen-Tawil syndrome, are described. Finally, specific subtypes of the long QT syndrome, characterized by high malignancy and frequent failure of available therapies, such as calmodulin-related LQTS and the triadin knockout syndrome, are also reviewed.

L. Crotti (✉)
IRCCS Istituto Auxologico Italiano, Center for Cardiac Arrhythmias of Genetic Origin and Laboratory of Cardiovascular Genetics, Milan, Italy

Department of Cardiovascular, Neural and Metabolic Sciences, San Luca Hospital, IRCCS Istituto Auxologico Italiano, Milan, Italy

Department of Medicine and Surgery, University of Milano-Bicocca, Milan, Italy
e-mail: l.crotti@auxologico.it

M.-C. Kotta · S. Castelletti
IRCCS Istituto Auxologico Italiano, Center for Cardiac Arrhythmias of Genetic Origin and Laboratory of Cardiovascular Genetics, Milan, Italy
e-mail: m.kotta@auxologico.it; s.castelletti@auxologico.it

D. Thomas, C. A. Remme (eds.), *Channelopathies in Heart Disease*, Cardiac and Vascular Biology, https://doi.org/10.1007/978-3-319-77812-9_7

7.1 Long QT Syndrome

7.1.1 Definition

The long QT syndrome (LQTS) is an inherited cardiac channelopathy characterized by QT interval prolongation on the electrocardiogram (ECG) (Merri et al. 1989) and by an increased risk of life-threatening arrhythmias in patients with a structurally normal heart (Vincent and Abildskov 1974; Schwartz et al. 1975). The typical life-threatening arrhythmia is a stress-induced torsade de pointes (TdP) that can be self-limited or can degenerate into ventricular fibrillation (VF) leading to cardiac arrest (CA) or sudden cardiac death (SCD).

7.1.2 Historical Perspective

The disease was formally first reported by Anton Jervell and Fred Lange-Nielsen in 1957 (Jervell and Lange-Nielsen 1957). They described a family in which four deaf-mute children all suffered attacks of fainting and presented a markedly prolonged QT interval, while three of them died suddenly (Jervell and Lange-Nielsen 1957). After the exclusion of known causes of QT prolongation and heart abnormalities at autopsy of one of the children, Jervell and Nielsen concluded that the clinical and electrocardiographic features were consistent with the same heart disease, the nature of which was to them obscure (Jervell and Lange-Nielsen 1957). Six years later, Cesarino Romano and co-workers described a similar condition without hearing loss in a paediatric patient (Romano et al. 1963), but it was 1 year later that Owen Connor Ward reported on this entity as a new familial condition (Ward 1964). In the following decade, more than 200 cases were published in the literature, collectively reported by Schwartz et al. (1975).

The strong association between the syncopal episodes and physical or emotional stress pointed to an autonomic system imbalance in the genesis of the arrhythmias and led to the use of beta blockers as first-line therapy. The first support of the sympathetic imbalance theory came in 1966 by Yanotwitz et al., who showed a QT prolongation after either right stellectomy or left ganglion stimulation in anaesthetized dogs (Yanowitz et al. 1966). Subsequently, this was further supported by the observation that several LQTS patients have reproducible episodes of T-wave alternation upon either unilateral stimulation of the left stellate ganglion or bilateral stimulation of both stellate ganglia but with a greater intensity applied to the left (Schwartz and Malliani 1975). These observations provided the rationale to perform left cardiac sympathetic denervation to prevent arrhythmic events in LQTS patients (Schwartz and Malliani 1975; Moss and McDonald 1971).

In the 1980s, it was suggested that some patients might be affected by LQTS and nonetheless have a normal QT interval on the ECG (Schwartz 1983; Moss et al. 1985). Based on that, the possibility of an unknown intracardiac abnormality as underlying cause of the disease was proposed, which would initiate VF when coupled with a sudden increase in sympathetic activity (Schwartz 1985). The actual hypothesis that LQTS may be underlined by a cardiac repolarization abnormality due to a genetic defect impacting on the outward K^+ current was first presented by isolated reports in the mid to late 1980s (Schwartz 1986; Attwell and Lee 1988). However, it was only in 1995 that Curran et al. (1995) and Wang et al. (1995) confirmed this hypothesis providing evidence that mutations in the genes encoding the cardiac ion channels conducting the outward rapid delayed rectifier current I_{Kr} and the fast inward Na^+ current I_{Na} are responsible for LQTS.

7.1.3 Molecular Genetic Basis

The ECG reflects on the body surface level the sequence of electrical activities that, on the cellular level, orchestrate the cycle of cardiac excitation and rest. Excitation is achieved through changes in cardiomyocytes' transmembrane potential from a resting potential to an action potential and back. The duration and shape of the action potential are defined by the interplay of depolarizing inward and repolarizing outward electric currents. The inward and outward flow of ions occurs through ion channels, macromolecular protein complexes of the cell membrane. Ion channels, in response to changes in transmembrane potential, alter their stereochemical structure in a way so as to serve a particular electrochemical gradient. These stereochemical changes, though small, lead to the opening of a gate from which, in fractions of a second, millions of ions enter or leave the cell resulting in an electric current, I, of some picoamperes (Grant 2009).

The cardiac action potential progresses in five phases: phase 0 of rapid depolarization, phase 1 of initial repolarization, phase 2 of plateau, phase 3 of rapid repolarization and phase 4 that represents the resting potential phase. In ventricular myocytes, depolarization is mainly orchestrated by the fast inward Na^+ current I_{Na} and the inward long-lasting Ca^{+2} current $I_{Ca\text{-}L}$, while the main repolarizing K^+ currents are the transient outward current I_{to}, the outward slow delayed rectifier current I_{Ks}, the outward rapid delayed rectifier current I_{Kr} and the inward rectifier current I_{K1} (Grant 2009; Conrath and Opthof 2006).

With current knowledge in hand, it seems evident that even minor alterations of these ionic currents can affect substantially the duration of the ventricular action potential and lead to anomalous ventricular repolarization and, hence, arrhythmia. In fact, QT interval prolongation, that is a delay of ventricular repolarization, actually expresses the prolongation of the action potential duration in ventricular myocytes. Today we know that LQTS is an inherited cardiac arrhythmia caused mainly by autosomal dominant heterozygous mutations in the genes encoding cardiac ion channels, or their accessory/interacting proteins, and belongs to the wider disease family of cardiac channelopathies. The most common forms are related to mutations in genes encoding the ion channels conducting I_{Kr}, I_{Ks} and I_{Na}. Depending on the gene involved, LQTS is divided into subtypes 1–17, which are denoted as LQT1, LQT2 and so forth (Schwartz et al. 2012; Crotti et al. 2008; Schwartz and Crotti 2017). It is worth noting that although LQT4 and LQT7 were originally described and considered as true subtypes of LQTS, they have proven to represent complex clinical entities in which QT prolongation may be moderate, or overestimated in initial descriptions, and for which declassification as LQTS has been suggested

(Schwartz et al. 2012; Zhang et al. 2005). These two clinical entities are discussed separately further below.

7.1.3.1 The Major LQTS Genes: K^+- and Na^+-Associated Disease

The main LQTS subtypes are those of LQT1, LQT2, LQT3, LQT5 and LQT6. LQT1 is associated with mutations in the *KCNQ1* gene encoding the α subunit of the cardiac K^+ channel Kv7.1 (Wang et al. 1996a), while LQT5 is associated with mutations in the *KCNE1* gene, encoding the channel's β subunit mink (Splawski et al. 1997b). These two ion channel proteins co-assemble and conduct the I_{Ks} current (Barhanin et al. 1996; Sanguinetti et al. 1996b). Accordingly, LQT2 and LQT6 are associated with mutations in the *KCNH2* and *KCNE2* genes that encode for the α (Kv11.1) and β (MiRP1) subunits, respectively, of the cardiac K^+ channel conducting I_{Kr} (Curran et al. 1995; Abbott et al. 1999). Mutations in the *SCN5A* gene encoding the α subunit of the cardiac Na^+ channel (Nav1.5), conducting I_{Na}, constitute the LQT3 subtype of the syndrome (Wang et al. 1995) (Table 7.1).

Mutation Types

Mutations responsible for LQTS are mainly heterozygous point mutations following an autosomal dominant inheritance pattern. However, there has been a significant number of reports in the literature of compound heterozygous (Larsen et al. 1999; Westenskow et al. 2004) as well as homozygous mutation carriers (Lupoglazoff et al. 2001; Piippo et al. 2000) that present with a malignant phenotype and prognosis. An identical mutation in homozygous state produces, as expected, a severe phenotype (Lupoglazoff et al. 2001; Piippo et al. 2000; Johnson Jr et al. 2003) and has also been indicated as a cause of stillbirth (Hoorntje et al. 1999). Most commonly identified are single-base substitutions leading to missense or nonsense amino acid changes but also in frame deletions and insertions, frameshifts as well as splicing mutations (Zhang et al. 2004; Tester et al. 2005; Crotti et al. 2009a). Copy number variants have also been described (Koopmann et al. 2006; Eddy et al. 2008), but they appear to be an infrequent cause of the syndrome (Barc et al. 2011).

Mutation Effects

In the LQT1 and LQT2 forms, mutations confer a loss-of-function effect, often by interfering with the ability of the α subunits of the K^+ channel to assemble into a tetramer. This defect results in a ~50% reduction of available functional channels with a concomitant reduction in the respective ionic current and is defined as happloinsufficiency (Keating and Sanguinetti 2001). Another loss-of-function mechanism relates to mutations that allow the mutant and wild-type subunits to co-assemble but then either modify the tetrameric channel's gating or permeability properties or lead eventually to early degradation of the protein complex. In any case, this is defined as dominant-negative suppression and is a very common mechanism in LQTS (Sanguinetti et al. 1996a; Bianchi et al. 1999, 2000). An example of early degradation is that of mutations in the N-terminal intracellular part of the Kv7.1 channel (LQT1) that contains highly conserved signal sequences for channel targeting to the cell membrane. Mutant and wild-type α subunit heterotetramers do not pass the endoplasmic reticulum's quality control and are

Table 7.1 Genes associated to long and short QT syndromes

Subtype	Gene	Chromosome	Protein (subunit[a])	Effect	Frequency (%)
The long QT syndrome genes					
LQT1	***KCNQ1***	11p15.5	K_v7.1 (α)	$\downarrow I_{Ks}$	40–45
LQT2	*KCNH2*	7q35-36	K_v11.1 (α)	$\downarrow I_{Kr}$	30–45
LQT3	*SCN5A*	3p21-23	Na_v1.5 (α)	$\uparrow I_{Na}$	5–10
LQT4	***ANK2***	4q25-27	Ankyrin B	$\downarrow I_{Na\text{-}Ca}$	<1
LQT5	***KCNE1***	21q22.1-22.2	MinK (β)	$\downarrow I_{Ks}$	1–2
LQT6	*KCNE2*	21q22.1-22.2	MiRP1 (β)	$\downarrow I_{Kr}$	0.5–1.5
LQT7	***KCNJ2***	17q23.1-24.2	Kir2.1 (α)	$\downarrow I_{K1}$	<1
LQT8	***CACNA1C***	12p13.3	Ca_v1.2 (α_{1c})	$\uparrow I_{Ca\text{-}L}$	<1
LQT9	*CAV3*	3p25	Caveolin-3	$\uparrow I_{Na}$	<1
LQT10	*SCN4B*	11q23.3	Na_v1.5 (β_4)	$\uparrow I_{Na}$	<1
LQT11	*AKAP9*	7q21-22	Yotiao	$\downarrow I_{Ks}$	<1
LQT12	*SNTA1*	20q11.2	α1-Syntrophin	$\uparrow I_{Na}$	<1
LQT13	*KCNE3*	11q13-14	MiRP2 (β)	$\downarrow I_{Ks}$	<1
LQT14	*CALM1*	14q32.11	Calmodulin 1	$\uparrow I_{Ca\text{-}L}$	unknown
LQT15	*CALM2*	2p21	Calmodulin 2	$\uparrow I_{Ca\text{-}L}$	unknown
LQT16	*CALM3*	19q13.32	Calmodulin 3	$\uparrow I_{Ca\text{-}L}$	unknown
LQT17	*TRDN*	6q22.31	Triadin	$\uparrow I_{Ca\text{-}L}$	unknown
The short QT syndrome genes					
SQT1	*KCNH2*	7q35-36	K_v11.1 (α)	$\uparrow I_{Kr}$	unknown
SQT2	*KCNQ1*	11p15.5	K_v7.1 (α)	$\uparrow I_{Ks}$	unknown
SQT3	*KCNJ2*	17q23.1-24.2	Kir2.1 (α)	$\uparrow I_{K1}$	unknown

LQTS genes with additional and/or extracardiac phenotypes are highlighted in bold
↑ denotes increase, and ↓ denotes decrease of the ionic current that is affected in case of a mutation in the respective gene
[a]Main or accessory subunit of the ion channel, where applicable

therefore not allowed to reach the membrane surface (Dahimène et al. 2006). Another example is the case of wild-type α and mutant β subunit channel complexes that are also being retained in the endoplasmic reticulum (Harmer and Tinker 2007). Trafficking deficiency and endoplasmic reticulum retainment are the most frequent mechanism in LQT2 (Delisle et al. 2004).

Many K^+ channel mutations (especially in the S6 transmembrane segment and in the N-terminus) have been shown to alter the biophysical properties of the α subunits through shifting of either their inactivation voltage dependence threshold to more negative values or their activation voltage dependence threshold to more positive values, thereby increasing inactivation or decreasing activation duration, respectively (Chouabe et al. 1997). Mutations in the C-terminal part of the Kv7.1 (LQT1) channel, that contains calmodulin-binding, phosphorylation, trafficking and assembly sequence motifs, have also been described that interfere with both proper channel kinetics and correct folding, trafficking and assembly (Haitin and Attali 2008; Shamgar et al. 2006). In particular, mutations that modify crucial phosphorylation sequence motifs, such as the C-terminal leucine zipper, may do even more harm. Because the sympathetic nervous system exerts its control on heart rate through β-adrenergic receptor-activated Kv7.1 channel phosphorylation, that normally increases I_{Ks} and decreases action potential duration in a rate-dependent manner, these mutations may in fact partially disconnect the channel from neural control (Marx et al. 2002). Particular mutations in the channel's β subunit minK (LQT5) have also been described that affect the phosphorylated Kv7.1 channel's further response to these neural signals (Kurokawa et al. 2003).

In the case of Na^+ channel mutations (*SCN5A*, LQT3), a gain-of-function effect is conferred through either incomplete, destabilized or delayed inactivation of the Na^+ channel. In any case, the result is the presence of late I_{Na} depolarizing currents during the plateau phase of the action potential that, although of low magnitude, prolong its duration (Bennett et al. 1995; Wang et al. 1996b). Mutations in crucial linker sequences, such as the one linking DIII with DIV that contains the inactivation IFM motif (isoleucine-phenylalanine-methionine) where phenylalanine moves and physically occludes the passage towards the cytoplasm, lead to inactivation inhibition (Stuhmer et al. 1989), while others, such as the ΔKPQ, may result in persistent I_{Na} through more than one defective biophysical mechanisms (Dumaine et al. 1996).

LQTS Genetic Testing

Genetic testing in LQTS should be performed whenever there is a strong or reasonable clinical suspicion in an index patient, as well as in their family members upon identification of the disease-causing mutation in the proband (class I recommendation) (Ackerman et al. 2011). Genetic testing of the proband is always performed through comprehensive mutation scanning of the LQTS genes since distinct genetic mutations, often characterized as 'private', are found within single families. In a study by Tester et al. (2005), 79% of the total 211 mutations that were detected in an LQTS patient cohort were unique, and, thus far, hundreds have been described dispersed throughout the coding sequence or within the intron-exon boundaries of the genes.

Between 2000 and 2006, three large studies provided crucial information regarding the frequency of what we now consider as main subtypes and the respective yield of genetic testing in patients with a definite LQTS clinical diagnosis (Splawski et al. 2000; Napolitano et al. 2005; Tester et al. 2006). It then became evident that, firstly, the LQT1 and LQT2 forms represent the largest part of the syndrome's genetic substrate and, secondly, that genetic testing of the five major LQTS genes (*KCNQ1*, *KCNH2*, *SCN5A*, *KCNE1* and *KCNE2*) yields a positive result in approximately 70% of clinically positive cases. For a syndrome as genetically heterogeneous as LQTS, these results have had a significant impact, underlying the fact that the majority of patients can be successfully genotyped through routine mutation scanning of these five genes. This has made feasible the transition of LQTS genetic testing from the research to the diagnostic setting in recent years, and with the advent of more robust genetic technologies, the respective yield of genetic testing may now reach 75–80% among patients with a definite clinical diagnosis.

7.1.3.2 The Minor LQTS Genes: K^+- and Na^+-Associated Disease

The LQT9 to LQT13 forms of the syndrome are rare and thus far only preliminary reports are available. LQT9 is associated with mutations in the *CAV3* gene that encodes for caveolin-3, a protein involved in the compartmentalization and regulation of various ion channels (Vatta et al. 2006). In the myocardium, it has been shown to co-localize with the Na^+ channel in the cell membrane, creating a macromolecular complex that regulates further signal transduction. Functional studies of the *CAV3* mutations originally identified showed that they exert a functional effect through increase of I_{Na} (Vatta et al. 2006).

An increase of the I_{Na} current is the final common mechanism also among LQT10 and LQT12, that relate to mutations in the *SCN4B* gene, encoding the β4 subunit of the Na^+ channel, and in the *SNTA1* gene, encoding α1-syntrophin, a natural interacting partner of the Na^+ channel's α subunit, respectively (Medeiros-Domingo et al. 2007; Ueda et al. 2008). LQT11 is linked to mutations in the *AKAP9* gene encoding A-kinase anchor protein 9 that binds to the endoplasmic C-terminal tail of the Kv7.1 channel (LQT1) and regulates its protein kinase A-dependent phosphorylation (Chen et al. 2007). The wild-type Kv7.1-mutant AKAP-9 complex has been shown to result in decreased channel phosphorylation and reduction of I_{Ks} (Chen et al. 2007). Finally, LQT13 involves mutations in the *KCNJ5* gene encoding the inwardly rectifying K^+ channel subunit Kir3.4 (Yang et al. 2010) (Table 7.1).

7.1.3.3 The Ca^{2+}-Associated LQTS Genes

LQT8: Timothy Syndrome and Beyond

The subtype LQT8 was originally described by Marks et al. as an entity that combined significant QT interval prolongation, syndactyly and a high risk of SCD (Marks et al. 1995a, b). Almost 10 years later, Splawski et al. (2004, 2005) reported on Timothy syndrome, a malignant LQTS variant form with additional extracardiac findings that may range from syndactyly to autism. Specific mutations arising de novo in the *CACNA1C* gene, which encodes the α1c subunit of the Ca^{2+} channel Cav1.2, were described (Splawski et al. 2004, 2005; Gillis et al. 2012); however, parental mosaicism has been later demonstrated to incorrectly exclude parental transmission in some cases (Etheridge et al. 2011). Mutations affect channel inactivation, thereby producing sustained $I_{Ca\text{-}L}$ currents during the plateau phase of the action potential.

Although initially LQT8 had been considered to deviate from classical LQTS and it was more of a hallmark of Timothy syndrome, it has recently been shown that mutations of the Ca^{2+} channel may underlie 'pure' LQTS without any features of Timothy syndrome (Boczek et al. 2013) or even cases of idiopathic VF (IVF) (Leinonen et al. 2018). It has been suggested that mutations in particular 'hot-spot' regions within the *CACNA1C* gene may result in a Timothy syndrome phenotype, instead of an LQTS (Landstrom et al. 2016). Timothy syndrome's clinical features are discussed in more detail in a later section.

LQT14–LQT16: Calmodulinopathy

In 2013, we described a variant of LQTS with otherwise typical clinical features of the syndrome but with quite severe clinical manifestations occurring very early in life (Crotti et al. 2013a). Infants born to healthy parents, with structurally normal hearts and presenting with extreme QT interval prolongation (QT > 600 ms), intermittent 2:1 atrioventricular block and T-wave alternans, suffered recurrent cardiac arrest episodes. Genetic testing of the major LQTS genes was negative, and whole-exome sequencing was employed to identify the underlying genetic cause. This approach led to the identification of de novo mutations in the *CALM1* and *CALM2* genes encoding calmodulin that were further ascertained in an unrelated cohort of genotype-negative LQTS patients (Fig. 7.1). Mutations in the *CALM3* gene were later also identified in similar clinical cases (Reed et al. 2015; Boczek et al. 2016).

Calmodulin is a Ca^{2+}-binding signal transducer messenger protein that greatly influences the activity of ion channels, among other targets. In particular, the cardiac K^+ channel Kv7.1 (*KCNQ1*-LQT1), the Ca^{2+} channel Cav1.2 (*CACNA1C*-LQT8) as well as several types of Na^+ channels are known to be regulated in a Ca^{2+}-dependent manner (Shamgar et al. 2006; Zühlke et al. 1999; Kim et al. 2004). In the context of LQTS, mutant calmodulin has been demonstrated to have a reduced affinity for Ca^{2+} (Crotti et al. 2013a; Makita et al. 2014; Pipilas et al. 2016) and to impair Ca^{2+}-dependent inactivation of the L-type Ca^{2+} channel Cav1.2 (Pipilas et al. 2016; Yin et al. 2014), with a strong dominant-negative effect, as demonstrated in patient-specific induced pluripotent stem cell-derived cardiomyocytes (iPSC-CMs) in which the physiological balance of mutant and wild-type proteins is respected (Rocchetti et al. 2017). Indeed, three different genes, in three different chromosomes, are known to encode the same calmodulin protein (Crotti et al. 2013a). Therefore, despite the presence of five alleles encoding a wild-type protein, the only defective allele is able to significantly impair inactivation of the L-type Ca^{2+} channel, resulting in increased inward $I_{Ca\text{-}L}$ current during the plateau phase and prolongation of repolarization, which has been shown to be adrenergic stimulation-sensitive (Rocchetti et al. 2017).

After the original description of this new clinical entity of severe LQTS in infants (Crotti et al. 2013a), it was shown that calmodulin mutations may be the underlying genetic cause in cases of less severe, but still malignant, LQTS (Boczek et al. 2016; Makita et al. 2014), thereby further justifying the assignment of *CALM1*, *CALM2* and *CALM3* mutations as true LQT14, LQT15 and LQT16 subtypes. Furthermore, calmodulin mutations have also been associated with cases of catecholaminergic polymorphic ventricular tachycardia (CPVT) (Nyegaard et al. 2012) and IVF (Marsman et al. 2014), again with an early occurrence of malignant symptoms. To better understand the phenotypic manifestations and possible therapies for patients affected by what has been come to know as 'calmodulinopathy' (George 2015; Crotti et al. 2016c), we have recently established an international calmodulinopathy patient registry aiming at elucidating the particular clinical and genetic features of this severe disease entity (Crotti et al. 2016c).

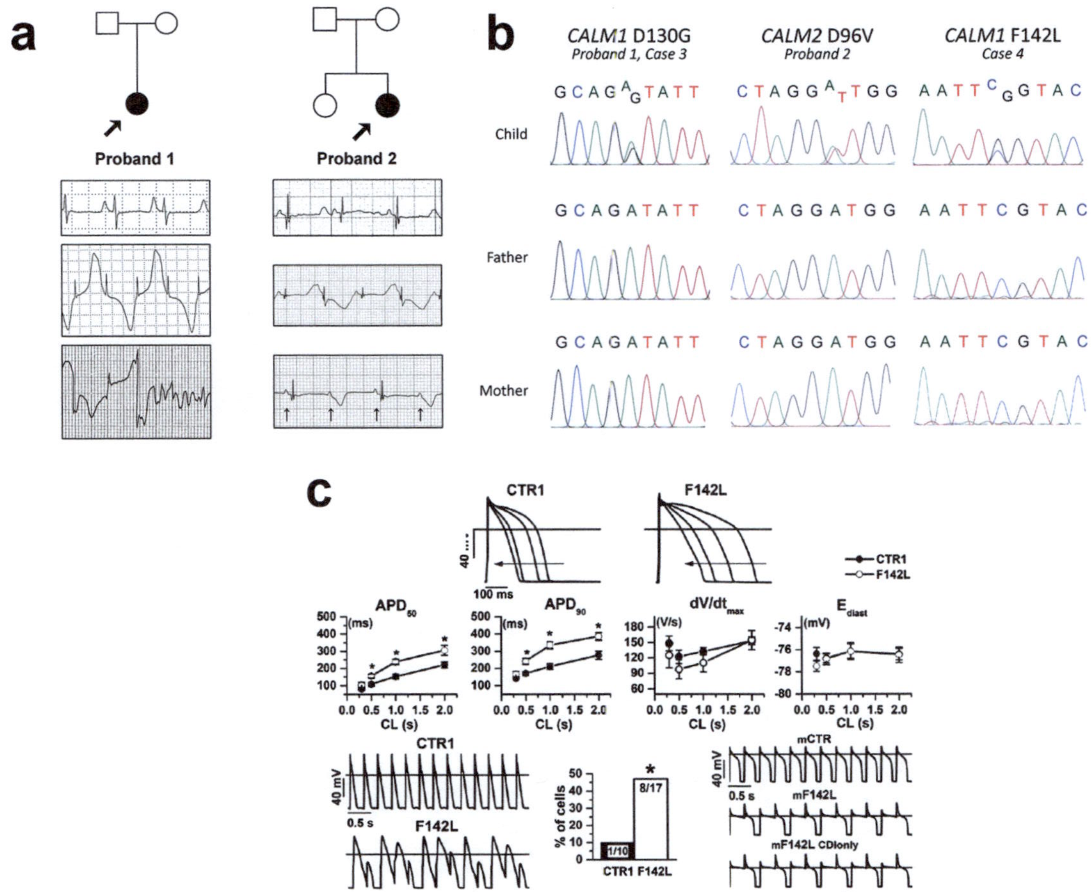

Fig. 7.1 (**a**) Representative electrocardiographic recordings from two probands with early onset, life-threatening cardiac arrhythmias. Upper trace represents baseline ECG for both probands. Middle trace illustrates T-wave alternans. Lower trace for proband 1 illustrates onset of ventricular fibrillation following a

LQT17: Triadin Knockout Syndrome

The most recently identified LQTS subtype, LQT17, pertains to homozygous or compound heterozygous frameshift mutations in the *TRDN* gene which encodes triadin (Altmann et al. 2015). Triadin is a major anchoring protein in cardiomyocytes, responsible for the structural integrity and crosstalk of sarcoplasmic reticulum Ca^{2+}-release channels, Ca^{2+}-sensitive proteins and L-type Ca^{2+} channels (Chopra and Knollmann 2013). From the thus far available descriptions, it seems that LQT17 patients mostly present with a notable QTc interval prolongation, recurrent stress-induced syncope and CA at a young age (Altmann et al. 2015). Studies in *Trdn*-null mice have shown that triadin ablation affects Ca^{2+}-dependent inactivation of the L-type Ca^{2+} channel resulting in Ca^{2+} overload and propensity for ventricular arrhythmias upon β adrenergic receptor stimulation (Chopra et al. 2009) (Table 7.1).

7.1.4 Prevalence

At variance with other inherited cardiac disorders for which only estimates exist, the prevalence of LQTS has been established with a large prospective study (Schwartz et al. 2009b). This involved 44,596 neonatal ECG data complemented by genetic testing of infants 3–4 weeks of age. A markedly prolonged QTc, defined according to the European Task Force on Neonatal Electrocardiography (Schwartz et al. 2002) as a QTc $\geq$ 470 ms, was present in 0.7/1000 infants ($n = 31$). Among these, an LQTS disease-causing mutation was found in 46% of the 28 that underwent genetic testing. As 1.4% of infants had a QTc between 440 and 469 ms, and some of them are also affected (unpublished data), it follows that the prevalence of the disease should be close to 1/2000 live births. This does not include the 'silent mutation carriers' that are affected subjects with a QTc $\leq$ 440 ms (Napolitano et al. 2005). The prevalence of this group of subjects is less precise and has been estimated to be up to 36% according to data from cascade family screening: 36% of LQT1, 19% of LQT2 and 10% of LQT3 patients (Priori et al. 2003). Despite their normal QT interval, this group of patients may have as much as a 10% risk of experiencing a major cardiac event by the age of 40 years. In particular, they have a 4% probability of CA that, although lower than those with a prolonged QT, it is ten times higher than that of unaffected individuals (Goldenberg et al. 2011). First proposed in the 1980s (Schwartz 1985) and then supported by the proven low penetrance of the disease (Priori et al. 1999), the concept that the spectrum of the disease can include patients with a normal QT interval has important practical and medicolegal consequences as

---

Fig. 7.1 (continued) period of T-wave alternans. Lower trace for proband 2 illustrates 2:1 AV block (arrows mark p-waves coincident with atrial depolarization). (**b**) DNA sequence traces indicating heterozygous calmodulin gene mutations identified in the four LQTS cases (the first mutation is in common between two cases). (**c**) Isolated *CALM1*-p.F142L iPSC-CMs show prolonged action potential duration and failure of the action potential to shorten at high pacing rates, a finding reproduced by numerical simulations

it does not allow to discharge as unaffected a sibling with a normal QTc without performing a cascade genetic screening.

7.1.5 Clinical Presentation

The clinical presentation of LQTS is very broad, ranging from no symptoms to sudden death that may even represent the first manifestation of the disease in 12% of untreated patients. However, the mortality rate among untreated symptomatic patients, originally reported as high as 5% per year (Schwartz et al. 1975), is significantly decreased by therapy (Moss et al. 1985, 2000).

The most typical presentation of the disease is a loss of consciousness occurring after a sudden increase in sympathetic activity, as under emotional or physical stress. It is due to a TdP ventricular tachycardia often degenerating into VF.

The trigger of the arrhythmic event differs according to the underlying genotype (Schwartz et al. 2001). In LQT1, events typically occur upon exercise or stress, as the impairment of the I_{Ks} current prevents the shortening of the QT during increases in heart rate. In this subgroup of patients, swimming has been identified as the triggering event in 33% of cases (Schwartz et al. 1991, 2001; Moss et al. 1999). Conversely, LQT2 and LQT3 patients have a preserved I_{Ks} current; therefore, they are at lower risk during exercise and experience events typically at rest. In LQT2 patients, 63% of events occur during emotional stress, and sudden auditory stimuli are typical triggers (Schwartz et al. 2001). In this subgroup of patients, moreover, hormonal changes increase the risk of events in women during menopause transition and postmenopausal periods (Buber et al. 2011). Hormonal changes together with sleep disruption are probably also the underlying causes of increased risk in the postpartum period in LQT2 patients (Schwartz et al. 2012). LQT3 patients typically experience events while asleep or at rest, while in the context of particular mutations, overlapping phenotypes of Brugada syndrome, sinus node dysfunction and cardiac conduction defects may be observed (Makita et al. 2008).

7.1.6 Electrocardiographic Features

Basal ECG, exercise stress test and 12-lead 24-hour Holter monitoring are all extremely important to identify the main electrocardiographic features of the disease.

7.1.6.1 Surface ECG

Typical features on surface ECG include (1) QTc prolongation, (2) peculiar T-wave morphology, (3) T-wave alternans and (4) low heart rate for the age.

QT prolongation is the main feature of the disease. Normal QTc values are up to 440 ms for men and 460 ms for women after puberty (Merri et al. 1989), calculated with the Bazett's formula (Bazett 1997). A gender difference in QTc is absent at birth (Stramba-Badiale et al. 1995) probably due to hormonal reasons (Carnethon

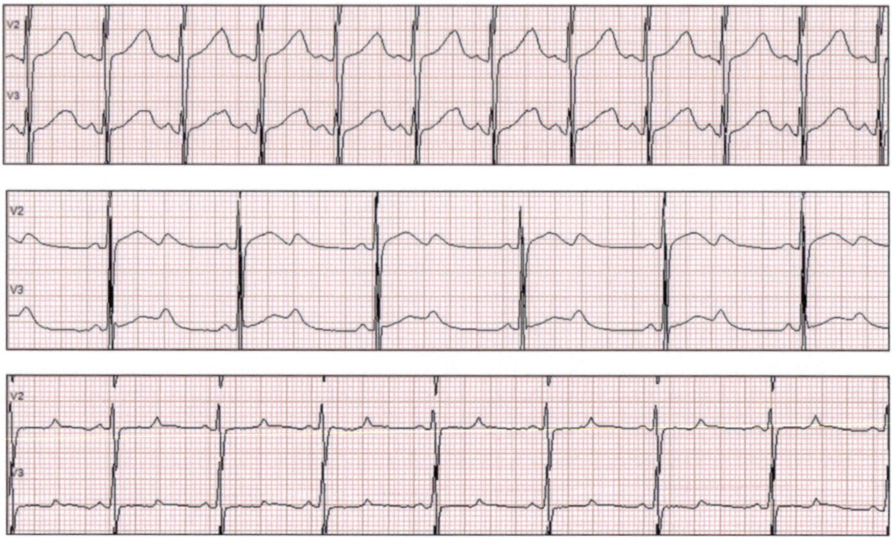

Fig. 7.2 Representative electrocardiographic recordings from LQT1 (upper panel), LQT2 (middle panel) and LQT3 (lower panel) patients. Upper panel shows broad-based T waves, characteristic of LQT1; middle panel shows biphasic and notched T waves, characteristic of LQT2; lower panel shows prolonged ST segment with sharpened T waves, characteristic of LQT3

et al. 2003; Kadish et al. 2004; Saito et al. 2009). Indeed, in puberty hormonal factors come into play and start introducing gender differences, since androgens contribute towards QT shortening (Kääb et al. 2004; Bai et al. 2005), while estrogens may modulate cardiac ion channel expression (Drici et al. 1996). In general, the longer the QT, the greater the risk for malignant arrhythmias (Moss et al. 1991). However, also individuals with a normal QTc may in fact be affected and have a 4% probability of CA (Goldenberg et al. 2011).

Another striking feature of LQTS patients, besides the QT prolongation, is the morphology of the T waves that can be biphasic, bifid or notched, typically in V2–V5 (Schwartz 1985). Gene-specific ECG patterns have been described (i.e. LQT1 patients tend to have broad-based T waves, LQT2 biphasic and notched T waves, LQT3 prolonged ST with sharpened T waves), but even within families an extreme heterogeneity can be present (Fig. 7.2).

T-wave alternans is a beat-to-beat alternation of the polarity and/or amplitude of the T wave, and it has been reported since 1975 (Schwartz and Malliani 1975). This peculiar feature represents a marker of major electrical instability. It usually appears during stress but also briefly at rest, and its observation should prompt reassessment of therapy. Due to its transient nature, it is more easily detected during 24-hour Holter ECG monitoring.

7.1.6.2 Exercise Stress Test

Exercise stress test represents a useful tool for LQTS diagnosis. Both the inability of QT shortening during exercise and the QT prolongation during the recovery phase have been identified as typical features of affected patients since the early descriptions of the disease (Locati et al. 1988; Vincent et al. 1991). Many studies have then supported this initial finding, recognizing also a gene-specific behaviour: during exercise, LQT3 patients shorten their QT interval much more than LQT2 patients (Schwartz et al. 1995) that shorten their QT interval more than LQT1 patients (Swan et al. 1999); during recovery, an exaggeration of QT prolongation distinguishes LQTS patients, particularly LQT1, from controls (Horner et al. 2011). However, only recently a clear and easily applicable QTc cut-off has been validated in more than one cohort through a three-step algorithm (Sy et al. 2011) leading to the inclusion of a QTc $\geq$ 480 ms at the fourth minute of recovery as a diagnostic criterion of the disease (Schwartz and Crotti 2011).

7.1.6.3 12-Lead 24-Hour Holter Recording

Besides the detection of QT prolongation and repolarization abnormalities, 12-lead 24-hour Holter monitoring provides information about heart rate and its reflex control. Indeed, heart rate is lower among LQTS patients than unaffected subjects (Schwartz et al. 1975). Also, sudden sinus pauses followed by accentuated repolarization abnormalities are sometimes observed in LQTS patients. These pauses are isolated and not cyclic, therefore different from the sinus arrhythmias (Malfatto et al. 1994). Therefore, 24-hour Holter ECG can support the diagnostic process but can also provide prognostic information (i.e. identification of T-wave alternans, important QT prolongation and frequent sudden pauses with abnormal repolarization abnormalities in the subsequent beat are all negative prognostic factors). Finally, Holter monitoring is useful while increasing the posology of beta-blocker therapy to the maximum tolerated dose for the patient.

7.1.7 Echocardiographic Features

The finding that LQTS patients show a rapid early contraction and an extended plateau phase well visible at M-mode Doppler before the rapid relaxation (Nador et al. 1991) casted doubt on LQTS as a pure electrical disease. The observation that these abnormalities were abolished by calcium blockers (De Ferrari et al. 1994) and the validation of these findings in a large Norwegian study (Haugaa et al. 2009) has definitely led to the conclusion that mechanical abnormalities do exist in LQTS patients (De Ferrari and Schwartz 2009) and may represent the mechanical counterparts of early afterdepolarizations (EADs) (ter Bekke et al. 2015; De Ferrari and Schwartz 2015).

7.1.8 Multisystem Disorders and QT Prolongation

7.1.8.1 The Jervell and Lange-Nielsen Syndrome

The J–LN syndrome is a very severe variant of LQTS combined with congenital deafness (Jervell and Lange-Nielsen 1957) due to the presence of homozygous or compound heterozygous mutations in the *KCNQ1* (LQT1) and *KCNE1* (LQT5) genes and is inherited in an autosomal recessive manner (Splawski et al. 1997a; Neyroud et al. 1997; Schulze-Bahr et al. 1997; Schwartz et al. 2006). Studies in *KCNE1*(−/−) and *KCNQ1*(−/−) transgenic mice have shown that due to the concomitant expression of these genes in the inner ear, mutations may result in deafness through disturbance of the ionic balance within endolymph and of the development of the cochlear and vestibular end organs (Vetter et al. 1996; Lee et al. 2000). Patients with J–LN syndrome also show clear differences compared to the other types of LQTS patients, including LQT1 and LQT5 patients with whom they share an impairment of the I_{Ks} current. J–LN syndrome patients present a severe phenotype: 90% of them are symptomatic and the QT is markedly prolonged (on average 557 ± 65 ms) (Schwartz et al. 2006). In this group of patients, beta blockers and left cardiac sympathetic denervation appear of limited efficacy, making the therapeutic approach complex, even more considering that 50% of them become symptomatic by the age of 3 years. Thus, despite their young age, an ICD should be seriously considered in these patients. However, within J–LN patients, there are subgroups at lower risk that are those with a QTc < 500 ms, those asymptomatic during the first year of life and especially those with *KCNE1* mutations (Schwartz et al. 2006).

7.1.8.2 The Timothy Syndrome

First named in 1992 as the 'heart and hand syndrome' for its two main features, syndactyly and QT prolongation (Reichenbach et al. 1992), the Timothy syndrome (TS) is a disorder characterized by multisystem cardiac, neurological, facial and hand dysfunction and developmental defects. It is caused by mutations in *CACNA1C*, the gene encoding for the alpha subunit of the calcium channel $Ca_v1.2$ (Splawski et al. 2004). Usually, besides markedly prolonged QT, the patient may manifest arrhythmias such as life-threatening ventricular arrhythmias but also foetal bradycardia and atrioventricular conduction block (Splawski et al. 2004, 2005). In 70% of patients, cardiac congenital defects such as patent foramen ovale, patent ductus arteriosus, ventricular septal defects and also tetralogy of Fallot can be observed, while in 25% of cases, hypertrophic or dilated cardiomyopathy may be present (Splawski et al. 2004; Lo-A-Njoe et al. 2005).

Neurological signs and symptoms occur in approximately 80% of patients and range from developmental delays (language, motor and generalized cognitive impairment) to autism spectrum disorder (Splawski et al. 2004; Gillis et al. 2012). Intractable seizures, cortical blindness, myopathy and profound developmental delay have been reported only in one case (Gillis et al. 2012). The most typical dysmorphic feature is the hand and feet syndactyly, present in almost 90% of cases

(Marks et al. 1995a, b). Other dysmorphologies include flattened noses, low-set ears, premaxillary underdevelopment, thin vermilion of the upper lip, round face, small, widely spaced teeth and poor dental enamel with severe dental cavities (Marks et al. 1995a, b; Splawski et al. 2005; Napolitano et al. 1993). Other more rare extracardiac features are frequent infections secondary to altered immune responses, joint contracture and intermittent hypoglycaemia (Gillis et al. 2012).

In typical cases the ECG is characterized by a long ST segment with a small T wave, but also giant negative T waves in precordial leads can be observed (Napolitano et al. 1993; Napolitano and Antzelevitch 2011). The QTc interval range is from 480 to 700 ms, and macroscopic T-wave alternans can be observed during Holter monitoring (Splawski et al. 2004, 2005; Napolitano and Antzelevitch 2011). The cardiac rhythm disturbances often occur early in life, prompting the diagnosis within the first few days of life. Occasionally it can be suspected prenatally because of foetal bradycardia or foetal distress secondary to heart rhythm disturbances. Rarely diagnosis is made after the age of 4 (Splawski et al. 2004, 2005; Marks et al. 1995a, b).

A milder phenotype of TS does exist, it lacks syndactyly and it is characterized by musculoskeletal problems, atrial fibrillation and more facial deformities (Splawski et al. 2005). In almost 80% of cases, death occurs at 14–15 years, and it is due to ventricular tachycardia or ventricular fibrillation without a specific trigger (Splawski et al. 2004, 2005; Reichenbach et al. 1992). As for the majority of LQTS forms, treatment includes beta blockers and left cardiac sympathetic denervation (LCSD). In some cases, pacemakers can be implanted during the first days of life to avoid pause-induced TdP; however, an implantable cardioverter defibrillator (ICD) should be considered in all affected individuals.

7.1.8.3 Ankyrin-B Syndrome (LQT4)

In 2003, a loss-of-function mutation in the *ANK2* gene encoding ankyrin B (ANK-B) was identified as being responsible for LQT4 (Mohler et al. 2003). It however soon became evident that the phenotypic manifestations of the LQT4 subtype deviated substantially from classical LQTS (Mohler et al. 2004). Besides the presence of clinical features such as severe sinus node dysfunction, atrial fibrillation, severe bradycardia, isorhythmic atrioventricular dissociation, heart rate variability, delayed cardiac conduction, conduction block, IVF and exercise-induced ventricular tachycardia (Mohler et al. 2004, 2007; Le Scouarnec et al. 2008), the hallmark feature of QT prolongation was in many patient cases absent (Zhang et al. 2005; Mohler et al. 2004). This wide spectrum of clinical phenotypes has led to the use of the term 'sick sinus syndrome with bradycardia' or 'ankyrin-B syndrome' in order to distinguish this subgroup of patients in which a QT prolongation may be or may not be present (Zhang et al. 2005; Mohler et al. 2004; Le Scouarnec et al. 2008) as part of a more broad cardiac disease. Recently, a mutation in *ANK2* has been identified in two large multigenerational families affected with ankyrin-B syndrome and structural heart disease (Swayne et al. 2017).

7.1.8.4 Andersen-Tawil Syndrome (LQT7)

After the first description in 1963 (Klein et al. 1963), the triad of symptoms characterizing this syndrome was first reported by Andersen in 1971 (Andersen et al. 1971), and then the disease was further characterized by Tawil et al. (1994). It was initially labelled as LQT7 due to the erroneous measurements of the QT interval that was including the prominent U wave (Tristani-Firouzi et al. 2002). Today, the name 'Andersen-Tawil syndrome, ATS' (Donaldson et al. 2003), in recognition of these two clinicians' work, is widely accepted. So far, the only gene linked to this syndrome is *KCNJ2* that encodes the α subunit of the K^+ ion channel Kir1.2 conducting the inward rectifier repolarizing current I_{K1} (Plaster et al. 2001). There is significant heterogeneity in phenotypic expression even among family members sharing the same disease-causing mutation (Plaster et al. 2001). Affected patients may present mild or prominent dysmorphic features which include short stature, low-set ears, wide-set eyes, small mandible, fifth-digit clinodactyly, syndactyly and broad forehead. Other less common features are bilateral ptosis, thin hair and scoliosis (Nguyen et al. 2013). Periodic paralysis attacks and muscle weakness usually predate the cardiac symptoms by at least 2 years as the attacks can arise even at 8 months of age and be as frequent as three per month on average (Yoon et al. 2006). The attacks are more frequently associated with hypokalaemia but may occur also in hyperkalaemia or normokalemia, and no clear trigger has been identified (Yoon et al. 2006; Sansone et al. 1997; Canűn et al. 1999). Ventricular arrhythmias in ATS range from premature ventricular beats to short runs of polymorphic ventricular tachycardia leading rarely to sudden cardiac death (Zhang et al. 2005; Yoon et al. 2006). Less manifest during pregnancy (Subbiah et al. 2008), no specific triggers of the arrhythmias have been identified (Nguyen et al. 2013). Other cardiac manifestations include prominent U waves in the anterior leads of ECG (Zhang et al. 2005; Tristani-Firouzi et al. 2002) with a QTc that is only marginally prolonged if the U wave is not included in the measurement (Nguyen et al. 2013). Diagnosis is made according to the diagnostic criteria formulated by Venance et al. (2006). However, the clinical variability of the disease can make difficult the diagnosis as some patients may be non-penetrant and some may show one, two or three of the classic triad symptoms in different combinations (Nguyen et al. 2013). Beta blockers in combination with flecainide have shown an optimal control of cardiac symptoms (Sansone and Tawil 2007; Pellizzón et al. 2008; Delannoy et al. 2013).

7.1.9 Diagnosis

The diagnosis of LQTS can be performed when a patient receives a so-called Schwartz score of $\geq$3.5 points (Table 7.2); alternatively it can be made in presence of an unequivocally pathogenic mutation and/or in presence of a QTc $\geq$ 500 ms in repeated 12-lead ECGs (Priori et al. 2013). The Schwartz score is assigned according to recently updated and validated criteria based on the electrocardiographic features, familial and personal history (Schwartz and Crotti 2011; Schwartz et al. 1993; Hayashi et al. 2016). It does not permit to identify the silent mutation carriers, but

Table 7.2 Diagnostic criteria for long QT syndrome

	Points
Electrocardiographic findings[a]	
QTc[b]	
≥480 ms	3
460–479 ms	2
450–459 ms (in males)	1
QTc fourth minute of recovery from exercise stress test ≥480 ms	1
Torsade de pointes[c]	2
T-wave alternans	1
Notched T wave in three leads	1
Low heart rate for age[d]	0.5
Clinical history	
Syncope[c]	
With stress	2
Without stress	1
Congenital deafness	0.5
Family history	
Family members with definite LQTS[e]	1
Unexplained SCD below age 30 among immediate family members[e]	0.5

Score: ≤1 point, low probability of LQTS; 1.5–3 points, intermediate probability of LQTS; ≥3 points, high probability of LQTS

[a]In the absence of medications or disorders known to affect these electrocardiographic features

[b]QTc calculated by Bazett's formula

[c]Mutually exclusive

[d]Resting heart rate below the second percentile for age

[e]The same family member cannot be counted in both (Adapted from Schwartz PJ and Crotti L. Circulation 2011;124:2181–4)

it is useful during the first contact with the patient to assess the probability of disease and therefore to decide whether to start a therapy.

7.1.10 Differential Diagnosis

The main differential diagnosis is the so-called acquired long QT syndrome (aLQTS); under this term are included all cases in which a prolonged QTc, often accompanied by TdP, is caused either by drugs, bradycardia or hypokalaemia.

First proposed by Schwartz and Moss (1982), the hypothesis that some cases of drug-induced arrhythmias depend on genetic predisposition was later confirmed in 2000 by the evidence of a LQT1 variant in a patient with an episode of VF on therapy with cisapride and normal QTc at baseline (Napolitano et al. 2000) and recently validated in a large cohort study (Itoh et al. 2016). In this study, 188 probands with aLQTS were screened for the five major LQTS genes, and genetic results and QTc values were compared with 2379 members of 1010 genotyped congenital LQTS families (cLQTS), including 1938 mutation carriers and

441 noncarriers, providing useful insight into differential diagnosis: aLQTS patients have a baseline QTc that is intermediate between that of cLQTS and of controls, while LQT2-causing mutations are significantly more frequent than LQT1-causing mutations in the aLQTS group compared to the cLQTS. These data were used to develop a score, based on QTc, age and symptoms, which allows the identification of the aLQTS subjects more likely to carry LQTS-causing mutations (Itoh et al. 2016). Of note, the prevalence of LQTS-causing mutations found in the aLQTS cohort was 28%, higher than that previously reported (Itoh et al. 2016).

Recently (Crotti et al. 2016a), a prolonged QT has been observed in young athletes performing competitive sports. In this subset of patients, genetic testing is usually negative in those with no family history and in those in whom the QTc normalizes after detraining. However, more studies are needed to better understand the underlying mechanism and perform a correct differential diagnosis.

7.1.11 Risk Stratification and Prognosis

Classically, risk stratification in LQTS is based on sex, age, QTc and genotype (Priori et al. 2003). However, since 2003, when a scheme for risk stratification was proposed (Priori et al. 2003), advances in molecular genetics have more and more contributed to risk stratification (Moss et al. 2002, 2007), making a mutation-specific risk assessment also possible in specific cases (Crotti et al. 2007). Furthermore, data so far published on modifier genes (Schwartz 2011; Crotti et al. 2005, 2009b, 2016b; Nof et al. 2010; Amin et al. 2012; Duchatelet et al. 2013; de Villiers et al. 2014; Earle et al. 2014; Arking et al. 2006, 2014; Tomás et al. 2010; Kolder et al. 2015; Schwartz et al. 2008) pave the way for including in the near future additional genetic markers in the risk stratification process. On the clinical side, some well-known ECG features are associated with increased risk of cardiac events such as the T-wave alternans, 2:1 functional AV block, extreme QT prolongation and the presence of sinus pauses followed by particularly prolonged QT. Furthermore, some autonomic parameters have been associated with arrhythmic risk in selected populations.

7.1.11.1 Genetic Risk Stratification

The identification of the pathological reference genes underlying LQTS not only makes molecular diagnosis possible but also provides useful information for risk stratification. In fact, it is known that particular mutations cause a milder (*forme fruste*) (Donger et al. 1997; Berthet et al. 1999) or more severe (Crotti et al. 2007) LQTS phenotype.

The type of mutation, as this is defined by its topology within the gene and type and degree of afflicted functionality, has been shown to correlate with disease risk (Moss et al. 2002, 2007; Nagaoka et al. 2008). In particular, for the *KCNQ1* gene, transmembrane, missense or dominant-negative suppression mutations have been shown to correlate with significantly increased risk of cardiac events of any type, independently of other factors, with respect to C-terminal, nonsense and happloinsufficiency mutations (Moss et al. 2007). As far as mutations in *KCNH2* are concerned, it is well known that those affecting the channel's pore or voltage sensor are associated with a more severe phenotype (Moss et al. 2002; Nagaoka et al. 2008).

On the other hand, considering, for instance, the increased phenotypic malignancy of the transmembrane, missense, dominant-negative *KCNQ1*-p.A341V mutation, compared to other mutations of this type, this classification still proves as an inadequate tool to be fully applied in patient management. Therefore, mutation-specific risk stratification strategies, which take into account the particular behavioural motif of each mutation, have been proposed (Crotti et al. 2007).

7.1.11.2 Modifier Genes

The evidence that even within families a very different clinical outcome may be observed among siblings carrying the same disease-causing mutation has stimulated the search for secondary factors that may shape the final clinical outcome. In fact, the degree of disease penetrance (i.e. the percentage of genetic carriers that clinically manifest the disease) and the variability of disease expressivity (i.e. the different ways or the extent the disease is expressed among those that do manifest it) seem to be defined by factors that extend beyond the main pathological substrate. Although the latter is undoubtedly the main determinant of phenotypic manifestation, it has long been recognized that modifying factors also come into play, whether environmental or genetic. Interpersonal variation among identical mutation carriers is principally attributed to genetic modifying factors, such as single nucleotide polymorphisms (SNPs) (Schwartz 2011). A modifying effect may be exerted from a SNP within the same or different gene than the disease gene that may attenuate or aggravate the pathological manifestation of the principal disease-causing mutation, thereby conferring protection or increased risk, respectively.

The first modifier of clinical severity in LQTS, the common *KCNH2* polymorphism p.K897T, was identified in 2005 (Crotti et al. 2005). Functional studies have shown that in the presence of a clinically latent C-terminal LQT2 (*KCNH2*) mutation, p.K897T accentuates and unmasks the I_{Kr} current loss produced by the disease-causing mutation (Crotti et al. 2005). This finding has already been validated (Nof et al. 2010). To date, other common SNPs in the LQTS genes, causing either subtle changes (increase or decrease) of magnitude of the main ionic currents or differential ion channel expression, have been described, such as SNPs in the 3′ UTR region of *KCNQ1* (Amin et al. 2012), the intronic *KCNQ1* rs2074238 SNP (Duchatelet et al. 2013) as well as intronic SNPs in *AKAP9* (LQT11) (de Villiers et al. 2014). Although some of these modifier SNPs have preliminarily produced conflicting results upon replication in heterogeneous patient cohorts (Crotti et al. 2016b; Earle et al. 2014) while others await independent validation, overall their identification has provided proof of concept of how common genetic variants in the LQTS genes may confer a detrimental as well as a protective effect on LQTS patients' arrhythmic risk.

Genetic modifiers may also lie beyond the immediate family of the pathological reference genes, and their identification is even more demanding. GWAS studies that aimed at the identification of SNPs influencing QT interval duration in the general population also contributed to the search and identification of modifier genes in

LQTS (Arking et al. 2014; Newton-Cheh et al. 2009; Pfeufer et al. 2009). Since the QT interval is by itself an inherited quantitative trait (Newton-Cheh et al. 2005), genes that genetically define it are natural candidates as modifiers. In fact, we have shown that two common variants in the nitric oxide synthase 1 adaptor protein gene *NOS1AP*, previously identified as a gene modifying QT interval duration in the general population (Arking et al. 2006), act as modifiers in the *KCNQ1*-p.A341V LQTS founder population by almost doubling the risk of life-threatening arrhythmias (Crotti et al. 2009b). The modifier role of *NOS1AP* has subsequently been validated in an independent, heterogeneous population of LQTS patients (Tomás et al. 2010). In addition, a more recent multicentre analysis focusing on more than 600 LQT2 patients and testing all major SNPs identified by GWAS studies as associated to QT interval duration confirmed the major role played by *NOS1AP* (Kolder et al. 2015). Thus, *NOS1AP* has consistently appeared as a strong modifying factor of clinical severity in LQTS. Another example is an earlier study by Schwartz et al. in the same *KCNQ1*-p.A341V LQTS founder population (Schwartz et al. 2008) that showed that particular SNPs in the α_2 and β_1 adrenergic receptor genes act synergistically to increase heart rate and baroreflex sensitivity, thereby exposing the already compromised *KCNQ1*-p.A341V LQTS patients to increased risk of cardiac events.

The identification and comprehension of the modifying factors that influence the pathological substrate and the final clinical outcome are principal areas of ongoing research that may hopefully one day be integrated into clinical practice. Nonetheless, the strength of the *NOS1AP* data is such that the presence of these deleterious SNPs should, even at present, indicate a more aggressive therapeutic strategy. In the future, not only mutation-specific (Crotti et al. 2007) but also SNP-specific (Duchatelet et al. 2013) stratification strategies are expected to contribute to a more precise and refined definition of risk.

7.1.11.3 Autonomic Parameters

In the large South African LQT1 founder population, in which all the affected members carry the *KCNQ1*-p.A341V mutation, Brink et al. (2005) and Schwartz et al. (2008) provided novel evidence that faster basal heart rates and brisk autonomic responses are associated with a greater probability of being symptomatic. Whereas among patients with a major arrhythmogenic substrate (QTc $>$ 500 ms) basal heart rate is rather unimportant, among patients with a QTc $\leq$ 500 ms, those in the lower tertile of heart rate were more frequently asymptomatic. Furthermore, relatively low values of baroreflex sensitivity—an index of the ability to respond with brisk increases in either vagal or sympathetic activity—were associated with a reduced probability of being symptomatic (Schwartz et al. 2008). However, BRS is not frequently used in clinical practice, and therefore we assessed the value of another and simpler marker of reflex vagal activation, i.e. the heart rate reduction during the first minute of recovery from an exercise stress test (Crotti et al. 2012). This is a parameter strongly correlated with BRS ($r = 0.64$, $p = 0.001$) (Crotti et al. 2012). We observed that symptomatic LQT1 patients reduced their HR during the first minute of recovery significantly more than asymptomatic patients and those with marked heart reductions had three times greater risk of suffering cardiac events (OR 3.28, 95%CI 1.3–8.3, $p = 0.012$). By striking contrast, the phenomenon so clearly evident among LQT1 patients is totally absent among LQT2 and LQT3 patients; this difference was expected, given that LQT2 and LQT3 patients have a normal I_{Ks} current and therefore are less prone to the onset of life-threatening arrhythmias related to rapid changes in heart rate, especially increases (Crotti et al. 2012).

In LQTS patients an exercise stress test should always be performed, not only to analyse repolarization changes useful for the diagnosis but also to aid risk stratification of LQT1 subjects. Another practical implication is that intense exercise training, which potentiates vagal reflexes, should be discouraged in LQT1 patients.

7.1.11.4 Pregnancy and Arrhythmic Risk

Pregnancy is characterized by a variety of physiological cardiovascular and hormonal changes that may influence the risk of cardiac events in patients with an underlying pro-arrhythmic substrate as LQTS.

Rashba et al. reported the first evidence that in LQTS patients the post-partum period, but not the pregnancy interval, is associated with an increased risk of cardiac events, that is reduced by beta-blocker therapy (Rashba et al. 1998). Subsequent studies showed that the women with an increased risk are mainly LQT2 women. This is probably because the post-partum period is characterized by sleep deprivation and sudden noises in the middle of the night (i.e. the baby who starts crying when the mother is sleeping), typical triggers for LQT2 patients (Seth et al. 2007). The pregnancy-related cardiovascular risk in LQT1 patients is relatively small (Heradien et al. 2006), while studies on LQT3 pregnant females are not available.

7.1.11.5 Events in the First Year of Life

Long QT syndrome patients with life-threatening arrhythmias in the first year of life represent a small subset of Romano-Ward patients (in LQTS International Registry 2% of 3323 subjects) (Spazzolini et al. 2009); however, their risk for subsequent cardiac arrest/sudden cardiac death in the following 10 years is particularly high (HR 23.4, $p < 0.01$), and their response to beta-blocker therapy is poor (Spazzolini et al. 2009).

This subgroup of patients, independently of genotype, should be regarded as a subgroup of LQTS patients in whom traditional treatments are usually not effective and in whom an aggressive strategy is usually needed, at variance with the vast majority of LQTS patients.

Unfortunately, sudden cardiac death can be the first clinical manifestation of the disease, and when this occurs very early in life, it may result in sudden infant death syndrome (SIDS) (Arnestad et al. 2007) or stillbirth (Crotti et al. 2013b).

This hypothesis has been validated by the evidence that 9.5% of SIDS victims carry an LQTS-associated functional variant: eight mutations and seven rare variants with functional effect were found in 19 of 201 cases as likely contributors to sudden death (9.5%, 95% CI 5.8–14.4%), while none was identified among the 187 controls (Arnestad et al. 2007). These data support the concept of neonatal ECG screening as a cost-effective method to prevent those sudden deaths due to unrecognized LQTS (Saul et al. 2014; Schwartz 2004).

Based on the hypothesis that LQTS, causing death during the first months of life, could cause death also shortly before birth (Saul et al. 2014; Schwartz 2004; Quaglini et al. 2006), a genetic study was performed in 97 stillbirths (Crotti et al. 2013b): 3.3% of cases indeed carried an LQTS-causing mutation, and 8.8% carried a rare missense variant with a potential pro-arrhythmic role (Crotti et al. 2013b).

7.1.12 Therapy

The trigger for most life-threatening events is a sudden increase in sympathetic activity (Schwartz et al. 1975). Accordingly, antiadrenergic therapies provide the main protection. Among these, beta blockers represent the first-line therapy since the 1970s. Further therapies for symptomatic patients on beta blockers include left sympathetic cardiac denervation, automatic defibrillator implantation and, in selected cases, sodium channel blockers.

7.1.12.1 Beta Blockers

The first study demonstrating the efficacy of beta blockers was published in 1985. It compared 233 symptomatic untreated patients with treated patients either with beta blockers or with non-antiadrenergic interventions (Schwartz 1985). Response to beta-blocker therapy can be different according to the different genetic subtypes. Specifically, LQT1 patients are those better protected by beta-blocker therapy, with failures mainly due to non-compliance or concomitant use of QT-prolonging drugs (Vincent et al. 2009). LQT2 patients on beta blockers have a 6–7% rate of resuscitated CA events (Priori et al. 2004), while those that appeared to be less protected were LQT3 patients (Schwartz and Crotti 2017). These data, despite being based on very small numbers, have led to the incorrect notion that beta blockers were useless in LQT3 patients, and therefore ICD implantation had to be considered even in paediatric cohorts (Etheridge et al. 2007). Subsequent studies showed that beta blockers are effective also in LQT3 patients when excluding those with life-threatening arrhythmias in the first year of life (Schwartz et al. 2009a; Wilde et al. 2016).

Among all beta blockers, the most extensively studied is propranolol, which has been shown to either decrease or prevent an increase in transmural dispersion of repolarization in response to strong sympathetic stimulation (Shimizu et al. 2002). Propranolol is the most widely used drug, at 2–3 mg/kg/day. Also nadolol is widely used since its longer half-life allows once- or twice-a-day administration, usually at 1–2 mg/kg/day. Not all beta blockers are equally effective in preventing arrhythmic events, and indeed a multicentre study has shown a greater risk of recurrences of cardiac events with metoprolol compared to propranolol or nadolol (Chockalingam et al. 2012).

7.1.12.2 Left Cardiac Sympathetic Denervation

LCSD is a surgical approach requiring the ablation of the first four thoracic ganglia (T1–T4), leaving intact the cephalic portion of the left stellate ganglion to avoid the Horner's syndrome, expected in 1–2% of patients (Odero et al. 2010). The procedure is traditionally performed in a retroclavicular approach; a thoracoscopic approach makes thoracotomy unnecessary but seems to be associated with a higher incidence of transient post-operative neurological pain (Odero et al. 2010; Collura et al. 2009). The rationale for LCSD is largely based on its antifibrillatory effect, on the major reduction of norepinephrine release at ventricular level in the absence of post-denervation supersensitivity (Schwartz 2014). In a cohort of 147 LQTS patients that underwent LCSD, the procedure showed a 91% reduction in cardiac events during a mean follow-up of 8 years (Schwartz et al. 2004). The population included 99% of symptomatic patients, with a mean QTc of 563 $\pm$ 65 ms, 48% with previous CA and 75% with syncope despite full-dose beta blockers. LCSD produced a mean QTc shortening of 39 ms, pointing to an action on the substrate as well as on the trigger, while a post-surgery QTc $<$ 500 ms predicted a very favourable outcome. It is worth noting that in five patients who underwent LCSD due to multiple ICD shocks and electrical storms, there was a 95% decrease in the number of shocks (from an average of 29 shocks/year) during a 4-year follow-up (Schwartz et al. 2004). Current indications for LCSD include (1) patients with appropriate VF-terminating ICD shocks, (2) patients with recurrence of syncope while on adequate drug therapy, (3) patients who do not tolerate beta-blocker therapy and (4) as a bridge to ICD for young patients at particularly high risk (Schwartz and Ackerman 2013).

7.1.12.3 Cardiac Pacing

In some patients, cardiac pacing can be used as an adjunct to beta-blocker therapy to allow increases in dosage. Its use is justified by the observation that pauses usually precede TdP. However, if a device is needed, one should seriously consider to implant a transvenous ICD that also has a pacing mode. Exceptions could be very small children in whom pacing may represent a transient bridge to ICD. It is important to bear in mind that fast heart rates, while on one hand may have antiarrhythmic effect, on the other hand may produce a tachycardia-mediated cardiomyopathy.

7.1.12.4 Implantable Cardioverter Defibrillator

The largest ICD-LQTS Registry published so far, including 233 patients, has given deep disquieting insights into ICD implantation indication in LQTS patients: the majority of implanted patients had not suffered a CA and, even worse, had not failed beta-blocker therapy (Schwartz et al. 2010). Accordingly to genotype, the majority of the asymptomatic patients implanted were LQT3 (45%), reflecting the incorrect concept that LQT3 patients are not adequately protected by beta blockers (Schwartz et al. 2009a). During a mean follow-up of 4.6 years, 28% of patients experienced at least an appropriate shock, and adverse events occurred in 25% (in total 31% of the patients, including 6% of inappropriate shocks). These observations have allowed the creation of a score, M-FACT, that is able to predict on the basis of simple clinical variables the appropriate ICD therapies. Main variables include age <20 years at implantation, QTc > 500 ms, prior CA and CA despite therapy. Within 7 years, 70% of patients with all these factors experienced an appropriate shock, while no patients with none of these factors had appropriate ICD shocks (Schwartz et al. 2010).

Current indications for ICD implantation are (1) previous CA and (2) recurrences despite full antiadrenergic therapy (possibly including LCSD). Asymptomatic patients should not be implanted with an ICD, unless despite optimal therapy a QTc > 550 ms is present together with sign of high electrical instability (i.e. T-wave alternans, 2:1 AV block, very long sinus pauses with subsequent beats showing aberrant T-wave morphology and further QT prolongation).

7.1.12.5 Gene-Specific Therapy and Management

The amazing unravelling of the underlying genetic component has made LQTS the first cardiac disease with a potential gene-specific management.

LQT1 patients are at higher risk during sympathetic activation; therefore, these patients should avoid competitive sports activity (Johnson and Ackerman 2012), particularly swimming which is a well-known trigger of arrhythmic events in this subset of patients (99% of arrhythmic events are associated with swimming) (Choi et al. 2004).

LQT2 patients are at higher risk when aroused from sleep or rest by a sudden noise. They are also particularly sensitive to serum potassium levels (Schwartz et al. 2001). Thus, telephone and alarm clocks should be removed from the bedrooms, and a combination with potassium-sparing agents should be considered in those patients with difficulties to maintain reasonable levels of potassium (Schwartz and Ackerman 2013).

In LQT3 patients, mexiletine may be used as a possible adjuvant (Schwartz et al. 1995). This is based on the demonstration that *SCN5A* LQTS-causing mutations have a 'gain-of-function' effect (Bennett et al. 1995), suggesting the use of sodium channel blockers. As the effect of mexiletine is mutation-specific (Ruan et al. 2007), its effectiveness should be tested by the acute oral drug test technique (half the daily dose during continuous ECG monitoring): within 90 min the peak plasma concentration is reached, and if the QTc is shortened by more than 40 ms without PR prolongation or QRS widening, then mexiletine can be added to the therapy (Fig. 7.3). Limited data are available about ranolazine, a sodium channel blocker specific for the late sodium current (Moss et al. 2008). Given its direct late sodium current blocking properties, propranolol is the preferred beta blocker in these patients.

7.1.12.6 Asymptomatic LQTS Patients and Patients with Normal QTc

As the first manifestation of LQTS in approximately 13% of cases is SCD, beta-blocker treatment ideally should be initiated in all LQTS patients with evident QT prolongation including those still asymptomatic. Possible exception could be (1) asymptomatic LQT1 men who were diagnosed after 40 years because they seldom have a first event after this age and (2) any asymptomatic LQTS patient who is older than 50 years and has a resting QTc < 480 ms. LQT2 women remain at

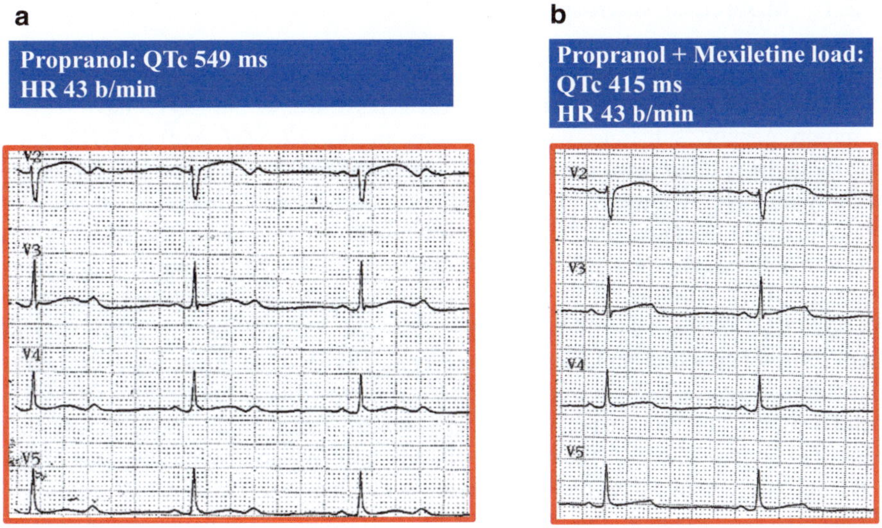

Fig. 7.3 (**a**) Major QTc prolongation (QTc 549 ms) in a 32-year-old female patient on beta-blocker therapy. (**b**) Oral acute load of mexiletine shown to dramatically reduce the length of the QT interval (QTc 415 ms)

risk throughout life, and therefore with few exceptions they should be always treated (Schwartz and Ackerman 2013).

Mutation carrier patients have a significantly lower risk of life-threatening arrhythmias when QTc is normal (<440 ms) compared to the phenotypically affected patients; however, it is still ten times greater than negative family members (Goldenberg et al. 2011). Therefore, the decision to start therapy should be carefully individualized.

7.2 Short QT Syndrome

The short QT syndrome (SQTS) is a malignant cardiac disease characterized by a short QT interval at the basal ECG and by the presence of ventricular tachyarrhythmias leading to syncope and SCD. It was in the early 1990s that an association between a short QT interval and SCD was proven, but it was only at the beginning of the new millennium that the existence of this new syndrome was postulated after the description of a family with paroxysmal atrial fibrillation (AF), constantly short QT interval (Gussak et al. 2000) and multiple cases of SCD (Gaita et al. 2003). Since then, other cases affected by SQTS have been reported (Bellocq et al. 2004; Priori et al. 2005; Kirilmaz et al. 2005; Hong et al. 2005; Lu et al. 2006); however, the available descriptions are still quite few, thereby limiting the possibility of forming a generalized clinical picture and estimation of the syndrome's prevalence.

7.2.1 Molecular Genetic Basis

In contrast to LQTS whose genetic substrate has largely been elucidated, in SQTS our knowledge is still largely scanty. Short QT syndrome and LQTS have partly a common genetic substrate with mutations in the *KCNQ1* and *KCNH2* genes encoding the α subunits of the cardiac K^+ channels Kv7.1 and Kv11.1, respectively, underlying both syndromes. In fact, while loss-of-function mutations in *KCNQ1* underlie LQT1 (Wang et al. 1996a), gain-of-function mutations that augment I_{Ks} and result in abbreviation of action potential duration, and hence QT interval shortening, underlie SQT2 (Bellocq et al. 2004). Accordingly, gain-of-function mutations in *KCNH2* that augment I_{Kr} underlie SQT1 (Brugada et al. 2004), in contrast to the LQT2 loss-of-function mutations (Curran et al. 1995). In addition, gain-of-function mutations in the *KCNJ2* gene encoding the α subunit of the K^+ ion channel Kir1.2 lie behind SQT3, thereby increasing the I_{K1} current (Priori et al. 2005) (Table 7.1). Short QT intervals, but in the context of an overall Brugada syndrome phenotype, have been described due to mutations in the genes encoding the α and β2 subunits of the cardiac Ca^{2+} channel, but these most likely form part of a distinct clinical entity (Antzelevitch et al. 2007).

Genetic testing of the three main genes responsible thus far for SQTS may be considered (class IIb recommendation) for patients with a strong clinical suspicion, while it is recommended for the family members of a proband in whom a disease-causing mutation has already been identified (class I recommendation) (Ackerman et al. 2011). The yield of genetic testing is still relatively low, ranging from 11 to 20% of cases (Ackerman et al. 2011; Mazzanti et al. 2014).

7.2.2 Clinical Presentation

The clinical presentation of SQTS in the available cohorts (Mazzanti et al. 2014; Giustetto et al. 2011; Gollob et al. 2011) tends to be quite severe, with a high incidence of CA and SCD. The age of onset of symptoms is highly variable, ranging from first manifestations in utero or at age 70 (Mazzanti et al. 2014; Giustetto et al. 2011; Gollob et al. 2011). Sometimes SQTS manifests as SIDS (Arnestad et al. 2007). However, the peak of occurrence of life-threatening arrhythmias seems to be the first year of life and then between age 14 and 40 (Mazzanti et al. 2014; Giustetto et al. 2011; Gollob et al. 2011). Atrial fibrillation has been observed in several patients and is probably related to short atrial refractory periods (Mazzanti et al. 2014; Giustetto et al. 2011; Gollob et al. 2011).

SQTS is an autosomal dominant disease and is expected to be equally prevalent in male and female patients; however, data from the European SQTS Registry suggest a higher prevalence of the disease among males with a mean age at diagnosis between 20 and 30 years (Giustetto et al. 2011). Among the 53 patients with SQTS reported, there was a high clinical heterogeneity, with 62% of patients being symptomatic at presentation: 4 had died suddenly, 13 had an aborted SCD (range 3 months to 62 years), 8 had syncope and 13 had palpitations (6 of whom had documented AF or flutter) (Giustetto et al. 2011). Differences have been observed between the two genders; although CA had a similar prevalence, more than 90% of males had CA between 14 and 40 years of age, while the events were widely distributed across the years among females (Giustetto et al. 2011).

7.2.3 Diagnosis

The diagnosis of SQTS is still debated as there is no consensus on the QTc value to be used as a cut-off for the diagnosis due to the considerable overlap of the QTc of affected and healthy individuals. Indeed, the SQTS patients described so far had a QTc in the range of 250–380 ms (Mazzanti et al. 2014; Giustetto et al. 2011; Gollob et al. 2011), and large studies in healthy populations (altogether >28,000 individuals) showed that a QTc in the lowest 0.5 percentile of the normal distribution (QTc $\leq$ 330 ms) (Gallagher et al. 2006) or below 340 ms (Anttonen et al. 2007) and 320 ms (Anttonen et al. 2007; Dhutia et al. 2016) was not associated with an increased risk of SCD. The 2015 ESC Guidelines for the prevention of SCD (Priori et al. 2015) suggest that a diagnosis should be reached in all patients with a QTc $\leq$ 340 ms, while when the QTc is between 340 and 360 ms, SQTS can be diagnosed only in the presence of at least one additional criterion that could be either presence of a pathogenic mutation, family history of SQTS, family history of SCD below age 40 or survival of a VT/VF episode in the absence of heart disease. The HRS/EHRA/APHRS expert consensus statement on the diagnosis and management of patients with inherited primary arrhythmia syndromes (Priori et al. 2013) has suggested a slightly lower cut-off (QTc $\leq$ 330 ms).

Besides QTc values, other ECG characteristics can be useful to suspect or diagnose the disease. For instance, the morphology of the T wave can be informative. Affected patients also have a short or absent ST segment, with the T wave initiating immediately from the S wave. SQT1 patients present tall, sharp, narrow, fine and symmetrical T waves, especially in leads V2–V4 (Schimpf et al. 2005), with a relatively prolonged T_{peak}–T_{end} interval, suggesting increased transmural dispersion of repolarization.

SQT2 patients show less narrow symmetrical T waves (Bellocq et al. 2004), while SQT3 patients present an asymmetrical pattern with a less steep ascending section of the T wave followed by a rapid descending terminal phase (Priori et al. 2005). Patients carrying Ca^{2+} channel mutations present with ECGs that may coincide with a Brugada-type ST elevation in V1–V2 either at baseline or after ajmaline administration (Antzelevitch et al. 2007). A significantly shorter J point–T peak interval and a shorter T_{peak}–T_{end}/QT ratio have been reported as features that may help in distinguishing healthy people from SQTS patients (Anttonen et al. 2007; Watanabe et al. 2010). Furthermore, early repolarization has been described in 65% of SQTS patients and associated with arrhythmic events (Watanabe et al. 2010).

An additional helpful element for the diagnosis of SQTS is the analysis of the QT adaptation during 24-h ECG recordings and exercise stress test (Giustetto et al. 2015). Indeed, the QT interval does not shorten physiologically in response to heart rate increases, and therefore the slope of the QT-RR relationship is usually less steep (Giustetto et al. 2015). Also the analysis of the PQ interval could provide some insights. Indeed, most SQTS patients have a PQ depression $\geq$0.05 mV, rarely observed in healthy individuals, that may be due to the heterogeneous abbreviation of atrial repolarization (Tülümen et al. 2014).

7.2.4 Differential Diagnosis

Main differential diagnoses are reversible causes of a short QTc such as hypercalcaemia, hyperkalaemia, digital toxicity, acidosis and hyperthermia. Also, androgens use and increased vagal tone may shorten the QT interval. Therefore, all possible causes of transient short QT should be ruled out to reach a correct diagnosis (Schimpf et al. 2005; Wolpert et al. 2005; Liu et al. 2003).

7.2.5 Therapy

The management of SQTS patients is quite controversial, above all in asymptomatic patients, as risk stratification remains challenging and a high rate of inappropriate shocks due to T-wave oversensing is reported when an ICD is implanted (Schimpf et al. 2003). While in survivors of a CA or in patients with spontaneous sustained VT an ICD is clearly indicated, for asymptomatic patients there are no data supporting the role of invasive electrophysiological study with programmed ventricular stimulation to stratify risk (Priori et al. 2015).

Among all available pharmacological treatments, hydroquinidine (HQ) appears to induce normalization of the QT interval and of the effective refractory period in patients with *KCNH2* mutations, whereas in those without a mutation, the effect appears weaker and significantly variable (Giustetto et al. 2011; Gaita et al. 2004). Sotalol is considered a possible alternative to hydroquinidine (Priori et al. 2015). When HQ was tested in 41% of the patients of the European SQTS Registry, it was found to prevent the induction of ventricular arrhythmias during electrophysiological study in all studied patients, while none of the treated patients had arrhythmic events during follow-up (Giustetto et al. 2011). Therapy was interrupted only by 9% of the patients because of gastroenteric intolerance or dermatitis (Giustetto et al. 2011). The incidence of arrhythmic events during follow-up (64 $\pm$ 27 months) was 4.9% per year in pharmacologically untreated patients, while no arrhythmias occurred in patients on HQ. Ventricular fibrillation occurred in already symptomatic untreated patients (Giustetto et al. 2011). These data support the use of HQ as an alternative option to ICD in those patients who would qualify for an ICD but present contraindications or refusal (Priori et al. 2015). The use of HQ or sotalol may be considered also in asymptomatic patients with a family history of SCD (Priori et al. 2015) and in those who already have an ICD to prevent the occurrence of multiple arrhythmic events.

Compliance with Ethical Standards

Conflict of Interest LC declares that she has no conflict of interest. MCK declares that she has no conflict of interest. SC declares that she has no conflict of interest.

Ethical Approval All procedures performed in studies involving human participants were in accordance with the ethical standards of the institutional and/or national research committee and with the 1964 Helsinki declaration and its later amendments or comparable ethical standards. This article does not contain any studies with animals performed by any of the authors.

References

Abbott GW, Sesti F, Splawski I, et al. MiRP1 forms IKr potassium channels with HERG and is associated with cardiac arrhythmia. Cell. 1999;97(2):175–87.

Ackerman MJ, Priori SG, Willems S, et al. HRS/EHRA expert consensus statement on the state of genetic testing for the channelopathies and cardiomyopathies: this document was developed as a partnership between the Heart Rhythm Society (HRS) and the European Heart Rhythm Association (EHRA). Europace. 2011;13(8):1077–109. Erratum in: Europace. 2012 Feb;14(2):277.

Altmann HM, Tester DJ, Will ML, et al. Homozygous/compound heterozygous triadin mutations associated with autosomal-recessive long-QT syndrome and pediatric sudden cardiac arrest: elucidation of the triadin knockout syndrome. Circulation. 2015;131(23):2051–60.

Amin AS, Giudicessi JR, Tijsen AJ, et al. Variants in the 3' untranslated region of the KCNQ1-encoded Kv7.1 potassium channel modify disease severity in patients with type 1 long QT syndrome in an allele-specific manner. Eur Heart J. 2012;33(6):714–23.

Andersen ED, Krasilnikoff PA, Overvad H. Intermittent muscular weakness, extrasystoles, and multiple developmental anomalies. A new syndrome? Acta Paediatr Scand. 1971;60(5):559–64.

Anttonen O, Junttila MJ, Rissanen H, et al. Prevalence and prognostic significance of short QT interval in a middle-aged Finnish population. Circulation. 2007;116(7):714–20.

Antzelevitch C, Pollevick GD, Cordeiro JM, et al. Loss-of-function mutations in the cardiac calcium channel underlie a new clinical entity characterized by ST-segment elevation, short QT intervals, and sudden cardiac death. Circulation. 2007;115(4):442–9.

Arking DE, Pfeufer A, Post W, et al. A common genetic variant in the NOS1 regulator NOS1AP modulates cardiac repolarization. Nat Genet. 2006;38(6):644–51.

Arking DE, Pulit SL, Crotti L, et al. Genetic association study of QT interval highlights role for calcium signaling pathways in myocardial repolarization. Nat Genet. 2014;46(8):826–36.

Arnestad M, Crotti L, Rognum TO, et al. Prevalence of long-QT syndrome gene variants in sudden infant death syndrome. Circulation. 2007;115(3):361–7.

Attwell D, Lee JA. A cellular basis for the primary long Q-T syndromes. Lancet. 1988;1 (8595):1136–9.

Bai CX, Kurokawa J, Tamagawa M, et al. Nontranscriptional regulation of cardiac repolarization currents by testosterone. Circulation. 2005;112(12):1701–10.

Barc J, Briec F, Schmitt S, et al. Screening for copy number variation in genes associated with the long QT syndrome: clinical relevance. J Am Coll Cardiol. 2011;57(1):40–7.

Barhanin J, Lesage F, Guillemare E, et al. K(V)LQT1 and lsK (minK) proteins associate to form the I(Ks) cardiac potassium current. Nature. 1996;384(6604):78–80.

Bazett HC. An analysis of the time-relations of electrocardiograms. Ann Noninvasive Electrocardiol. 1997;2(2):177–94.

Bellocq C, van Ginneken AC, Bezzina CR, et al. Mutation in the KCNQ1 gene leading to the short QT-interval syndrome. Circulation. 2004;109(20):2394–7.

Bennett PB, Yazawa K, Makita N, et al. Molecular mechanism for an inherited cardiac arrhythmia. Nature. 1995;376(6542):683–5.

Berthet M, Denjoy I, Donger C, et al. C-terminal HERG mutations: the role of hypokalemia and a KCNQ1-associated mutation in cardiac event occurrence. Circulation. 1999;99(11):1464–70.

Bianchi L, Shen Z, Dennis AT, et al. Cellular dysfunction of LQT5-minK mutants: abnormalities of IKs, IKr and trafficking in long QT syndrome. Hum Mol Genet. 1999;8(8):1499–507.

Bianchi L, Priori SG, Napolitano C, et al. Mechanisms of I(Ks) suppression in LQT1 mutants. Am J Physiol Heart Circ Physiol. 2000;279(6):H3003–11.

Boczek NJ, Best JM, Tester DJ, et al. Exome sequencing and systems biology converge to identify novel mutations in the L-type calcium channel, CACNA1C, linked to autosomal dominant long QT syndrome. Circ Cardiovasc Genet. 2013;6(3):279–89.

Boczek NJ, Gomez-Hurtado N, Ye D, et al. Spectrum and prevalence of CALM1-, CALM2-, and CALM3-encoded calmodulin variants in long QT syndrome and functional characterization of a novel long QT syndrome-associated calmodulin missense variant, E141G. Circ Cardiovasc Genet. 2016;9(2):136–46.

Brink PA, Crotti L, Corfield V, et al. Phenotypic variability and unusual clinical severity of congenital long QT syndrome in a founder population. Circulation. 2005;112:2602–10.

Brugada R, Hong K, Dumaine R, et al. Sudden death associated with short-QT syndrome linked to mutations in HERG. Circulation. 2004;109(1):30–5.

Buber J, Mathew J, Moss AJ, et al. Risk of recurrent cardiac events after onset of menopause in women with congenital long-QT syndrome types 1 and 2. Circulation. 2011;123(24):2784–91.

Canún S, Pérez N, Beirana LG. Andersen syndrome autosomal dominant in three generations. Am J Med Genet. 1999;85(2):147–56.

Carnethon MR, Anthony MS, Cascio WE, et al. A prospective evaluation of the risk of QT prolongation with hormone replacement therapy: the atherosclerosis risk in communities study. Ann Epidemiol. 2003;13(7):530–6.

Chen L, Marquardt ML, Tester DJ, et al. Mutation of an A-kinase-anchoring protein causes long-QT syndrome. Proc Natl Acad Sci USA. 2007;104(52):20990–5.

Chockalingam P, Crotti L, Girardengo G, et al. Not all beta-blockers are equal in the management of long QT syndrome types 1 and 2. J Am Coll Cardiol. 2012;60(20):2092–9.

Choi G, Kopplin LJ, Tester DJ, et al. Spectrum and frequency of cardiac channel defects in swimming-triggered arrhythmia syndromes. Circulation. 2004;110(15):2119–24.

Chopra N, Knollmann BC. Triadin regulates cardiac muscle couplon structure and microdomain Ca (2+) signalling: a path towards ventricular arrhythmias. Cardiovasc Res. 2013;98(2):187–91.

Chopra N, Yang T, Asghari P, et al. Ablation of triadin causes loss of cardiac Ca2+ release units, impaired excitation-contraction coupling, and cardiac arrhythmias. Proc Natl Acad Sci U S A. 2009;106(18):7636–41.

Chouabe C, Neyroud N, Guicheney P, et al. Properties of KvLQT1 K+ channel mutations in Romano-Ward and Jervell and Lange-Nielsen inherited cardiac arrhythmias. EMBO J. 1997;116(17):5472–9.

Collura CA, Johnson JN, Moir C, et al. Left cardiac sympathetic denervation for the treatment of long QT syndrome and catecholaminergic polymorphic ventricular tachycardia using video-assisted thoracic surgery. Heart Rhythm. 2009;6(6):752–9.

Conrath CE, Opthof T. Ventricular repolarization: an overview of (patho)physiology, sympathetic effects and genetic aspects. Prog Biophys Mol Biol. 2006;92(3):269–307.

Crotti L, Lundquist AL, Insolia R, et al. KCNH2-K897T is a genetic modifier of latent congenital long-QT syndrome. Circulation. 2005;112(9):1251–8.

Crotti L, Spazzolini C, Schwartz PJ, et al. The common long-QT syndrome mutation KCNQ1/A341V causes unusually severe clinical manifestations in patients with different ethnic backgrounds: toward a mutation-specific risk stratification. Circulation. 2007;116(21):2366–75.

Crotti L, Celano G, Dagradi F, et al. Congenital long QT syndrome. Orphanet J Rare Dis. 2008;3:18.

Crotti L, Lewandowska MA, Schwartz PJ, et al. A KCNH2 branch point mutation causing aberrant splicing contributes to an explanation of genotype-negative long QT syndrome. Heart Rhythm. 2009a;6(2):212–8.

Crotti L, Monti MC, Insolia R, et al. NOS1AP is a genetic modifier of the long-QT syndrome. Circulation. 2009b;120(17):1657–63.

Crotti L, Spazzolini C, Porretta AP, et al. Vagal reflexes following an exercise stress test: a simple clinical tool for gene-specific risk stratification in the long QT syndrome. J Am Coll Cardiol. 2012;60:2515–224.

Crotti L, Johnson CN, Graf E, et al. Calmodulin mutations associated with recurrent cardiac arrest in infants. Circulation. 2013a;127(9):1009–17.

Crotti L, Tester DJ, White WM, et al. Long QT syndrome-associated mutations in intrauterine fetal death. JAMA. 2013b;309(14):1473–82.

Crotti L, Dossena C, Spazzolini C, et al. LQTS diagnosis in genotype-negative athletes with a long QT interval. A different clinical entity? Eur Heart J. 2016a;37(Abstract Suppl):207.

Crotti L, Lahtinen AM, Spazzolini C, et al. Genetic modifiers for the long-QT syndrome: how important is the role of variants in the 3' untranslated region of KCNQ1? Circ Cardiovasc Genet. 2016b;9(4):330–9.

Crotti L, Spazzolini C, Boczek NJ, et al. International Calmodulinopathy Registry (ICaMR). Circulation. 2016c;134:A14840.

Curran ME, Splawski I, Timothy KW, et al. A molecular basis for cardiac arrhythmia: HERG mutations cause long QT syndrome. Cell. 1995;80(5):795–803.

Dahimène S, Alcoléa S, Naud P, et al. The N-terminal juxtamembranous domain of KCNQ1 is critical for channel surface expression: implications in the Romano-Ward LQT1 syndrome. Circ Res. 2006;99(10):1076–83.

De Ferrari GM, Schwartz PJ. Long QT syndrome, a purely electrical disease? Not anymore. Eur Heart J. 2009;30(3):253–5.

De Ferrari GM, Schwartz PJ. Vox clamantis in deserto. We spoke but nobody was listening: echocardiography can help risk stratification of the long-QT syndrome. Eur Heart J. 2015;36 (3):148–50.

De Ferrari GM, Nador F, Beria G, et al. Effect of calcium channel block on the wall motion abnormality of the idiopathic long QT syndrome. Circulation. 1994;89(5):2126–32.

de Villiers CP, van der Merwe L, Crotti L, et al. AKAP9 is a genetic modifier of congenital long-QT syndrome type 1. Circ Cardiovasc Genet. 2014;7(5):599–606.

Delannoy E, Sacher F, Maury P, et al. Cardiac characteristics and long-term outcome in Andersen-Tawil syndrome patients related to KCNJ2 mutation. Europace. 2013;15(12):1805–11.

Delisle BP, Anson BD, Rajamani S, et al. Biology of cardiac arrhythmias: ion channel protein trafficking. Circ Res. 2004;94(11):1418–28.

Dhutia H, Malhotra A, Parpia S, et al. The prevalence and significance of a short QT interval in 18,825 low-risk individuals including athletes. Br J Sports Med. 2016;50(2):124–9.

Donaldson MR, Jensen JL, Tristani-Firouzi M, et al. PIP2 binding residues of Kir2.1 are common targets of mutations causing Andersen syndrome. Neurology. 2003;60(11):1811–6.

Donger C, Denjoy I, Berthet M, et al. KVLQT1 C-terminal missense mutation causes a forme fruste long-QT syndrome. Circulation. 1997;96(9):2778–81.

Drici MD, Burklow TR, Haridasse V, et al. Sex hormones prolong the QT interval and downregulate potassium channel expression in the rabbit heart. Circulation. 1996;94(6):1471–4.

Duchatelet S, Crotti L, Peat RA, et al. Identification of a KCNQ1 polymorphism acting as a protective modifier against arrhythmic risk in long-QT syndrome. Circ Cardiovasc Genet. 2013;6(4):354–61.

Dumaine R, Wang Q, Keating MT, et al. Multiple mechanisms of Na+ channel-linked long-QT syndrome. Circ Res. 1996;78:916–24.

Earle N, Yeo Han D, Pilbrow A, et al. Single nucleotide polymorphisms in arrhythmia genes modify the risk of cardiac events and sudden death in long QT syndrome. Heart Rhythm. 2014;11 (1):76–82.

Eddy CA, MacCormick JM, Chung SK, et al. Identification of large gene deletions and duplications in KCNQ1 and KCNH2 in patients with long QT syndrome. Heart Rhythm. 2008;5(9):1275–81.

Etheridge SP, Sanatani S, Cohen MI, et al. Long QT syndrome in children in the era of implantable defibrillators. J Am Coll Cardiol. 2007;50(14):1335–40.
Etheridge SP, Bowles NE, Arrington CB, et al. Somatic mosaicism contributes to phenotypic variation in Timothy syndrome. Am J Med Genet A. 2011;155A(10):2578–83.
Gaita F, Giustetto C, Bianchi F, et al. Short QT Syndrome: a familial cause of sudden death. Circulation. 2003;108(8):965–70.
Gaita F, Giustetto C, Bianchi F, et al. Short QT syndrome: pharmacological treatment. J Am Coll Cardiol. 2004;43(8):1494–9.
Gallagher MM, Magliano G, Yap YG, et al. Distribution and prognostic significance of QT intervals in the lowest half centile in 12,012 apparently healthy persons. Am J Cardiol. 2006;98(7):933–5.
George AL Jr. Calmodulinopathy: a genetic trilogy. Heart Rhythm. 2015;12(2):423–4.
Gillis J, Burashnikov E, Antzelevitch C, et al. Long QT, syndactyly, joint contractures, stroke and novel CACNA1C mutation: expanding the spectrum of Timothy syndrome. Am J Med Genet A. 2012;158A(1):182–7.
Giustetto C, Schimpf R, Mazzanti A, et al. Long-term follow-up of patients with short QT syndrome. J Am Coll Cardiol. 2011;58(6):587–95.
Giustetto C, Scrocco C, Schimpf R, et al. Usefulness of exercise test in the diagnosis of short QT syndrome. Europace. 2015;17(4):628–34.
Goldenberg I, Horr S, Moss AJ, et al. Risk for life-threatening cardiac events in patients with genotype-confirmed long-QT syndrome and normal-range corrected QT intervals. J Am Coll Cardiol. 2011;57(1):51–9.
Gollob MH, Redpath CJ, Roberts JD. The short QT syndrome: proposed diagnostic criteria. J Am Coll Cardiol. 2011;57(7):802–12.
Grant AO. Cardiac ion channels. Circ Arrhythm Electrophysiol. 2009;2(2):185–94.
Gussak I, Brugada P, Brugada J, et al. Idiopathic short QT interval: a new clinical syndrome? Cardiology. 2000;94(2):99–102.
Haitin Y, Attali B. The C-terminus of Kv7 channels: a multifunctional module. J Physiol. 2008;586 (7):1803–10.
Harmer SC, Tinker A. The role of abnormal trafficking of KCNE1 in long QT syndrome 5. Biochem Soc Trans. 2007;35(Pt 5):1074–6.
Haugaa KH, Edvardsen T, Leren TP, et al. Left ventricular mechanical dispersion by tissue Doppler imaging: a novel approach for identifying high-risk individuals with long QT syndrome. Eur Heart J. 2009;30(3):330–7.
Hayashi K, Konno T, Fujino N, et al. Impact of updated diagnostic criteria for long QT syndrome on clinical detection of diseased patients. JACC Clin Electrophysiol. 2016;2(3):279–87.
Heradien MJ, Goosen A, Crotti L, et al. Does pregnancy increase cardiac risk for LQT1 patients with the KCNQ1-A341V mutation? J Am Coll Cardiol. 2006;48:1410–5.
Hong K, Bjerregaard P, Gussak I, et al. Short QT syndrome and atrial fibrillation caused by mutation in KCNH2. J Cardiovasc Electrophysiol. 2005;16(4):394–6.
Hoorntje T, Alders M, van Tintelen P, et al. Homozygous premature truncation of the HERG protein: the human HERG knockout. Circulation. 1999;100(12):1264–7.
Horner JM, Horner MM, Ackerman MJ. The diagnostic utility of recovery phase QTc during treadmill exercise stress testing in the evaluation of long QT syndrome. Heart Rhythm. 2011;8 (11):1698–704.
Itoh H, Crotti L, Aiba T, et al. The genetics underlying acquired long QT syndrome: impact for genetic screening. Eur Heart J. 2016;37(18):1456–64.
Jervell A, Lange-Nielsen F. Congenital deaf-mutism, functional heart disease with prolongation of the Q-T interval, and sudden death. Am Heart J. 1957;54(1):59–68.
Johnson JN, Ackerman MJ. Competitive sports participation in athletes with congenital long QT syndrome. JAMA. 2012;308(8):764–5.

Johnson WH Jr, Yang P, Yang T, et al. Clinical, genetic, and biophysical characterization of a homozygous HERG mutation causing severe neonatal long QT syndrome. Pediatr Res. 2003;53(5):744–8.

Kääb S, Pfeufer A, Hinterseer M, et al. Long QT syndrome. Why does sex matter? Z Kardiol. 2004;93(9):641–5.

Kadish AH, Greenland P, Limacher MC, et al. Estrogen and progestin use and the QT interval in postmenopausal women. Ann Noninvasive Electrocardiol. 2004;9(4):366–74.

Keating MT, Sanguinetti MC. Molecular and cellular mechanisms of cardiac arrhythmias. Cell. 2001;104(4):569–80.

Kim J, Ghosh S, Liu H, et al. Calmodulin mediates Ca2+ sensitivity of sodium channels. J Biol Chem. 2004;279:45004–12.

Kirilmaz A, Ulusoy RE, Kardesoglu E, et al. Short QT interval syndrome: a case report. J Electrocardiol. 2005;38(4):371–4.

Klein R, Ganelin R, Marks JF, et al. Periodic paralysis with cardiac arrhythmia. J Pediatr. 1963;62(3):371–85.

Kolder IC, Tanck MW, Postema PG, et al. Analysis for genetic modifiers of disease severity in patients with long-QT syndrome type 2. Circ Cardiovasc Genet. 2015;8(3):447–56.

Koopmann TT, Alders M, Jongbloed RJ, et al. Long QT syndrome caused by a large duplication in the KCNH2 (HERG) gene undetectable by current polymerase chain reaction-based exon-scanning methodologies. Heart Rhythm. 2006;3(1):52–5.

Kurokawa J, Chen L, Kass RS. Requirement of subunit expression for cAMP mediated regulation of a heart potassium channel. Proc Natl Acad Sci U S A. 2003;100(4):2122–7.

Landstrom AP, Boczek NJ, Ye D, et al. Novel long QT syndrome-associated missense mutation, L762F, in CACNA1C-encoded L-type calcium channel imparts a slower inactivation tau and increased sustained and window current. Int J Cardiol. 2016;220:290–8.

Larsen LA, Fosdal I, Andersen PS, et al. Recessive Romano-Ward syndrome associated with compound heterozygosity for two mutations in the KVLQT1 gene. Eur J Hum Genet. 1999;7(6):724–8.

Le Scouarnec S, Bhasin N, Vieyres C, et al. Dysfunction in ankyrin-B-dependent ion channel and transporter targeting causes human sinus node disease. Proc Natl Acad Sci U S A. 2008;105(40):15617–22.

Lee MP, Ravenel JD, Hu RJ, et al. Targeted disruption of the Kvlqtl gene causes deafness and gastric hyperplasia in mice. J Clin Invest. 2000;106(12):1447–55.

Leinonen JT, Crotti L, Djupsjöbacka A, et al. The genetics underlying idiopathic ventricular fibrillation: A special role for catecholaminergic polymorphic ventricular tachycardia? Int J Cardiol. 2018;250:139–45.

Liu X-K, Katchman A, Whitfield BH, et al. In vivo androgen treatment shortens the QT interval and increases the densities of inward and delayed rectifier potassium currents in orchiectomized male rabbits. Cardiovasc Res. 2003;57(1):28–36.

Lo-A-Njoe SM, Wilde AA, van Erven L, et al. Syndactyly and long QT syndrome (CaV1.2 missense mutation G406R) is associated with hypertrophic cardiomyopathy. Heart Rhythm. 2005;2(12):1365–8.

Locati EH, Pancaldi A, Pala M, et al. Exercise-induced electrocardiographic changes in patients with the long QT syndrome. Circulation. 1988;78(Suppl II):42.

Lu LX, Zhou W, Zhang X, et al. Short QT syndrome: a case report and review of literature. Resuscitation. 2006;71(1):115–21.

Lupoglazoff JM, Cheav T, Baroudi G, et al. Homozygous SCN5A mutation in long-QT syndrome with functional two-to-one atrioventricular block. Circ Res. 2001;89(2):E16–21.

Makita N, Behr E, Shimizu W, et al. The E1784K mutation in SCN5A is associated with mixed clinical phenotype of type 3 long QT syndrome. J Clin Invest. 2008;118(6):2219–29.

Makita N, Yagihara N, Crotti L, et al. Novel calmodulin mutations associated with congenital arrhythmia susceptibility. Circ Cardiovasc Genet. 2014;7(4):466–74.

Malfatto G, Beria G, Sala S, et al. Quantitative analysis of T wave abnormalities and their prognostic implications in the idiopathic long QT syndrome. J Am Coll Cardiol. 1994;23(2):296–301.

Marks ML, Trippel DL, Keating MT. Long QT syndrome associated with syndactyly identified in females. Am J Cardiol. 1995a;76(10):744–5.

Marks ML, Whisler SL, Clericuzio C, et al. A new form of long QT syndrome associated with syndactyly. J Am Coll Cardiol. 1995b;25(1):59–64.

Marsman RF, Barc J, Beekman L, et al. A mutation in CALM1 encoding calmodulin in familial idiopathic ventricular fibrillation in childhood and adolescence. J Am Coll Cardiol. 2014;63 (3):259–66.

Marx SO, Kurokawa J, Reiken S, et al. Requirement of a macromolecular signaling complex for beta adrenergic receptor modulation of the KCNQ1-KCNE1 potassium channel. Science. 2002;295(5554):496–9.

Mazzanti A, Kanthan A, Monteforte N, et al. Novel insight into the natural history of short QT syndrome. J Am Coll Cardiol. 2014;63(13):1300–8.

Medeiros-Domingo A, Kaku T, Tester DJ, et al. SCN4B-encoded sodium channel beta4 subunit in congenital long-QT syndrome. Circulation. 2007;116(2):134–42.

Merri M, Benhorin J, Alberti M, et al. Electrocardiographic quantitation of ventricular repolarization. Circulation. 1989;80(5):1301–8.

Mohler PJ, Schott JJ, Gramolini AO, et al. Ankyrin-B mutation causes type 4 long-QT cardiac arrhythmia and sudden cardiac death. Nature. 2003;421(6923):634–9.

Mohler PJ, Splawski I, Napolitano C, et al. A cardiac arrhythmia syndrome caused by loss of ankyrin-B function. Proc Natl Acad Sci U S A. 2004;101(24):9137–42.

Mohler PJ, Le Scouarnec S, Denjoy I, et al. Defining the cellular phenotype of 'ankyrin-B syndrome' variants: human ANK2 variants associated with clinical phenotypes display a spectrum of activities in cardiomyocytes. Circulation. 2007;115(4):432–41.

Moss AJ, McDonald J. Unilateral cervicothoracic sympathetic ganglionectomy for the treatment of long QT interval syndrome. N Engl J Med. 1971;285(16):903–4.

Moss AJ, Schwartz PJ, Crampton RS, et al. The long QT syndrome: a prospective international study. Circulation. 1985;71(1):17–21.

Moss AJ, Schwartz PJ, Crampton RS, et al. The long QT syndrome. Prospective longitudinal study of 328 families. Circulation. 1991;84(3):1136–44.

Moss AJ, Robinson JL, Gessman L, et al. Comparison of clinical and genetic variables of cardiac events associated with loud noise versus swimming among subjects with the long QT syndrome. Am J Cardiol. 1999;84(8):876–9.

Moss AJ, Zareba W, Hall WJ, et al. Effectiveness and limitations of blocker therapy in congenital long-QT syndrome. Circulation. 2000;101(6):616–23.

Moss AJ, Zareba W, Kaufman ES, et al. Increased risk of arrhythmic events in long-QT syndrome with mutations in the pore region of the human ether-a-gogo-related gene potassium channel. Circulation. 2002;105(7):794–9.

Moss AJ, Shimizu W, Wilde AA, et al. Clinical aspects of type-1 long-QT syndrome by location, coding type, and biophysical function of mutations involving the KCNQ1 gene. Circulation. 2007;115(19):2481–9.

Moss AJ, Zareba W, Schwarz KQ, et al. Ranolazine shortens repolarization in patients with sustained inward sodium current due to type-3 long-QT syndrome. J Cardiovasc Electrophysiol. 2008;19(12):1289–93.

Nador F, Beria G, De Ferrari GM, et al. Unsuspected echocardiographic abnormality in the long QT syndrome. Diagnostic, prognostic, and pathogenetic implications. Circulation. 1991;84 (4):1530–42.

Nagaoka I, Shimizu W, Itoh H, et al. Mutation site dependent variability of cardiac events in Japanese LQT2 form of congenital long-QT syndrome. Circ J. 2008;72(5):694–9.

Napolitano C, Antzelevitch C. Phenotypical manifestations of mutations in the genes encoding subunits of the cardiac voltage-dependent L-type calcium channel. Circ Res. 2011;108 (5):607–18.

Napolitano C, Splawski I, Timothy KW, et al. Timothy syndrome. In: Adam MP, Ardinger HH, Pagon RA, et al., editors. GeneReviews®. Seattle, WA: University of Washington. Seattle; 1993-2018; 2006 Feb 15 [Updated 2015 Jul 16].

Napolitano C, Schwartz PJ, Brown AM, et al. Evidence for a cardiac ion channel mutation underlying drug-induced QT prolongation and life-threatening arrhythmias. J Cardiovasc Electrophysiol. 2000;11(6):691–6.

Napolitano C, Priori SG, Schwartz PJ, et al. Genetic testing in the long QT syndrome: development and validation of an efficient approach to genotyping in clinical practice. JAMA. 2005;294 (23):2975–80.

Newton-Cheh C, Larson MG, Corey DC, et al. QT interval is a heritable quantitative trait with evidence of linkage to chromosome 3 in a genome-wide linkage analysis: The Framingham Heart Study. Heart Rhythm. 2005;2(3):277–84.

Newton-Cheh C, Eijgelsheim M, Rice KM, et al. Common variants at ten loci influence QT interval duration in the QTGEN Study. Nat Genet. 2009;41(4):399–406.

Neyroud N, Tesson F, Denjoy I, et al. A novel mutation in the potassium channel gene KVLQT1 causes the Jervell and Lange-Nielsen cardioauditory syndrome. Nat Genet. 1997;15(2):186–9.

Nguyen HL, Pieper GH, Wilders R. Andersen–Tawil syndrome: clinical and molecular aspects. Int J Cardiol. 2013;170(1):1–16.

Nof E, Cordeiro JM, Pérez GJ, et al. A common single nucleotide polymorphism can exacerbate long-QT type 2 syndrome leading to sudden infant death. Circ Cardiovasc Genet. 2010;3 (2):199–206.

Nyegaard M, Overgaard MT, Søndergaard MT, et al. Mutations in calmodulin cause ventricular tachycardia and sudden cardiac death. Am J Hum Genet. 2012;91(4):703–12.

Odero A, Bozzani A, De Ferrari GM, et al. Left cardiac sympathetic denervation for the prevention of life-threatening arrhythmias: the surgical supraclavicular approach to cervicothoracic sympathectomy. Heart Rhythm. 2010;7(8):1161–5.

Pellizzón OA, Kalaizich L, Ptácek LJ, et al. Flecainide suppresses bidirectional ventricular tachycardia and reverses tachycardia-induced cardiomyopathy in Andersen-Tawil syndrome. J Cardiovasc Electrophysiol. 2008;19(1):95–7.

Pfeufer A, Sanna S, Arking DE, et al. Common variants at ten loci modulate the QT interval duration in the QTSCD Study. Nat Genet. 2009;41(4):407–14.

Piippo K, Laitinen P, Swan H, et al. Homozygosity for a HERG potassium channel mutation causes a severe form of long QT syndrome: identification of an apparent founder mutation in the Finns. J Am Coll Cardiol. 2000;35(7):1919–25.

Pipilas DC, Johnson CN, Webster G, et al. Novel calmodulin mutations associated with congenital long QT syndrome affect calcium current in human cardiomyocytes. Heart Rhythm. 2016;13 (10):2012–9.

Plaster NM, Tawil R, Tristani-Firouzi M, et al. Mutations in Kir2.1 cause the developmental and episodic electrical phenotypes of Andersen's syndrome. Cell. 2001;105(4):511–9.

Priori SG, Napolitano C, Schwartz PJ. Low penetrance in the long-QT syndrome: clinical impact. Circulation. 1999;99(4):529–33.

Priori SG, Schwartz PJ, Napolitano C, et al. Risk stratification in the long-QT syndrome. N Engl J Med. 2003;348(19):1866–74.

Priori SG, Napolitano C, Schwartz PJ, et al. Association of long QT syndrome loci and cardiac events among patients treated with beta-blockers. JAMA. 2004;292(11):1341–4.

Priori SG, Pandit SV, Rivolta I, et al. A novel form of short QT syndrome (SQT3) is caused by a mutation in the KCNJ2 gene. Circ Res. 2005;96(7):800–7.

Priori SG, Wilde AA, Horie M, et al. HRS/EHRA/APHRS expert consensus statement on the diagnosis and management of patients with inherited primary arrhythmia syndromes. Heart Rhythm. 2013;10(12):1932–63.

Priori SG, Blomström-Lundqvist C, Mazzanti A, et al. ESC Guidelines for the management of patients with ventricular arrhythmias and the prevention of sudden cardiac death: the task force for the management of patients with ventricular arrhythmias and the prevention of sudden cardiac death of the European Society of Cardiology (ESC). Endorsed by: Association for European Paediatric and Congenital Cardiology (AEPC). Eur Heart J. 2015;36:2793–867.

Quaglini S, Rognoni C, Spazzolini C, et al. Cost-effectiveness of neonatal ECG screening for the long QT syndrome. Eur Heart J. 2006;27(15):1824–32.
Rashba EJ, Zareba W, Moss AJ, et al. Influence of pregnancy on the risk for cardiac events in patients with hereditary long QT syndrome. Circulation. 1998;97(5):451–6.
Reed GJ, Boczek NJ, Etheridge SP, et al. CALM3 mutation associated with long QT syndrome. Heart Rhythm. 2015;12(2):419–22.
Reichenbach H, Meister EM, Theile H. The heart-hand syndrome. A new variant of disorders of heart conduction and syndactylia including osseous changes in hands and feet. Kinderarztl Prax. 1992;60(2):54–6.
Rocchetti M, Sala L, Dreizehnter L, et al. Elucidating the arrhythmogenic mechanism of long QT syndrome caused by the CALM1-F142L mutation using patient-specific induced pluripotent stem cell-derived cardiomyocytes. Cardiovasc Res. 2017;113(5):531–41.
Romano C, Gemme G, Pongiglione R. Rare cardiac arrhythmias of the pediatric age. ii. Syncopal attacks due to paroxysmal ventricular fibrillation (presentation of 1st case in italian pediatric literature). Clin Pediatr. 1963;45:656–83.
Ruan Y, Liu N, Bloise R, et al. Gating properties of SCN5A mutations and the response to mexiletine in long-QT syndrome type 3 patients. Circulation. 2007;116(10):1137–44.
Saito T, Ciobotaru A, Bopassa JC, et al. Estrogen contributes to gender differences in mouse ventricular repolarization. Circ Res. 2009;105(4):343–52.
Sanguinetti MC, Curran ME, Spector PS, et al. Spectrum of HERG K+-channel dysfunction in an inherited cardiac arrhythmia. Proc Natl Acad Sci U S A. 1996a;93(5):2208–12.
Sanguinetti MC, Curran ME, Zou A, et al. Coassembly of K(V)LQT1 and minK (IsK) proteins to form cardiac I(Ks) potassium channel. Nature. 1996b;384(6604):80–3.
Sansone V, Tawil R. Management and treatment of Andersen-Tawil syndrome (ATS). Neurotherapeutics. 2007;4(2):233–7.
Sansone V, Griggs RC, Meola G, et al. Andersen's syndrome: a distinct periodic paralysis. Ann Neurol. 1997;42(3):305–12.
Saul JP, Schwartz PJ, Ackerman MJ, et al. Rationale and objectives for ECG screening in infancy. Heart Rhythm. 2014;11(12):2316–21.
Schimpf R, Wolpert C, Bianchi F, et al. Congenital short QT syndrome and implantable cardioverter defibrillator treatment: inherent risk for inappropriate shock delivery. J Cardiovasc Electrophysiol. 2003;14(12):1273–7.
Schimpf R, Wolpert C, Gaita F, et al. Short QT syndrome. Cardiovasc Res. 2005;67(3):357–66.
Schulze-Bahr E, Wang Q, Wedekind H, et al. KCNE1 mutations cause Jervell and Lange-Nielsen syndrome. Nat Genet. 1997;17(3):267–8.
Schwartz PJ. The idiopathic long QT syndrome: the need for a prospective registry. Eur Heart J. 1983;4(8):529–31.
Schwartz PJ. Idiopathic long QT syndrome: progress and questions. Am Heart J. 1985;109 (2):399–411.
Schwartz PJ. Prevention of the arrhythmias in the long QT syndrome. In: Kulbertus HE, editor. Medical management of cardiac arrhythmias. Edinburgh: Churchill Livingstone; 1986. p. 153–61.
Schwartz PJ. Stillbirths, sudden infant deaths, and long-QT syndrome: puzzle or mosaic, the pieces of the Jigsaw are being fitted together. Circulation. 2004;109(24):2930–2.
Schwartz PJ. Sudden cardiac death, founder populations, and mushrooms: what is the link with gold mines and modifier genes? Heart Rhythm. 2011;8(4):548–50.
Schwartz PJ. Cardiac sympathetic denervation to prevent life-threatening arrhythmias. Nat Rev Cardiol. 2014;11(6):346–53.
Schwartz PJ, Ackerman MJ. The long QT syndrome: a transatlantic clinical approach to diagnosis and therapy. Eur Heart J. 2013;34(40):3109–16.
Schwartz PJ, Crotti L. QTc behavior during exercise and genetic testing for the long-QT syndrome. Circulation. 2011;124(20):2181–4.

Schwartz PJ, Crotti L. Long QT and short QT syndromes. In: Zipes DP, Jalife J, editors. Cardiac electrophysiology: from cell to bedside. 7th ed. Philadelphia: Elsevier/Saunders; 2017. p. 893–904. ISBN: 9780323447331.

Schwartz PJ, Malliani A. Electrical alternation of the T-wave: clinical and experimental evidence of its relationship with the sympathetic nervous system and with the long Q-T syndrome. Am Heart J. 1975;89(1):45–50.

Schwartz PJ, Moss AJ. Prolonged QT interval: what does it mean? J Cardiovasc Med. 1982;7:1317.

Schwartz PJ, Periti M, Malliani A. The long Q-T syndrome. Am Heart J. 1975;89(3):378–90.

Schwartz PJ, Zaza A, Locati E, et al. Stress and sudden death. The case of the long QT syndrome. Circulation. 1991;83(4 Suppl II):71–80.

Schwartz PJ, Moss AJ, Vincent GM, et al. Diagnostic criteria for the long QT syndrome. An update. Circulation. 1993;88(2):782–4.

Schwartz PJ, Priori SG, Locati EH, et al. Long QT syndrome patients with mutations of the SCN5A and HERG genes have differential responses to Na+ channel blockade and to increases in heart rate: implications for gene-specific therapy. Circulation. 1995;92(12):3381–6.

Schwartz PJ, Priori SG, Spazzolini C, et al. Genotype-phenotype correlation in the long-QT syndrome: gene-specific triggers for life-threatening arrhythmias. Circulation. 2001;103 (1):89–95.

Schwartz PJ, Garson A, Paul T, et al. Guidelines for the interpretation of the neonatal electrocardiogram. A task force of the European Society of Cardiology. Eur Heart J. 2002;23 (17):1329–44.

Schwartz PJ, Priori SG, Cerrone M, et al. Left cardiac sympathetic denervation in the management of high-risk patients affected by the long-QT syndrome. Circulation. 2004;109(15):1826–33.

Schwartz PJ, Spazzolini C, Crotti L, et al. The Jervell and Lange-Nielsen syndrome: natural history, molecular basis, and clinical outcome. Circulation. 2006;113(6):783–90.

Schwartz PJ, Vanoli E, Crotti L, et al. Neural control of heart rate is an arrhythmia risk modifier in long QT syndrome. J Am Coll Cardiol. 2008;51(9):920–9.

Schwartz PJ, Spazzolini C, Crotti L. All LQT3 patients need an ICD: true or false? Heart Rhythm. 2009a;6:113–20.

Schwartz PJ, Stramba-Badiale M, Crotti L, et al. Prevalence of the congenital long-QT syndrome. Circulation. 2009b;120(18):1761–7.

Schwartz PJ, Spazzolini C, Priori SG, et al. Who are the long-QT syndrome patients who receive an implantable cardioverter-defibrillator and what happens to them? Data from the European Long-QT Syndrome Implantable Cardioverter-Defibrillator (LQTS ICD) Registry. Circulation. 2010;122(13):1272–82.

Schwartz PJ, Crotti L, Insolia R. Long-QT syndrome: from genetics to management. Circ Arrhythm Electrophysiol. 2012;5(4):868–77.

Seth R, Moss AJ, McNitt S, et al. Long QT syndrome and pregnancy. J Am Coll Cardiol. 2007;49 (10):1092–8.

Shamgar L, Ma L, Schmitt N, et al. Calmodulin is essential for cardiac IKS channel gating and assembly: impaired function in long-QT mutations. Circ Res. 2006;98(8):1055–63.

Shimizu W, Tanabe Y, Aiba T, et al. Differential effects of beta-blockade on dispersion of repolarization in the absence and presence of sympathetic stimulation between the LQT1 and LQT2 forms of congenital long QT syndrome. J Am Coll Cardiol. 2002;39(12):1984–91.

Spazzolini C, Mullally J, Moss AJ, et al. Clinical implications for patients with long QT syndrome who experience a cardiac event during infancy. J Am Coll Cardiol. 2009;54(9):832–7.

Splawski I, Timothy KW, Vincent GM, et al. Molecular basis of the long-QT syndrome associated with deafness. N Engl J Med. 1997a;336(22):1562–7.

Splawski I, Tristani-Firouzi M, Lehmann MH, et al. Mutations in the hminK gene cause long QT syndrome and suppress IKs function. Nat Genet. 1997b;17(3):338–40.

Splawski I, Shen J, Timothy KW, et al. Spectrum of mutations in long-QT syndrome genes. KVLQT1, HERG, SCN5A, KCNE1, and KCNE2. Circulation. 2000;102(10):1178–85.

Splawski I, Timothy KW, Sharpe LM, et al. Ca(V)1.2 calcium channel dysfunction causes a multisystem disorder including arrhythmia and autism. Cell. 2004;119(1):19–31.

Splawski I, Timothy KW, Decher N, et al. Severe arrhythmia disorder caused by cardiac L-type calcium channel mutations. Proc Natl Acad Sci U S A. 2005;102(23):8089–96.

Stramba-Badiale M, Spagnolo D, Bosi G, et al. Are gender differences in QTc present at birth? MISNES investigators. Multicenter Italian study on neonatal electrocardiography and sudden infant death syndrome. Am J Cardiol. 1995;75(17):1277–8.

Stuhmer W, Conti F, Suzuki H, et al. Structural parts involved in activation and inactivation of the sodium channel. Nature. 1989;339(6226):597–603.

Subbiah RN, Gula LJ, Skanes AC, et al. Andersen-Tawil syndrome: management challenges during pregnancy, labor, and delivery. J Cardiovasc Electrophysiol. 2008;19(9):987–9.

Swan H, Viitasalo M, Piippo K, et al. Sinus node function and ventricular repolarization during exercise stress test in long QT syndrome patients with KvLQT1 and HERG potassium channel defects. J Am Coll Cardiol. 1999;34(3):823–9.

Swayne LA, Murphy NP, Asuri S, et al. Novel variant in the ANK2 membrane-binding domain is associated with ankyrin-B syndrome and structural heart disease in a first nations population with a high rate of long QT syndrome. Circ Cardiovasc Genet. 2017;10(1):e001537.

Sy RW, van der Werf C, Chattha IS, et al. Derivation and validation of a simple exercise-based algorithm for prediction of genetic testing in relatives of LQTS probands. Circulation. 2011;124 (20):2187–94.

Tawil R, Ptacek LJ, Pavlakis SG, et al. Andersen's syndrome: potassium-sensitive periodic paralysis, ventricular ectopy, and dysmorphic features: Andersen's syndrome. Ann Neurol. 1994;35 (3):326–30.

ter Bekke RMA, Haugaa KH, van den Wijngaard A, et al. Electromechanical window negativity in genotyped long-QT syndrome patients: relation to arrhythmia risk. Eur Heart J. 2015;36 (3):179–86.

Tester DJ, Will ML, Haglund CM, et al. Compendium of cardiac channel mutations in 541 consecutive unrelated patients referred for long QT syndrome genetic testing. Heart Rhythm. 2005;2 (5):507–17.

Tester DJ, Will ML, Haglund CM, et al. Effect of clinical phenotype on yield of long QT syndrome genetic testing. J Am Coll Cardiol. 2006;47(4):764–8.

Tomás M, Napolitano C, De Giuli L, et al. Polymorphisms in the NOS1AP gene modulate QT interval duration and risk of arrhythmias in the long QT syndrome. J Am Coll Cardiol. 2010;55 (24):2745–52.

Tristani-Firouzi M, Jensen JL, Donaldson MR, et al. Functional and clinical characterization of KCNJ2 mutations associated with LQT7 (Andersen syndrome). J Clin Invest. 2002;110 (3):381–8.

Tülümen E, Giustetto C, Wolpert C, et al. PQ segment depression in patients with short QT syndrome: a novel marker for diagnosing short QT syndrome? Heart Rhythm. 2014;11 (6):1024–30.

Ueda K, Valdivia C, Medeiros-Domingo A, et al. Syntrophin mutation associated with long QT syndrome through activation of the nNOS-SCN5A macromolecular complex. Proc Natl Acad Sci U S A. 2008;105(27):9355–60.

Vatta M, Ackerman MJ, Ye B, et al. Mutant caveolin-3 induces persistent late sodium current and is associated with long-QT syndrome. Circulation. 2006;114(20):2104–12.

Venance SL, Cannon SC, Fialho D, et al. The primary periodic paralyses: diagnosis, pathogenesis and treatment. Brain J Neurol. 2006;129(Pt 1):8–17.

Vetter DE, Mann JR, Wangemann P, et al. Inner ear defects induced by null mutation of the isk gene. Neuron. 1996;17(6):1251–64.

Vincent GM, Abildskov JA, Burgess MJ. MJ Q-T interval syndromes. Prog Cardiovasc Dis. 1974;16(6):523–30.

Vincent GM, Jaiswal D, Timothy KW. Effects of exercise on heart rate, QT, QTc and QT/QS2 in the Romano-Ward inherited long QT syndrome. Am J Cardiol. 1991;68(5):498–503.

Vincent GM, Schwartz PJ, Denjoy I, et al. High efficacy of beta-blockers in long-QT syndrome type 1: contribution of noncompliance and QT-prolonging drugs to the occurrence of beta-blocker treatment 'failures'. Circulation. 2009;119:215–21.

Wang Q, Shen J, Splawski I, et al. SCN5A mutations associated with an inherited cardiac arrhythmia, long QT syndrome. Cell. 1995;80(5):805–11.

Wang Q, Curran ME, Splawski I, et al. Positional cloning of a novel potassium channel gene: KVLQT1 mutations cause cardiac arrhythmias. Nat Genet. 1996a;12(1):17–23.

Wang DW, Yazawa K, George AL Jr, et al. Characterization of human cardiac Na+ channel mutations in the congenital long QT syndrome. Proc Natl Acad Sci U S A. 1996b;93 (23):13200–5.

Ward OC. A new familial cardiac syndrome in children. J Ir Med Assoc. 1964;54:103–6.

Watanabe H, Makiyama T, Koyama T, et al. High prevalence of early repolarization in short QT syndrome. Heart Rhythm. 2010;7(5):647–52.

Westenskow P, Splawski I, Timothy KW, et al. Compound mutations: a common cause of severe long-QT syndrome. Circulation. 2004;109(15):1834–41.

Wilde AA, Moss AJ, Kaufman ES, et al. Clinical aspects of type 3 long-QT syndrome: an international multicenter study. Circulation. 2016;134(12):872–82.

Wolpert C, Schimpf R, Giustetto C, et al. Further insights into the effect of quinidine in short QT syndrome caused by a mutation in HERG. J Cardiovasc Electrophysiol. 2005;16(1):54–8.

Yang Y, Yang Y, Liang B, et al. Identification of a Kir3.4 mutation in congenital Long QT Syndrome. Am J Hum Genet. 2010;86:872–80.

Yanowitz F, Preston JB, Abildskov JA. Functional distribution of right and left stellate innervation to the ventricles. Production of neurogenic electrocardiographic changes by unilateral alteration of sympathetic tone. Circ Res. 1966;18(4):416–28.

Yin G, Hassan F, Haroun AR, et al. Arrhythmogenic calmodulin mutations disrupt intracellular cardiomyocyte Ca2+ regulation by distinct mechanisms. J Am Heart Assoc. 2014;3(3):e000996.

Yoon G, Oberoi S, Tristani-Firouzi M, et al. Andersen-Tawil syndrome: prospective cohort analysis and expansion of the phenotype. Am J Med Genet A. 2006;140A(4):312–21.

Zhang L, Vincent GM, Baralle M, et al. An intronic mutation causes long QT syndrome. J Am Coll Cardiol. 2004;44(6):1283–91.

Zhang L, Benson DW, Tristani-Firouzi M, et al. Electrocardiographic features in Andersen-Tawil syndrome patients with KCNJ2 mutations: characteristic T-U-wave patterns predict the KCNJ2 genotype. Circulation. 2005;111(21):2720–6.

Zühlke RD, Pitt GS, Deisseroth K, et al. Calmodulin supports both inactivation and facilitation of L-type calcium channels. Nature. 1999;399:159–62.

Brugada Syndrome: Current Perspectives

8

Apichai Khongphatthanayothin and Koonlawee Nademanee

Abstract

Brugada syndrome was first reported as a distinct entity in 1992. It is diagnosed by signature EKG changes including at least 2-mm J-point elevation with coved-type ST elevation and T-wave inversion in at least one right precordial leads (type I Brugada marker). Initially thought of as a rare entity, Brugada syndrome is now widely recognized as a common cause of natural death among young men as a result of ventricular arrhythmia occurring at rest, particularly during sleep. The etiology of this disease is likely multifactorial, with genetic predisposition playing an important role in the pathogenesis. Mutation in the sodium channel gene *SCN5A* is seen in up to 20–25% of patients with at least 17 additional genes were reported to be associated with the disease. However, mutation in a single gene could be implicated in less than 30% of patients with Brugada syndrome, and a recent study showing association of this disease with common single nucleotide polymorphism of *SCN5A*, *SCN10A*, and *HEY2* pointed toward polygenic or oligogenic pattern of inheritance rather than a single gene defect. Despite initial reports of normal structural heart in majority of patients, recent reports showed frequent minor structural abnormalities, especially in the right ventricular outflow tract (RVOT) in these patients. Brugada syndrome is now viewed as a spectrum of cardiomyopathy as well as channelopathy. The pathophysiologic underlying cardiac arrhythmia in Brugada syndrome is still unresolved with continued debate on depolarization versus repolarization defect. Treatment is largely dependent on the symptom with implantable cardioverter-defibrillator

A. Khongphatthanayothin
Chulalongkorn University, Bangkok, Thailand

K. Nademanee (✉)
Chulalongkorn University, Bangkok, Thailand

Pacific Rim Electrophysiology Research Institute, Los Angeles, CA, USA
e-mail: wee@pacificrimep.com

D. Thomas, C. A. Remme (eds.), *Channelopathies in Heart Disease*, Cardiac and Vascular Biology, https://doi.org/10.1007/978-3-319-77812-9_8

(ICD) implantation indicated in patients with severe symptoms. Avoiding medications and/or conditions that predispose the patient to ventricular arrhythmia is advised in all patients. Quinidine, by blocking transient outward current, has been used with some success. RF ablation of the epicardial substrate in the RVOT has been shown to prevent recurrence of ventricular arrhythmia in severe cases.

8.1 Introduction

In 1992, Brugada and Brugada described eight patients who had the following distinct clinical characteristics: (1) the presence of coved ST-segment elevation followed by a negative T wave in the right precordial leads and (2) life-threatening ventricular arrhythmias that could lead to sudden cardiac death, cardiac arrest, or symptoms caused by spontaneous self-terminating ventricular tachycardia (VT) or ventricular fibrillation (VF) episodes, i.e., syncope, agonal respiration, and seizure (Brugada and Brugada 1992). The entity is later widely recognized as "Brugada syndrome."

The Brugada syndrome (BrS) has also been linked with sudden unexpected death syndrome (SUDS) that usually occurs at night in young Southeast Asian men with a structurally normal heart (Nademanee et al. 1997; Vatta et al. 2002). BrS patients often have a history of unexplained sudden cardiac death in the family, and the syndrome is a well-known autosomal dominant inherited arrhythmia disorder and associated with gene mutations that are predominantly confined to the *SCN5A* gene, which encodes for the α-subunit of the cardiac sodium channel, causing loss of sodium current (INa) (Vatta et al. 2002; Wilde et al. 2002; Antzelevitch et al. 2005, 2016; Priori et al. 2013). However, the causal role of these genetic variants remains controversial.

SUDS has the same phenotype as BrS (Nademanee et al. 1997), and shortly thereafter we reported that both syndromes also shared the same genetic and biophysical basis (Vatta et al. 2002). In the past two and a half decades, we have witnessed major progress toward better understanding of the syndrome and gained knowledge in epidemiology, genetic aspects of the syndrome, pathophysiology, and patient management (Wilde et al. 2002, 2010; Antzelevitch et al. 2005, 2016; Priori et al. 2013; Probst et al. 2009; Nademanee et al. 2011; Veerakul and Nademanee 2012). Four consensus reports were published to help in defining diagnostic criteria, risk stratification, and management of BrS patients (Wilde et al. 2002; Antzelevitch et al. 2005, 2016; Priori et al. 2013). However, there are controversies, especially on the role of sodium channel mutations and electrophysiologic mechanisms underlying the syndrome(Wilde et al. 2010), risk stratification (Veerakul and Nademanee 2012; Wilde et al. 2010; Priori et al. 2012; Brugada et al. 2002; Probst et al. 2010; Kamakura et al. 2009; Paul et al. 2007; Eckardt et al. 2005; Takagi et al. 2007; Giustetto et al. 2009), and treatment of asymptomatic patients (Priori et al. 2012; Takagi et al. 2007; Giustetto et al. 2009; Viskin and Rogowski 2007; Veerakul et al. 2008).

8.2 Clinical Presentation and Diagnostic Criteria

The clinical spectrum of BrS patients ranges from asymptomatic to sudden cardiac death (Antzelevitch et al. 2005; Priori et al. 2013; Veerakul and Nademanee 2012). Patients may have a late onset of VT/VF despite having had an abnormal ECG pattern for decades (Nademanee et al. 1997; Antzelevitch et al. 2005). Syncope or seizures due to self-terminating VT/VF episodes are also common as well as agonal respiration and difficulty to arouse at night time again due to self-terminating VF episodes (Nademanee et al. 1997; Antzelevitch et al. 2005).

Life-threatening ventricular tachyarrhythmias in patients with BrS usually consist of VF or a rapid polymorphic VT although rare cases of monomorphic VT have been reported (Boersma et al. 2001; Shimada et al. 1996; Rodríguez-Mañero et al. 2016). In a study looking at the stored electrogram of the implantable cardioverter-defibrillator (ICD), polymorphic VT was often preceded by ventricular premature beats (VPB) with the same morphology as the one that initiated the VT (Kakishita et al. 2000). The VPB typically occurred at the end of the T wave and only rarely followed a long-short sequence. Site of earliest ventricular activation in patients with VT induced in the electrophysiologic lab almost always involves the right ventricular outflow tract (RVOT), at least for the initial beats. VF and sudden cardiac death mostly occur at rest or during sleep or at night in the early morning hours (Nademanee et al. 1997; Matsuo et al. 1999; Aizawa et al. 2016).

Supraventricular tachycardia, including AV nodal reentry tachycardia, atrial flutter, and atrial fibrillation, is common (Eckardt et al. 2001; Morita et al. 2002a; Rodríguez-Mañero et al. 2013). Depolarization abnormalities including prolonged P wave, PR interval, and QRS duration are also frequently observed (Antzelevitch et al. 2005; Takagi et al. 2007; Maury et al. 2013).

Diagnosis of BrS relies on the signature marker (type I Brugada pattern) on the ECG. The two original Brugada consensus (Wilde et al. 2002; Antzelevitch et al. 2005) reports classified the Brugada ECG pattern into three types (Fig. 8.1):

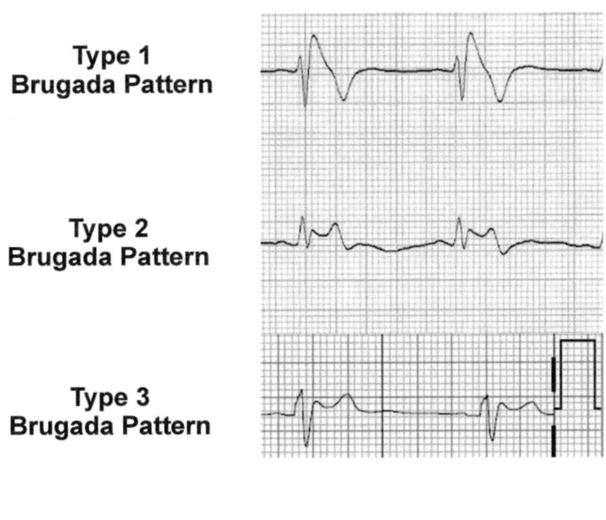

Fig. 8.1 Three types of Brugada ECG pattern. Type 1 is a coved-type pattern and type 2 is a "saddle-back" type which has the ST elevation >2 mm without T-wave inversion. Type 3 pattern is a J-point elevation without ST elevation >1 mm

1. Type 1 pattern has ST elevation >2 mm giving rise to a coved-type ST segment, in electrical continuity with a negative T wave and without a separating isoelectric line.
2. Type 2 has a high take-off ST-segment elevation. In this variant, the J-point elevation (>2 mm) gives rise to a gradually descending elevated ST segment (remaining >1 mm above the baseline) and a positive or biphasic T wave. This ST-T segment morphology is referred to as the saddle-back type.
3. Type 3 is the coved or saddle-back type with <1 mm ST-segment elevation.

One has to be cognizant that the Brugada ECG pattern often is wax and wane. Sodium channel blockers, ajmaline, procainamide, and flecainide, could be used to unmask the ECG pattern—since the details of how to perform a drug challenge test to unmask the Brugada ECG pattern have been nicely reviewed elsewhere, we shall not repeat here(Wilde et al. 2002; Antzelevitch et al. 2005, 2016). In recent years, it has become clear that the right ventricular outflow tract (RVOT) is the likely arrhythmogenic substrate site. And the RVOT is the only cardiac structure lying just beneath the third and second intercostal space. We and others have demonstrated that placement of right precordial lead ECG recordings over the higher intercostal spaces (third and second intercostal space) significantly increases diagnostic yield in bringing the Brugada ECG pattern (Veerakul et al. 2000; Shimizu et al. 2000). In our institution, we always routinely record right precordial lead ECG (V1–V3) from fourth, third, and second intercostal space in every patient suspected of BrS. Figure 8.2 shows an example of the ECG tracings from a patient with BrS: the Brugada ECG pattern is absent in the conventional fourth intercostal space lead placement but become apparent in the higher intercostal spaces (third and second).

The original criteria for the diagnosis of BrS mandate that the patient must have type 1 Brugada ECG pattern in at least two leads with or without a sodium channel blocker challenge test and one of the following clinical manifestations: (1) a history of spontaneous VT/VF episodes or aborted sudden cardiac death, (2) a family history of sudden cardiac death or coved-type ECG, (3) agonal respiration during sleep, or (4) inducibility of VT/VF by programmed electrical stimulation. However, since

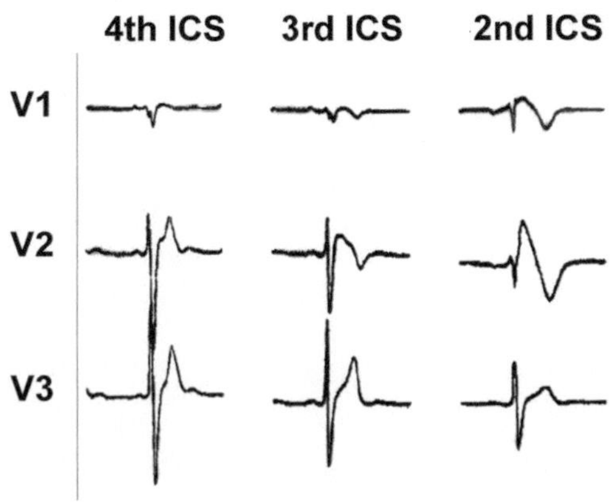

Fig. 8.2 An example of how placement of right precordial leads in higher intercostal spaces unmasks the Brugada ECG pattern in a BrS patient. The higher intercostal spaces (ICS), third and second ICS of V2 in this patients showed distinct coved-type Brugada ECG pattern (type 1)

Table 8.1 Conditions which can mimic Brugada ECG pattern (Brugada phenocopy) (Antzelevitch et al. 2016)

Atypical right bundle branch block (RBBB)
Left ventricular hypertrophy
Early depolarization Acute pericarditis
Acute myocardial ischemia or infarction
Acute stroke
Pulmonary embolism Prinzmetal angina
Dissecting aortic aneurysm
Various central and autonomic nervous system abnormalities
Duchenne muscular dystrophy
Friedreich ataxia
Spinobulbar muscular dystrophy Myoclonic dystrophy
Arrhythmogenic right ventricular cardiomyopathy (ARVC) Hypothermia
Mechanical compression of the right ventricular outflow tract (RVOT) as occurs in pectus excavatum, mediastinal tumor, or hemopericardium

several potential flaws exist in these criteria, they have since then been refined. While the clinical manifestations mentioned above remain important in recommending treatment and risk stratification, they are no longer listed among the diagnostic criteria, and type I Brugada ECG pattern in at least one lead (recorded from either the second, third, or fourth intercostal space) is now required for the diagnosis (Priori et al. 2013). Recently, another guideline was proposed utilizing a scoring system for the diagnosis of Brugada syndrome in the same fashion as the system for long QT syndrome, the so-called Shanghai score (Antzelevitch et al. 2016). Unfortunately, the benefit and the accuracy of the Shanghai score remain unclear, and one may use the old criteria and merely categorize the patients into two groups: symptomatic or asymptomatic BrS.

A Brugada ECG pattern can also be seen in several other conditions. Thus, the diagnosis can be made in a patient showing the Brugada ECG pattern with demonstration of a grossly normal heart by cardiological tests and in the absence of other conditions that can mimic the Brugada ECG pattern (Table 8.1).

8.3 Epidemiology

The prevalence of type I Brugada ECG pattern was estimated to be 1:2000 based on combined studies encompassing >400,000 patients from North America, Europe, and Asia (Postema 2012). The prevalence, however, varies widely among countries with low prevalence in North America (0.005–0.1%) and Western countries including Europe (0–0.017%) compared to Asia (0.15–0.27% in Japan, 0.18% in the Philippines) (Probst et al. 2007). Interestingly, several epidemiologic studies in Thailand have demonstrated that the prevalence of type I Brugada electrocardiogram (ECG) pattern is higher in Thailand than in other Asian countries and even much higher than that in the European countries and the USA. The incidence is ranging from 0.8% to 1.8% in nonfebrile patients to 4% in febrile patients [fever is a well-known precipitating factor for BrS (Rattanawong et al. 2016)] compared to a much lower incidence in Europe and in the USA.

The majority of BrS patients are relatively young between 20 and 40 years (Antzelevitch et al. 2005; Probst et al. 2007) although cases have been reported in children as young as 2 days old and in elderly up until the age of 84 years (Antzelevitch et al. 2005; Probst et al. 2007). Despite what appears to be an autosomal dominant inheritance pattern, BrS has up to tenfold higher prevalence in males with greater severity (Nademanee et al. 1997; Benito et al. 2008). World-wide, the syndrome is probably responsible for 4–12% of all sudden deaths and at least 20% of sudden deaths in patients with structurally normal hearts (Wilde et al. 2002; Antzelevitch et al. 2005).

8.4 Genetics of BrS

Because of occurrence in siblings and family history of sudden death in affected individuals, genetics was thought to play an important part of BrS since the first description in 1992 (Brugada and Brugada 1992). Transmission in familial cases is consistent with autosomal dominant with incomplete penetrance, especially in females whose incidence is up to 10 times lower than in male. In 1998, Chen et al. reported the first mutation, linked to BrS, in the *SCN5A* gene which encodes for the α-subunit of the sodium channel (Chen et al. 1998). Since then, more than 300 *SCN5A* mutations have been reported in BrS and represent currently the most common genotype. Functional studies demonstrate that *SCN5A* mutations in BrS cause loss of function of the sodium channel due to decreased expression of the sodium channel protein (Nav1.5) on the sarcolemma (Valdivia et al. 2004), expression of nonfunctional channels (Kyndt et al. 2001), or altered gating properties (delayed activation, earlier inactivation, faster inactivation, enhanced slow inactivation, and delayed recovery from inactivation) (Bezzina et al. 1999; Dumaine et al. 1999; Akai et al. 2000; Amin et al. 2005; Nakajima et al. 2015; Kinoshita et al. 2016). The loss of function of the sodium channel results in a decrease in sodium current and in turn impairs the fast upstroke of phase 0 of the action potential causing a slow conduction in the heart.

Even though *SCN5A* mutations are the most common defect found in 11–28% of BrS probands, the genetics of BrS have become heterogeneous. In addition to *SCN5A* mutations, more mutations are found in gene encoding protein of potassium and calcium channels. There have been now mutations of at least 18 genes that have been associated with BrS phenotype (Table 8.2). The contribution of the other 17 genes (except *SCN5A*) as the etiology of BrS, however, is small, and the presence of rare variants in these genes as the sole cause of BrS has been called into question due to similar percentage of normal individual carrying rare variants in these genes compared to BrS patients (Le Scouarnec et al. 2015).

Table 8.2 Gene mutations associated with Brugada syndrome (Antzelevitch et al. 2016)

Subtype	Locus	Gene/protein	Ion channel	Percent of probands
BrS1	3p21	*SCN5A*, Nav1.5	↓Ina	11–28%
BrS2	3p24	*GPD1L*	↓Ina	Rare
BrS3	12p13.3	*CACNA1C*, Cav1.2	↓Ica	6.6%
BrS4	10p12.33	*CACNB2b*, Cavβ2b	↓Ica	4.8%
BrS5	19q13.1	*SCN1B*, Navβ1	↓Ina	1.1%
BrS6	11q13-14	*KCNE3*, MiRP2	↑Ito	Rare
BrS7	11q23.3	*SCN3B*, Navβ3	↓Ina	Rare
BrS8	12p11.23	*KCNJ8*, Kir6.1	↑IK-ATP	2%
BrS9	7q21.11	*CACNA2D1*, Cav a2δ1	↓Ica	1.8%
BrS10	1p13.2	*KCND3*, Kv4.3	↑Ito	Rare
BrS11	17p13.1	*RANGRF*, MOG1	↓INa	Rare
BrS12	3p21.2-p14.3	*SLMAP*	↓Ina	Rare
BrS13	12p12.1	*ABCC9*, SUR2A	↑IK-ATP	Rare
BrS14	11q23	*SCN2B*, Navβ2	↓INa	Rare
BrS15	12p11	*PKP2*, Plakophillin-2	↓INa	Rare
BrS16	3q28	*FGF12*, FHAF1	↓INa	Rare
BrS17	3p22.2	*SCN10A*, Nav1.8	↓INa	5–16.7%
BrS18	6q	*HEY2* (Transcriptional factor)	↑Ina	Rare

Note: Upward and downward arrows show increase and decrease

Despite the fact that many genes have been identified and linked to the syndrome, gene mutation alone cannot explain the phenotype in full. Kapplinger et al. found nearly 300 *SCN5A* mutations in 211 unrelated probands (Kapplinger et al. 2010). The *SCN5A* mutations were found in 21% of patients with the Brugada phenotype and 2–5% of healthy controls, respectively. These findings suggest an important role of *SCN5A* mutations causing loss-of-function sodium channel in the phenotype manifestation. However, 80% of these mutations were only present in a single individual or one family, and a causal role of these mutations in the Brugada syndrome is far from clearly established.

Probst et al. studied 13 large families with *SCN5A* mutations and revealed the following intriguing findings: Many of the mutation carriers did not have the Brugada signature sign on ECG nor could it be provoked by sodium channel blockers (Probst et al. 2009). Moreover, in 5 of the 13 families with more than five clinically affected individuals, there were one or two affected individuals with the Brugada phenotype who did not have the familial *SCN5A* mutations. Furthermore, the Brugada ECG pattern was induced in eight mutation-negative patients (Probst et al. 2009). These findings along with a report of a case of identical twins carrying a *SCN5A* mutation of which only one displayed the phenotype suggest that *SCN5A* mutations may act as modifiers rather than causative mutations (Sakabe et al. 2002a).

SCN5A mutations may cause not only BrS but other diseases as well. Indeed, *SCN5A* mutations have also been associated with long QT syndrome (Bezzina et al.

1999; Wang et al. 1995), cardiac conduction disease (Tan et al. 2001), sick sinus syndrome (Benson et al. 2003), atrial fibrillation (Darbar et al. 2008; Olson et al. 2005), and dilated cardiomyopathy with overlap syndromes identified in specific families (Meregalli et al. 2009).

Because of the seemingly complex inheritance pattern in BrS, a genome-wide association (GWAS) study was done to explore the contribution of common single nucleotide polymorphisms (SNP) in this disease. In this study, the investigators could demonstrate association of three common SNP in *SCN5A*, *SCN10A*, and *HEY2* with BrS. Each "risk allele" was associated with an odd ratio of 1.6–2.8 for having Brugada syndrome (Bezzina et al. 2013). Risk was progressively higher with presence of more risk alleles and was as high as 21.5 if 5 or 6 risk alleles were present in a single individual. This study suggests that the genetic basis of BrS and susceptibility to VF therein in the individual patient is not caused by a single major genetic mutation (classical Mendelian view) but rather by inheritance of multiple susceptibility genetic variants (oligogenic) acting in concert through one or more mechanistic pathways. This finding partially explains why simple genetic testing may have easily missed these rare genetic variants in many of the BrS patients.

Because of the complexity of genetic mechanism in this disease, genetic testing for the diagnosis of BrS is not as helpful as some other monogenic diseases associated with sudden cardiac death like the long QT syndrome or catecholaminergic polymorphic VT (CPVT). Given the background rate of rare variants of unknown significance (VUS) in *SCN5A* of approximately 2% in normal population and 20% yield in patients with BrS (signal-to-noise ratio of 10:1), the HRS/EHRS Expert Consensus Guidelines written in 2011 listed genetic testing for the diagnosis of Brugada syndrome as "can be useful" (Class IIa) in a proband with clinical diagnosis of BrS (Ackerman et al. 2011). The test is probabilistic rather than deterministic, and interpretation of an abnormal test must consider all the clinical and molecular genetic findings. Due to presence of rare genetic variants in the normal population, a false-positive result is possible. For example, a recent study of 870 whole exome sequencing data and 6161 genotype array data in general population found that of 28 variants associated with BrS, none of whom had Brugada marker on the ECG. Syncope, ventricular arrhythmia, and mortality were also not significantly different between those who had the variants and those who did not (control group) (Ghouse et al. 2017).

The role of genetic testing in risk stratification remains unclear. Crotti et al. reported their findings of a comprehensive mutational analysis of all 12 BrS genes for a single large cohort of BrS patients (Crotti et al. 2012). They found putative pathogenic mutations in 21% of their BrS cohort. Similar to other reports, 78% of the mutations in BrS were still confined to *SCN5A*. Interestingly, in male patients, the yield of positive testing varied from 11% in those older than 40 years of age to 21% in male patients 20–40 years of age and to 83% in male patients younger than 20 years of age. The BrS patients with prolonged PR interval >200 ms had very high incidence of *SCN5A* mutations (39%) compared to those with normal PR interval. The yield of identifying the mutations is similar between those who have only the typical type 1 Brugada ECG pattern (asymptomatic cases) and those with symptoms and/or family history of sudden cardiac death, so presence of identifiable mutation did not appear to correlate with the symptoms or prognosis. Based on their findings, the authors recommended genetic testing for all patients who just have the type 1 Brugada ECG pattern (asymptomatic cases) as well as symptomatic cases. Due to their findings of a low prevalence of non-*SCN5A* mutations, they suggested that it was reasonable to test most patients for *SCN5A* mutations alone first, with further testing for the other minor BrS genes only in special circumstances. This recommendation concurs with the position paper of the Canadian Cardiovascular Society but not with consensus of the Heart Rhythm Society/European Heart Rhythm Association which states that either comprehensive genetic testing or target testing for *SCN5A* can be used (Ackerman et al. 2011; Gollob et al. 2011; Kaufman 2012).

8.5 Pathophysiology

Using arterially perfused wedge preparation of canine right ventricle (RV), Antzelevitch and his colleagues proposed the repolarization theory as the electrophysiologic abnormality underlying BrS (Yan and Antzelevitch 1999). In their experimental studies, they observed the transmembrane voltage gradient between the RV epicardium and endocardium due to the loss of the action potential (AP) dome only in the epicardium but not in the endocardium; RV epicardium is well known to have abundant Ito. Upon exposure to sodium channel blockers in combination with acetylcholine, this area then developed a notch and dome appearance of the epicardium AP leading to a coved-type ST-segment elevation in the right precordial leads. When the loss of the AP dome was further accentuated, it caused marked shortening of the epicardial AP in certain regions causing pronounced heterogeneity of transmembrane voltage potentials and, in turn, causing phase 2 reentry and triggered VF (Yan and Antzelevitch 1999). However, thus far, there have not been clear clinical relevant data in humans to support this theory. Perhaps, the observational study showing that quinidine, a strong Ito blocker, is effective in treating BrS patients could be inferred as weak-indirect evidence that supports the repolarization theory (Belhassen et al. 2004). While the repolarization theory enjoyed its popularity early on, the lack of strong clinical relevant findings to convincingly support the concept and subsequent clinical evidence led to the other theory, that of depolarization disorder.

Using an electrical guidewire to record an epicardial electrogram from a conus branch of the right coronary artery, Negase et al. were the first to show abnormal electrograms characterized by late potentials following the QRS which were recorded from the free wall of RVOT epicardium in BrS patients (Nagase et al. 2002). Their findings suggest conduction delay in the RVOT epicardium. Two studies conducted in an explanted heart in addition to biopsies showing (ultra-) structural changes in the right ventricular outflow tract of BrS patients demonstrated conduction disorder in these patients. The explanted hearts showed no evidence of repolarization abnormality; instead they found evidence of interstitial fibrosis causing conduction delay in one heart (Coronel et al. 2005) and right ventricular

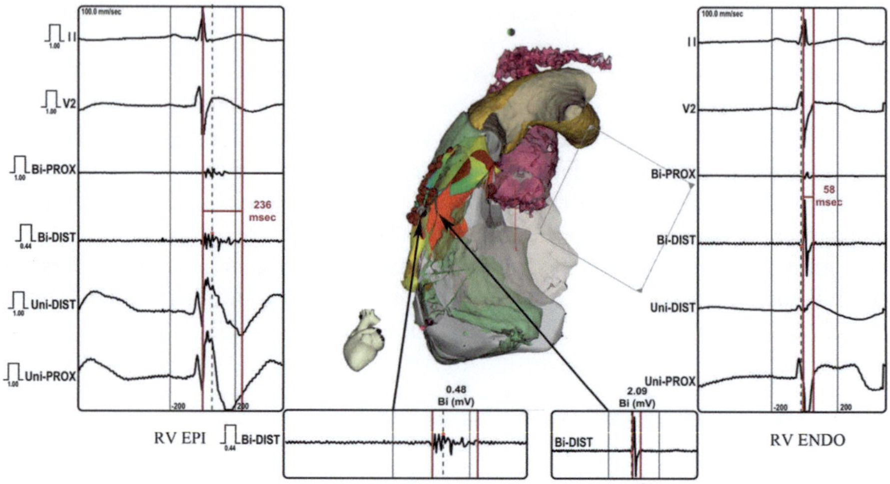

Fig. 8.3 A left lateral view of the right ventricular outflow tract (RVOT) displays the difference in ventricular electrograms between the endocardial and epicardial site of the anterior RVOT of a BrS patient with electrical storm. The left and right insets display bipolar and unipolar electrograms recorded from the epicardium and endocardium from the same site of the RVOT, respectively. *Bi-DIST* bipolar distal, *Bi-PROX* bipolar proximal, *Uni-DIST* unipolar distal, *Uni-PROX* unipolar proximal. Reproduced with permission from Nademanee K, Veerakul G, Chandanamattha P, et al. Prevention of ventricular fibrillation episodes in Brugada syndrome by catheter ablation over the anterior right ventricular outflow tract epicardium. Circulation. 2011;123:1270–9

excitation failure and activation delay by current-to-load mismatch in the sub-epicardium in the other heart (Hoogendijk et al. 2010). As a result, the delay in the AP of the RVOT causes the electrical gradient from the more positive RV to RVOT, leading to the ST elevation of the right precordial leads—similar to the situation of a myocardial injury at the RVOT—and as the RVOT depolarizes later (during repolarization of the RV), this gradient is reversed, and the net current flows toward the RV, resulting in a negative T wave in the same right precordial leads. The experiment from the same group in this explanted heart also showed that this site is the arrhythmogenic site during programmed stimulation-induced VF.

Perhaps, the most compelling findings to support depolarization disorder came from our study (Nademanee et al. 2011). Our group carried out a study to determine the substrate sites and arrhythmogenic mechanisms in this disease. We found that all our BrS patients had abnormal low-voltage, fractionated late potentials exclusively clustering in the anterior aspect of the RVOT epicardium and not seen anywhere else. Figures 8.3 and 8.4 show an example of low-voltage fractionated electrograms recorded from the anterior RVOT epicardium of a patient who presented with electrical storm. Ablation at this area normalized the Brugada ECG pattern and prevented recurrent VF episodes. As shown in Fig. 8.3, the endocardial site (arrow) displays a single potential of 2.09 mV, with a duration of 58 ms, and did not extend beyond the QRS compared to the epicardial counterpart that showed low-voltage late potential (0.48 mV), with a duration of 236 ms with late potential extended beyond

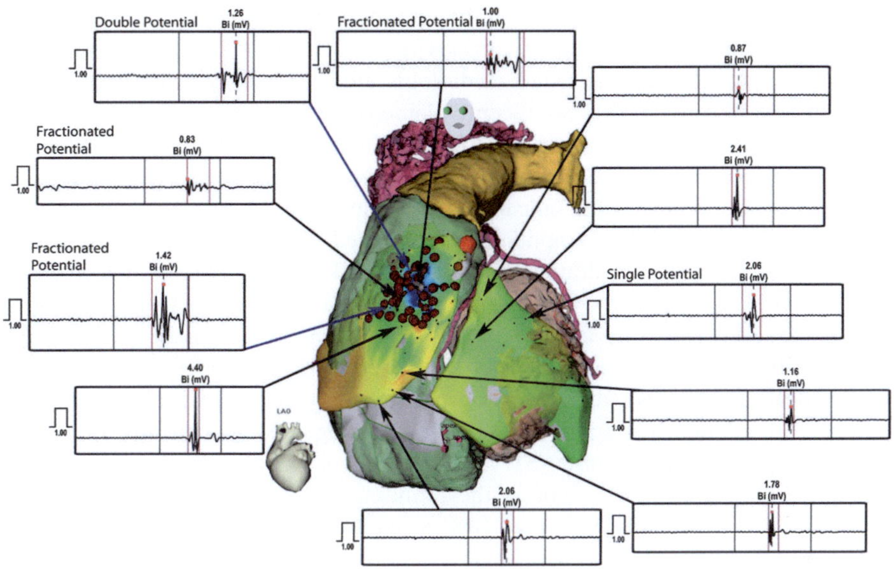

Fig. 8.4 Comparison of ventricular electrograms recorded from different sites in both the left ventricle (LV) and right ventricle (RV) of the same patient as in Fig. 8.3. Reproduced with permission from Nademanee K, Veerakul G, Chandanamattha P, et al. Prevention of ventricular fibrillation episodes in Brugada syndrome by catheter ablation over the anterior right ventricular outflow tract epicardium. Circulation. 2011;123:1270–9

the QRS. Figure 8.4 shows epicardial electrograms recorded from various sites of the epicardium in both the LV and RV epicardium. Note that abnormal fractionated electrograms and double potential electrograms are only localized in the anterior aspect of the RVOT epicardium. Similar observations were found in all our study patients, and these findings clearly provide the strongest clinical evidence that the delayed depolarization at the anterior aspect of the RVOT is the most likely underlying electrophysiologic mechanism underlying BrS.

Recently, Szél and Antzelevitch also observed fractionated electrograms in the RV epicardium, but they suggested that this is due to a heterogeneous epicardial loss of dome and local re-excitation via a concealed phase 2 reentry, rejecting the possibility of abnormal depolarization or structural abnormalities (Szél and Antzelevitch 2014). However, in our patients another mechanism seems in operation. Fractionated signals were clearly associated with diastolic potentials in the BrS patients; we observed that the fractionation was present immediately after activation, arguing against reactivation of calcium channels (which would require a minimum time delay). Fractionated activity and diastolic potentials have been shown to be associated with reentrant arrhythmias and are observed in BrS patients. Nevertheless, while it is quite apparent that depolarization disorder is likely to be the main mechanism underlying the BrS, one must be mindful that repolarization abnormality could contribute to the arrhythmogenesis of BrS patients, along with genetic mutations of ionic channel and other precipitating factors.

Brugada syndrome was classically thought of as primary electrical disease with no structural defect of the heart (Antzelevitch et al. 2005). Despite no structural abnormality detected by the usual cardiologic studies such as echocardiography, angiography, or cardiac MRI, we recently demonstrated minor yet similar pathological findings in six autopsy hearts of SADS victims who had a family history of BrS and six biopsy specimens in patients with BrS (Nademanee et al. 2015). Increased collagen in the RVOT with epicardial and interstitial fibrosis was found with corresponding area of low-amplitude, fractionated electrogram in biopsy cases. Connexin-43 signals were diminished in the RVOT of these patients which raises the possibility of cardiomyocyte electrical uncoupling as one of the pathophysiology in this disease. These findings point to the likelihood of combined structural abnormalities and ion channel defects as the basis of ventricular arrhythmia and sudden death in patients with BrS. A recent MRI study of cases with BrS compared to age- and sex-matched control demonstrated a slight but statistically significant larger RV end-systolic volume (31 vs. 28 mL/m^2, $p = 0.038$) and lower ejection fraction (61% vs. 64%, $p = 0.004$); these findings attest to these minor structural abnormalities which likely make BrS a cardiomyopathic entity as well as channelopathy (Bastiaenen et al. 2017).

8.6 Modulating and Precipitating Factors

As mentioned above, the Brugada ECG pattern is often concealed but can be unmasked or modulated by sodium channel blockers, a febrile state, vagotonic agents, autonomic nervous system changes, tricyclic or tetracyclic antidepressants, first-generation antihistamines (dimenhydrinate), a combination of glucose and insulin, hyperkalemia, hypokalemia, hypercalcemia, and alcohol and cocaine toxicity.

8.6.1 Autonomic Nervous System

The effect of sympathetic stimulation by isoproterenol infusion, resulting in normalization of the BrS pattern, suggests that sympathetic activity could modify the VF substrate (Miyazaki et al. 1996). The presence of the Brugada ECG pattern is probably a prerequisite for the increased risk of SCD, and normalization of the ECG pattern is associated with a decreased risk (Antzelevitch et al. 2005). This concept is strengthened by the fact that some patients with "VF storms" associated with BrS can be effectively treated with isoproterenol infusion (Tanaka et al. 2001). On the other hand, increased vagal tone could be arrhythmogenic in BrS patients. Increased vagal tone, as well as acute β-blockade, was found to promote VF induction in the electrophysiology laboratory (Kasanuki et al. 1997). Abe et al. found that fluctuations in late potentials on signal-averaged ECG (SAECG) occurred predominantly at night, suggesting that conduction delay and, by inference, the arrhythmogenic substrate are autonomically modulated (Abe et al. 2012). Therefore, it is plausible that at night during sleep, when vagal tone is usually increased and associated with the withdrawal of sympathetic activity, the VF substrate is modulated and more susceptible to arrhythmogenesis. Kasanuki et al. also showed a sudden increase in vagal activity, as measured by heart rate variability (HRV), just before episodes of VF in a patient with BS (Kasanuki et al. 1997). However, Krittayaphong et al. studied HRV from 24-h Holter data of SUDS patients with the Brugada ECG marker, aiming to determine the circadian pattern of sympathetic and parasympathetic activity (Krittayaphong et al. 2003). Surprisingly, they found decreased HRV at night in SUDS patients when compared with the control group and suggested that these patients had an abnormal increase in sympathetic activity or decrease in the vagal tone at night. Although the explanation for the different findings of Kasanuki et al. and that of Krittayaphong et al. is unknown, it is clear that the sympathovagal balance in the BrS patients plays a significant role in the circadian variation of VF occurrence. However, further studies are needed to clearly define the complex interplay between the autonomic nervous system and the arrhythmic mechanisms of BrS.

8.6.2 Hypokalemia

Hypokalemia has been implicated as a contributing cause for the prevalence of SUDS in the northeastern region of Thailand where potassium deficiency is endemic (Nimmannit et al. 1991). Serum potassium levels in the northeastern population are significantly lower than those of the population in Bangkok, which lies in the central part of Thailand where potassium is abundant in food.

Hypokalemia is a well-known predisposing factor to ventricular arrhythmias. Furthermore, it has been shown that there is commonly a shift of serum potassium into the muscular compartment between midnight and 7 am, decreasing the amount of serum potassium (Andres et al. 1957). If this phenomenon indeed occurs in BS/SUDS patients in Thailand, then it is likely that low serum potassium is a key factor that precipitates VF at night in these patients.

8.6.3 Sleep, Heavy Meals, and Alcohol

Because the majority of VF episodes occur at night, the question is whether a sleep disorder is a trigger of VF. Thus far, none of our sleep studies in BrS patients has found any evidence of a sleep disorder, including sleep apnea. One theory that many SUDS researchers have informally discussed as a possible precipitating factor is eating a heavy meal at dinnertime before retiring to bed. A Thai Ministry of Public Health Report (1990) suggested that a large meal of glutinous rice ("sticky rice") or carbohydrates ingested on the night of death precipitated SUDS attacks. Both carbohydrates and glutinous rice have been shown to shift potassium into cells and thus lower the serum potassium level. Postprandial increased ST segment elevation in lead V2 has been seen in patients with symptomatic BrS on a Holter study

(Mizumaki et al. 2007). A study by Nogami et al. showed that glucose and insulin could unmask the Brugada ECG marker or accentuate the J-junction elevation of the ST segment (Nogami et al. 2003). They observed a slight decrease in the serum potassium levels of their study patients, but it did not reach statistical significance. Nevertheless, these findings bode well for a heavy carbohydrate meal being a precipitating factor for sudden death in SUDS patients. Alcohol has also been implicated in triggering recurrent VF in a patient with Brugada syndrome (Ohkubo et al. 2013).

8.6.4 Body Temperature and Febrile Illness

Dumaine et al. discovered that the *T1620M* missense mutation causes accelerated inactivation of the sodium channel at physiologic body temperature but not at room temperature (Dumaine et al. 1999). Identification of this temperature-sensitive mutation that precipitates the net loss of sodium current prompted investigators to recognize that a hot climate and body temperature may be important modulating factors. Indeed, several case reports have emerged recently, demonstrating that febrile illness or external heat could unmask BrS and/or precipitate VF occurrence (Antzelevitch and Brugada 2002; Morita et al. 2002b; Saura et al. 2002; Porres et al. 2002; Kum et al. 2002; Canpolat et al. 2017; Chung et al. 2017; De Marco et al. 2012; Skinner et al. 2007). In children, ventricular arrhythmia triggered by fever can be mistaken for episodes of febrile seizure (Skinner et al. 2007). We have encountered a case of a young male patient who died suddenly after a spiking fever of 40 °C after abdominal surgery. Upon review of the ECG, the patient had the typical BrS pattern, but had not had a prior medical problem and had been asymptomatic. The northeastern part of Thailand where SUDS is prevalent is well known for its hot climate, with temperatures reaching as high as 41 °C. It is again unclear how much climate influences the occurrence of SUDS in Thailand, but a study is underway. It is entirely possible that high climatic temperatures or a febrile state could modulate the functional expression of mutant channels in other genes responsible for BrS. In the meantime, physicians should factor in temperature as a cause for arrhythmogenesis in BrS. They should be cognizant of the association between temperature and BrS during diagnosis and treatment, advising patients to promptly treat fevers.

Figure 8.5 proposes pathophysiology of BrS and demonstrates how these modulating and predisposing factors can affect the arrhythmia and the clinical outcomes in many ways: (1) modifying the VF substrates, (2) affecting the gene expression of the ion channel defects, (3) affecting the triggering PVCs and the initiating process of VF, and (4) influencing the sustaining process of the VF episodes.

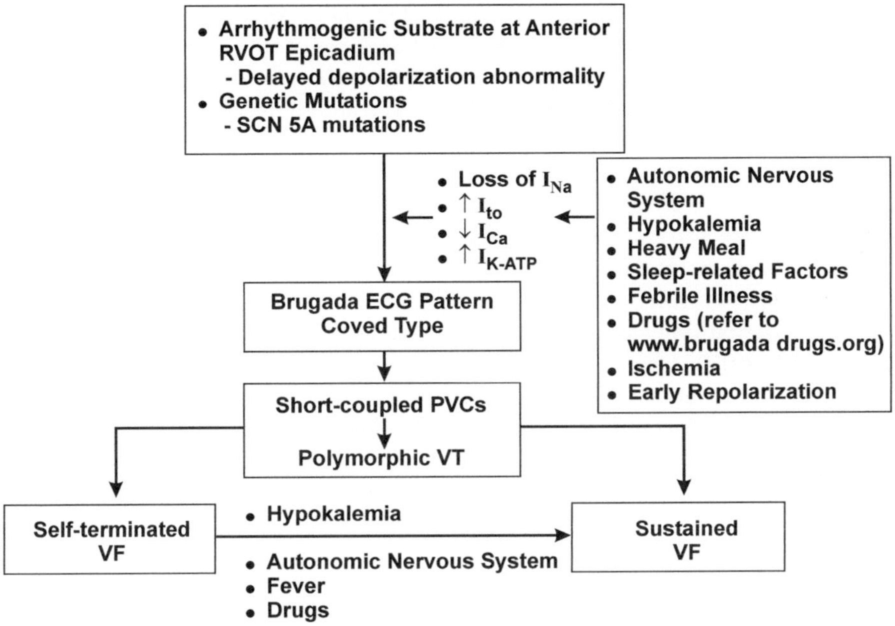

Fig. 8.5 Proposed pathophysiologic mechanisms of BrS with respect to predisposing factors

8.7 Combined Syndromes

In addition to the above precipitating factors, BrS patients often have other concomitant arrhythmias or arrhythmic syndromes. Atrial fibrillation is one of the common arrhythmias in BrS. Incidence of combined early repolarization syndrome in the BrS patients occurred in 15%, and in this subset (Letsas et al. 2008), the incidence of recurrent VF episodes is significantly higher than that of the BrS alone. Similarly, combined syndromes of a progressive conduction defect and BrS and long QT syndrome and BrS are occasionally observed (Bezzina et al. 1999; Shirai et al. 2002).

8.8 Risk Stratification

There is little to debate that BrS patients who survived an out-of-hospital cardiac arrest are at high risk of recurrent VF episodes and need ICD treatment. Likewise, symptomatic patients with recurrent syncope, agonal respiration at night during sleep, or unknown seizure are at risk of dying suddenly without protection and have Class I indication for ICD treatment. One large study involving 1029 BrS patients in 11 European centers found annual cardiac events rate of 7.7% in patients who presented with sudden cardiac arrest, 1.9% in patients with syncope, and 0.5% in asymptomatic patients (Probst et al. 2010). The heated debate is more on how one best identifies high-risk patients for sudden death and the need of ICD treatment in asymptomatic BrS patients. At the beginning, the Brugada registry reported a significantly high risk of asymptomatic patients with positive VT inducibility by programmed electrical stimulation (Antzelevitch et al. 2005; Brugada et al. 2011). However, more recent studies found a much lower incidence of sudden death or VF in this group and questioned the specificity of programmed electrical stimulation (PES) in risk stratifying asymptomatic BrS patients (Veerakul and Nademanee 2012; Priori et al. 2012). Our own experience of asymptomatic patients shows that the annual cardiac event rate (VF or death) is so low (0.25% per year) that it will be very unlikely for any risk stratification strategy to be able to identify high-risk patients for ICD treatment as a primary prevention (Veerakul et al. 2008).

Electrocardiographic risk factors include presence of spontaneous type I Brugada pattern which carries a higher risk of arrhythmic events compared to drug-induced Brugada pattern (Adler 2016). QRS fragmentation (Priori et al. 2012; Morita et al. 2008; Tokioka et al. 2014), early repolarization pattern (Tokioka et al. 2014; Kawata et al. 2013; Kaneko et al. 2014), a significant S wave in lead I (≥0.1 mV and/or ≥40 ms) (Calò et al. 2016), and prolonged Tpeak-Tend or Tpeak-Tend/QTc (Zumhagen et al. 2016; Maury et al. 2015) were reported to be associated with high risk for ventricular arrhythmia. Other studies found exercise testing (Makimoto et al. 2010; Amin et al. 2009), signal-averaged ECG (Huang et al. 2009), and shortening of the ventricular refractory period (<200 ms) (Priori et al. 2012) to be valuable tools for identifying high-risk patients. However, it is unclear how useful any of these parameters would be in identifying asymptomatic BrS for ICD treatment. In our own asymptomatic cohorts of 115 patients, we found that only two patients after 10 years of follow-up had either VF or sudden death (one EPS positive and the other EPS negative) (Veerakul et al. 2008). With this low event rate and after a decade of follow-up, it is very clear to us that any risk stratification strategy would be very unlikely to be of any value in selecting patients for an ICD treatment in our population.

8.9 Treatment

BrS patients should be informed of various modulating and precipitating factors—as discussed above—that could bring about malignant arrhythmias: fever, electrolyte abnormalities, alcohol consumption, and a whole host of drugs as listed in www.brugadadrugs.org (Postema et al. 2009). VF and sudden death in BrS usually occur at rest and at night. Therefore, one has to be cognizant of the circadian variation of sympathovagal balance, hormones, and other metabolic factors which are likely to contribute to this circadian pattern (Wilde et al. 2002; Antzelevitch et al. 2005).

8.9.1 Anti-arrhythmic Drugs

Quinidine has been the only drug that consistently shows benefits in preventing recurrent VF episodes. The drug has been shown to suppress inducibility of VF on PES and reduce the number of appropriate ICD shocks (Belhassen et al. 2004, 2015; Hermida et al. 2004; Mizusawa et al. 2006). Blockade of Ito is thought to be the mechanism by which quinidine is effective. Unfortunately, there are two major problems with quinidine: (1) Only two-third of patients could tolerate the drug, and serious side effects such as thrombocytopenia could be very serious. (2) Quinidine is not available in many countries (Viskin et al. 2007). In Thailand, there is no supply of the drug, and the only drug we could use is amiodarone with variable success. Bepridil—another Ito blocker—which is only available in Japan has been used in BrS patients to suppress VF (Kaneko et al. 2014; Ohgo et al. 2007; Murakami et al. 2010). Cilostazol—an oral phosphodiesterase inhibitor—has also been shown to be of benefit in preventing recurrent VF in Brugada syndrome (Ohgo et al. 2007; Tsuchiya et al. 2002).

However, there is very little data to objectively assess the safety and efficacy of both Bepridil and Cilostazol in BrS patients at this time.

8.9.2 Implantable Cardioverter-Defibrillator

Symptomatic BrS patients (with a past history of VT/VF or syncope) have a Class I indication for ICD treatment (Wilde et al. 2002; Antzelevitch et al. 2005, 2016; Priori et al. 2013). In our DEBUT study (Tsuchiya et al. 2002)—Defibrillator versus β-blocker in Unexplained Death in Thailand: A Randomized Clinical Trial—we found that ICDs provided full protection from death related to primary VF in the study population which included 59% of patients with BrS. However, we also found that unwanted effects of the ICD were also frequent (30%). Most of the complications were minor; they included defibrillation discharges caused by supraventricular tachycardia or sinus tachycardia and T-wave over-sensing. All of the complications were corrected by reprogramming the devices without major intervention. However, one patient had pocket erosion with infection that required removal of the ICD, and one patient needed to have his ICD lead replaced because of an insulation break. Other studies also show similar results that long-term follow-up of Brugada ICD patients has a high complication rate up to a third of patients (Nademanee et al. 2003; Sacher et al. 2006; Sarkozy et al. 2007; Steven et al. 2011). The majority of the complications, similar to our study, were mostly inappropriate shocks occurring in 36% of patients at follow-up; however, one registry recorded an 18% rate of serious vents including pericardial effusion, lead fracture, infection, and subclavian vein thrombosis (Steven et al. 2011).

With the above relatively high complication rate of ICD in BrS patients, one has to be extremely cautious to use ICD in asymptomatic BrS patients. The fact that the event rates in the asymptomatic BrS population are quite low in most series makes ICD treatment in this subset questionable with respect to whether ICD benefits

Table 8.3 Indication for ICD implantation in patients with Brugada syndrome (Priori et al. 2013)

Class I indication (ICD is recommended):
Symptomatic patients displaying the type 1 Brugada ECG (either spontaneously or after sodium channel blockade) who present with aborted sudden death should receive an ICD
Similar patients presenting with related symptoms such as syncope, seizure, or nocturnal agonal respiration and having documented ventricular fibrillation or tachycardia should also undergo ICD implantation. Electrophysiologic study (EPS) is recommended in symptomatic patients only for the assessment of supraventricular arrhythmia.
Class IIa (ICD could be useful):
In symptomatic patients with type 1 pattern, in whom syncope was likely caused by VT/VF
Class IIb (ICD may be considered):
In asymptomatic patients inducible by programmed electrical stimulation (PES)
Class III indication:
ICDs are not indicated in asymptomatic patients with drug-induced type I ECG and on the basis of a family history of SCD alone

would outweigh the risk. As mentioned in the risk stratification section, thus far there have not been any convincing methods of risk stratification to identify high-risk asymptomatic patients for ICD therapy. One could follow the guideline established by the HRS/EHRA/APHRS expert consensus statement (Priori et al. 2013) recommendations for ICD implantation, as summarized in Table 8.3.

The recent approval of leadless ICD, which has been shown to be quite effective in terminating VT/VF episode, is a welcome addition to therapeutic modalities for BrS patients 116. Early users reported issues with sensing and higher sensing screening failure rate compared to other inherited arrhythmia syndrome due to the dynamic nature of QRS and T-wave morphologies in patients with BrS (Kamakura et al. 2017; Conte et al. 2017; Olde Nordkamp et al. 2016). Further studies and clinical trials to determine efficacy and safety are warranted in the high-risk BrS population.

8.9.3 Catheter Ablation

The early attempt of catheter ablation in treating BrS syndrome patients was limited to a few report cases of patients with electrical storms. The initial approach was designed to target initiating PVCs that trigger VF, which were found to come from the RVOT (Haïssaguerre et al. 2003; Nakagawa et al. 2008; Darmon et al. 2004). The ablation was performed on the endocardial site of the RVOT. However, this approach has not been widely successful largely because patients with BrS rarely had frequent PVCs to be mapped, and therefore it was quite difficult to identify precise targets for ablation and clearly assess the acute outcomes of the ablation. We have reported our epicardial approach for substrate ablations that was indeed safe and effective (Nademanee et al. 2011). We identified and proved that anterior RVOT epicardium is the most common arrhythmogenic substrate sites for our BrS patients. However, subsequently, in a significant number of patients, the RV body and the inferolateral aspect and the area near the tricuspid valve are also infrequently involved. These sites consistently have abnormal late potentials and low-voltage fractionate ventricular electrograms; these abnormal electrograms tended to cluster exclusively in this area but not anywhere else. After ablations at this RVOT epicardial site, the Brugada ECG pattern normalizes, and VT/VF episode subsides. We have now performed 54 BrS patients with frequent ICD discharges. Long-term outcomes (median 30 months) have been excellent with no recurrent VT/VF in all patients off medication. More recent studies ranging from individual case reports as well as collective collaborative studies have confirmed our findings that BrS arrhythmogenic substrates are ubiquitous in the RVOT epicardium and catheter ablations are beneficial in treating symptomatic BrS (Brugada et al. 2015; Zhang et al. 2016). Whether ablation would substitute for an ICD in high-risk BrS patients remains unknown. Further studies clearly need to be done to assess values and limitations of catheter ablation in patients with BrS.

8.10 Conclusions and Future Perspective

The two decades of BrS research have witnessed an impressive progress of our understanding of several aspects of the syndrome with respect to role of genetics, electrophysiological mechanisms, and clinical characteristics. However, many questions and controversies remain regarding the role of genetic background including polymorphisms as well as rare variants and gene-gene or gene-environmental interaction and other confounding factors such as fever and gender in this population. The cause and the role of fibrosis and anatomical abnormality in the RVOT in arrhythmogenesis will likely be the subject of intense investigation in the coming years. It's certain that the debate will continue regarding the role of repolarization in BrS and whether our findings of abnormal delayed depolarization in the anterior RVOT epicardium is seen in other centers and more specifically other population besides ours. How does one best treat asymptomatic BrS patients? Will there be any better risk stratification strategy to identify high-risk groups? Why is the anterior RVOT epicardium the arrhythmogenic sites for these patients? Why is there such a male preponderance and why do most of the VF episodes usually occur at night time? Whole genome sequencing may identify rare and common variants in genes modulating ion channels that may combine as multiple hits to cause abnormal conduction in the RVOT areas and in turn increase susceptibility to VF. Further research will continue to answer these questions. Meanwhile, refinement of treatment is needed. We will need to know the efficacy and safety of quinidine in the randomized trial studies which currently are being conducted. In the near future, more study of the subcutaneous leadless ICD will be forthcoming. The expanding role of catheter ablation of the epicardial substrate beyond the population of BrS with frequent VF needs to be also evaluated by assessing the risks and benefits of the procedures, especially with respect to complications related to the epicardial ablation approach. We indeed anticipate with excitement that the next decades will have these answers and in turn advance our ability to care for our BrS patients.

Acknowledgement The authors would like to thank the National Research Council of Thailand (NRCT) for support on the study of Brugada syndrome.

Compliance with Ethical Standards

Conflict of Interest Author Apichai Khongphatthanayothin declares that he has no conflict of interest. Author Koonlawee Nademanee declares that the following conflict of interests: Research grants from Medtronic Inc & Royalty from Biosense Cordis Webster Inc.

Ethical Approval All procedures performed in studies involving human participants were in accordance with the ethical standards of the institutional and/or national research committee and with the 1964 Helsinki declaration and its later amendments or comparable ethical standards. Inform consent was obtained from all individual participants included in the study.

References

Abe A, Kobayashi K, Yuzawa H, Sato H, Fukunaga S, Fujino T, Okano Y, Yamazaki J, Miwa Y, Yoshino H, Ikeda T. Comparison of late potentials for 24 hours between Brugada syndrome and arrhythmogenic right ventricular cardiomyopathy using a novel signal-averaging system based on Holter ECG. Circ Arrhythm Electrophysiol. 2012;5:789–95.

Ackerman MJ, Priori SG, Willems S, Berul C, Brugada R, Calkins H, Camm AJ, Ellinor PT, Gollob M, Hamilton R, Hershberger RE, Judge DP, Le Marec H, McKenna WJ, Schulze-Bahr E, Semsarian C, Towbin JA, Watkins H, Wilde A, Wolpert C, Zipes DP, HRS, EHRA. HRS/EHRA expert consensus statement on the state of genetic testing for the channelopathies and cardiomyopathies: this document was developed as a partnership between the Heart Rhythm Society (HRS) and the European Heart Rhythm Association (EHRA). Europace. 2011;13:1077–109.

Adler A. Brugada syndrome: diagnosis, risk stratification, and management. Curr Opin Cardiol. 2016;31:37–45.

Aizawa Y, Takatsuki S, Kaneko Y, Noda T, Katsumata Y, Nishiyama T, Kimura T, Nishiyama N, Fukumoto K, Niwano S, Kurita T, Mitsuhashi T, Kamakura S, Shimizu A, Horie M, Fukuda K. Comparison of circadian, weekly, and seasonal variations of electrical storms and single events of ventricular fibrillation in patients with Brugada syndrome. Int J Cardiol Heart Vasc. 2016;11:104–10.

Akai J, Makita N, Sakurada H, Shirai N, Ueda K, Kitabatake A, Nakazawa K, Kimura A, Hiraoka M. A novel SCN5A mutation associated with idiopathic ventricular fibrillation without typical ECG findings of Brugada syndrome. FEBS Lett. 2000;479:29–34.

Amin AS, Verkerk AO, Bhuiyan ZA, Wilde AA, Tan HL. Novel Brugada syndrome-causing mutation in ion-conducting pore of cardiac Na+ channel does not affect ion selectivity properties. Acta Physiol Scand. 2005;185:291–301.

Amin AS, de Groot EA, Ruijter JM, Wilde AA, Tan HL. Exercise-induced ECG changes in Brugada syndrome. Circ Arrhythm Electrophysiol. 2009;2:531–9.

Andres R, Cader G, Goldman P, Zierler KL. Net potassium movement between resting muscle and plasma in man in the basal state and during the night. J Clin Invest. 1957;36:723–9.

Antzelevitch C, Brugada R. Fever and Brugada syndrome. Pacing Clin Electrophysiol. 2002;25:1537–9.

Antzelevitch C, Brugada P, Borggrefe M, Brugada J, Brugada R, Corrado D, Gussak I, LeMarec H, Nademanee K, Perez Riera AR, Shimizu W, Schulze-Bahr E, Tan H, Wilde A. Brugada syndrome: report of the second consensus conference: endorsed by the Heart Rhythm Society and the European Heart Rhythm Association. Circulation. 2005;111:659–70.

Antzelevitch C, Yan GX, Ackerman MJ, Borggrefe M, Corrado D, Guo J, Gussak I, Hasdemir C, Horie M, Huikuri H, Ma C, Morita H, Nam GB, Sacher F, Shimizu W, Viskin S, Wilde AA. J-Wave syndromes expert consensus conference report: emerging concepts and gaps in knowledge. Heart Rhythm. 2016;13:e295–324.

Bastiaenen R, Cox AT, Castelletti S, Wijeyeratne YD, Colbeck N, Pakroo N, Ahmed H, Bunce N, Anderson L, Moon JC, Prasad S, Sharma S, Behr ER. Late gadolinium enhancement in Brugada syndrome: a marker for subtle underlying cardiomyopathy? Heart Rhythm. 2017;14:583–9.

Belhassen B, Glick A, Viskin S. Efficacy of quinidine in high-risk patients with Brugada syndrome. Circulation. 2004;110:1731–7.

Belhassen B, Rahkovich M, Michowitz Y, Glick A, Viskin S. Management of Brugada syndrome: thirty-three-year experience using electrophysiologically guided therapy with class 1A antiarrhythmic drugs. Circ Arrhythm Electrophysiol. 2015;8:1393–402.

Benito B, Sarkozy A, Mont L, Henkens S, Berruezo A, Tamborero D, Arzamendi D, Berne P, Brugada R, Brugada P, Brugada J. Gender differences in clinical manifestations of Brugada syndrome. J Am Coll Cardiol. 2008;52:1567–73.

Benson DW, Wang DW, Dyment M, Knilans TK, Fish FA, Strieper MJ, Rhodes TH, George AL. Congenital sick sinus syndrome caused by recessive mutations in the cardiac sodium channel gene (SCN5A). J Clin Invest. 2003;112:1019–28.

Bezzina C, Veldkamp MW, van Den Berg MP, Postma AV, Rook MB, Viersma JW, van Langen IM, Tan-Sindhunata G, Bink-Boelkens MT, van Der Hout AH, Mannens MM, Wilde AA. A single Na(+) channel mutation causing both long-QT and Brugada syndromes. Circ Res. 1999;85:1206–13.

Bezzina CR, Barc J, Mizusawa Y, Remme CA, Gourraud JB, Simonet F, Verkerk AO, Schwartz PJ, Crotti L, Dagradi F, Guicheney P, Fressart V, Leenhardt A, Antzelevitch C, Bartkowiak S, Borggrefe M, Schimpf R, Schulze-Bahr E, Zumhagen S, Behr ER, Bastiaenen R, Tfelt-Hansen J, Olesen MS, Kääb S, Beckmann BM, Weeke P, Watanabe H, Endo N, Minamino T, Horie M, Ohno S, Hasegawa K, Makita N, Nogami A, Shimizu W, Aiba T, Froguel P, Balkau B, Lantieri O, Torchio M, Wiese C, Weber D, Wolswinkel R, Coronel R, Boukens BJ, Bézieau S, Charpentier E, Chatel S, Despres A, Gros F, Kyndt F, Lecointe S, Lindenbaum P, Portero V, Violleau J, Gessler M, Tan HL, Roden DM, Christoffels VM, Le Marec H, Wilde AA, Probst V, Schott JJ, Dina C, Redon R. Common variants at SCN5A-SCN10A and HEY2 are associated with Brugada syndrome, a rare disease with high risk of sudden cardiac death. Nat Genet. 2013;45:1044–9.

Boersma LV, Jaarsma W, Jessurun ER, Van Hemel NH, Wever EF. Brugada syndrome: a case report of monomorphic ventricular tachycardia. Pacing Clin Electrophysiol. 2001;24:112–5.

Brugada P, Brugada J. Right bundle branch block, persistent ST segment elevation and sudden cardiac death: a distinct clinical and electrocardiographic syndrome. A multicenter report. J Am Coll Cardiol. 1992;20:1391–6.

Brugada J, Brugada R, Antzelevitch C, Towbin J, Nademanee K, Brugada P. Long-term follow-up of individuals with the electrocardiographic pattern of right bundle-branch block and ST-segment elevation in precordial leads V1 to V3. Circulation. 2002;105:73–8.

Brugada J, Brugada R, Brugada P. Electrophysiologic testing predicts events in Brugada syndrome patients. Heart Rhythm. 2011;8:1595–7.

Brugada J, Pappone C, Berruezo A, et al. Brugada syndrome phenotype elimination by epicardial substrate ablation. Circ Arrhythm Electrophysiol. 2015;8:1373–81.

Calò L, Giustetto C, Martino A, Sciarra L, Cerrato N, Marziali M, Rauzino J, Carlino G, de Ruvo E, Guerra F, Rebecchi M, Lanzillo C, Anselmino M, Castro A, Turreni F, Penco M, Volpe M, Capucci A, Gaita F. A new electrocardiographic marker of sudden death in Brugada syndrome: the S-wave in lead I. J Am Coll Cardiol. 2016;67:1427–40.

Canpolat U, Bayazit Y, Aytemir K. Brugada syndrome unmasked by heat exhaustion. Ann Noninvasive Electrocardiol. 2017;22. https://doi.org/10.1111/anec.12356.

Chen Q, Kirsch GE, Zhang D, Brugada R, Brugada J, Brugada P, Potenza D, Moya A, Borggrefe M, Breithardt G, Ortiz-Lopez R, Wang Z, Antzelevitch C, O'Brien RE, Schulze-

Bahr E, Keating MT, Towbin JA, Wang Q. Genetic basis and molecular mechanism for idiopathic ventricular fibrillation. Nature. 1998;392:293–6.

Chung FP, Raharjo SB, Lin YJ, Chang SL, Lo LW, Hu YF, Tuan TC, Chao TF, Liao JN, Lin CY, Chang YT, Hung Y, Te A, Yamada S, Tasaka H, Wang CT, Chen SA. A novel method to enhance phenotype, epicardial functional substrates, and ventricular tachyarrhythmias in Brugada syndrome. Heart Rhythm. 2017;14:508–17.

Conte G, Kawabata M, de Asmundis C, Taravelli E, Petracca F, Ruggiero D, Caputo ML, Regoli F, Chierchia GB, Chiodini A, Del Bufalo A, Moccetti T, Goya M, Hirao K, Vicentini A, De Ferrari GM, Brugada P, Auricchio A. High rate of subcutaneous implantable cardioverter-defibrillator sensing screening failure in patients with Brugada syndrome: a comparison with other inherited primary arrhythmia syndromes. Europace. 2017. https://doi.org/10.1093/europace/eux009.

Coronel R, Casini S, Koopmann TT, Wilms-Schopman FJ, Verkerk AO, de Groot JR, Bhuiyan Z, Bezzina CR, Veldkamp MW, Linnenbank AC, van der Wal AC, Tan HL, Brugada P, Wilde AA, de Bakker JM. Right ventricular fibrosis and conduction delay in a patient with clinical signs of Brugada syndrome: a combined electrophysiological, genetic, histopathologic, and computational study. Circulation. 2005;112:2769–77.

Crotti L, Marcou CA, Tester DJ, Castelletti S, Giudicessi JR, Torchio M, Medeiros-Domingo A, Simone S, Will ML, Dagradi F, Schwartz PJ, Ackerman MJ. Spectrum and prevalence of mutations involving BrS1- through BrS12-susceptibility genes in a cohort of unrelated patients referred for Brugada syndrome genetic testing: implications for genetic testing. J Am Coll Cardiol. 2012;60:1410–8.

Darbar D, Kannankeril PJ, Donahue BS, Kucera G, Stubblefield T, Haines JL, George AL, Roden DM. Cardiac sodium channel (SCN5A) variants associated with atrial fibrillation. Circulation. 2008;117:1927–35.

Darmon JP, Bettouche S, Deswardt P, Tiger F, Ricard P, Bernasconi F, Saoudi N. Radiofrequency ablation of ventricular fibrillation and multiple right and left atrial tachycardia in a patient with Brugada syndrome. J Interv Card Electrophysiol. 2004;11:205–9.

De Marco S, Giannini C, Chiavaroli V, De Leonibus C, Chiarelli F, Mohn A. Brugada syndrome unmasked by febrile illness in an asymptomatic child. J Pediatr. 2012;161:769–769.e761.

Dumaine R, Towbin JA, Brugada P, Vatta M, Nesterenko DV, Nesterenko VV, Brugada J, Brugada R, Antzelevitch C. Ionic mechanisms responsible for the electrocardiographic phenotype of the Brugada syndrome are temperature dependent. Circ Res. 1999;85:803–9.

Eckardt L, Kirchhof P, Loh P, Schulze-Bahr E, Johna R, Wichter T, Breithardt G, Haverkamp W, Borggrefe M. Brugada syndrome and supraventricular tachyarrhythmias: a novel association? J Cardiovasc Electrophysiol. 2001;12:680–5.

Eckardt L, Probst V, Smits JP, Bahr ES, Wolpert C, Schimpf R, Wichter T, Boisseau P, Heinecke A, Breithardt G, Borggrefe M, LeMarec H, Böcker D, Wilde AA. Long-term prognosis of individuals with right precordial ST-segment-elevation Brugada syndrome. Circulation. 2005;111:257–63.

Ghouse J, Have CT, Skov MW, Andreasen L, Ahlberg G, Nielsen JB, Skaaby T, Olesen SP, Grarup N, Linneberg A, Pedersen O, Vestergaard H, Haunsø S, Svendsen JH, Hansen T, Kanters JK, Olesen MS. Numerous Brugada syndrome-associated genetic variants have no effect on J-point elevation, syncope susceptibility, malignant cardiac arrhythmia, and all-cause mortality. Genet Med. 2017;19:521–8.

Giustetto C, Drago S, Demarchi PG, Dalmasso P, Bianchi F, Masi AS, Carvalho P, Occhetta E, Rossetti G, Riccardi R, Bertona R, Gaita F, Section IAoAaCA-P. Risk stratification of the patients with Brugada type electrocardiogram: a community-based prospective study. Europace. 2009;11:507–13.

Gollob MH, Blier L, Brugada R, Champagne J, Chauhan V, Connors S, Gardner M, Green MS, Gow R, Hamilton R, Harris L, Healey JS, Hodgkinson K, Honeywell C, Kantoch M, Kirsh J, Krahn A, Mullen M, Parkash R, Redfearn D, Rutberg J, Sanatani S, Woo A. Recommendations for the use of genetic testing in the clinical evaluation of inherited cardiac arrhythmias associated with sudden cardiac death: Canadian Cardiovascular Society/Canadian Heart Rhythm Society joint position paper. Can J Cardiol. 2011;27:232–45.

Haïssaguerre M, Extramiana F, Hocini M, Cauchemez B, Jaïs P, Cabrera JA, Farré J, Farre G, Leenhardt A, Sanders P, Scavée C, Hsu LF, Weerasooriya R, Shah DC, Frank R, Maury P, Delay M, Garrigue S, Clémenty J. Mapping and ablation of ventricular fibrillation associated with long-QT and Brugada syndromes. Circulation. 2003;108:925–8.

Hermida JS, Denjoy I, Clerc J, Extramiana F, Jarry G, Milliez P, Guicheney P, Di Fusco S, Rey JL, Cauchemez B, Leenhardt A. Hydroquinidine therapy in Brugada syndrome. J Am Coll Cardiol. 2004;43:1853–60.

Hoogendijk MG, Potse M, Linnenbank AC, Verkerk AO, den Ruijter HM, van Amersfoorth SC, Klaver EC, Beekman L, Bezzina CR, Postema PG, Tan HL, Reimer AG, van der Wal AC, Ten Harkel AD, Dalinghaus M, Vinet A, Wilde AA, de Bakker JM, Coronel R. Mechanism of right precordial ST-segment elevation in structural heart disease: excitation failure by current-to-load mismatch. Heart Rhythm. 2010;7:238–48.

Huang Z, Patel C, Li W, Xie Q, Wu R, Zhang L, Tang R, Wan X, Ma Y, Zhen W, Gao L, Yan GX. Role of signal-averaged electrocardiograms in arrhythmic risk stratification of patients with Brugada syndrome: a prospective study. Heart Rhythm. 2009;6:1156–62.

Kakishita M, Kurita T, Matsuo K, Taguchi A, Suyama K, Shimizu W, Aihara N, Kamakura S, Yamamoto F, Kobayashi J, Kosakai Y, Ohe T. Mode of onset of ventricular fibrillation in patients with Brugada syndrome detected by implantable cardioverter defibrillator therapy. J Am Coll Cardiol. 2000;36:1646–53.

Kamakura S, Ohe T, Nakazawa K, Aizawa Y, Shimizu A, Horie M, Ogawa S, Okumura K, Tsuchihashi K, Sugi K, Makita N, Hagiwara N, Inoue H, Atarashi H, Aihara N, Shimizu W, Kurita T, Suyama K, Noda T, Satomi K, Okamura H, Tomoike H, Japan BSIi. Long-term prognosis of probands with Brugada-pattern ST-elevation in leads V1-V3. Circ Arrhythm Electrophysiol. 2009;2:495–503.

Kamakura T, Wada M, Ishibashi K, Inoue YY, Miyamoto K, Okamura H, Nagase S, Noda T, Aiba T, Yasuda S, Kusano K. Impact of electrocardiogram screening during drug challenge test for the prediction of T-wave oversensing by a subcutaneous implantable cardioverter defibrillator in patients with Brugada syndrome. Heart Vessels. 2017;32:1277–83.

Kaneko Y, Horie M, Niwano S, Kusano KF, Takatsuki S, Kurita T, Mitsuhashi T, Nakajima T, Irie T, Hasegawa K, Noda T, Kamakura S, Aizawa Y, Yasuoka R, Torigoe K, Suzuki H, Ohe T, Shimizu A, Fukuda K, Kurabayashi M. Electrical storm in patients with Brugada syndrome is associated with early repolarization. Circ Arrhythm Electrophysiol. 2014;7:1122–8.

Kapplinger JD, Tester DJ, Alders M, Benito B, Berthet M, Brugada J, Brugada P, Fressart V, Guerchicoff A, Harris-Kerr C, Kamakura S, Kyndt F, Koopmann TT, Miyamoto Y, Pfeiffer R, Pollevick GD, Probst V, Zumhagen S, Vatta M, Towbin JA, Shimizu W, Schulze-Bahr E, Antzelevitch C, Salisbury BA, Guicheney P, Wilde AA, Brugada R, Schott JJ, Ackerman MJ. An international compendium of mutations in the SCN5A-encoded cardiac sodium channel in patients referred for Brugada syndrome genetic testing. Heart Rhythm. 2010;7:33–46.

Kasanuki H, Ohnishi S, Ohtuka M, Matsuda N, Nirei T, Isogai R, Shoda M, Toyoshima Y, Hosoda S. Idiopathic ventricular fibrillation induced with vagal activity in patients without obvious heart disease. Circulation. 1997;95:2277–85.

Kaufman ES. Genetic testing in Brugada syndrome. J Am Coll Cardiol. 2012;60:1419–20.

Kawata H, Morita H, Yamada Y, Noda T, Satomi K, Aiba T, Isobe M, Nagase S, Nakamura K, Fukushima Kusano K, Ito H, Kamakura S, Shimizu W. Prognostic significance of early repolarization in inferolateral leads in Brugada patients with documented ventricular fibrillation: a novel risk factor for Brugada syndrome with ventricular fibrillation. Heart Rhythm. 2013;10:1161–8.

Kinoshita K, Takahashi H, Hata Y, Nishide K, Kato M, Fujita H, Yoshida S, Murai K, Mizumaki K, Nishida K, Yamaguchi Y, Kano M, Tabata T, Nishida N. SCN5A(K817E), a novel Brugada syndrome-associated mutation that alters the activation gating of NaV1.5 channel. Heart Rhythm. 2016;13:1113–20.

Krittayaphong R, Veerakul G, Nademanee K, Kangkagate C. Heart rate variability in patients with Brugada syndrome in Thailand. Eur Heart J. 2003;24:1771–8.

Kum LC, Fung JW, Sanderson JE. Brugada syndrome unmasked by febrile illness. Pacing Clin Electrophysiol. 2002;25:1660–1.

Kyndt F, Probst V, Potet F, Demolombe S, Chevallier JC, Baro I, Moisan JP, Boisseau P, Schott JJ, Escande D, Le Marec H. Novel SCN5A mutation leading either to isolated cardiac conduction defect or Brugada syndrome in a large French family. Circulation. 2001;104:3081–6.

Le Scouarnec S, Karakachoff M, Gourraud JB, Lindenbaum P, Bonnaud S, Portero V, Duboscq-Bidot L, Daumy X, Simonet F, Teusan R, Baron E, Violleau J, Persyn E, Bellanger L, Barc J, Chatel S, Martins R, Mabo P, Sacher F, Haïssaguerre M, Kyndt F, Schmitt S, Bézieau S, Le Marec H, Dina C, Schott JJ, Probst V, Redon R. Testing the burden of rare variation in arrhythmia-susceptibility genes provides new insights into molecular diagnosis for Brugada syndrome. Hum Mol Genet. 2015;24:2757–63.

Letsas KP, Sacher F, Probst V, Weber R, Knecht S, Kalusche D, Haïssaguerre M, Arentz T. Prevalence of early repolarization pattern in inferolateral leads in patients with Brugada syndrome. Heart Rhythm. 2008;5:1685–9.

Makimoto H, Nakagawa E, Takaki H, Yamada Y, Okamura H, Noda T, Satomi K, Suyama K, Aihara N, Kurita T, Kamakura S, Shimizu W. Augmented ST-segment elevation during recovery from exercise predicts cardiac events in patients with Brugada syndrome. J Am Coll Cardiol. 2010;56:1576–84.

Matsuo K, Kurita T, Inagaki M, Kakishita M, Aihara N, Shimizu W, Taguchi A, Suyama K, Kamakura S, Shimomura K. The circadian pattern of the development of ventricular fibrillation in patients with Brugada syndrome. Eur Heart J. 1999;20:465–70.

Maury P, Rollin A, Sacher F, Gourraud JB, Raczka F, Pasquié JL, Duparc A, Mondoly P, Cardin C, Delay M, Derval N, Chatel S, Bongard V, Sadron M, Denis A, Davy JM, Hocini M, Jaïs P, Jesel L, Haïssaguerre M, Probst V. Prevalence and prognostic role of various conduction disturbances in patients with the Brugada syndrome. Am J Cardiol. 2013;112:1384–9.

Maury P, Sacher F, Gourraud JB, Pasquié JL, Raczka F, Bongard V, Duparc A, Mondoly P, Sadron M, Chatel S, Derval N, Denis A, Cardin C, Davy JM, Hocini M, Jaïs P, Jesel L, Carrié D, Galinier M, Haïssaguerre M, Probst V, Rollin A. Increased Tpeak-Tend interval is highly and independently related to arrhythmic events in Brugada syndrome. Heart Rhythm. 2015;12:2469–76.

Meregalli PG, Tan HL, Probst V, Koopmann TT, Tanck MW, Bhuiyan ZA, Sacher F, Kyndt F, Schott JJ, Albuisson J, Mabo P, Bezzina CR, Le Marec H, Wilde AA. Type of SCN5A mutation determines clinical severity and degree of conduction slowing in loss-of-function sodium channelopathies. Heart Rhythm. 2009;6:341–8.

Miyazaki T, Mitamura H, Miyoshi S, Soejima K, Aizawa Y, Ogawa S. Autonomic and antiarrhythmic drug modulation of ST segment elevation in patients with Brugada syndrome. J Am Coll Cardiol. 1996;27:1061–70.

Mizumaki K, Fujiki A, Nishida K, Iwamoto J, Sakamoto T, Sakabe M, Tsuneda T, Sugao M, Inoue H. Postprandial augmentation of bradycardia-dependent ST elevation in patients with Brugada syndrome. J Cardiovasc Electrophysiol. 2007;18:839–44.

Mizusawa Y, Sakurada H, Nishizaki M, Hiraoka M. Effects of low-dose quinidine on ventricular tachyarrhythmias in patients with Brugada syndrome: low-dose quinidine therapy as an adjunctive treatment. J Cardiovasc Pharmacol. 2006;47:359–64.

Morita H, Kusano-Fukushima K, Nagase S, Fujimoto Y, Hisamatsu K, Fujio H, Haraoka K, Kobayashi M, Morita ST, Nakamura K, Emori T, Matsubara H, Hina K, Kita T, Fukatani M, Ohe T. Atrial fibrillation and atrial vulnerability in patients with Brugada syndrome. J Am Coll Cardiol. 2002a;40:1437–44.

Morita H, Nagase S, Kusano K, Ohe T. Spontaneous T wave alternans and premature ventricular contractions during febrile illness in a patient with Brugada syndrome. J Cardiovasc Electrophysiol. 2002b;13:816–8.

Morita H, Kusano KF, Miura D, Nagase S, Nakamura K, Morita ST, Ohe T, Zipes DP, Wu J. Fragmented QRS as a marker of conduction abnormality and a predictor of prognosis of Brugada syndrome. Circulation. 2008;118:1697–704.

Murakami M, Nakamura K, Kusano KF, Morita H, Nakagawa K, Tanaka M, Tada T, Toh N, Nishii N, Nagase S, Hata Y, Kohno K, Miura D, Ohe T, Ito H. Efficacy of low-dose bepridil for prevention of ventricular fibrillation in patients with Brugada syndrome with and without SCN5A mutation. J Cardiovasc Pharmacol. 2010;56:389–95.

Nademanee K, Veerakul G, Nimmannit S, Chaowakul V, Bhuripanyo K, Likittanasombat K, Tunsanga K, Kuasirikul S, Malasit P, Tansupasawadikul S, Tatsanavivat P. Arrhythmogenic marker for the sudden unexplained death syndrome in Thai men. Circulation. 1997;96:2595–600.

Nademanee K, Veerakul G, Mower M, Likittanasombat K, Krittayapong R, Bhuripanyo K, Sitthisook S, Chaothawee L, Lai MY, Azen SP. Defibrillator versus beta-blockers for unexplained death in Thailand (DEBUT): a randomized clinical trial. Circulation. 2003;107:2221–6.

Nademanee K, Veerakul G, Chandanamattha P, Chaothawee L, Ariyachaipanich A, Jirasirirojanakorn K, Likittanasombat K, Bhuripanyo K, Ngarmukos T. Prevention of ventricular fibrillation episodes in Brugada syndrome by catheter ablation over the anterior right ventricular outflow tract epicardium. Circulation. 2011;123:1270–9.

Nademanee K, Raju H, de Noronha SV, Papadakis M, Robinson L, Rothery S, Makita N, Kowase S, Boonmee N, Vitayakritsirikul V, Ratanarapee S, Sharma S, van der Wal AC, Christiansen M, Tan HL, Wilde AA, Nogami A, Sheppard MN, Veerakul G, Behr ER. Fibrosis, Connexin-43, and conduction abnormalities in the Brugada syndrome. J Am Coll Cardiol. 2015;66:1976–86.

Nagase S, Kusano KF, Morita H, Fujimoto Y, Kakishita M, Nakamura K, Emori T, Matsubara H, Ohe T. Epicardial electrogram of the right ventricular outflow tract in patients with the Brugada syndrome: using the epicardial lead. J Am Coll Cardiol. 2002;39:1992–5.

Nakagawa E, Takagi M, Tatsumi H, Yoshiyama M. Successful radiofrequency catheter ablation for electrical storm of ventricular fibrillation in a patient with Brugada syndrome. Circ J. 2008;72:1025–9.

Nakajima T, Kaneko Y, Saito A, Ota M, Iijima T, Kurabayashi M. Enhanced fast-inactivated state stability of cardiac sodium channels by a novel voltage sensor SCN5A mutation, R1632C, as a cause of atypical Brugada syndrome. Heart Rhythm. 2015;12:2296–304.

Nimmannit S, Malasit P, Chaovakul V, Susaengrat W, Vasuvattakul S, Nilwarangkur S. Pathogenesis of sudden unexplained nocturnal death (lai tai) and endemic distal renal tubular acidosis. Lancet. 1991;338:930–2.

Nogami A, Nakao M, Kubota S, Sugiyasu A, Doi H, Yokoyama K, Yumoto K, Tamaki T, Kato K, Hosokawa N, Sagai H, Nakamura H, Nitta J, Yamauchi Y, Aonuma K. Enhancement of J-ST-segment elevation by the glucose and insulin test in Brugada syndrome. Pacing Clin Electrophysiol. 2003;26:332–7.

Ohgo T, Okamura H, Noda T, Satomi K, Suyama K, Kurita T, Aihara N, Kamakura S, Ohe T, Shimizu W. Acute and chronic management in patients with Brugada syndrome associated with electrical storm of ventricular fibrillation. Heart Rhythm. 2007;4:695–700.

Ohkubo K, Nakai T, Watanabe I. Alcohol-induced ventricular fibrillation in a case of Brugada syndrome. Europace. 2013;15:1058.

Olde Nordkamp LR, Conte G, Rosenmöller BR, Warnaars JL, Tan HL, Caputo ML, Regoli F, Moccetti T, Auricchio A, Knops RE, Wilde AA. Brugada syndrome and the subcutaneous implantable cardioverter-defibrillator. J Am Coll Cardiol. 2016;68:665–6.

Olson TM, Michels VV, Ballew JD, Reyna SP, Karst ML, Herron KJ, Horton SC, Rodeheffer RJ, Anderson JL. Sodium channel mutations and susceptibility to heart failure and atrial fibrillation. JAMA. 2005;293:447–54.

Paul M, Gerss J, Schulze-Bahr E, Wichter T, Vahlhaus C, Wilde AA, Breithardt G, Eckardt L. Role of programmed ventricular stimulation in patients with Brugada syndrome: a meta-analysis of worldwide published data. Eur Heart J. 2007;28:2126–33.

Porres JM, Brugada J, Urbistondo V, García F, Reviejo K, Marco P. Fever unmasking the Brugada syndrome. Pacing Clin Electrophysiol. 2002;25:1646–8.

Postema PG. About Brugada syndrome and its prevalence. Europace. 2012;14:925–8.

Postema PG, Wolpert C, Amin AS, Probst V, Borggrefe M, Roden DM, Priori SG, Tan HL, Hiraoka M, Brugada J, Wilde AA. Drugs and Brugada syndrome patients: review of the literature, recommendations, and an up-to-date website (www.brugadadrugs.org). Heart Rhythm. 2009;6:1335–41.

Priori SG, Gasparini M, Napolitano C, Della Bella P, Ottonelli AG, Sassone B, Giordano U, Pappone C, Mascioli G, Rossetti G, De Nardis R, Colombo M. Risk stratification in Brugada syndrome: results of the PRELUDE (PRogrammed ELectrical stimUlation preDictive valuE) registry. J Am Coll Cardiol. 2012;59:37–45.

Priori SG, Wilde AA, Horie M, Cho Y, Behr ER, Berul C, Blom N, Brugada J, Chiang CE, Huikuri H, Kannankeril P, Krahn A, Leenhardt A, Moss A, Schwartz PJ, Shimizu W, Tomaselli G, Tracy C. HRS/EHRA/APHRS expert consensus statement on the diagnosis and management of patients with inherited primary arrhythmia syndromes: document endorsed by HRS, EHRA, and APHRS in May 2013 and by ACCF, AHA, PACES, and AEPC in June 2013. Heart Rhythm. 2013;10:1932–63.

Probst V, Denjoy I, Meregalli PG, Amirault JC, Sacher F, Mansourati J, Babuty D, Villain E, Victor J, Schott JJ, Lupoglazoff JM, Mabo P, Veltmann C, Jesel L, Chevalier P, Clur SA, Haissaguerre M, Wolpert C, Le Marec H, Wilde AA. Clinical aspects and prognosis of Brugada syndrome in children. Circulation. 2007;115:2042–8.

Probst V, Wilde AA, Barc J, Sacher F, Babuty D, Mabo P, Mansourati J, Le Scouarnec S, Kyndt F, Le Caignec C, Guicheney P, Gouas L, Albuisson J, Meregalli PG, Le Marec H, Tan HL, Schott JJ. SCN5A mutations and the role of genetic background in the pathophysiology of Brugada syndrome. Circ Cardiovasc Genet. 2009;2:552–7.

Probst V, Veltmann C, Eckardt L, Meregalli PG, Gaita F, Tan HL, Babuty D, Sacher F, Giustetto C, Schulze-Bahr E, Borggrefe M, Haissaguerre M, Mabo P, Le Marec H, Wolpert C, Wilde AA. Long-term prognosis of patients diagnosed with Brugada syndrome: results from the FINGER Brugada Syndrome Registry. Circulation. 2010;121:635–43.

Rattanawong P, Vutthikraivit W, Charoensri A, Jongraksak T, Prombandankul A, Kanjanahattakij N, et al. Fever-induced Brugada syndrome is more common than previously suspected: a cross-sectional study from an endemic area. Ann Noninvasive Electrocardiol. 2016;21(2):136–41.

Rodríguez-Mañero M, Namdar M, Sarkozy A, Casado-Arroyo R, Ricciardi D, de Asmundis C, Chierchia GB, Wauters K, Rao JY, Bayrak F, Van Malderen S, Brugada P. Prevalence, clinical characteristics and management of atrial fibrillation in patients with Brugada syndrome. Am J Cardiol. 2013;111:362–7.

Rodríguez-Mañero M, Sacher F, de Asmundis C, Maury P, Lambiase PD, Sarkozy A, Probst V, Gandjbakhch E, Castro-Hevia J, Saenen J, Fukushima Kusano K, Rollin A, Arbelo E, Valderrábano M, Arias MA, Mosquera-Pérez I, Schilling R, Chierchia GB, García-Bolao I, García-Seara J, Hernandez-Ojeda J, Kamakura T, Martínez-Sande L, González-Juanatey JR, Haïssaguerre M, Brugada J, Brugada P. Monomorphic ventricular tachycardia in patients with Brugada syndrome: a multicenter retrospective study. Heart Rhythm. 2016;13:669–82.

Sacher F, Probst V, Iesaka Y, Jacon P, Laborderie J, Mizon-Gérard F, Mabo P, Reuter S, Lamaison D, Takahashi Y, O'Neill MD, Garrigue S, Pierre B, Jaïs P, Pasquié JL, Hocini M, Salvador-Mazenq M, Nogami A, Amiel A, Defaye P, Bordachar P, Boveda S, Maury P, Klug D, Babuty D, Haïssaguerre M, Mansourati J, Clémenty J, Le Marec H. Outcome after implantation of a cardioverter-defibrillator in patients with Brugada syndrome: a multicenter study. Circulation. 2006;114:2317–24.

Sakabe M, Fujiki A, Tsuneda T, Nishida K, Sugao M, Mizumaki K, Inoue H. Brugada syndrome occurring in an identical twin: a case report. J Cardiol. 2002a;40:111–5.

Sakabe M, Fujiki A, Tsuneda T, Nishida K, Sugao M, Mizumaki K, Inoue H. Brugada syndrome occurring in an identical twin: a case report. J Cardiol. 2002b;40:111–5.

Sarkozy A, Boussy T, Kourgiannides G, Chierchia GB, Richter S, De Potter T, Geelen P, Wellens F, Spreeuwenberg MD, Brugada P. Long-term follow-up of primary prophylactic implantable cardioverter-defibrillator therapy in Brugada syndrome. Eur Heart J. 2007;28:334–44.

Saura D, García-Alberola A, Carrillo P, Pascual D, Martínez-Sánchez J, Valdés M. Brugada-like electrocardiographic pattern induced by fever. Pacing Clin Electrophysiol. 2002;25:856–9.

Shimada M, Miyazaki T, Miyoshi S, Soejima K, Hori S, Mitamura H, Ogawa S. Sustained monomorphic ventricular tachycardia in a patient with Brugada syndrome. Jpn Circ J. 1996;60:364–70.

Shimizu W, Matsuo K, Takagi M, Tanabe Y, Aiba T, Taguchi A, Suyama K, Kurita T, Aihara N, Kamakura S. Body surface distribution and response to drugs of ST segment elevation in Brugada syndrome: clinical implication of eighty-seven-lead body surface potential mapping and its application to twelve-lead electrocardiograms. J Cardiovasc Electrophysiol. 2000;11:396–404.

Shirai N, Makita N, Sasaki K, Yokoi H, Sakuma I, Sakurada H, Akai J, Kimura A, Hiraoka M, Kitabatake A. A mutant cardiac sodium channel with multiple biophysical defects associated with overlapping clinical features of Brugada syndrome and cardiac conduction disease. Cardiovasc Res. 2002;53:348–54.

Skinner JR, Chung SK, Nel CA, Shelling AN, Crawford JR, McKenzie N, Pinnock R, French JK, Rees MI. Brugada syndrome masquerading as febrile seizures. Pediatrics. 2007;119:e1206–11.

Steven D, Roberts-Thomson KC, Inada K, Seiler J, Koplan BA, Tedrow UB, Sweeney MO, Epstein LE, Stevenson WG. Long-term follow-up in patients with presumptive Brugada syndrome treated with implanted defibrillators. J Cardiovasc Electrophysiol. 2011;22:1115–9.

Szél T, Antzelevitch C. Abnormal repolarization as the basis for late potentials and fractionated electrograms recorded from epicardium in experimental models of Brugada syndrome. J Am Coll Cardiol. 2014;63:2037–45.

Takagi M, Yokoyama Y, Aonuma K, Aihara N, Hiraoka M, Investigators JIVFSJ-I. Clinical characteristics and risk stratification in symptomatic and asymptomatic patients with Brugada syndrome: multicenter study in Japan. J Cardiovasc Electrophysiol. 2007;18:1244–51.

Tan HL, Bink-Boelkens MT, Bezzina CR, Viswanathan PC, Beaufort-Krol GC, van Tintelen PJ, van den Berg MP, Wilde AA, Balser JR. A sodium-channel mutation causes isolated cardiac conduction disease. Nature. 2001;409:1043–7.

Tanaka H, Kinoshita O, Uchikawa S, Kasai H, Nakamura M, Izawa A, Yokoseki O, Kitabayashi H, Takahashi W, Yazaki Y, Watanabe N, Imamura H, Kubo K. Successful prevention of recurrent ventricular fibrillation by intravenous isoproterenol in a patient with Brugada syndrome. Pacing Clin Electrophysiol. 2001;24:1293–4.

Tokioka K, Kusano KF, Morita H, Miura D, Nishii N, Nagase S, Nakamura K, Kohno K, Ito H, Ohe T. Electrocardiographic parameters and fatal arrhythmic events in patients with Brugada syndrome: combination of depolarization and repolarization abnormalities. J Am Coll Cardiol. 2014;63:2131–8.

Tsuchiya T, Ashikaga K, Honda T, Arita M. Prevention of ventricular fibrillation by cilostazol, an oral phosphodiesterase inhibitor, in a patient with Brugada syndrome. J Cardiovasc Electrophysiol. 2002;13:698–701.

Valdivia CR, Tester DJ, Rok BA, Porter CB, Munger TM, Jahangir A, Makielski JC, Ackerman MJ. A trafficking defective, Brugada syndrome-causing SCN5A mutation rescued by drugs. Cardiovasc Res. 2004;62:53–62.

Vatta M, Dumaine R, Varghese G, Richard TA, Shimizu W, Aihara N, Nademanee K, Brugada R, Brugada J, Veerakul G, Li H, Bowles NE, Brugada P, Antzelevitch C, Towbin JA. Genetic and biophysical basis of sudden unexplained nocturnal death syndrome (SUNDS), a disease allelic to Brugada syndrome. Hum Mol Genet. 2002;11:337–45.

Veerakul G, Nademanee K. Brugada syndrome: two decades of progress. Circ J. 2012;76:2713–22.

Veerakul G, Chaothawee L, Nademanee K. Usefulness of positioning ECG leads at V1-3 at higher intercostal spaces to detect Brugada syndrome. Circulation. 2000;102(Suppl):18.

Veerakul G, Camblock J, Schwab M, Nademanee K. Low mortality rate among asymptomatic Brugada syndrome patients: a multi-center control-randomized study comparing ICD versus no-ICD treatment. Circulation. 2008;118:S982.

Viskin S, Rogowski O. Asymptomatic Brugada syndrome: a cardiac ticking time-bomb? Europace. 2007;9:707–10.

Viskin S, Antzelevitch C, Márquez MF, Belhassen B. Quinidine: a valuable medication joins the list of 'endangered species'. Europace. 2007;9:1105–6.

Wang Q, Shen J, Splawski I, Atkinson D, Li Z, Robinson JL, Moss AJ, Towbin JA, Keating MT. SCN5A mutations associated with an inherited cardiac arrhythmia, long QT syndrome. Cell. 1995;80:805–11.

Wilde AA, Antzelevitch C, Borggrefe M, Brugada J, Brugada R, Brugada P, Corrado D, Hauer RN, Kass RS, Nademanee K, Priori SG, Towbin JA, Cardiology SGotMBoAotESo. Proposed diagnostic criteria for the Brugada syndrome: consensus report. Circulation. 2002;106:2514–9.

Wilde AA, Postema PG, Di Diego JM, Viskin S, Morita H, Fish JM, Antzelevitch C. The pathophysiological mechanism underlying Brugada syndrome: depolarization versus repolarization. J Mol Cell Cardiol. 2010;49:543–53.

Yan GX, Antzelevitch C. Cellular basis for the Brugada syndrome and other mechanisms of arrhythmogenesis associated with ST-segment elevation. Circulation. 1999;100:1660–6.

Zhang P, Tung R, Zhang Z, et al. Characterization of the epicardial substrate for catheter ablation of Brugada syndrome. Heart Rhythm. 2016;11:2151–8.

Zumhagen S, Zeidler EM, Stallmeyer B, Ernsting M, Eckardt L, Schulze-Bahr E. Tpeak-Tend interval and Tpeak-Tend/QT ratio in patients with Brugada syndrome. Europace. 2016;18:1866–72.

Sinus Node Disease and Cardiac Conduction Disease

9

Patrick A. Schweizer

Abstract

Primary sinus node disease (SND) and cardiac conduction defect (CCD) are frequent clinical entities with significant morbidity and mortality, which are major indications for the implantation of electronic pacemakers. Throughout the previous two decades, pathogenetic mechanisms underlying both disorders have been investigated in detail, and it has been demonstrated that distinct genetic defects and/or predisposing genetic constellations play important roles in a considerable number of cases. Furthermore it has been shown that both entities often are related to a broader clinical spectrum including overlapping arrhythmia syndromes and structural cardiac abnormalities, indicating that specified genetic defects are key to distinct clinical phenotypes. This book chapter summarizes the work, which most profoundly influences the current understanding of primary excitation and conduction disorders of the heart. The novel mechanistic insight into important pathogenetic aspects of these disorders may lay the groundwork for more mechanism-based, individually tailored clinical management of patients with primary SND and CCD in the future.

9.1 Sinus Node Disease

9.1.1 Clinical Aspects of Sinus Node Disease (SND)

Loss or dysfunction of sinoatrial nodal cells results in sinus node disease (SND), a term commonly used for disorders associated with failure in rate initiation or conduction from the sinoatrial node (SAN) to the atrium, comprising sinus

P. A. Schweizer (✉)
Department of Cardiology, Heidelberg Center for Rhythm Disorders (HCR), Medical University Hospital Heidelberg, Heidelberg, Germany
e-mail: patrick.schweizer@med.uni-heidelberg.de

D. Thomas, C. A. Remme (eds.), *Channelopathies in Heart Disease*, Cardiac and Vascular Biology, https://doi.org/10.1007/978-3-319-77812-9_9

bradycardia, SAN block or arrest, and bradycardia-tachycardia syndrome (Birchfield et al. 1957; Kaplan et al. 1973). In many cases, SND leads to symptoms like dizziness, fatigue, pre-syncope/syncope, or collapse, and the implantation of an electronic pacemaker is currently the only effective therapy (Jensen et al. 2014; Lamas et al. 2000). In 30–50% of all electronic pacemaker implantations, SND is the primary indication, resulting in more than 100,000 pacemakers in the USA in the year 2000, costing more than US$2 billion. With the aging of the population, the number of patients with SND will increase dramatically over the next 50 years. Based on a recent population study, it was calculated that there were approximately 78,000 incident cases of SND in 2012 in the USA, and this number would increase to almost 172,000 per year by 2060 resulting in a major public health burden (Jensen et al. 2014).

In the majority of the cases, SND is a primary "idiopathic" disorder which occurs clearly age-dependent and equally among gender and was found to be associated with risk factors like greater body mass index, height, N-terminal pro-B type natriuretic peptide, cystatin C, and additional cardiovascular disease (Jensen et al. 2014).

Furthermore, primary SND has been related to inherited forms, and specified genes have been linked to SND and shown to be mutated in affected family members (Choudhury et al. 2015). Interestingly, abnormalities of SAN function are particularly common in heart failure and cardiomyopathies (structural, ischemic, or inflammatory) indicating a genetic and/or mechanistic link between electrical and structural dysfunction (Sanders et al. 2004a, b; Zicha et al. 2005). In addition, there are secondary causes of SND like drug intake, myocardial infarction/ischemia, or heart surgery (Monfredi and Boyett 2015). In particular mentionable is the association of SND with the generation of atrial tachyarrhythmias, mostly atrial fibrillation (AF), as these tachycardias affect around 50% of patients with SND (Gomes et al. 1981), leading to the term "bradycardia-tachycardia syndrome." Clinical data suggest that both conditions lay the groundwork for the development and perpetuation of each other (Sairaku et al. 2012).

Many genetic and epigenetic factors constitute the clinical phenotype of primary SND, which represents a complex and heterogenous disease entity. The following sections will provide a comprehensive overview of the different factors that contribute to the clinical development of SND.

9.1.2 The Sinoatrial Node

The sinoatrial node (SAN) is a complex and heterogeneous tissue, which constitutes the primary pacemaker of the heart (Dobrzynski et al. 2005). It is located at the entrance of the *superior* vena cava to the right atrium and is thought to consist of <10,000 highly specialized cells, capable of automatically depolarizing, and by this pace ~5 billion working cardiomyocytes downstream of the SAN (Cho 2015). Automaticity is modulated by the central nervous system via sympathetic and parasympathetic stimulation and thus can be adapted to the physiological needs of the body. Interestingly, at embryonic stages, early myocardial cells possess the capability to spontaneous excitation (Yasui et al. 2001; Schweizer et al. 2009). Later, at postnatal stages, automaticity is restricted to specialized cells of the sinoatrial node (SAN) and the conduction system, while tissue of the working type myocardium remains quiescent if not activated by the neighboring cell (Kurata et al. 2005).

The spontaneous excitation originates from the center of the SAN and is then propagated from the leading pacemaker site to the periphery, where it connects the SAN to the atrial muscle of the crista terminalis and right atrial free wall (Boyett et al. 2000, 2006). The SAN center has little electrical coupling to protect it from the inhibitory hyperpolarizing influence of surrounding cardiac muscle and is characterized by unique ionic currents appropriate for pacemaking. By contrast, the SAN periphery, although capable of spontaneous depolarization, achieves to drive the surrounding atrial muscle by a large inward sodium current (consequently, an action potential with a rapid upstroke) to generate sufficient depolarizing current and pronounced electrical coupling to deliver current to the atrial muscle (Boyett et al. 2000; Fedorov et al. 2010). Thus, molecular mechanisms underlying specified excitation and conduction are crucial for proper pacemaking and constitute critical pathomechanistic components in SND.

9.1.3 Sinoatrial Node Remodeling

SND was originally attributed to idiopathic fibrosis, cell atrophy, or ischemia. However, whether chronic ischemia is a cause of SND remains unresolved as postmortem studies could not establish a definite association of the grade of SAN artery disease with symptomatic SND (Evans and Shaw 1977; Shaw et al. 1987). Recent evidence is accumulating that changes in the electrophysiology of the SAN, known as electrical remodeling, may contribute importantly to SND (Choudhury et al. 2015). In this context, patterns of SAN remodeling between different predisposing diseases/states, i.e., heart failure, aging, diabetes, atrial fibrillation, and endurance sports, are diverse but are suspected to lay a molecular groundwork for the common end point of sinus node disease (Choudhury et al. 2015) (Fig. 9.1). The following examples highlight this view: with respect to age-related changes, various studies demonstrated remodeling processes of the SAN, going along with a specific footprint of ion-channel downregulation, including hallmark pacemaker channels HCN1, HCN4, Cav1.2, and Nav1.5 (Hao et al. 2011; Tellez et al. 2011; Larson et al. 2013) (Table 9.1).

Furthermore, endurance training is associated with marked sinus bradycardia. Athletes more often show symptomatic SND and AF later in life, compared to control groups (Baldesberger et al. 2008). Originally, bradycardia was considered a result of high vagal tone, in terms of a neural response to exercise, thought to be fully reversible after cessation of excessive training. However "intrinsic heart rate," investigated by complete pharmacological vagal blockade has been shown to be lower in trained individuals (Boyett et al. 2013). Recently a study in rodents demonstrated downregulation of *HCN4* and *TBX3* in trained animals (D'Souza et al. 2014). Furthermore it was shown by the same group that miR-423-5p contributes to training-induced bradycardia by targeting *HCN4* (D'Souza et al. 2017). These

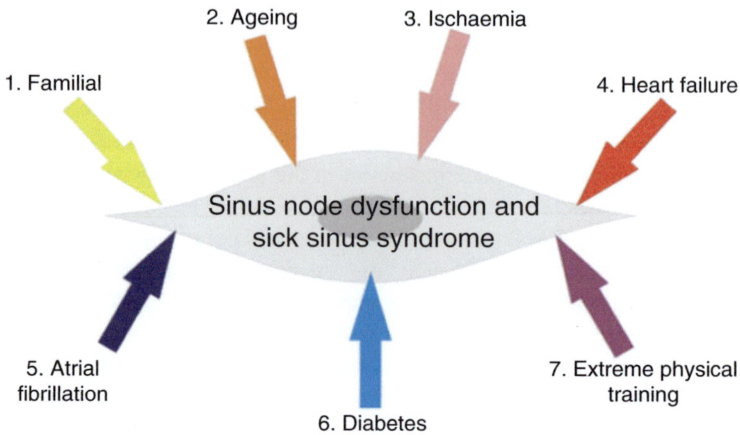

Fig. 9.1 Illustration of the most important etiologies of SND (modified from Monfredi et al. 2010)

Table 9.1 Genes and mechanisms involved in electrical remodeling of the SAN in different causes of SND. The inherited genes are mutations found in patients affected by familial SND, while data on other causes of electrical remodeling were observed in animal models. Downward arrows mean downregulation (modified from Choudhury et al. 2015)

Cause of SND	Ion channels and genes involved
Familial/inherited	HCN4, SCN5A, RYR2, CASQ2, ANKB, MYH6, CACNA1D
	KCNQ1, CASQ2, GIRK1, GIRK4
Aging	↓Nav1.5, ↓Cx43, ↓RYR2, ↓HCN1, ↓HCN4
Heart failure	↓HCN4
Exercise training	↓HCN4, ↓TBX3
Atrial tachyarrhythmia	↓HCN2, ↓HCN4

data suggest electrical remodeling of the SAN as a key mechanism for exercise-induced bradycardia rather than high vagal tone, pointing to molecular changes associated with endurance sports that may aggravate in a subset of patients leading to SND later in life.

9.1.4 Sinus Node Disease and Atrial Tachyarrhythmias

Regarding known genetic pathomechanisms of SND, it is interesting to note that most disease genes for SND also associate with AF. As SND is increasingly recognized not simply to be a disease of the SAN but also including the conduction system and the atrial myocardium (Sanders et al. 2004a, b), electrical and structural remodeling of these structures lay the groundwork for the development of AF as well (Monfredi and Boyett 2015). Concomitant bradycardia further facilitates the development of AF through an increased probability of atrial ectopic activity and a greater dispersion of refractoriness, which both are established pathomechanisms of AF (Amasyali et al. 2014). The other way around, AF and other supraventricular tachycardias are known to compromise the SAN by the fast rate leading to SAN remodeling and dysfunction. In this context, it has been reported that atrial tachyarrhythmias cause alterations of Ca^{2+} cycling, as well as reduced I_f and I_{Ks} currents due to downregulation of HCN2, HCN4, and minK channels within the SAN, respectively (Yeh et al. 2009). Thus, it becomes obvious that bradycardia and atrial tachyarrhythmias in SND are not incoherent processes; rather they are the result of the same underlying pathomechanisms and reinforce each other (Monfredi and Boyett 2015). Therefore treating the one might co-effect the other, although limited data exist upon this relationship.

9.1.5 Genetic Findings of Familial Sinus Node Disease

Primary sinus node dysfunction has been related to inherited, familial forms of the disease (Spellberg 1971). Several genes have been associated with the disorder (Table 9.2). "Loss-of-function" mutations within those genes were related to either congenital SND or to phenotypes that developed throughout life with variable penetrance in families. The findings facilitated not only novel insight into SAN pathophysiology but also uncovered SND as a primary Mendelian disorder in a subset of cases. Among the genes associated with the syndrome, loss-of-function mutations of the *SCN5A* gene underlying the cardiac sodium channel alpha subunit are an established pathomechanism (OMIM sick sinus syndrome 1; Benson et al. 2003). Based on electrophysiological studies and computational modeling, mutated channels were demonstrated to cause either abnormally slow pacemaking or to produce sinus exit block (Butters et al. 2010). In addition, *SCN5A* mutations associate with multiple arrhythmic disorders including Brugada syndrome, long QT syndrome, and dilated cardiomyopathy (Remme 2013), but little is known about the mechanisms underlying phenotypic specification. However, the possibility of multiple overlapping symptoms, also summarized as "sodium channel disease," requires particular attention in the management of such patients.

Furthermore, mutations in *HCN4* underlying a significant proportion of the pacemaker current I_f in the SAN have been demonstrated to cause hereditary SND (OMIM sick sinus syndrome 2). Although initially linked to rather asymptomatic sinus bradycardia (Milanesi et al. 2006; Nof et al. 2007; Schweizer et al. 2010), a significant number of *HCN4* mutations were associated with symptomatic bradycardia requiring pacemaker implantation (Schulze-Bahr et al. 2003; Duhme et al. 2013; Schweizer et al. 2014). Moreover, *HCN4* loss-of-function mutations were shown to facilitate bradycardia-tachycardia syndrome and atrial fibrillation, indicating that I_f-channel dysfunction also contributes to the development of atrial tachyarrhythmias and in particular AF (Ellinor et al. 2012; Duhme et al. 2013; Macri et al. 2014). Recently, the phenotypic spectrum of *HCN4* mutations was expanded to a combined electromechanical phenotype of sinus bradycardia and noncompaction cardiomyopathy (Schweizer et al. 2014; Milano et al. 2014), and it

Table 9.2 Genes linked to human SND

Causative gene	Mechanism mutation	Rhythm disorder	Additional phenotype	Reference
SCN5A	Loss-of-function	SB, SA block, SArr, CD, AF, BrS	DCM	Benson et al. (2003), Remme (2013) and Haas et al. (2015)
	Gain-of-function	LQT-3, SB	–	Makita et al. (2008)
HCN4	Loss-of-function	SB, CI, AVB, AF, APC, VPC	NCCM, MVP, DAA	Schulze-Bahr et al. (2003), Schweizer et al. (2010, 2014), Vermeer et al. (2016)
	Gain-of-function	IST	–	Baruscotti et al. (2017)
MYH6	Loss-of-function	SB, AVB, SSS	DCM, HCM	Holm et al. (2011)
ANK2	Loss-of-function	SB, AF, CD, LQT-4	–	Le Scouarnec et al. (2008)
CACNA1D	Loss-of-function	SB, AVB (autosomal recessive)	Inner ear deafness	Baig et al. (2011)
GNB2	Loss-of-function	SB, AVB	–	Stallmeyer et al. (2017)
GNB5	Loss-of-function	SB	Eye, gastric, and neural disease	Lodder et al. (2016)
KCNE2	Loss-of-function	SB, LQT-6	–	Nawathe et al. (2013)
KCNQ1	Loss-of-function	SB, AF, LQT-1	–	Henrion et al. (2012)
	Gain-of-function	SB, SQT, AF	–	Ki et al. (2014))
RYR2	Loss-of-function	SB, CPVT, ARVC	NCCM, HCM	Postma et al. (2005)
CASQ2	Loss-of-function	SB, CPVT (autosomal recessive)	HCM	Postma et al. (2002)
GIRK1	Unknown	SB, SA block	–	Holmegard et al. (2010)
GIRK4	Unknown	SB, SA block	–	Holmegard et al. (2010)

AF atrial fibrillation, *APC* atrial premature contraction, *ARVC* arrhythmogenic right ventricular cardiomyopathy, *BrS* Brugada syndrome, *CD* conduction defect, *CPVT* catecholaminergic polymorphic ventricular tachycardia, *DAA* dilatation of the ascending aorta, *DCM* dilated cardiomyopathy, *HCM* hypertrophic cardiomyopathy, *IST* idiopathic sinus tachycardia, *LQT* long QT syndrome, *NCCM* noncompaction cardiomyopathy, *SA-block* sinoatrial block, *SArr* sinus arrest, *SB* sinus bradycardia, *SQT* short QT syndrome, *SSS* sick sinus syndrome, *VPC* ventricular premature contraction

was shown that defects of the mitral valve and dilation of the ascending aorta also form part of this symptom complex (Vermeer et al. 2016). A very recent work identified HCN4 "gain-of-function" to be associated with inappropriate sinus tachycardia in a familial trait (Baruscotti et al. 2017), providing a novel molecular mechanism underlying this previously unresolved disorder.

Furthermore, the non-ion channel genes *MYH6* have been demonstrated to contribute to SND in humans as well (OMIM sick sinus syndrome 3; Holm et al. 2011). *MYH6* was shown to be importantly involved in SND pathogenesis, as a common missense variant (allelic frequency 0.38% in Icelanders) increases the lifetime risk of developing SND to ~50%, although pathomechanisms remained unresolved yet (Holm et al. 2011).

Other genes have been associated with combined phenotypes of SND together with additional cardiac (i.e., *KCNQ1* and *ANK2* mutations linked to long QT syndrome) or noncardiac (i.e., *CACNA1D* mutations associate with deafness) symptoms (Ki et al. 2014; Le Scouarnec et al. 2008; Baig et al. 2011), in agreement with the view that rhythm genes often are crucial for other physiological processes as well (Akhirome and Jay 2015). Accordingly, dysfunction of calcium-handling proteins (RYR2, Postma et al. 2005 and CASQ2, Postma et al. 2002) has been reported to cause sinus bradycardia in addition to catecholaminergic polymorphic ventricular tachycardia (CPVT) and various structural phenotypes (hypertrophic cardiomyopathy, noncompaction cardiomyopathy). Apart from the clinical scenario, experimental animal models suggested various ion channel and transcriptional/regulatory proteins to be implicated in SND, although the clinical relevance of these mechanisms needs to be determined.

9.2 Isolated and Progressive Cardiac Conduction Defects

9.2.1 Clinical Aspects and Classification of Cardiac Conduction Defect

Cardiac conduction defect (CCD) as failure in the propagation of the cardiac impulse along the specialized electrical system is a primary disorder, if not explained by other pathophysiological states like congenital, ischemic or structural heart disease, infection, drug intake, or disturbed metabolic states. Isolated cardiac conduction defect (ICCD) constitutes a heterogeneous group of disease-causing mechanisms resulting in potentially life-threatening heart block.

Usually patients present with exercise intolerance and/or dyspnea due to compromised AV conduction, which later results in pre-syncope or syncope due to periods of ventricular asystole caused by high-grade AV block. At disease onset, patients are mostly asymptomatic and only rarely show hemodynamic disturbance due to prolonged AV conduction.

The disorder was first described in 1964 by Lenegre and Lev and thus carries the synonym Morbus Lev-Lenegre (Lev 1964; Lenegre 1964). Both authors independently reported from patients with diseased cardiac conduction that is AV block or left and/or right bundle branch block resulting in symptoms like dizziness, syncope, and sudden cardiac death. Postmortem investigations revealed distinct fibrosis of the cardiac conduction system, providing the initial pathogenetic hypothesis of the disease mechanism. Further, it was shown that the disorder progressed in an age-related manner (Probst et al. 2003). With respect to its pathophysiological mechanisms, two forms of primary CCD are distinguished: a senile form with late onset (age > 50 years), pointing to age-related fibrotic degeneration and remodeling of the conduction system similar to the SAN, and a hereditary form, which more often has an early onset (age < 50 years) and goes along with a family history of CCD, sudden cardiac death, congenital heart disease, and/or cardiomyopathy originating from an underlying pathogenic mutation in susceptibility genes. As the mechanism of hereditary CCD is in part an accelerated degeneration of the conduction system, progress of degeneration might occur much faster in such patients compared to others affected by the senile form of the disease.

9.2.2 Genetic Findings of CCD

The identification of mutations in the depolarizing cardiac ion channel gene *SCN5A* in patients affected by CCD, for the first time, offered a plausible explanation for the inheritance of this idiopathic disorder (Schott et al. 1999; Tan et al. 2001). Since then ICCD has been associated with multiple different *SCN5A* mutations (OMIM progressive cardiac conduction defect 1a), which constitutes the most important disease gene for the disorder. More recently, mutations in other genes have been reported but are less frequently identified among patients with ICCD than *SCN5A* mutations. Recent data suggested the yield of genetic testing in CCD to ~37%, with a single recurrent SCN5A mutation (c.2582_2583delTT) being the predominant genetic hit (Hofman et al. 2013). However, this single center study was confined to mutation scanning of single genes. Thus, genetic distribution among larger populations using modern sequencing techniques remains to be explored.

Clinically, patients carrying *SCN5A* mutations usually present with bradycardia, a prolonged PR interval, wide QRS, and left-axis deviation. Generally, it is important to note that patients with a pathogenic mutation in *SCN5A* should be advised to avoid drugs with sodium-channel-blocking effects (please refer to https://www.brugadadrugs.org/). With respect to the possibility of an overlap syndrome associated with "sodium channel disease," some patients requiring pacemaker therapy may benefit from an implantable cardioverter defibrillator (ICD), which should be carefully evaluated prior implantation.

Another important disease gene linked to isolated CCD is *TRPM4* (OMIM progressive cardiac conduction defect 1b), which encodes a Ca^{2+}-sensitive unselective cation channel that is highly expressed in the Purkinje system. Critical mutations in *TRPM4* linked to CCD were shown to cause attenuated deSUMOylation of the TRPM4 channel resulting in increased expression at the cytoplasmic membrane (Kruse et al. 2009; Liu et al. 2010). Consequently, TRPM4 gain-of-function results in membrane depolarization, which reduces availability of Nav1.5 and therefore leads to conduction disturbance. Mutations segregated with multiple families and clinical phenotype are typically characterized by a right bundle branch block that progresses to complete heart block. Based on recent studies with small cohorts, the estimated yield of *TRPM4* mutations in progressive CCD is up to 15% (Stallmeyer et al. 2012; Daumy et al. 2016).

Mutations in *LMNA* are associated with a broad phenotypic spectrum, known as laminopathies, including Hutchinson-Gilford progeria, autosomal recessive Charcot-Marie-Tooth syndrome, and Emery-Dreifuss muscular dystrophy. Importantly, mutations in *LMNA* are linked to dilated cardiomyopathy. Disease onset is typically preceded by marked CCD (Brodt et al. 2013). Carriers of *LMNA* mutations have a considerable risk of malignant ventricular arrhythmias, even with preserved left ventricular ejection fraction (Kumar et al. 2016). Thus, on the basis of a pacemaker indication, ICD implantation should be considered (Priori et al. 2013), especially in the presence of additional risk factors such as male sex, nonsustained VT, left ventricular ejection fraction <45%, and the presence of a non-missense *LMNA* mutation (van Rijsingen et al. 2012).

NKX2–5 encodes a transcription factor involved in cardiomyogenesis and formation of cardiac structure, as well as development of the conduction system. Accordingly mutations in *NKX2–5* were found to cause congenital heart defects with autosomal dominant inheritance. Mostly, CCD characterized by different degrees of AV block was reported to be accompanied by ostium secundum atrial septal defects (ASD) (Schott et al. 1998; Stallmeyer et al. 2010).

Furthermore CCD is an essential symptom of Holt-Oram syndrome (HOS), an autosomal dominant.

disease, which is caused by mutations in the transcription factor *TBX5*. Affected individuals have skeletal anomalies involving the radius, carpal, or hand bones. In addition, patients display congenital heart defects, typically a secundum ASD or ventricular septal defect (VSD). Progressive CCD also forms part of the symptom complex often requiring pacemaker implantation. With respect to the disease mechanism of CCD, TBX5 is critical for normal cardiac development in prenatal life, while it controls SCN5A expression making it important in regulating cardiac conduction in postnatal life (Arnolds et al. 2012). Accordingly, genome-wide association studies (GWAS) identified common variations of *TBX5* associated with both PR and QRS durations, again underlining its relevance for cardiac conduction (Sotoodehnia et al. 2010).

Moreover, muscular dystrophies commonly involve the cardiac muscle as well and can cause CCD (Groh 2012). Among the muscular dystrophies with frequent involvement of the conduction system, the autosomal dominant myotonic dystrophies caused by repeat expansions in *DMPK* (type 1) or *CNBP* (type 2) are most prevalent. In particular in myotonic dystrophy type I (also called Steinert's disease), the majority of patients develop CCP, which is the major cause of sudden death (Groh et al. 2008). Thus, pacemaker implantation should be considered at low threshold.

Other neuromuscular disorders associated with CCD are Emery-Dreifuss muscular dystrophy, limb-girdle muscular dystrophy type IB, and myofibrillar myopathy (Groh 2012). The most common form with cardiac involvement is the autosomal dominant desmin-related myopathy, which commonly associates with DCM and CCD but also causes supraventricular and ventricular arrhythmias (van Spaendonck-Zwarts et al. 2011). Given the increased risk of SCD in many neuromuscular disorders, clinical guidelines suggest a more aggressive approach than with other CCD patients, including a lower threshold to implant an ICD in patients with pacemaker indication. Other genes that have been described in association with familial CCD are the sodium channel β-subunit gene *SCN1B* (Watanabe et al. 2008), *GJA5* encoding the gap junction protein Cx40 (Makita et al. 2012), and *PRKAG2*. The latter is involved in hypertrophic cardiomyopathy and the WPW syndrome as well (Gollob et al. 2001a, b).

Taken together, many factors determine the clinical phenotype of SND and CCD, representing complex and heterogeneous disorders. There is growing evidence that genetic disposition plays an important role in the pathogenesis of SND and CCD, and mutations in specified genes have been shown to cause hereditary forms in a subset of cases. Particularly, certain genetic defects were linked to distinct clinical profiles, which may pave the way for better diagnosis and surveillance of patients in the future. Thus, evidence of a genetic form of SND and/or CCD may have the potential to improve clinical stratification of patients as genetic changes can underlie specified clinical pathways that may also point to overlapping cardiac phenotypes or syndromic comorbidity.

Acknowledgments This work was supported in parts by grants from the Molecular Medicine Partnership Unit, Heidelberg (Senior Career Fellowship to P.A.S.).

Compliance with Ethical Standards

Conflict of Interest Dr. Schweizer indicates no potential conflicts of interest.

Ethical Approval This article does not contain any studies with human participants or animals performed by any of the authors.

References

Akhirome E, Jay PY. Rhythm genes sing more than one tune: noncanonical functions of cardiac ion channels. Circ Arrhythm Electrophysiol. 2015;8:261–2.

Amasyali B, Kilic A, Kilit C. Sinus node dysfunction and atrial fibrillation: which one dominates? Int J Cardiol. 2014;175:379–80.

Arnolds DE, Liu F, Fahrenbach JP, Kim GH, Schillinger KJ, Smemo S, McNally EM, Nobrega MA, Patel VV, Moskowitz IP. TBX5 drives Scn5a expression to regulate cardiac conduction system function. J Clin Invest. 2012;122:2509–18.

Baig SM, Koschak A, Lieb A, Gebhart M, Dafinger C, Nürnberg G, Ali A, Ahmad I, Sinnegger-Brauns MJ, Brandt N, Engel J, Mangoni ME, Farooq M, Khan HU, Nürnberg P, Striessnig J,

Bolz HJ. Loss of Ca(v)1.3 (CACNA1D) function in a human channelopathy with bradycardia and congenital deafness. Nat Neurosci. 2011;14:77–84.

Baldesberger S, Bauersfeld U, Candinas R, Seifert B, Zuber M, Ritter M, Jenni R, Oechslin E, Luthi P, Scharf C, Marti B, Attenhofer Jost CH. Sinus node disease and arrhythmias in the long-term follow-up of former professional cyclists. Eur Heart J. 2008;29:71–8.

Baruscotti M, Bucchi A, Milanesi R, Paina M, Barbuti A, Gnecchi-Ruscone T, Bianco E, Vitali-Serdoz L, Cappato R, DiFrancesco D. A gain-of-function mutation in the cardiac pacemaker HCN4 channel increasing cAMP sensitivity is associated with familial inappropriate sinus tachycardia. Eur Heart J. 2017;38:280–8.

Benson DW, Wang DW, Dyment M, Knilans TK, Fish FA, Strieper MJ, Rhodes TH, George AL Jr. Congenital sick sinus syndrome caused by recessive mutations in the cardiac sodium channel gene (SCN5A). J Clin Invest. 2003;112:1019–28.

Birchfield RI, Menefee EE, Bryant GD. Disease of the sinoatrial node associated with bradycardia, asystole, syncope, and paroxysmal atrial fibrillation. Circulation. 1957;16:20–6.

Boyett MR, Honjo H, Kodama I. The sinoatrial node, a heterogeneous pacemaker structure. Cardiovasc Res. 2000;47:658–87.

Boyett MR, Inada S, Yoo S, Li J, Liu J, Tellez JO, Greener ID, Honjo H, Billeter R, Lei M, Zhang H, Efimov IR, Dobrzynski H. Connexins in the sinoatrial and atrioventricular nodes. Adv Cardiol. 2006;42:175–97.

Boyett MR, D'Souza A, Zhang H, Morris GM, Dobrzynski H, Monfredi O. Viewpoint: Is the resting bradycardia in athletes the result of remodeling of the sinoatrial node rather than high vagal tone? J Appl Physiol (1985). 2013;114:1351–5.

Brodt C, Siegfried JD, Hofmeyer M, Martel J, Rampersaud E, Li D, Morales A, Hershberger RE. Temporal relationship of conduction system disease and ventricular dysfunction in LMNA cardiomyopathy. J Card Fail. 2013;19:233–9.

Butters TD, Aslanidi OV, Inada S, Boyett MR, Hancox JC, Lei M, Zhang H. Mechanistic links between Na+ channel (SCN5A) mutations and impaired cardiac pacemaking in sick sinus syndrome. Circ Res. 2010;107:126–37.

Cho HC. Pacing the heart with genes: recent progress in biological pacing. Curr Cardiol Rep. 2015; 17:65.

Choudhury M, Boyett MR, Morris GM. Biology of the sinus node and its disease. Arrhythm Electrophysiol Rev. 2015;4:28–34.

Daumy X, Amarouch MY, Lindenbaum P, Bonnaud S, Charpentier E, Bianchi B, et al. Targeted resequencing identifies TRPM4 as a major gene predisposing to progressive familial heart block type I. Int J Cardiol. 2016;207:349–58.

Dobrzynski H, Li J, Tellez J, Greener ID, Nikolski VP, Wright SE, Parson SH, Jones SA, Lancaster MK, Yamamoto M, Honjo H, Takagishi Y, Kodama I, Efimov IR, Billeter R, Boyett MR. Computer three dimensional reconstruction of the sinoatrial node. Circulation. 2005;111: 846–54.

D'Souza A, Bucchi A, Johnsen AB, Logantha SJ, Monfredi O, Yanni J, Prehar S, Hart G, Cartwright E, Wisloff U, Dobryznski H, DiFrancesco D, Morris GM, Boyett MR. Exercise training reduces resting heart rate via downregulation of the funny channel HCN4. Nat Commun. 2014;13:3775.

D'Souza A, Pearman CM, Wang Y, Nakao S, Logantha SJRJ, Cox C, Bennett H, Zhang Y, Johnsen AB, Linscheid N, Poulsen PC, Elliott J, Coulson J, McPhee J, Robertson A, da Costa Martins PA, Kitmitto A, Wisløff U, Cartwright EJ, Monfredi O, Lundby A, Dobrzynski H, Oceandy D, Morris GM, Boyett MR. Targeting miR-423-5p reverses exercise training-induced HCN4 channel remodeling and sinus bradycardia. Circ Res. 2017;121:1058–68.

Duhme N, Schweizer PA, Thomas D, Becker R, Schröter J, Schlichting I, Bahrends T, Draguhn A, Bruehl C, Katus HA, Koenen M. Altered HCN4 channel C-linker interaction is associated with familial tachycardia-bradycardia syndrome and atrial fibrillation. Eur Heart J. 2013;34:2768–75.

Ellinor PT, Lunetta KL, Albert CM, Glazer NL, Ritchie MD, Smith AV, Arking DE, Müller-Nurasyid M, Krijthe BP, Lubitz SA, Bis JC, Chung MK, Dörr M, Ozaki K, Roberts JD, Smith JG,

Pfeufer A, Sinner MF, Lohman K, Ding J, Smith NL, Smith JD, Rienstra M, Rice KM, Van Wagoner DR, Magnani JW, Wakili R, Clauss S, Rotter JI, Steinbeck G, Launer LJ, Davies RW, Borkovich M, Harris TB, Lin H, Völker U, Völzke H, Milan DJ, Hofman A, Boerwinkle E, Chen LY, Soliman EZ, Voight BF, Li G, Chakravarti A, Kubo M, Tedrow UB, Rose LM, Ridker PM, Conen D, Tsunoda T, Furukawa T, Sotoodehnia N, Xu S, Kamatani N, Levy D, Nakamura Y, Parvez B, Mahida S, Furie KL, Rosand J, Muhammad R, Psaty BM, Meitinger T, Perz S, Wichmann HE, Witteman JC, Kao WH, Kathiresan S, Roden DM, Uitterlinden AG, Rivadeneira F, McKnight B, Sjögren M, Newman AB, Liu Y, Gollob MH, Melander O, Tanaka T, Stricker BH, Felix SB, Alonso A, Darbar D, Barnard J, Chasman DI, Heckbert SR, Benjamin EJ, Gudnason V, Kääb S. Meta-analysis identifies six new susceptibility loci for atrial fibrillation. Nat Genet. 2012;44:670–5.

Evans R, Shaw D. Pathological studies in sinoatrial disorder (sick sinus syndrome). Br Heart J. 1977;39:778–86.

Fedorov VV, Glukhov AV, Chang R, Kostecki G, Aferol H, Hucker WJ, Wuskell JP, Loew LM, Schuessler RB, Moazami N, Efimov IR. Optical mapping of the isolated coronary-perfused human sinus node. J Am Coll Cardiol. 2010;56:1386–94.

Gollob MH, Green MS, Tang AS, Gollob T, Karibe A, Ali Hassan AS, Ahmad F, Lozado R, Shah G, Fananapazir L, Bachinski LL, Roberts R. Identification of a gene responsible for familial Wolff-Parkinson-white syndrome. N Engl J Med. 2001a;344:1823–31.

Gollob MH, Seger JJ, Gollob TN, Tapscott T, Gonzales O, Bachinski L, Roberts R. Novel PRKAG2 mutation responsible for the genetic syndrome of ventricular preexcitation and conduction system disease with childhood onset and absence of cardiac hypertrophy. Circulation. 2001b;104:3030–3.

Gomes JA, Kang PS, Matheson M, Gough Jr WB, El-Sherif N. Coexistence of sick sinus rhythm and atrial flutter-fibrillation. Circulation. 1981;63:80–6.

Groh WJ. Arrhythmias in the muscular dystrophies. Heart Rhythm. 2012;9:1890–5.

Groh WJ, Groh MR, Saha C, Kincaid JC, Simmons Z, Ciafaloni E, Pourmand R, Otten RF, Bhakta D, Nair GV, Marashdeh MM, Zipes DP, Pascuzzi RM. Electrocardiographic abnormalities and sudden death in myotonic dystrophy type 1. N Engl J Med. 2008;358:2688–97.

Haas J, Frese KS, Peil B, Kloos W, Keller A, Nietsch R, Feng Z, Müller S, Kayvanpour E, Vogel B, Sedaghat-Hamedani F, Lim WK, Zhao X, Fradkin D, Köhler D, Fischer S, Franke J, Marquart S, Barb I, Li DT, Amr A, Ehlermann P, Mereles D, Weis T, Hassel S, Kremer A, King V, Wirsz E, Isnard R, Komajda M, Serio A, Grasso M, Syrris P, Wicks E, Plagnol V, Lopes L, Gadgaard T, Eiskjær H, Jørgensen M, Garcia-Giustiniani D, Ortiz-Genga M, Crespo-Leiro MG, Deprez RH, Christiaans I, van Rijsingen IA, Wilde AA, Waldenstrom A, Bolognesi M, Bellazzi R, Mörner S, Bermejo JL, Monserrat L, Villard E, Mogensen J, Pinto YM, Charron P, Elliott P, Arbustini E, Katus HA, Meder B. Atlas of the clinical genetics of human dilated cardiomyopathy. Eur Heart J. 2015;36:1123–35.

Hao X, Zhang Y, Zhang X, Nirmalan M, Davies L, Konstantinou D, Yin F, Dobrzynski H, Wang X, Grace A, Zhang H, Boyett M, Huang CL, Lei M. TGF-β1-mediated fibrosis and ion channel remodeling are key mechanisms in producing the sinus node dysfunction associated with SCN5A deficiency and aging. Circ Arrhythm Electrophysiol. 2011;4:397–406.

Henrion U, Zumhagen S, Steinke K, Strutz-Seebohm N, Stallmeyer B, Lang F, Schulze-Bahr E, Seebohm G. Overlapping cardiac phenotype associated with a familial mutation in the voltage sensor of the KCNQ1 channel. Cell Physiol Biochem. 2012;29:809–18.

Hofman N, Tan HL, Alders M, Kolder I, de Haij S, Mannens MM, Lombardi MP, Dit Deprez RH, van Langen I, Wilde AA. Yield of molecular and clinical testing for arrhythmia syndromes: report of 15 years' experience. Circulation. 2013;128:1513–21.

Holm H, Gudbjartsson DF, Sulem P, Masson G, Helgadottir HT, Zanon C, Magnusson OT, Helgason A, Saemundsdottir J, Gylfason A, Stefansdottir H, Gretarsdottir S, Matthiasson SE, Thorgeirsson GM, Jonasdottir A, Sigurdsson A, Stefansson H, Werge T, Rafnar T, Kiemeney LA, Parvez B, Muhammad R, Roden DM, Darbar D, Thorleifsson G, Walters GB, Kong A,

Thorsteinsdottir U, Arnar DO, Stefansson K. A rare variant in MYH6 is associated with high risk of sick sinus syndrome. Nat Genet. 2011;43:316–20.

Holmegard HN, Theilade J, Benn M, Duno M, Haunso S, Svendsen JH. Genetic variation in the inwardly rectifying K channel subunits KCNJ3 (GIRK1) and KCNJ5 (GIRK4) in patients with sinus node dysfunction. Cardiology. 2010;115:176–81.

Jensen PN, Gronroos NN, Chen LY, Folsom AR, deFilippi C, Heckbert SR, Alonso A. Incidence of and risk factors for sick sinus syndrome in the general population. J Am Coll Cardiol. 2014;64:531–8.

Kaplan BM, Langendorf R, Lev M, Pick A. Tachycardia-bradycardia syndrome (so-called "sick sinus syndrome"). Pathology, mechanisms and treatment. Am J Cardiol. 1973;31:497–508.

Ki CS, Jung CL, Kim HJ, Baek KH, Park SJ, On YK, Kim KS, Noh SJ, Youm JB, Kim JS, Cho H. A KCNQ1 mutation causes age-dependant bradycardia and persistent atrial fibrillation. Pflugers Arch. 2014;466:529–40.

Kruse M, Schulze-Bahr E, Corfield V, Beckmann A, Stallmeyer B, Kurtbay G, Ohmert I, Schulze-Bahr E, Brink P, Pongs O. Impaired endocytosis of the ion channel TRPM4 is associated with human progressive familial heart block type I. J Clin Invest. 2009;119:2737–44.

Kumar S, Baldinger SH, Gandjbakhch E, Maury P, Sellal JM, Androulakis AF, Waintraub X, Charron P, Rollin A, Richard P, Stevenson WG, Macintyre CJ, Ho CY, Thompson T, Vohra JK, Kalman JM, Zeppenfeld K, Sacher F, Tedrow UB, Lakdawala NK. Long-term arrhythmic and nonarrhythmic outcomes of Lamin a/C mutation carriers. J Am Coll Cardiol. 2016;68:2299–307.

Kurata Y, Hisatome I, Matsuda H, Shibamoto T. Dynamical mechanisms of pacemaker generation in IK1-downregulated human ventricular myocytes: insights from bifurcation analyses of a mathematical model. Biophys J. 2005;89:2865–87.

Lamas GA, Lee K, Sweeny M, et al. The mode selection trial (MOST) in sinus node dysfunction: design, rationale, and baseline characteristics of the first 1000 patients. Am Heart J. 2000;140:541–51.

Larson ED, St Clair JR, Sumner WA, Bannister RA, Proenza C. Depressed pacemaker activity of sinoatrial node myocytes contributes to the age-dependent decline in maximum heart rate. Proc Natl Acad Sci U S A. 2013;110:18011–6.

Le Scouarnec S, Bhasin N, Vieyres C, Hund TJ, Cunha SR, Koval O, Marionneau C, Chen B, Wu Y, Demolombe S, Song LS, Le Marec H, Probst V, Schott JJ, Anderson ME, Mohler PJ. Dysfunction in ankyrin-B-dependent ion channel and transporter targeting causes human sinus node disease. Proc Natl Acad Sci U S A. 2008;105:15617–22.

Lenegre J. Etiology and pathology of bilateral bundle branch block in relation to complete heart block. Prog Cardiovasc Dis. 1964;6:409–44.

Lev M. The pathology of complete atrioventricular block. Prog Cardiovasc Dis. 1964;6:317–26.

Liu H, El Zein L, Kruse M, Guinamard R, Beckmann A, Bozio A, Kurtbay G, Mégarbané A, Ohmert I, Blaysat G, Villain E, Pongs O, Bouvagnet P. Gain-of-function mutations in TRPM4 cause autosomal dominant isolated cardiac conduction disease. Circ Cardiovasc Genet. 2010;3:374–85.

Lodder EM, De Nittis P, Koopman CD, Wiszniewski W, Moura de Souza CF, Lahrouchi N, Guex N, Napolioni V, Tessadori F, Beekman L, Nannenberg EA, Boualla L, Blom NA, de Graaff W, Kamermans M, Cocciadiferro D, Malerba N, Mandriani B, Akdemir ZH, Fish RJ, Eldomery MK, Ratbi I, Wilde AA, de Boer T, Simonds WF, Neerman-Arbez M, Sutton VR, Kok F, Lupski JR, Reymond A, Bezzina CR, Bakkers J, Merla G. GNB5 mutations cause an autosomal-recessive multisystem syndrome with sinus Bradycardia and cognitive disability. Am J Hum Genet. 2016;99:704–10.

Macri V, Mahida SN, Zhang ML, Sinner MF, Dolmatova EV, Tucker NR, McLellan M, Shea MA, Milan DJ, Lunetta KL, Benjamin EJ, Ellinor PT. A novel trafficking-defective HCN4 mutation is associated with early-onset atrial fibrillation. Heart Rhythm. 2014;11:1055–62.

Makita N, Behr E, Shimizu W, Horie M, Sunami A, Crotti L, Schulze-Bahr E, Fukuhara S, Mochizuki N, Makiyama T, Itoh H, Christiansen M, McKeown P, Miyamoto K, Kamakura S,

Tsutsui H, Schwartz PJ, George AL Jr, Roden DM. The E1784K mutation in SCN5A is associated with mixed clinical phenotype of type 3 long QT syndrome. J Clin Invest. 2008; 18:2219–29.

Makita N, Seki A, Sumitomo N, Chkourko H, Fukuhara S, Watanabe H, et al. A connexin40 mutation associated with a malignant variant of progressive familial heart block type I. Circ Arrhythm Electrophysiol. 2012;5:163–72.

Milanesi R, Baruscotti M, Gnecchi-Ruscone T, DiFrancesco D. Familial sinus bradycardia associated with a mutation in the cardiac pacemaker channel. N Engl J Med. 2006;354:151–7.

Milano A, Vermeer AM, Lodder EM, Barc J, Verkerk AO, Postma AV, van der Bilt IA, Baars MJ, van Haelst PL, Caliskan K, Hoedemaekers YM, Le Scouarnec S, Redon R, Pinto YM, Christiaans I, Wilde AA, Bezzina CR. HCN4 mutations in multiple families with bradycardia and left ventricular noncompaction cardiomyopathy. J Am Coll Cardiol. 2014;64:745–56.

Monfredi O, Boyett MR. Sick sinus syndrome and atrial fibrillation in older persons—a view from the sinoatrial nodal myocyte. J Mol Cell Cardiol. 2015;83:88–100.

Monfredi O, Dobrzynski H, Mondal T, Boyett MR, Morris GM. The anatomy and physiology of the sinoatrial node—a contemporary review. Pacing Clin Electrophysiol. 2010;33:1392–406.

Nawathe PA, Kryukova Y, Oren RV, Milanesi R, Clancy CE, Lu JT, Moss AJ, Difrancesco D, Robinson RB. An LQTS6 MiRP1 mutation suppresses pacemaker current and is associated with sinus bradycardia. J Cardiovasc Electrophysiol. 2013;24:1021–7.

Nof E, Luria D, Brass D, Marek D, Lahat H, Reznik-Wolf H, Pras E, Dascal N, Eldar M, Glikson M. Point mutation in the HCN4 cardiac ion channel pore affecting synthesis, trafficking, and functional expression is associated with familial asymptomatic sinus bradycardia. Circulation. 2007;116:463–70.

Postma AV, Denjoy I, Hoorntje TM, et al. Absence of calsequestrin 2 causes severe forms of catecholaminergic polymorphic ventricular tachycardia. Circ Res. 2002;91:e21–6.

Postma AV, Denjoy I, Kamblock J, et al. Catecholaminergic polymorphic ventricular tachycardia: RYR2 mutations, bradycardia, and follow up of the patients. J Med Genet. 2005;42:863–70.

Priori SG, Wilde AA, Horie M, Cho Y, Behr ER, Berul C, et al. Executive summary: HRS/EHRA/APHRS expert consensus statement on the diagnosis and management of patients with inherited primary arrhythmia syndromes. Europace. 2013;15:1389–406.

Probst V, Kyndt F, Potet F, Trochu J-N, Mialet G, Demolombe S, Schott J-J, Baró I, Escande D, Le Marec H. Haploinsufficiency in combination with aging causes SCN5A-linked hereditary Lenègre disease. J Am Coll Cardiol. 2003;41:643–52.

Remme CA. Cardiac sodium channelopathy associated with SCN5A mutations: electrophysiological, molecular and genetic aspects. J Physiol. 2013;591:4099–116.

Sairaku A, Nakano Y, Oda N, Makita Y, Kajihara K, Tokuyama T, et al. Prediction of sinus node dysfunction in patients with persistent atrial flutter using the flutter cycle length. Europace. 2012;14:380–7.

Sanders P, Kistler PM, Morton JB, Spence SJ, Kalman JM. Remodeling of sinus node function in patients with congestive heart failure: reduction in sinus node reserve. Circulation. 2004a;110: 897–903.

Sanders P, Morton JB, Kistler PM, Spence SJ, Davidson NC, Hussin A, Vohra JK, Sparks PB, Kalman JM. Electrophysiological and electroanatomic characterization of the atria in sinus node disease: evidence of diffuse atrial remodeling. Circulation. 2004b;109:1514–22.

Schott JJ, Benson DW, Basson CT, Pease W, Silberbach GM, Moak JP, et al. Congenital heart disease caused by mutations in the transcription factor NKX2-5. Science. 1998;281:108–11.

Schott JJ, Alshinawi C, Kyndt F, Probst V, Hoorntje TM, Hulsbeek M, Wilde AAM, Escande D, Mannens MM, Le Marec H. Cardiac conduction defects associate with mutations in SCN5A. Nat Genet. 1999;23:20–1.

Schulze-Bahr E, Neu A, Friederich P, Kaupp UB, Breithardt G, Pongs O, Isbrandt D. Pacemaker channel dysfunction in a patient with sinus node disease. J Clin Invest. 2003;111:1537–45.

Schweizer PA, Yampolsky P, Malik R, Thomas D, Zehelein J, Katus HA, Koenen M. Transcription profiling of HCN-channel isotypes throughout mouse cardiac development. Basic Res Cardiol. 2009;104:621–9.

Schweizer PA, Duhme N, Thomas D, Becker R, Zehelein J, Draguhn A, Bruehl C, Katus HA, Koenen M. cAMP sensitivity of HCN pacemaker channels determines basal heart rate but is not critical for autonomic rate control. Circ Arrhythm Electrophysiol. 2010;3:542–52.

Schweizer PA, Schröter J, Greiner S, Haas J, Yampolsky P, Mereles D, Buss SJ, Seyler C, Bruehl C, Draguhn A, Koenen M, Meder B, Katus HA, Thomas D. The symptom complex of familial sinus node dysfunction and myocardial non-compaction is associated with mutations in the HCN4 channel. J Am Coll Cardiol. 2014;64:757–67.

Shaw DB, Linker NJ, Heaver PA, et al. Chronic sinoatrial disorder (sick sinus syndrome): a possible result of cardiac ischaemia. Br Heart J. 1987;58:598–607.

Sotoodehnia N, Isaacs A, de Bakker PI, Dorr M, Newton-Cheh C, Nolte IM, et al. Common variants in 22 loci are associated with QRS duration and cardiac ventricular conduction. Nat Genet. 2010;42:1068–76.

Spellberg RD. Familial sinus node disease. Chest. 1971;60:246–51.

Stallmeyer B, Fenge H, Nowak-Gottl U, Schulze-Bahr E. Mutational spectrum in the cardiac transcription factor gene NKX2.5 (CSX) associated with congenital heart disease. Clin Genet. 2010;78:533–40.

Stallmeyer B, Zumhagen S, Denjoy I, Duthoit G, Hebert JL, Ferrer X, et al. Mutational spectrum in the Ca(2+)-activated cation channel gene TRPM4 in patients with cardiac conductance disturbances. Hum Mutat. 2012;33:109–17.

Stallmeyer B, Kuß J, Kotthoff S, Zumhagen S, Vowinkel K, Rinné S, Matschke LA, Friedrich C, Schulze-Bahr E, Rust S, Seebohm G, Decher N, Schulze-Bahr E. A mutation in the G-protein gene GNB2 causes familial sinus node and atrioventricular conduction dysfunction. Circ Res. 2017;120:e33–44.

Tan HL, Bink-Boelkens MT, Bezzina CR, Viswanathan PC, Beaufort-Krol GC, van Tintelen PJ, van den Berg MP, Wilde AAM, Balser JR. A sodium-channel mutation causes isolated cardiac conduction disease. Nature. 2001;409:1043–7.

Tellez JO, Mczewski M, Yanni J, Sutyagin P, Mackiewicz U, Atkinson A, Inada S, Beresewicz A, Billeter R, Dobrzynski H, Boyett MR. Ageing-dependent remodelling of ion channel and Ca2+ clock genes underlying sino-atrial node pacemaking. Exp Physiol. 2011;96:1163–78.

van Rijsingen IA, Arbustini E, Elliott PM, Mogensen J, Hermansvan Ast JF, van der Kooi AJ, et al. Risk factors for malignant ventricular arrhythmias in Lamin a/c mutation carriers a European cohort study. J Am Coll Cardiol. 2012;59:493–500.

van Spaendonck-Zwarts KY, van Hessem L, Jongbloed JD, de Walle HE, Capetanaki Y, van der Kooi AJ, van Langen IM, van den Berg MP, van Tintelen JP. Desmin-related myopathy. Clin Genet. 2011;80:354–66.

Vermeer AMC, Lodder EM, Thomas D, Duijkers FAM, Marcelis C, van Gorselen EOF, Fortner P, Buss SJ, Mereles D, Katus HA, Wilde AA, Bezzina CR, Boekholdt SM, Schweizer PA, Christiaans I. Dilatation of the aorta ascendens forms part of the clinical spectrum of HCN4 mutations. J Am Coll Cardiol. 2016;67:2313–5.

Watanabe H, Koopmann TT, Le Scouarnec S, Yang T, Ingram CR, Schott JJ, Demolombe S, Probst V, Anselme F, Escande D, Wiesfeld AC, Pfeufer A, Kääb S, Wichmann HE, Hasdemir C, Aizawa Y, Wilde AA, Roden DM, Bezzina CR. Sodium channel beta1 subunit mutations associated with Brugada syndrome and cardiac conduction disease in humans. J Clin Invest. 2008;118:2260–8.

Yasui K, Liu W, Opthof T, Kada K, Lee JK, Kamiya K, Kodama I. I(f) current and spontaneous activity in mouse embryonic ventricular myocytes. Circ Res. 2001;88:536–42.

Yeh YH, Burstein B, Qi XY, Sakabe M, Chartier D, Comtois P, Wang Z, Kuo CT, Nattel S. Funny current downregulation and sinus node dysfunction associated with atrial tachyarrhythmia: a molecular basis for tachycardia-bradycardia syndrome. Circulation. 2009;119:1576–85.

Zicha S, Fernández-Velasco M, Lonardo G, L'Heureux N, Nattel S. Sinus node dysfunction and hyperpolarization-activated (HCN) channel subunit remodeling in a canine heart failure model. Cardiovasc Res. 2005;66:472–81.

Catecholaminergic Polymorphic Ventricular Tachycardia 10

Riccardo Maragna and Carlo Napolitano

Abstract

Catecholaminergic polymorphic ventricular tachycardia (CPVT) can be broadly viewed as a genetically determined disorder of the cardiac calcium handling. Different genes have been causally linked to the disease. It is currently evident that authors tend to define as "CPVT" all the forms of cardiac arrhythmia triggered by adrenergic activation. However, to distinguish the clinical condition described by Philippe Coumel in the 1970s form, other CPVT-like phenotypes have value since these latter have more complex and less defined clinical manifestations.

The severe clinical manifestations of CPVT are now much better understood and treated, thanks to the insights provided by more than 15 years of research after the first CPVT gene was identified. Still, the clinical diagnosis can be elusive due to presence of an unremarkable resting ECG and structurally normal heart. Thus, it is not uncommon that the opportunity of prescribing lifesaving treatments is missed in the daily clinical activity. The awareness on the crucial role of early diagnosis and therapy should be part of the standard cardiologic knowledge despite the relative low prevalence of the disease (1–2:10,000). Additional hurdles for the clinical management can derive from the incomplete response to therapy and from limited patient's compliance to lifelong drug therapy. For this reason a remarkable effort is directed toward the development of novel therapies and particularly gene therapy. This strategy is attracting the attention of the medical community given the strong preclinical evidences of effectiveness. Here, we provide an update on CPVT from the clinical, genetic, and experimental

Figures have been originally prepared for this chapter.

R. Maragna · C. Napolitano (✉)
Molecular Cardiology, ICS Maugeri, IRCCS, Pavia, Italy
e-mail: riccardo.maragna01@universitadipavia.it; carlo.napolitano@icsmaugeri.it

D. Thomas, C. A. Remme (eds.), *Channelopathies in Heart Disease*, Cardiac and Vascular Biology, https://doi.org/10.1007/978-3-319-77812-9_10

point of view in the attempt to delineate the current scenario and the future directions.

10.1 Introduction and Definition

Catecholaminergic polymorphic ventricular tachycardia (CPVT) is an inherited arrhythmogenic disorder that presents with a distinguishing pattern of adrenergically mediated arrhythmias that may cause syncope and sudden cardiac death (SCD). Following the identification of anecdotal cases, CPVT was formally described by Coumel et al. (1978) and the main features where described in a subsequent follow-up paper by the same group in 1995 (Leenhardt et al. 1995). The clinical presentation encompasses exercise- or emotion-induced syncopal events that are the clinical sign of the onset of a distinctive ventricular arrhythmia, the so-called bidirectional ventricular tachycardia. The molecular pathophysiology of the disease has been progressively clarified and shown to be linked to defects in proteins controlling intracellular calcium handling in cardiac myocytes (Lahat et al. 2001; Priori et al. 2001). Here, the current knowledge of CPVT will be reviewed with a focus on the genetics, pathophysiology, and clinical management.

10.2 Clinical Presentation

10.2.1 Clinical Presentation and Epidemiology

CPVT is a lethal arrhythmogenic disease, whose main manifestation is represented by adrenergic-mediated ventricular arrhythmias (Coumel et al. 1978). CPVT typically leads the affected subjects to medical attention between 7 and 12 years of age after having experienced a syncopal episode related to a physical or emotional stress (Postma et al. 2005; Priori et al. 2002). Other cases are referred after a sudden death event triggered by exercise or emotion in one or more family members.

The presence of a high percentage of simplex (sporadic) cases and the high lethality at young age justifies the low prevalence of CPVT: the current estimated figure is in the order of 1:10,000 individuals (Napolitano et al. 2016; Priori et al. 2013).

However, indirect clues suggest that the prevalence of CPVT in the general population may be underestimated. Indeed, CPVT diagnosis can be elusive given the presence of a normal baseline ECG (Priori et al. 2002). For this reason, a history of exercise-induced syncope in young individuals is often attributed to non-cardiac conditions, thus leaving young patients untreated for long periods. In 2002 our group showed, in a cohort of 148 patients, a mean delay time of 2.5 years between first syncope and diagnosis (Priori et al. 2002). These data have been recently confirmed by a multicenter study of 226 CPVT patients, which, despite being set more than 10 years later, showed a mean delay of almost 1 year between the first syncope and diagnosis (Roston et al. 2015). This clinical evidence emphasizes the crucial importance of a thorough medical workout in a specialized center for all individuals experiencing exercise or emotion-related syncope.

CPVT diagnosis can also be missed in patients in whom SCD is the first manifestation of the disease. These may represent a relevant proportion of cases, since a family history of sudden cardiac death under 40 years of age is detectable in 30% of families (Leenhardt et al. 1995). A recent study (Jimenez-Jaimez et al. 2015), confirming previous reports, shows a prevalence of 14% of CPVT-related mutations in a cohort of 35 negative autopsies of young victims of unexplained sudden cardiac death.

Taken together, these observations point to the idea that CPVT prevalence is possibly higher than anticipated from the clinically diagnosed cases. The high lethality and the high rate of misdiagnosis outline the importance of a correct diagnostic approach and the remarkable clinical implications of systematic postmortem molecular analysis of SCD victims to prevent lethal events in family members (Priori et al. 2015).

10.2.2 CPVT Diagnosis

The diagnosis of CPVT can be established in the presence of a structurally normal heart, normal resting electrocardiogram, and reproducible exercise- or emotion-induced ventricular tachycardia (Priori et al. 2015). Thus, the most sensitive diagnostic tools are represented by the exercise stress testing and the 24-h Holter ECG monitoring.

The exercise stress test allows to identify the typical arrhythmic pattern: isolated ventricular or supraventricular beats at heart rate of 110–130 beats per minute; the complexity and frequency of arrhythmias progressively worsen with the increase of workload, from isolated premature beats to bigeminy and to ventricular tachycardia. When the exercise stops, arrhythmias gradually disappear (usually in less than 2 min), whereas if continued the duration of the VT progressively increases possibly leading to syncope.

The typical pattern of CPVT arrhythmias is called **bidirectional ventricular tachycardia** (Fig. 10.1), and it is characterized by an alternating QRS axis with a rotation of 180 degrees on a beat-to-beat basis. This highly specific pattern is not evident in a subset (minority) of CPVT patients in whom polymorphic VT is detected. It is important to note that the bidirectional pattern is not present in all ECG leads and there is no "preferred lead" to detect it. Thus, it is advisable to always obtain 12-lead recordings during exercise and Holter monitoring. This latter clinical test is useful for subjects in whom emotional stress is a stronger arrhythmogenic trigger than exercise. The adrenergic component with worsening of arrhythmias should be present in all cases to allow diagnosis. Ventricular arrhythmias may also be induced by isoproterenol infusion in patients who are unable to exercise, but the predictive value of this provocative test for risk stratification has been questioned (Marjamaa et al. 2012).

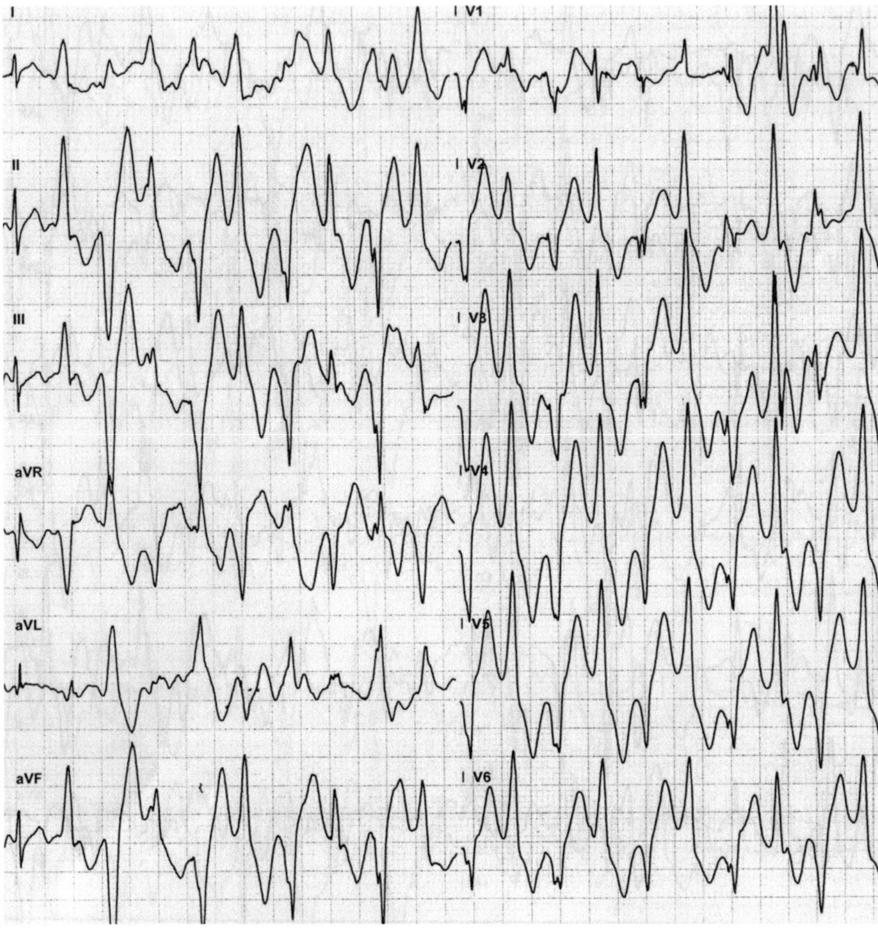

Fig. 10.1 Typical bidirectional ventricular tachycardia (VT). ECG showing the typical CPVT—pathognomonic ventricular arrhythmia, i.e., bidirectional VT: a ventricular tachycardia showing an alternating QRS axis with a rotation of 180° on a beat-to-beat basis

The resting ECG can show minor abnormalities (Fig. 10.2) that, albeit nonspecific, can still help the clinician in identifying affected individuals. In particular it can show sinus bradycardia (Postma et al. 2005) and the dynamic presence of prominent U waves (Leenhardt et al. 1995; Postma et al. 2005). The presence of a lower than normal heart rate probably reflects inhibition of the calcium clock (a pace-making mechanisms) in the sinus node cells (Neco et al. 2012). On the other hand, by mean of monophasic action potential recordings, Paavola et al. (2007) showed the concordance of DAD and U waves amplitude during adrenergic stimulation, suggesting that these latter can be a marker of propensity to arrhythmias (Viitasalo et al. 2008).

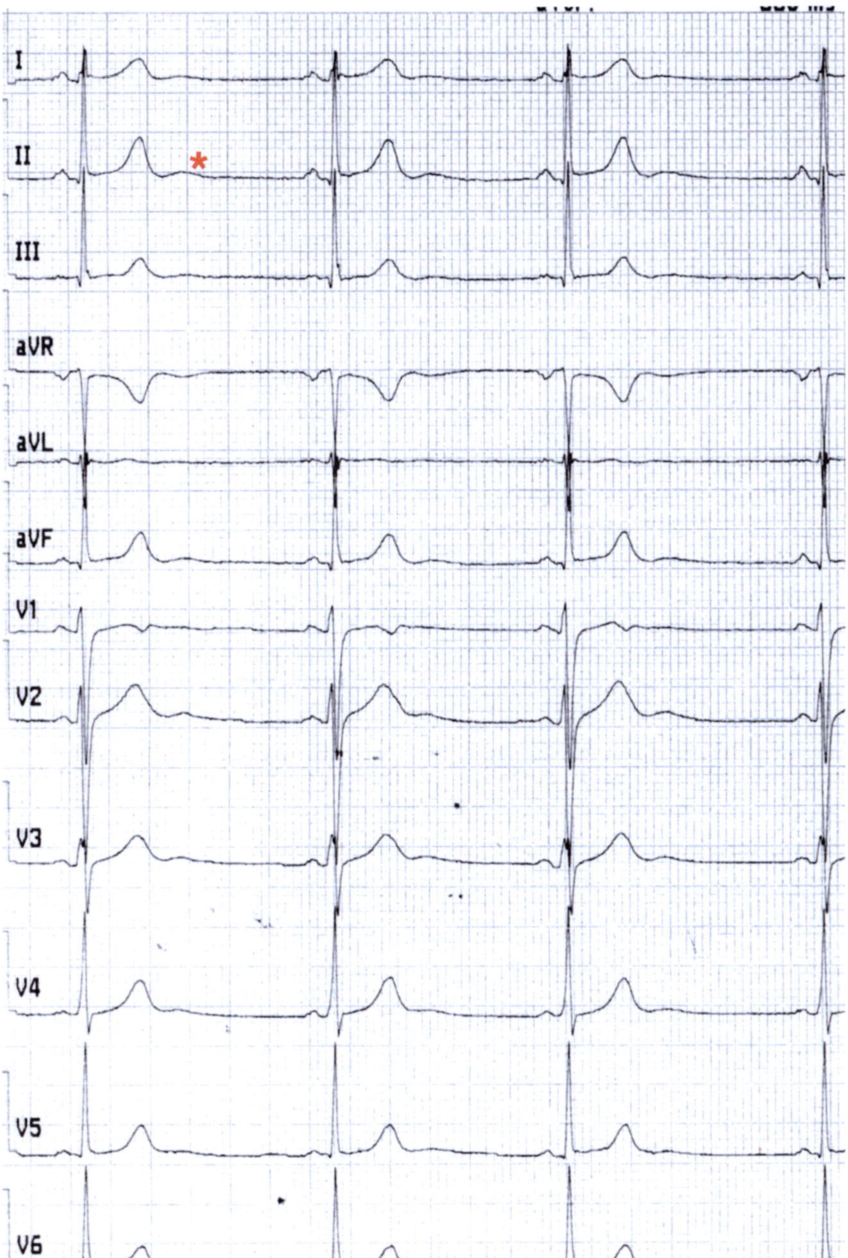

Fig. 10.2 ECG in CPVT patient. Baseline ECG of a 13-year-old CPVT patient who suffered two episodes of cardiac arrest during physical activity. The basal ECG trace shows common CPVT features: low heart rate (50 bpm) at rest and prominent U waves (see asterisk)

Supraventricular arrhythmias and tachycardia are also part of the picture. Isolated atrial ectopic beats, non-sustained supraventricular tachycardia, and short runs of atrial fibrillation are usually observed during exercise with a similar onset pattern as the one described for ventricular arrhythmias (Leenhardt et al. 1995). Since the propensity to triggered activity is heart rate dependent (the I_{Ti} current increases at fast stimulation rates), supraventricular arrhythmias can become a trigger for the development of ventricular arrhythmias.

Invasive electrophysiological testing with programmed electrical stimulation is of no value (either diagnostic or prognostic) in CPVT as the adrenergic-induced arrhythmias do not depend on reentrant circuits, but on DADs and triggered activity.

10.2.3 Natural History

CPVT is generally considered a severe condition among the other known inherited arrhythmias. When left untreated, CPVT exposes most affected individuals (60–80%) to life-threatening arrhythmias before the age of 40 (Priori et al. 2002). Moreover, in up to 30% of patients, the presenting symptom is ventricular fibrillation that may cause SCD (Roston et al. 2015).

Late-onset cases presenting the first event in the third or fourth decade of life have been described. A small study by Sy et al. (2011) has suggested that these cases are more likely to be women (80%) and genotype-negative individuals (i.e., those without mutations on the known genes) (Sy et al. 2011). There are also data supporting the idea of a time-dependent penetrance of CPVT with a reduced risk of events in the first years of life (van der Werf et al. 2012). This observation is important since the lack of inducible arrhythmias in mutation carriers at young age cannot be considered a marker of a benign condition.

Another factor that confirms the high lethality of the disease is the high prevalence of juvenile (<40 years) family history of sudden death in relatives of probands that is around 30% (Leenhardt et al. 1995; Priori et al. 2002). Instances of SIDS (sudden infant death syndrome) have also been associated with pathogenic variants in *RyR2* (Tester et al. 2007).

Overall, these data claim for the need of an early diagnosis and a prompt medical intervention both in symptomatic and asymptomatic individuals, when a diagnosis of CPVT is established, to interrupt the highly malignant natural course of the disease.

10.3 Pathophysiology

10.3.1 CPVT Genes and Phenotypes

CPVT is a cardiac calcium handling disorder (Fig. 10.3). More than a decade ago, only few highly specialized referral centers collected cohorts large enough to allow for linkage and candidate gene studies. These cohorts were carefully selected with a

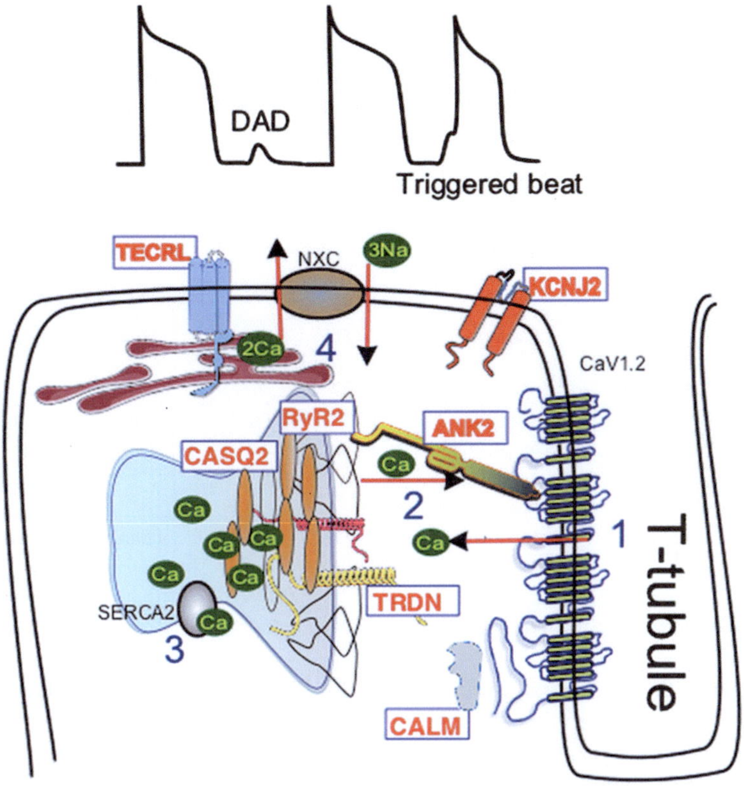

Fig. 10.3 CPVT genes and mechanisms. Cartoon showing a schematic representation with cellular localization of the proteins causing CPVT and the CPVT pathogenesis. Gene names are reported in red fonts. When calcium enters the cell through voltage-gated calcium channels (1), the RyR2 channels release an excess of Ca2+ during diastole (2); the SERCA pump (3) is unable to remove enough calcium especially during adrenergic activation that increases calcium release. The consequent activation of the NCX (4) causes DADs and triggered activity (see text for more details)

phenotype strictly adherent to that originally reported by Coumel et al. (1978). More recently, the growing efficiency of DNA sequencing technology and the inclusion of less neat phenotypes in cohort studies led to the identification of mutations of calcium handling proteins in patients with adrenergically mediated arrhythmias without typical CPVT (Table 10.1). These variants are less characterized both from the clinical and the mechanistic standpoint since CPVT genotypes other than those related to RyR2 mutations are rarely observed. Still, phenotypic differences among genotypes do exist, and their knowledge could support the clinical management. So far only RyR2 and CASQ2 are associated with typical CPVT (identical to that described in the original publication).

Ryanodine Receptor The first evidence of a calcium handling disorder as cause of the disease came in 2001 with identification of cardiac ryanodine receptor (*RyR2*)

Table 10.1 CPVT genes

Gene symbol	Chromosome	Gene	Phenotype	Inheritance	Prevalence
RyR2	1q43	Cardiac ryanodine receptor	CPVT, IVF, SCD	Dominant	60–70%
CASQ2	1p13.1	Cardiac calsequestrin	CPVT, SCD	Recessive[a]	3–5%
TRDN	6q22.31	Triadin	LQTS, polymorphic VT, SCD	Recessive	< 5%
CALM1, 2, 3	14q32.11, 2p21, 19q13.32	Calmodulin	LQTS, polymorphic VTs, SCD	Dominant	Unkn/rare
ANKB	4q25-q26	Ankyrin	CPVT, LQTS	Recessive	Unkn/rare
TECRL	7p22-p14	Trans-2,3-enoyl-CoA reductase	LQTS, polymorphic VTs, SCD	Recessive	Unkn/rare
KCNJ2	17q24.3	Inward rectifier	LQTS, bidirectional VT, periodic paralysis	Dominant	Unkn/rare

[a]One case of dominant pattern of transmission has been reported

mutations in the autosomal dominant CPVT1 (Priori et al. 2001). RyR2 is a large tetrameric channel spanning the membrane of the sarcoplasmic reticulum (SR) that controls the release of Ca^{2+} and the calcium-induced calcium release (CICR) process. As of today there are approximately 200 mutations reported in Human Gene Mutation Database and more than 500 variants (150 definitely pathogenic) in the ClinVar database (https://www.ncbi.nlm.nih.gov/clinvar/?term=ryr2%5Bgene%5D). These mutations are identified not only in typical CPVT but also in patients with adrenergically triggered ventricular fibrillation and at postmortem genetic testing in unexplained sudden death (SUDS) cases occurring in conditions of acute stress (Tester et al. 2004). These latter cases could represent CPVT with SCD and first manifestation (and not syncope as it happens in other cases).

RyR2 mutations tend to cluster in three specific areas of the protein (amino acids 77–466, 2246–2534, 3778–4967) with no overt differences in the clinical outcome according to mutation site. RyR2 mutations have been reported also in patients referred for idiopathic ventricular fibrillation and in relatives of SUDS cases. Therefore, RyR2 should be included routinely in the genetic testing panels of subjects with evidence of adrenergic arrhythmias. Overall *RyR2* is by far the most frequent CPVT gene, causing >60%.

More recently, cardiac ryanodine receptor mutations have been found in familial arrhythmias associated with cardiomyopathy: an exon 3 deletion was reported (Ohno et al. 2014) in subjects with adrenergic arrhythmias and dilated heart resembling left ventricular non-compaction. Additional *RyR2* mutations associated with cardiomyopathy were subsequently identified (Bhuiyan et al. 2007). Furthermore, preliminary evidence suggests that some mutations could also be associated with right ventricular cardiomyopathy (Roux-Buisson et al. 2014). The pathophysiology of RyR2-dependent cardiomyopathy is not well understood, and it is unclear why some *RyR2* mutations can produce structural abnormalities. A working hypothesis would be the presence of gain of function mutations (typical CPVT) vs. loss of function mutations (IVF and perhaps cardiomyopathy).In any case the practical consequence is the need of carrying out careful imaging investigations in the CPVT clinical workout.

CASQ2 The second identified CPVT gene is calsequestrin 2 (*CASQ2*). It was identified in a large Bedouin tribe with autosomal recessive inheritance through linkage analysis followed by candidate gene screening. CASQ2 is a component of CICR. It is located in the SR in close proximity of the ryanodine receptors (junctional SR—Fig. 10.3). CASQ2 acts as a calcium-buffering protein that regulates the SR Ca^{2+} concentration, and it modulates the RyR2 open probability. There are at least 24 reported mutations (http://www.hgmd.cf.ac.uk/ac/gene.php?gene=CASQ2), and they cause a very severe form of the disease. Fast polymorphic VTs, besides the typical bidirectional VT, can be observed. Interestingly anecdotal literature findings suggest that heterozygous CASQ2 mutation may rarely cause CPVT (Gray et al. 2016) with a dominant pattern of inheritance. This implies that CASQ2 should be always considered a candidate gene in RyR2 negative cases.

10.3.2 Genes Associated with Non-typical CPVT

At least four genes can cause adrenergic-induced arrhythmias that authors have classified as CPVT. However, there are peculiarities that distinguish the typical CPVT originally described by Coumel in 1970 that appears to be linked to RyR2 and CASQ2 to that caused by other genes. The CPVT-like variants are rarely observed in the clinical practice, and only few cases have been reported.

Calmodulin (CALM) In 2012 calmodulin 1 (*CALM1*) mutations in patients with adrenergic-dependent arrhythmias were identified. Mutations were also associated with idiopathic ventricular fibrillation by other authors (Marsman et al. 2014). Calmodulin is a small (140 amino acids) cytosolic protein that binds calcium and transduces calcium signaling to several intracellular targets, including RyR2 (Nyegaard et al. 2012).

More recently *a CALM3* mutation has been found in 1 out of 12 families with adrenergic-induced syncope and sudden death in the absence of typical CPVT arrhythmias (Gomez-Hurtado et al. 2016).

All three known calmodulin genes (*CALM1*, *CALM2*, and *CALM3*) generate identical proteins and are widely expressed in several tissues. It is important to note that *CALM1*, *CALM2,* or *CALM3* mutations can cause QT prolongation/long QT syndrome through an impairment of calcium-dependent inactivation of the voltage-dependent calcium current (I_{Ca}) and increased late inward sodium current

(I_{NaLate}) (Boczek et al. 2016), thus overlap phenotypes (CPVT/LQTS) are possible. Overall, calmodulin mutations appear to be rare.

Triadin (TRDN) Mutations were discovered in families with recurrent syncope and sudden death during exercise and no inducible typical bidirectional VT (Rooryck et al. 2015; Roux-Buisson et al. 2012). *TRDN* mutations have been found also in patients with unexplained cardiac arrest (IVF) (Walsh et al. 2016) and in a severe forms of long QT syndrome (Altmann et al. 2015).

TRDN mutations have recessive pattern of inheritance. Cardiac triadin gene encodes for multiple tissue specifically regulated splice variants. Cardiac triadin, also known as Trisk32, has a single transmembrane segment spanning the SR membrane where it participates in the formation of SR calcium-releasing macromolecular complex (that also includes RyR2, junctin, and calsequestrin), the so-called calcium release unites (CRU). *TRDN* knockout causes reduced SR Ca^{2+} release, impaired I_{Ca} inactivation, and intracellular Ca^{2+} overload. The reduced inactivation of the calcium channel is consistent with QT prolongation. Thus, *TRDN* generates a substrate with overlap CPVT/LQTS abnormalities. No clear typical CPVT has been so far found in TRDN mutation carriers.

Ankyrin (*ANKB*) ANKB is an adapter protein with key modulatory and membrane targeting function. It affects the activity of the cardiac sodium channel and the sodium calcium exchanger (NCX). Initially an Ankyrin-B mutation was identified in LQTS (LQT4). LQT4 is a non-typical LQTS variant with mild QT prolongation, sinus node dysfunctional arrhythmias (ventricular and atrial). Several mutations have been reported thereafter including mutations associated with adrenergically—induced arrhythmias and sudden death resembling—but not fully recapitulating—CPVT (Kline and Mohler 2014).

Inward Rectifier (*KCNJ2*) It is the gene of Andersen-Tawil syndrome (ATS or LQT7); it encodes for a two-transmembrane segment potassium channel (Kir2.1) that controls the IK1 (inward rectifier) currently active during phase 4 of cardiac action potential. Many ATS patients have bidirectional VTs although the arrhythmias are often unrelated with adrenergic stimulation and rarely cause SCD (Napolitano et al. 2012).

Trans-2,3-Enoyl-CoA Reductase- like Protein (*TECRL*) A recent whole-exome sequencing study in patients from three families with life-threatening arrhythmias, QT prolongation, and SCD identified an identical homozygous mutation variant in the *TECRL* gene which encodes the trans-2,3-enoyl-CoA reductase-like protein gene, a multi-pass membrane protein that resides in the endoplasmic reticulum (Devalla et al. 2016). There is no epidemiological information on prevalence, and the details on the phenotype(s) associated with TRCRL are scanty to draw clinically relevant conclusions that could be applied beyond the families reported in the paper from Devalla et al.

10.3.3 Genetic Testing in CPVT

CPVT genetic testing has an important role in confirming the diagnosis in cases of clinically overt disease and in identifying at-risk family members in the presymptomatic phase. However, caution is required when assessing the pathogenicity of the variants identified, as shown by recent data from Paludan-Muller et al. (2017), who showed that several variants reported in the literature can be found in the ExAC database with allelic frequency higher than the CPVT prevalence and could be classified as likely nonpathogenic. Furthermore, the presence of a growing number of "minor" and scarcely known genes can further complicate the interpretation of the results of genetic testing. Careful phenotyping of family members and familial co-segregation analysis should always be regarded as a compulsory step in these instances.

10.3.4 Arrhythmogenesis in CPVT

Extensive research in cellular models and in transgenic mice engineered with RyR2 and CASQ2 mutations has contributed to the understanding of the arrhythmogenesis in CPVT (Fig. 10.3). The final common abnormality is an excessive release ("leakage") of calcium ions from the sarcoplasmic reticulum (SR), which leads to a cytosolic Ca^{2+} overload. The cell reacts with the hyper-activation of the sodium-calcium exchanger (NCX), a ten transmembrane-domain transporter that contributes to stabilize the cytosolic [Ca^{2+}]. NCX can work in both directions, but in its "forward" mode (the preferred mode in the presence of elevated cytosolic [Ca^{2+}]), it extrudes one Ca^{2+} ion for three Na^{+} ions entering the cells, thus generating a depolarizing current, the so-called transient inward current (I_{Ti}). I_{Ti} is active during the phase 4 of cardiac action potential (electrical diastole) causing membrane depolarizations, the delayed afterdepolarizations (DADs), which can generate spontaneous ("triggered") arrhythmogenic action potentials when they are large enough to reactivate the sodium current (I_{Na}).

This arrhythmogenic mechanism is defined as triggered activity, and it is favored by catecholamines (adrenergic stimulation) that increase the SR calcium content and release (Priori and Chen 2011). The presence of DADs and triggered activity has been initially demonstrated by our group in a transgenic model of *RyR2*-CPVT (Cerrone et al. 2005; Liu et al. 2006) and subsequently shown in patient-derived IPS (Di Pasquale et al. 2013; Fatima et al. 2011) and in patients by means of intracardiac recordings (Paavola et al. 2007).

Further in vitro work has shown how a dysfunctional RyR2 or CASQ2 can lead to this chain of arrhythmogenic events. Thanks to the work of Wayne Chen, the concept of "store overload induced calcium release" (SOIRC) has been proposed and experimentally supported (Jiang et al. 2005; Priori and Chen 2011). There is a threshold of SR Ca^{2+} concentration enabling the release from SR (the Ca^{2+} store). Adrenergic stimulation increases the SR calcium content; therefore, the threshold for release ("spillover") is more readily reached. The SOIRC threshold is reduced by

CPVT mutations, thus leading to uncontrolled release and cytosolic Ca^{2+} overload, especially during adrenergic stimulation.

At molecular level CPVT mutations can affect SOICR through different mechanisms: altered RyR2 sensitivity to cytosolic or SR [Ca^{2+}], RyR2 channel instability (the channel "unzipping" mechanism) altered CASQ2 polymerization with increased free SR [Ca^{2+}] or impaired CASQ2/RyR2 interaction [for a detailed review of *RyR2* and *CASQ2* mutation molecular pathogenesis, see Priori and Chen (2011)]. The RyR2 leakage observed in CPVT is considered a gain-of-function effect. However, few *RyR2* mutations can cause a reduced SOICR (loss-of-function) due to reduced RyR2 calcium sensitivity. These mutations have been associated to idiopathic ventricular fibrillation and, possibly, to cardiomyopathy, but their arrhythmogenic mechanisms are not fully elucidated.

Besides the direct effect of *RyR2* and *CASQ2* mutations on SR calcium release, a role of co-player, in CPVT arrhythmogenesis, has been recently attributed to a specific subset of sodium channels, the so-called "neuronal" sodium channels. These are localized in the T-tubules in close proximity with the CRU. It has been suggested that beta-adrenergic stimulation induces late sustained sodium current in these channels that exacerbates the diastolic calcium release (Radwanski et al. 2016). This may represent an additional arrhythmogenic pathway in CPVT.

As pointed out earlier, *CALM* mutations are associated with non-typical CPVT. The pathophysiology of *CALM* mutations is not fully elucidated, and the mechanisms are complex since mutations in these genes also cause LQTS. In vitro expression of the CALM3-A103V mutation identified in a patient with borderline QTc (referred to medical attention for "atypical LQTS") showed reduced Ca^{2+} binding affinity, increased spark frequency, and increased Ca^{2+} release-triggering waves (Gomez-Hurtado et al. 2016). These effects are not substantially different from those demonstrated for *CALM2* and *CALM1* mutations (Hwang et al. 2014). On the contrary, the long QT syndrome-associated *CALM* mutations may (Hwang et al. 2014) or may not (Boczek et al. 2016) affect Ca^{2+} binding affinity, but they do not promote Ca^{2+} waves and affect RyR2 channel gating (Hwang et al. 2014); conversely they can cause a loss of inactivation in L-type calcium channel and a mild accentuation of sodium channel late current (Boczek et al. 2016; Limpitikul et al. 2014).

In the original study of Roux-Buisson, *TRDN* mutations expressed in COS-7 cells resulted in intracellular retention and degradation of the mutant protein triadin-1 overexpression in isolated myocytes. Terentyev et al. (2005) demonstrated that triadin sensitizes RyR2 channels and enhances SR Ca^{2+} release, while Chopra and colleagues (Chopra et al. 2009) suggested that triadin has a role in maintaining the calcium release unit structure by interacting with the ryanodine receptor, calsequestrin, and junctin. On this basis, it appears reasonable to speculate that the loss of triadin generates an arrhythmogenic substrate by disrupting calcium handling. *TRDN* mutations identified in patients with adrenergic arrhythmias lead to an almost complete loss of protein. Triadin knockout mice have a complex calcium handling abnormality with reduced SR Ca^{2+} release and impaired negative feedback on L-type calcium current. This would lead to uninhibited Ca^{2+} influx via I_{Ca}, the calcium-dependent inactivation of the transmembrane calcium current, with consequent Ca^{2+} overload and spontaneous diastolic SR releases during adrenergic stimulation leading to DADs (Chopra et al. 2009).

Interestingly, increased Ca^{2+} sparks and RyR2 leakage is also observed in association with *ANK2* mutations suggesting a pathophysiology like that of RyR2 mutations also in the case of ANK2-related CPVT (Camors et al. 2012).

Experiments performed in myocytes, derived from mutant-induced pluripotent stem cells obtained from patients with *TECRL* mutations, revealed smaller transient amplitudes, elevated diastolic [Ca^{2+}], and decreased SERCA and NCX activities and also showed prolonged action potentials (APs). It is currently unclear if and how these abnormalities can lead to CPVT, but in this cellular model increased propensity to DADs and triggered activity was suggested (Devalla et al. 2016).

10.4 Current Therapeutic Approach

In the latest years, thanks to the discovery of the causative genes and the understanding of the cellular mechanisms of CPVT, progress has been made in the use of available drugs and interventional therapies. A stepwise approach that exploits the reproducibility of arrhythmias for therapy optimization appears to be the best choice in most cases. The current version of the European Society of Cardiology guidelines (Priori et al. 2015) highlights this approach from lifestyle modifications to invasive treatments (Table 10.2).

STEP 1: Lifestyle changes—Avoidance of triggers (i.e., intense/sudden adrenergic activation). Intense sport activity and stressful environments are strongly contraindicated in all CPVT patients (class I indication). Only recreational physical activity may be allowed, provided good suppression of arrhythmias at exercise stress testing while on therapy.

STEP 2: Pharmacological measures—The block of β adrenergic receptors represents the first-line therapy for CPVT, due to the overt evidence of a direct connection between adrenergic activation and the onset of ventricular arrhythmias. Beta-blockers (BB) achieve a significant reduction of cardiac events (Hayashi et al. 2009; Priori et al. 2002). Nonselective BB and specifically nadolol are probably more effective (Leren et al. 2015). The typical approach includes nadolol, starting from 1 to 2 mg/kg per day. Propranolol (3–5 mg/kg TID) may represent an alternative, although therapy compliance may be harder to achieve. A possible explanation of the favorable effect of nadolol can be identified in the stronger negative chronotropic effect in comparison to other BB, which allows for longer/more intense exercise time. Selective BB should be used only in the case of strong contraindications toward nonselective agents (e.g., asthma).

It is important to underline that the carriers of a CPVT mutation ("silent" carriers) are at risk of events even in the absence of exercise-induced lethal arrhythmias. Therefore, BB therapy and the avoidance of intense exercise (Priori et al. 2015) are clinically appropriate in individuals carriers of pathogenic mutations also in the absence of clinical phenotype. This indication is supported

Table 10.2 Clinical management

Recommendation	Class of recommendation	Level of evidence
The following lifestyle changes are recommended in all CPVT patients with a diagnosis of CPVT: avoidance of competitive sports, strenuous exercise, and stressful environments	I	C
Beta-blockers are recommended in all patients with a clinical diagnosis of CPVT, based on the presence of documented spontaneous or stress-induced ventricular arrhythmias	I	C
ICD implantation in addition to beta-blockers with or without flecainide is recommended in patients with a diagnosis of CPVT who experience cardiac arrest, recurrent syncope, or polymorphic/bidirectional VT despite optimal therapy	I	C
Therapy with beta-blockers should be considered for genetically positive family members, even after a negative exercise test	IIa	C
Flecainide should be considered in addition to beta-blockers in patients with a diagnosis of CPVT who experience recurrent syncope or polymorphic/bidirectional VT while on beta-blockers, when there are risk/contraindications for the ICD, an ICD is not available or rejected by the patient	IIa	C
Flecainide should be considered in addition to beta-blockers in patients with a diagnosis of CPVT and in carriers of an ICD to reduce appropriate ICD shocks	IIa	C
Left cardiac sympathetic denervation (LCSD) may be considered in patients with a diagnosis of CPVT who experience recurrent syncope or polymorphic/bidirectional VT/several appropriate ICD shocks while on beta-blockers or beta-blockers plus flecainide and in patients who are intolerant or with contraindications to beta-blockers	IIb	C
Invasive EPS with PVS is not recommended for sudden cardiac death risk stratification	III	C

Adapted from the 2015 ESC guidelines for the prevention of ventricular arrhythmias and sudden cardiac death (Priori et al. 2015)

by data coming from the study of Hayashi et al. (2009), who showed that 13% of silent mutation carriers had a cardiac arrest during follow-up, in the absence of any therapy. Besides constantly adapting the therapy to the result of follow-up examinations, a crucial factor for therapeutic success in CPVT is represented by the compliance of patients to the treatment (Priori et al. 2015).

Remarkably, despite an optimal BB use, there is a relevant burden of recurrent life-threatening arrhythmias in approximately 30% of cases (Hayashi et al. 2009; Priori et al. 2002). Several hypotheses have been made regarding the limited efficacy of beta-blockers, such as the presence of a mutation-specific response to therapy (Priori et al. 2002) or a role of alpha adrenergic pathway in the adrenergic stimulation of the heart (Kurtzwald-Josefson et al. 2014), but further studies will be required to gather a complete picture.

Non-responders to maximally tolerated dose BB treatment should receive flecainide (Priori et al. 2015), a class IC antiarrhythmic agent, at a mean dose of 150 mg (3 mg/kg per day) with a class IIa indication (Priori et al. 2015). Initial evidence was provided in 2009, when Watanabe et al. (2009) suggested its in vitro ability to inhibit the RyR2-mediated calcium release from the sarcoplasmic reticulum. In the same study, clinical effectiveness was shown in two CPVT patients. Two studies have subsequently confirmed the efficacy of flecainide in suppressing ventricular arrhythmias in CPVT in up to 77% of patients non-responder to beta-blockers alone (Roston et al. 2015; van der Werf et al. 2011). The mechanism of action of flecainide is related to the block of the sodium current with consequent increase of the threshold for action potential triggering from DADs (Liu et al. 2011); a direct block of the RyR2 channels has been also postulated (Hilliard et al. 2009), but not confirmed by other authors (Bannister et al. 2015; Sikkel et al. 2013).

Recent data suggest that the antiarrhythmic effect of flecainide can also block the neuronal sodium channels residing in the T-tubules; these channels generate inward late sodium current upon adrenergic activation exacerbating the abnormal calcium release (Radwanski et al. 2016).

Clinically, Padfield et al. (2016) proposed the use of flecainide as monotherapy in CPVT patients who are intolerant to BB. There are multiple reasons however to consider this approach with caution (Napolitano 2016), due to potential side effects and narrow therapeutic window.

STEP 3: Invasive therapies left cardiac sympathetic denervation (LCSD) and implantable cardioverter defibrillations (ICD). CPVT patients unresponsive to BB and flecainide do exist. LCSD or ICD should be considered in this case. LCSD should be also considered in patients with ICD and multiple recurrences on optimal drug therapy in the attempt of reducing the number of shocks.

The role of LCSD has been supported in a cohort of 38 symptomatic CPVT patients (De Ferrari et al. 2015). However, LCSD is affected by potential complications, and recurrences are still possible in 20–30% of treated patients (De Ferrari et al. 2015; Hofferberth et al. 2014); therefore, it cannot be considered an alternative to ICD.

ICD implantation is a class I indication for those CPVT patients who experienced a cardiac arrest, a recurrence of syncope or a polymorphic VT despite optimal medical therapy (Priori et al. 2015). ICD therapy is associated with a risk of inappropriate shocks due to programming difficulties. Both appropriate and inappropriate ICD shocks can be painful and cause catecholamine release, which in turn can trigger further ventricular arrhythmias. Thus, ICD should not be regarded to as a stand-alone therapy, but it should be used in conjunction with optimal medical therapy. Careful device programming with a long delay to detection before shock and high cutoff rates for heart rate recognition can significantly reduce the risk of complications. Roses-Noguer et al. (Roses-Noguer et al. 2014) found that shocks delivered to VTs nearly always failed (1/40, 3%, effective) to restore sinus rhythm (Fig. 10.4), while shocks delivered to VF were usually successful (19/23, 83%, effective). Thus, turning off of VT therapy seems appropriate.

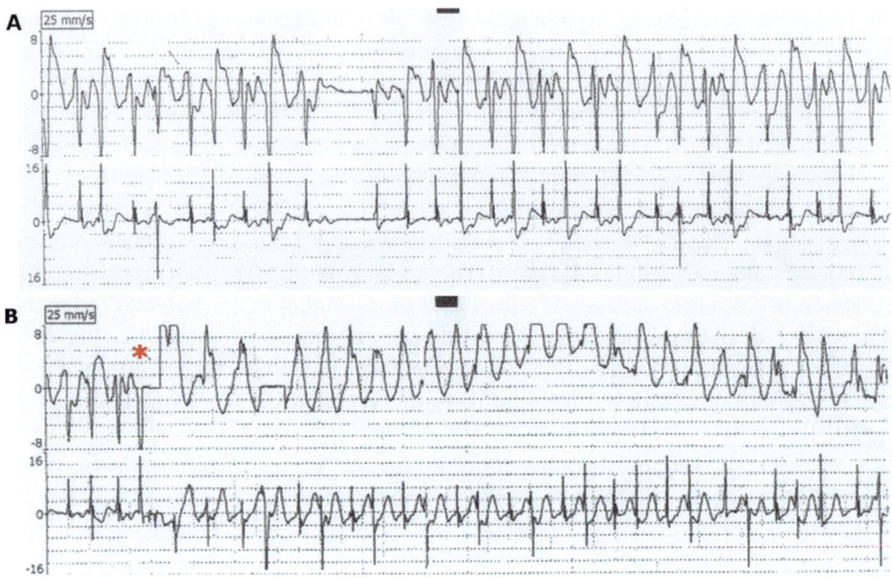

Fig. 10.4 Ineffective ICD in CPVT. ICD-stored electrograms of a bidirectional VT recognized (panel **a**) and correctly (according to device settings) treated (panel **b**). A 41 shock was delivered (asterisk) but it was unsuccessful in restoring sinus rhythm. The morphology of the arrhythmia changed after the ICD intervention

ICD and LCSD are not mutually exclusive; the two approaches can be combined in particularly severe cases with unsatisfactory response to drug therapy with ICD protecting from sudden death and LSCD reducing the risk of ICD shocks.

10.5 Experimental Therapies

10.5.1 Pharmacological Correction of SR Calcium Leakage

Control of RyR2 Channel Stability A well-described molecular mechanism for RyR2 dysfunction and reduced SOICR is channel instability (the channel “unzipping” mechanism). It has been proposed that in the wild-type RyR2, the N-terminal domain interacts with the central domain; RyR2 mutations in the N-terminal and central domains weaken this interaction (domain unzipping) (Ikemoto and Yamamoto 2000; Tateishi et al. 2009) and destabilize the closed state of the channel with increased leakage. It has been suggested that dantrolene, a drug used to treat acutely malignant hyperthermia due to RyR2 mutations, favors aberrant N-terminus intersubunit interactions of mutant cardiac RyR2, stabilizing the channels ("zipping") (Penttinen et al. 2015). At present there is no clinical experience with the use of dantrolene in CPVT.

CaMKII Inhibition There is emerging evidence of the role of the Ca^{2+}/calmodulin-dependent serine–threonine protein kinase II (CaMKII), in the modulation of calcium handling in response to adrenergic stimulation (Bers 2007), and there are theoretical reasons to think that CaMKII inhibition may result in additive/cooperative antiarrhythmic effect with beta-blockers. Indeed CaMKII acts not only in a PKA-dependent manner (the pathway inhibited by the beta-adrenergic receptor block) (Grimm and Brown 2010) but also through PKA-independent mechanisms (Oestreich et al. 2009) and autophosphorylation (Maier and Bers 2007; Napolitano et al. 2011).

We tested the hypothesis that the inhibition of CPVT CaMKII would be effective in an experimental model of CPVT in vivo (Liu et al. 2010). In our RyR2R4496C+/− knock-in mouse model of CPVT, CaMKII inhibition with KN-93 completely prevented catecholamine-induced sustained ventricular tachyarrhythmia blunted triggered activity and transient inward current induced by isoproterenol. Unfortunately, the available CaMKII inhibitors cannot be directly used in the clinical setting for several reasons including lack of cardiac specificity and potential unwanted effects. CaMKII plays several physiologic functions in the central nervous system (Chang et al. 1998; Tinsley et al. 2009) and in other tissues such as endothelial cells (Cui et al. 1996) and in the hormone secretion in parathyroid gland (Lu et al. 2011). In this context the development of CaMKII inhibitors with cardiac selectivity is highly desirable to achieve antiarrhythmic efficacy and to reduce the risk of adverse effects.

FKBP Binding and RyR2 Stability Several lines of evidence suggest that mutant RyR2 exhibits an increased sensitivity to cytosolic Ca^{2+} (Priori and Chen 2011). To explain such abnormal behavior, some authors have hypothesized the presence of abnormal/increased dissociation of FKBP12.6 from RyR2, a channel-stabilizing protein (Wehrens et al. 2003). During adrenergic stimulation, phosphorylation of RyR2 promotes Ca^{2+} leakage by abnormal dissociation of FKBP12.6 and consequent excessive increase of open probability. Accordingly, pharmacological enhancement/restoration of FKBB12.6 binding was tested. It has been shown that FKBP12.6 knockout mouse model develops stress-induced arrhythmias mainly due to increased RyR2 open probability (Wehrens et al. 2003). Unfortunately experiments from different authors provided conflicting evidence of efficacy (George et al. 2003; Jiang et al. 2005; Lehnart et al. 2008; Liu et al. 2006; Wehrens et al. 2003). Overall most of the results tend to dismiss a significant role of FKPB-RyR2 binding modulation as a CPVT therapeutic strategy.

10.5.2 Gene Therapy

The concept of using gene therapy to treat electrical dysfunction of the heart is not new (Donahue et al. 2000; Priori and Napolitano 2000), and the improvements of gene delivery technology with the use of adeno-associated vectors (Bongianino and Priori 2015) has made this approach for CPVT conceivable.

We reasoned that recessive CPVT would represent the ideal initial setting for developing and testing CPVT gene therapy. Calsequestrin mutation invariably cause a reduction of the expressed protein accompanied by a decrease in the ancillary proteins triadin and junctin (Denegri et al. 2012; Knollmann et al. 2006; Rizzi et al. 2008). This leads to a cascade of events: altered RyR2 gating and RyR2 calcium sensitivity (Raffaele di Barletta et al. 2006; Terentyev et al. 2006) with consequent diastolic Ca^{2+} overload, calcium wave fragmentation (Liu et al. 2013), delayed afterdepolarizations (DADs), adrenergically induced triggered arrhythmias (Liu et al. 2006), and ultrastructural abnormalities of the junctional sarcoplasmic reticulum (jSR) (Denegri et al. 2012).

We hypothesized that the normalization of the upstream abnormality—namely, reduced CASQ2 expression—could revert all the downstream molecular and physiological abnormalities resulting in a curative effect for recessive CPVT.

We subcloned wild-type *CASQ2* cDNA into an adeno-associated viral vector serotype 9 (AAV9) backbone (AAV9-*CASQ2*), and we infected calsequestrin null (knockout) mice at birth (Denegri et al. 2012). Five months after infection we demonstrated a complete normalization of both the electrophysiological and the ultrastructural abnormalities. Similar results were confirmed in a knock-in mouse harboring a CPVT missense mutation (R33Q). In this model, we showed that infection of adult mice could treat the manifestations of the disease, including life-threatening arrhythmias and ultrastructural abnormalities. This effect was still present up to 12 months after administration of a single dose (Denegri et al. 2014; Liu et al. 2013) (Fig. 10.5). The prevention of CPVT arrhythmias with AAV9-CASQ2 was obtained with a relatively low (40%) percentage of infected ventricular cells. This discrepancy between the nearly complete antiarrhythmic efficacy and the patchy and incomplete restoration of CASQ2 level can be due to the "source-sink" phenomenon (Xie et al. 2010). Cell-to-cell propagation is required for a DAD to trigger an action potential. It is only when adjacent myocytes develop synchronous DADs that their summation leads to a triggered action potential spreading to the ventricles. Thus, triggered arrhythmias cannot develop unless a "sufficient" number of the neighboring myocytes develop DADs (Xie et al. 2010). A recent independent work has replicated these results and showed that hearts with >33% of CASQ2 re-expression are protected from arrhythmias, while lower infection levels are insufficient to induce a therapeutic effect (Kurtzwald-Josefson et al. 2017).

Gene therapy is not a clinical reality yet, but the fast improving AAV vector technology and cardiac delivery strategies (Ishikawa et al. 2013) suggest that the first gene therapy clinical trial in CPVT could be planned in the next few years.

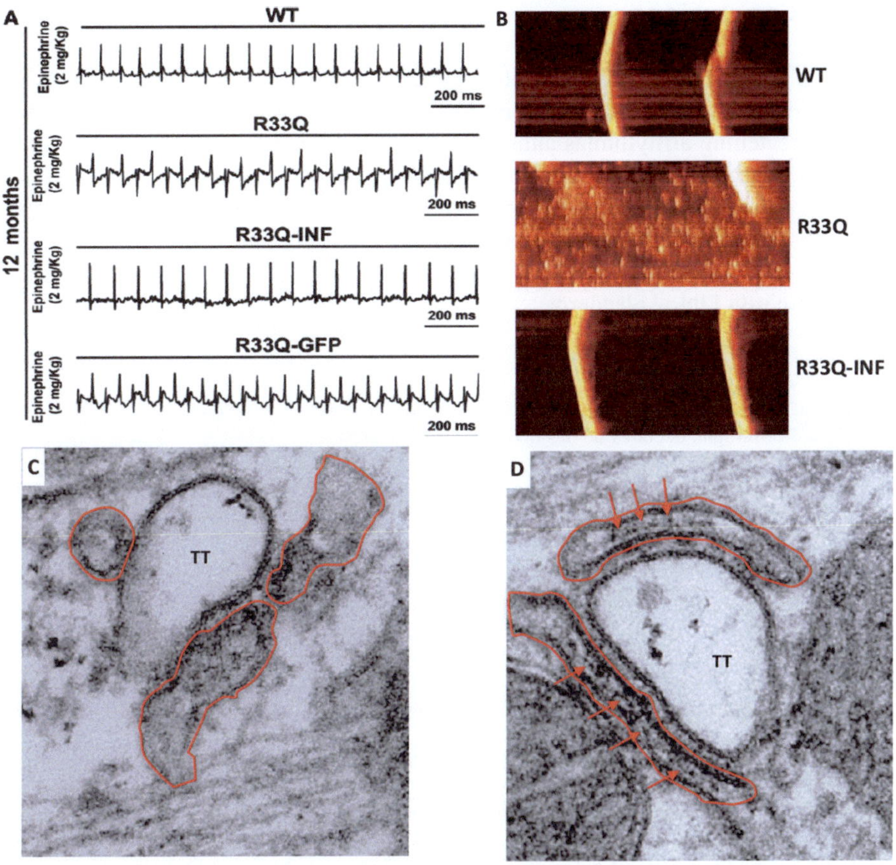

Fig. 10.5 Gene therapy in calsequestrin-dependent CPVT. The figure reports experimental data collected 1 year after a single injection of AAV9-CASQ2 gene therapy vector. (**a**) ECGs from wild-type (WT) and CASQ2-R33Q CPVT mice (first and second strip from top) during epineprine injection; CASQ2-R33Q mice develop typical bidirectional ventricular tachycardia, which is completely suppressed in mice infected with AAV-CASQ (third strip from top); mice infected with GFP only still show epinephrine inducible ventricular tachycardia. (**b**). Representative confocal line-scan images show spontaneous Ca^{2+} events in permeabilized R33Q, WT, and AAV-R33Q cells ($[Ca^{2+}]_i$ 100 nmol/L). R33Q cells did not show cell-wide waves but rather presented with chaotic and fragmented events and wavelets. WT (upper panel) and R33Q infected (R33Q-INF) with AAV9-CASQ2 myocytes exhibited regular spontaneous Ca^{2+} cell-wide waves. (**c**) Ultrastructural abnormalities in CASQ2-R33Q mice showing enlarged and fragmented junctional sarcoplasmic reticulum (encircled with red lines—jSR). (**d**) Mice treated with AAV9-CASQ show normal and regular jSR with clearly visible calsequestrin polymers (arrows). *TT* T-tubule

10.6 Conclusions

CPVT is a severe inherited arrhythmogenic condition that has been extensively studied in the last two decades. Currently we do have means to reduce the burden of life-threatening arrhythmias although they are ineffective in a subset of patients. Therefore, early diagnostic workout and optimal therapy management are essential, while many research groups are working to devise novel therapeutic strategies, including gene therapy that could be lead to a cure in the future.

Compliance with Ethical Standards

Sources of Funding None.

Conflict of Interest Carlo Napolitano declares he has no conflicts of interest.
Riccardo Maragna declares he has no conflicts of interest.

Ethical Approval This article does not contain any studies with human participants performed by any of the authors.

References

Altmann HM, Tester DJ, Will ML, Middha S, Evans JM, Eckloff BW, Ackerman MJ. Homozygous/compound heterozygous triadin mutations associated with autosomal-recessive long-QT syndrome and pediatric sudden cardiac arrest: elucidation of the triadin knockout syndrome. Circulation. 2015;131(23):2051–60.

Bannister ML, Thomas NL, Sikkel MB, Mukherjee S, Maxwell CE, MacLeod KT, George CH, Williams AJ. The mechanism of flecainide action in CPVT does not involve a direct effect on RyR2. Circ Res. 2015;116(8):1324–35.

Bers DM. Going to cAMP just got more complicated. J Physiol. 2007;583(Pt 2):415–6.

Bhuiyan ZA, van Den Berg MP, van Tintelen JP, Bink-Boelkens MT, Wiesfeld AC, Alders M, Postma AV, Van LI, Mannens MM, Wilde AA. Expanding spectrum of human RYR2-related disease. New electrocardiographic, structural, and genetic features. Circulation. 2007;116:1569–76.

Boczek NJ, Gomez-Hurtado N, Ye D, Calvert ML, Tester DJ, Kryshtal DO, Hwang HS, Johnson CN, Chazin WJ, Loporcaro CG, Shah M, Papez AL, Lau YR, Kanter R, Knollmann BC, Ackerman MJ. Spectrum and prevalence of CALM1-, CALM2-, and CALM3-encoded calmodulin variants in long QT syndrome and functional characterization of a novel long QT syndrome-associated calmodulin missense variant, E141G. Circ Cardiovasc Genet. 2016;9(2):136–46.

Bongianino R, Priori SG. Gene therapy to treat cardiac arrhythmias. Nat Rev Cardiol. 2015;12(9):531–46.

Camors E, Mohler PJ, Bers DM, Despa S. Ankyrin-B reduction enhances Ca spark-mediated SR Ca release promoting cardiac myocyte arrhythmic activity. J Mol Cell Cardiol. 2012;52(6):1240–8.

Cerrone M, Colombi B, Santoro M, di Barletta MR, Scelsi M, Villani L, Napolitano C, Priori SG. Bidirectional ventricular tachycardia and fibrillation elicited in a knock-in mouse model carrier of a mutation in the cardiac ryanodine receptor. Circ Res. 2005;96(10):e77–82.

Chang BH, Mukherji S, Soderling TR. Characterization of a calmodulin kinase II inhibitor protein in brain. Proc Natl Acad Sci U S A. 1998;95(18):10890–5.

Chopra N, Yang T, Asghari P, Moore ED, Huke S, Akin B, Cattolica RA, Perez CF, Hlaing T, Knollmann-Ritschel BE, Jones LR, Pessah IN, Allen PD, Franzini-Armstrong C, Knollmann BC. Ablation of triadin causes loss of cardiac Ca2+ release units, impaired excitation-contraction coupling, and cardiac arrhythmias. Proc Natl Acad Sci U S A. 2009;106 (18):7636–41.

Coumel P, Fidelle J, Lucet V, Attuel P, Bouvrain Y. Catecholamine-induced severe ventricular arrhythmias with Adams-stokes syndrome in children: report of four cases. Br Heart J. 1978;40:28–37.

Cui ZJ, Hidaka H, Dannies PS. KN-62, a calcium/calmodulin-dependent protein kinase II inhibitor, inhibits high potassium-stimulated prolactin secretion and intracellular calcium increases in anterior pituitary cells. Biochim Biophys Acta. 1996;1310(3):343–7.

De Ferrari GM, Dusi V, Spazzolini C, Bos JM, Abrams DJ, Berul CI, Crotti L, Davis AM, Eldar M, Kharlap M, Khoury A, Krahn AD, Leenhardt A, Moir CR, Odero A, Olde Nordkamp L, Paul T, Roses INF, Shkolnikova M, Till J, Wilde AA, Ackerman MJ, Schwartz PJ. Clinical management of catecholaminergic polymorphic ventricular tachycardia: the role of left cardiac sympathetic denervation. Circulation. 2015;131(25):2185–93.

Denegri M, Avelino-Cruz JE, Boncompagni S, De Simone SA, Auricchio A, Villani L, Volpe P, Protasi F, Napolitano C, Priori SG. Viral gene transfer rescues arrhythmogenic phenotype and ultrastructural abnormalities in adult calsequestrin-null mice with inherited arrhythmias. Circ Res. 2012;110(5):663–8.

Denegri M, Bongianino R, Lodola F, Boncompagni S, De Giusti VC, Avelino-Cruz JE, Liu N, Persampieri S, Curcio A, Esposito F, Pietrangelo L, Marty I, Villani L, Moyaho A, Baiardi P, Auricchio A, Protasi F, Napolitano C, Priori SG. Single delivery of an adeno-associated viral construct to transfer the CASQ2 gene to knock-in mice affected by catecholaminergic polymorphic ventricular tachycardia is able to cure the disease from birth to advanced age. Circulation. 2014;129(25):2673–81.

Devalla HD, Gelinas R, Aburawi EH, Beqqali A, Goyette P, Freund C, Chaix MA, Tadros R, Jiang H, Le Bechec A, Monshouwer-Kloots JJ, Zwetsloot T, Kosmidis G, Latour F, Alikashani A, Hoekstra M, Schlaepfer J, Mummery CL, Stevenson B, Kutalik Z, de Vries AA, Rivard L, Wilde AA, Talajic M, Verkerk AO, Al-Gazali L, Rioux JD, Bhuiyan ZA, Passier R. TECRL, a new life-threatening inherited arrhythmia gene associated with overlapping clinical features of both LQTS and CPVT. EMBO Mol Med. 2016;8(12):1390–408.

Di Pasquale E, Lodola F, Miragoli M, Denegri M, Avelino-Cruz JE, Buonocore M, Nakahama H, Portararo P, Bloise R, Napolitano C, Condorelli G, Priori SG. CaMKII inhibition rectifies arrhythmic phenotype in a patient-specific model of catecholaminergic polymorphic ventricular tachycardia. Cell Death Dis. 2013;4:e843.

Donahue JK, Heldman AW, Fraser H, McDonald AD, Miller JM, Rade JJ, Eschenhagen T, Marban E. Focal modification of electrical conduction in the heart by viral gene transfer. Nat Med. 2000;6(12):1395–8.

Fatima A, Xu G, Shao K, Papadopoulos S, Lehmann M, Arnaiz-Cot JJ, Rosa AO, Nguemo F, Matzkies M, Dittmann S, Stone SL, Linke M, Zechner U, Beyer V, Hennies HC, Rosenkranz S, Klauke B, Parwani AS, Haverkamp W, Pfitzer G, Farr M, Cleemann L, Morad M, Milting H, Hescheler J, Saric T. In vitro modeling of ryanodine receptor 2 dysfunction using human induced pluripotent stem cells. Cell Physiol Biochem. 2011;28(4):579–92.

George CH, Higgs GV, Lai FA. Ryanodine receptor mutations associated with stress-induced ventricular tachycardia mediate increased calcium release in stimulated cardiomyocytes. Circul Res. 2003;93(6):531–40.

Gomez-Hurtado N, Boczek NJ, Kryshtal DO, Johnson CN, Sun J, Nitu FR, Cornea RL, Chazin WJ, Calvert ML, Tester DJ, Ackerman MJ, Knollmann BC. Novel CPVT-associated calmodulin mutation in CALM3 (CALM3-A103V) activates arrhythmogenic Ca waves and sparks. Circ Arrhythm Electrophysiol. 2016;9(8)

Gray B, Bagnall RD, Lam L, Ingles J, Turner C, Haan E, Davis A, Yang PC, Clancy CE, Sy RW, Semsarian C. A novel heterozygous mutation in cardiac calsequestrin causes autosomal dominant catecholaminergic polymorphic ventricular tachycardia. Heart Rhythm. 2016;13 (8):1652–60.

Grimm M, Brown JH. Beta-adrenergic receptor signaling in the heart: role of CaMKII. J Mol Cell Cardiol. 2010;48(2):322–30.

Hayashi M, Denjoy I, Extramiana F, Maltret A, Buisson NR, Lupoglazoff JM, Klug D, Takatsuki S, Villain E, Kamblock J, Messali A, Guicheney P, Lunardi J, Leenhardt A. Incidence and risk factors of arrhythmic events in catecholaminergic polymorphic ventricular tachycardia. Circulation. 2009;119(18):2426–34.

Hilliard FA, Steele DS, Laver D, Yang Z, Le Marchand SJ, Chopra N, Piston DW, Huke S, Knollmann BC. Flecainide inhibits arrhythmogenic Ca2+ waves by open state block of ryanodine receptor Ca2+ release channels and reduction of Ca2+ spark mass. J Mol Cell Cardiol. 2009;48(2):293–301.

Hofferberth SC, Cecchin F, Loberman D, Fynn-Thompson F. Left thoracoscopic sympathectomy for cardiac denervation in patients with life-threatening ventricular arrhythmias. J Thorac Cardiovasc Surg. 2014;147(1):404–9.

Hwang HS, Nitu FR, Yang Y, Walweel K, Pereira L, Johnson CN, Faggioni M, Chazin WJ, Laver D, George AL Jr, Cornea RL, Bers DM, Knollmann BC. Divergent regulation of ryanodine receptor 2 calcium release channels by arrhythmogenic human calmodulin missense mutants. Circ Res. 2014;114(7):1114–24.

Ikemoto N, Yamamoto T. Postulated role of inter-domain interaction within the ryanodine receptor in Ca(2+) channel regulation. Trends Cardiovasc Med. 2000;10(7):310–6.

Ishikawa K, Aguero J, Naim C, Fish K, Hajjar RJ. Percutaneous approaches for efficient cardiac gene delivery. J Cardiovasc Transl Res. 2013;6(4):649–59.

Jiang D, Wang R, Xiao B, Kong H, Hunt DJ, Choi P, Zhang L, Chen SR. Enhanced store overload-induced Ca2+ release and channel sensitivity to luminal Ca2+ activation are common defects of RyR2 mutations linked to ventricular tachycardia and sudden death. Circ Res. 2005;97 (11):1173–81.

Jimenez-Jaimez J, Peinado R, Grima EZ, Segura F, Morina P, Sanchez Munoz JJ, Mazuelos F, Cozar R, Gimeno JR, Heras RP, Monserrat L, Domingo D, Ortiz-Genga M, Fernandez Pastor J, Alvarez M, Tercedor L. Diagnostic approach to unexplained cardiac arrest (from the FIVI-gen study). Am J Cardiol. 2015;116(6):894–9.

Kline CF, Mohler PJ. Defective interactions of protein partner with ion channels and transporters as alternative mechanisms of membrane channelopathies. Biochim Biophys Acta. 2014;1838 (2):723–30.

Knollmann BC, Chopra N, Hlaing T, Akin B, Yang T, Ettensohn K, Knollmann BE, Horton KD, Weissman NJ, Holinstat I, Zhang W, Roden DM, Jones LR, Franzini-Armstrong C, Pfeifer K. Casq2 deletion causes sarcoplasmic reticulum volume increase, premature Ca2+ release, and catecholaminergic polymorphic ventricular tachycardia. J Clin Invest. 2006;116(9):2510–20.

Kurtzwald-Josefson E, Hochhauser E, Bogachenko K, Harun-Khun S, Katz G, Aravot D, Seidman JG, Seidman CE, Eldar M, Shainberg A, Arad M. Alpha blockade potentiates CPVT therapy in calsequestrin-mutant mice. Heart Rhythm. 2014;11(8):1471–9.

Kurtzwald-Josefson E, Yadin D, Harun-Khun S, Waldman M, Aravot D, Shainberg A, Eldar M, Hochhauser E, Arad M. Viral delivered gene therapy to treat catecholamine dependent polymorphic ventricular tachycardia (CPVT2) in mouse models. Heart Rhythm. 2017;14 (7):1053–60.

Lahat H, Pras E, Olender T, Avidan N, Ben Asher E, Man O, Levy-Nissenbaum E, Khoury A, Lorber A, Goldman B, Lancet D, Eldar M. A missense mutation in a highly conserved region of CASQ2 is associated with autosomal recessive catecholamine-induced polymorphic ventricular tachycardia in Bedouin families from Israel. Am J Hum Genetics. 2001;69(6):1378–84.

Leenhardt A, Lucet V, Denjoy I, Grau F, Ngoc DD, Coumel P. Catecholaminergic polymorphic ventricular tachycardia in children. A 7-year follow-up of 21 patients. Circulation. 1995;91 (5):1512–9.

Lehnart SE, Mongillo M, Bellinger A, Lindegger N, Chen BX, Hsueh W, Reiken S, Wronska A, Drew LJ, Ward CW, Lederer WJ, Kass RS, Morley G, Marks AR. Leaky Ca2+ release channel/ ryanodine receptor 2 causes seizures and sudden cardiac death in mice. J Clin Invest. 2008;118 (6):2230–45.

Leren IS, Saberniak J, Majid E, Haland TF, Edvardsen T, Haugaa KH. Nadolol decreases the incidence and severity of ventricular arrhythmias during exercise stress testing compared with beta-selective beta-blockers in patients with Catecholaminergic polymorphic ventricular tachycardia. Heart Rhythm. 2015;15:S1547–5271.

Limpitikul WB, Dick IE, Joshi-Mukherjee R, Overgaard MT, George AL Jr, Yue DT. Calmodulin mutations associated with long QT syndrome prevent inactivation of cardiac L-type Ca(2+) currents and promote proarrhythmic behavior in ventricular myocytes. J Mol Cell Cardiol. 2014;74:115–24.

Liu N, Colombi B, Memmi M, Zissimopoulos S, Rizzi N, Negri S, Imbriani M, Napolitano C, Lai FA, Priori SG. Arrhythmogenesis in catecholaminergic polymorphic ventricular tachycardia: insights from a RyR2 R4496C knock-in mouse model. Circ Res. 2006;99(3):292–8.

Liu N, Denegri M, Ruan Y, Bachetti T, Seregni M, Napolitano C, Priori SG. Sodium channel blockers prevent triggered activity but not abnormal Ca2+ release in a knock-in mouse model with ryanodine receptor mutation R4496C. Circulation. 2010;122:A13707.

Liu N, Denegri M, Ruan Y, Avelino-Cruz JE, Perissi A, Negri S, Napolitano C, Coetzee WA, Boyden PA, Priori SG. Short communication: flecainide exerts an antiarrhythmic effect in a mouse model of Catecholaminergic polymorphic ventricular tachycardia by increasing the threshold for triggered activity. Circ Res. 2011;109(3):291–5.

Liu N, Denegri M, Dun W, Boncompagni S, Lodola F, Protasi F, Napolitano C, Boyden PA, Priori SG. Abnormal propagation of calcium waves and ultrastructural remodeling in recessive Catecholaminergic polymorphic ventricular tachycardia. Circul Res. 2013;113(2):142–52.

Lu M, Berglund E, Larsson C, Hoog A, Farnebo LO, Branstrom R. Calmodulin and calmodulin-dependent protein kinase II inhibit hormone secretion in human parathyroid adenoma. J Endocrinol. 2011;208(1):31–9.

Maier LS, Bers DM. Role of Ca2+/calmodulin-dependent protein kinase (CaMK) in excitation-contraction coupling in the heart. Cardiovasc Res. 2007;73(4):631–40.

Marjamaa A, Hiippala A, Arrhenius B, Lahtinen AM, Kontula K, Toivonen L, Happonen JM, Swan H. Intravenous epinephrine infusion test in diagnosis of catecholaminergic polymorphic ventricular tachycardia. J Cardiovasc Electrophysiol. 2012;23(2):194–9.

Marsman RF, Barc J, Beekman L, Alders M, Dooijes D, van den Wijngaard A, Ratbi I, Sefiani A, Bhuiyan ZA, Wilde AA, Bezzina CR. A mutation in CALM1 encoding calmodulin in familial idiopathic ventricular fibrillation in childhood and adolescence. J Am Coll Cardiol. 2014;63 (3):259–66.

Napolitano C. Flecainide monotherapy for Catecholaminergic polymorphic ventricular tachycardia: perspectives and limitations. Heart Rhythm. 2016;13(2):614–5.

Napolitano C, Priori SG, Bloise R. Catecholaminergic polymorphic ventricular tachycardia. In: Pagon RA, Adam MP, Ardinger HH, Wallace SE, Amemiya A, Bean LJH, Bird TD, Ledbetter N, Mefford HC, Smith RJH, Stephens K, editors. GeneReviews(R). Seattle: University of Wahington; 2016.

Napolitano C, Liu N, Priori SG. Role of calmodulin kinase in catecholaminergic polymorphic ventricular tachycardia. Heart Rhythm. 2011;8(10):1601–5.

Napolitano C, Bloise R, Monteforte N, Priori SG. Sudden cardiac death and genetic ion channelopathies: long QT, Brugada, short QT, Catecholaminergic polymorphic ventricular tachycardia, and idiopathic ventricular fibrillation. Circulation. 2012;125(16):2027–34.

Neco P, Torrente AG, Mesirca P, Zorio E, Liu N, Priori SG, Napolitano C, Richard S, Benitah JP, Mangoni ME, Gomez AM. Paradoxical effect of increased diastolic Ca2+ release and decreased sinoatrial node activity in a mouse model of catecholaminergic polymorphic ventricular tachycardia. Circulation. 2012;126(4):392–401.

Nyegaard M, Overgaard MT, Sondergaard MT, Vranas M, Behr ER, Hildebrandt LL, Lund J, Hedley PL, Camm AJ, Wettrell G, Fosdal I, Christiansen M, Borglum AD. Mutations in calmodulin cause ventricular tachycardia and sudden cardiac death. Am J Hum Genet. 2012;91(4):703–12.

Oestreich EA, Malik S, Goonasekera SA, Blaxall BC, Kelley GG, Dirksen RT, Smrcka AV. Epac and phospholipase Cepsilon regulate Ca2+ release in the heart by activation of protein kinase Cepsilon and calcium-calmodulin kinase II. J Biol Chem. 2009;284(3):1514–22.

Ohno S, Omura M, Kawamura M, Kimura H, Itoh H, Makiyama T, Ushinohama H, Makita N, Horie M. Exon 3 deletion of RYR2 encoding cardiac ryanodine receptor is associated with left ventricular non-compaction. Europace. 2014;16(11):1646–54.

Paavola J, Viitasalo M, Laitinen-Forsblom PJ, Pasternack M, Swan H, Tikkanen I, Toivonen L, Kontula K, Laine M. Mutant ryanodine receptors in Catecholaminergic polymorphic ventricular tachycardia generate delayed afterdepolarizations due to increased propensity to Ca2+ waves. Eur Heart J. 2007;28(9):1135–42.

Padfield GJ, AlAhmari L, Lieve KV, AlAhmari T, Roston TM, Wilde AA, Krahn AD, Sanatani S. Flecainide monotherapy is an option for selected patients with Catecholaminergic polymorphic ventricular tachycardia intolerant of beta-blockade. Heart Rhythm. 2016;13(2):609–13.

Paludan-Muller C, Ahlberg G, Ghouse J, Herfelt C, Svendsen JH, Haunso S, Kanters JK, Olesen MS. Integration of 60,000 exomes and ACMG guidelines question the role of Catecholaminergic polymorphic ventricular tachycardia-associated variants. Clin Genet. 2017;91(1):63–72.

Penttinen K, Swan H, Vanninen S, Paavola J, Lahtinen AM, Kontula K, Aalto-Setala K. Antiarrhythmic effects of dantrolene in patients with Catecholaminergic polymorphic ventricular tachycardia and replication of the responses using iPSC models. PLoS One. 2015;10(5): e0125366.

Postma AV, Denjoy I, Kamblock J, Alders M, Lupoglazoff JM, Vaksmann G, Dubosq-Bidot L, Sebillon P, Mannens MM, Guicheney P, Wilde AA. Catecholaminergic polymorphic ventricular tachycardia: RYR2 mutations, bradycardia, and follow up of the patients. J Med Genet. 2005;42(11):863–70.

Priori SG, Chen SR. Inherited dysfunction of sarcoplasmic reticulum Ca2+ handling and arrhythmogenesis. Circ Res. 2011;108(7):871–83.

Priori SG, Napolitano C. From catheters to vectors: the dawn of molecular electrophysiology. Nat Med. 2000;6(12):1316–8.

Priori SG, Napolitano C, Tiso N, Memmi M, Vignati G, Bloise R, Sorrentino V, Danieli GA. Mutations in the cardiac ryanodine receptor gene (hRyR2) underlie Catecholaminergic polymorphic ventricular tachycardia. Circulation. 2001;103(2):196–200.

Priori SG, Napolitano C, Memmi M, Colombi B, Drago F, Gasparini M, DeSimone L, Coltorti F, Bloise R, Keegan R, Cruz Filho FE, Vignati G, Benatar A, DeLogu A. Clinical and molecular characterization of patients with catecholaminergic polymorphic ventricular tachycardia. Circulation. 2002;106(1):69–74.

Priori SG, Wilde AA, Horie M, Cho Y, Behr ER, Berul C, Blom N, Brugada J, Chiang CE, Huikuri H, Kannankeril P, Krahn A, Leenhardt A, Moss A, Schwartz PJ, Shimizu W, Tomaselli G, Tracy C. HRS/EHRA/APHRS expert consensus statement on the diagnosis and management of patients with inherited primary arrhythmia syndromes: document endorsed by HRS, EHRA, and APHRS in may 2013 and by ACCF, AHA, PACES, and AEPC in June 2013. Heart Rhythm. 2013;10(12):1932–63.

Priori SG, Blomstrom-Lundqvist C, Mazzanti A, Blom N, Borggrefe M, Camm J, Elliott PM, Fitzsimons D, Hatala R, Hindricks G, Kirchhof P, Kjeldsen K, Kuck KH, Hernandez-Madrid A, Nikolaou N, Norekval TM, Spaulding C, Van Veldhuisen DJ. 2015 ESC guidelines for the management of patients with ventricular arrhythmias and the prevention of sudden cardiac death: the task force for the management of patients with ventricular arrhythmias and the prevention of sudden cardiac death of the European Society of Cardiology (ESC). Endorsed by: Association for European Paediatric and Congenital Cardiology (AEPC). Eur Heart J. 2015;36(41):2793–867.

Radwanski PB, Ho HT, Veeraraghavan R, Brunello L, Liu B, Belevych AE, Unudurthi SD, Makara MA, Priori SG, Volpe P, Armoundas AA, Dillmann WH, Knollmann BC, Mohler PJ, Hund TJ, Gyorke S. Neuronal Na+ channels are integral components of pro-arrhythmic Na+/Ca2+ signaling nanodomain that promotes cardiac arrhythmias during beta-adrenergic stimulation. JACC Basic Transl Sci. 2016;1(4):251–66.

Raffaele di Barletta M, Viatchenko-Karpinski S, Nori A, Memmi M, Terentyev D, Turcato F, Valle G, Rizzi N, Napolitano C, Gyorke S, Volpe P, Priori SG. Clinical phenotype and functional characterization of CASQ2 mutations associated with catecholaminergic polymorphic ventricular tachycardia. Circulation. 2006;114(10):1012–9.

Rizzi N, Liu N, Napolitano C, Nori A, Turcato F, Colombi B, Bicciato S, Arcelli D, Spedito A, Scelsi M, Villani L, Esposito G, Boncompagni S, Protasi F, Volpe P, Priori SG. Unexpected structural and functional consequences of the R33Q homozygous mutation in cardiac calsequestrin: a complex arrhythmogenic cascade in a knock in mouse model. Circ Res. 2008;103(3):298–306.

Rooryck C, Kyndt F, Bozon D, Roux-Buisson N, Sacher F, Probst V, Thambo JB. New family with catecholaminergic polymorphic ventricular tachycardia linked to the triadin gene. J Cardiovasc Electrophysiol. 2015;26(10):1146–50.

Roses-Noguer F, Jarman JW, Clague JR, Till J. Outcomes of defibrillator therapy in catecholaminergic polymorphic ventricular tachycardia. Heart Rhythm. 2014;11(1):58–66.

Roston TM, Vinocur JM, Maginot KR, Mohammed S, Salerno JC, Etheridge SP, Cohen M, Hamilton RM, Pflaumer A, Kanter RJ, Potts JE, LaPage MJ, Collins KK, Gebauer RA, Temple JD, Batra AS, Erickson C, Miszczak-Knecht M, Kubus P, Bar-Cohen Y, Kantoch M, Thomas VC, Hessling G, Anderson C, Young ML, Cabrera Ortega M, Lau YR, Johnsrude CL, Fournier A, Kannankeril PJ, Sanatani S. Catecholaminergic polymorphic ventricular tachycardia in children: analysis of therapeutic strategies and outcomes from an international multicenter registry. Circ Arrhythm Electrophysiol. 2015;8(3):633–42.

Roux-Buisson N, Cacheux M, Fourest-Lieuvin A, Fauconnier J, Brocard J, Denjoy I, Durand P, Guicheney P, Kyndt F, Leenhardt A, Le Marec H, Lucet V, Mabo P, Probst V, Monnier N, Ray PF, Santoni E, Tremeaux P, Lacampagne A, Faure J, Lunardi J, Marty I. Absence of triadin, a protein of the calcium release complex, is responsible for cardiac arrhythmia with sudden death in human. Hum Mol Genet. 2012;21(12):2759–67.

Roux-Buisson N, Gandjbakhch E, Donal E, Probst V, Deharo JC, Chevalier P, Klug D, Mansencal N, Delacretaz E, Cosnay P, Scanu P, Extramiana F, Keller D, Hidden-Lucet F, Trapani J, Fouret P, Frank R, Fressart V, Faure J, Lunardi J, Charron P. Prevalence and significance of rare RYR2 variants in arrhythmogenic right ventricular cardiomyopathy/dysplasia: results of a systematic screening. Heart Rhythm. 2014;11(11):1999–2009.

Sikkel MB, Collins TP, Rowlands C, Shah M, O'Gara P, Williams AJ, Harding SE, Lyon AR, MacLeod KT. Flecainide reduces Ca(2+) spark and wave frequency via inhibition of the sarcolemmal sodium current. Cardiovasc Res. 2013;98(2):286–96.

Sy RW, Gollob MH, Klein GJ, Yee R, Skanes AC, Gula LJ, Leong-Sit P, Gow RM, Green MS, Birnie DH, Krahn AD. Arrhythmia characterization and long-term outcomes in Catecholaminergic polymorphic ventricular tachycardia. Heart Rhythm. 2011;8(6):864–71.

Tateishi H, Yano M, Mochizuki M, Suetomi T, Ono M, Xu X, Uchinoumi H, Okuda S, Oda T, Kobayashi S, Yamamoto T, Ikeda Y, Ohkusa T, Ikemoto N, Matsuzaki M. Defective domain-domain interactions within the ryanodine receptor as a critical cause of diastolic Ca2+ leak in failing hearts. Cardiovasc Res. 2009;81(3):536–45.

Terentyev D, Cala SE, Houle TD, Viatchenko-Karpinski S, Gyorke I, Terentyeva R, Williams SC, Gyorke S. Triadin overexpression stimulates excitation-contraction coupling and increases predisposition to cellular arrhythmia in cardiac myocytes. Circ Res. 2005;96(6):651–8.

Terentyev D, Nori A, Santoro M, Viatchenko-Karpinski S, Kubalova Z, Gyorke I, Terentyeva R, Vedamoorthyrao S, Blom NA, Valle G, Napolitano C, Williams SC, Volpe P, Priori SG, Gyorke S. Abnormal interactions of calsequestrin with the ryanodine receptor calcium release channel complex linked to exercise-induced sudden cardiac death. Circ Res. 2006;98(9):1151–8.

Tester DJ, Spoon DB, Valdivia HH, Makielski JC, Ackerman MJ. Targeted mutational analysis of the RyR2-encoded cardiac ryanodine receptor in sudden unexplained death: a molecular autopsy of 49 medical examiner/coroner's cases. Mayo Clin Proc. 2004;79(11):1380–4.

Tester DJ, Dura M, Carturan E, Reiken S, Wronska A, Marks AR, Ackerman MJ. A mechanism for sudden infant death syndrome (SIDS): stress-induced leak via ryanodine receptors. Heart Rhythm. 2007;4(6):733–9.

Tinsley CJ, Narduzzo KE, Ho JW, Barker GR, Brown MW, Warburton EC. A role for calcium-calmodulin-dependent protein kinase II in the consolidation of visual object recognition memory. Eur J Neurosci. 2009;30(6):1128–39.

van der Werf C, Kannankeril PJ, Sacher F, Krahn AD, Viskin S, Leenhardt A, Shimizu W, Sumitomo N, Fish FA, Bhuiyan ZA, Willems AR, van der Veen MJ, Watanabe H, Laborderie J, Haissaguerre M, Knollmann BC, Wilde AA. Flecainide therapy reduces exercise-induced ventricular arrhythmias in patients with catecholaminergic polymorphic ventricular tachycardia. J Am Coll Cardiol. 2011;57(22):2244–54.

van der Werf C, Nederend I, Hofman N, van Geloven N, Ebink C, Frohn-Mulder IM, Alings AM, Bosker HA, Bracke FA, van den Heuvel F, Waalewijn RA, Bikker H, van Tintelen JP, Bhuiyan ZA, van den Berg MP, Wilde AA. Familial evaluation in catecholaminergic polymorphic ventricular tachycardia: disease penetrance and expression in cardiac ryanodine receptor mutation-carrying relatives. Circ Arrhythm Electrophysiol. 2012;5(4):748–56.

Viitasalo M, Oikarinen L, Vaananen H, Kontula K, Toivonen L, Swan H. U-waves and T-wave peak to T-wave end intervals in patients with catecholaminergic polymorphic ventricular tachycardia, effects of beta-blockers. Heart Rhythm. 2008;5(10):1382–8.

Walsh MA, Stuart AG, Schlecht HB, James AF, Hancox JC, Newbury-Ecob RA. Compound heterozygous triadin mutation causing cardiac arrest in two siblings. Pacing Clin Electrophysiol. 2016;39(5):497–501.

Watanabe H, Chopra N, Laver D, Hwang HS, Davies SS, Roach DE, Duff HJ, Roden DM, Wilde AA, Knollmann BC. Flecainide prevents catecholaminergic polymorphic ventricular tachycardia in mice and humans. Nat Med. 2009;15(4):380–3.

Wehrens XH, Lehnart SE, Huang F, Vest JA, Reiken SR, Mohler PJ, Sun J, Guatimosim S, Song LS, Rosemblit N, D'Armiento JM, Napolitano C, Memmi M, Priori SG, Lederer WJ, Marks AR. FKBP12.6 deficiency and defective calcium release channel (ryanodine receptor) function linked to exercise-induced sudden cardiac death. Cell. 2003;113(7):829–40.

Xie Y, Sato D, Garfinkel A, Qu Z, Weiss JN. So little source, so much sink: requirements for afterdepolarizations to propagate in tissue. Biophys J. 2010;99(5):1408–15.

Idiopathic Ventricular Fibrillation and Early Repolarization 11

Pieter G. Postema

Abstract

In this chapter, an overview is provided on idiopathic ventricular fibrillation (IVF) and early repolarization. Idiopathic ventricular fibrillation is a tragic and notoriously difficult disease entity to manage. The IVF patients in whom a cardiac arrest occurs are generally considered healthy and do not show any currently identifiable abnormalities that denote their increased risk for malignant and life-threatening arrhythmias. Importantly, IVF can be inheritable and tear through families. The number of patients who survive their first manifestation of the disease is low, and the recurrence rate of IVF is appreciable. Treatment in patients who survived their event is performed by implantation of a cardioverter defibrillator (ICD) and a decrease of the risk of IVF recurrence may be achieved by prescription of quinidine. During a VF storm, administration of isoproterenol can be essential, and possibly also sedation might be effective. Research into the origin and genetic underpinning of IVF is limited by its malignant character, yet has revealed genetic variants in the *DPP6* gene and *CALM1* gene that impact on cellular cardiac electrophysiology. In contrast, early repolarization is a description of electrocardiographic variants with elevation, slurring, or notching in the terminal QRS complex or early ST segment. Although considered a benign electrocardiographic characteristic for many decades, in the past decade associations have been made with a propensity to sudden cardiac arrest by VF. Importantly, early repolarization is of very common occurrence in many young and healthy individuals. The description of a rare malignant association has spurred scientific interest in this phenomenon. Fortunately, its benign prognosis is still valid for the vast majority of individuals who display an early repolarization pattern. The challenge for the future will be to delineate benign

P. G. Postema (✉)
Department of Clinical and Experimental Cardiology, Heart Centre, Academic Medical Centre, University of Amsterdam, Amsterdam, The Netherlands
e-mail: p.g.postema@amc.nl

D. Thomas, C. A. Remme (eds.), *Channelopathies in Heart Disease*, Cardiac and Vascular Biology, https://doi.org/10.1007/978-3-319-77812-9_11

from malignant variants of early repolarization, the development of aids in risk stratification and a better understanding underlying pathophysiology on cellular and genetic level, to further guide clinicians and patients.

11.1 Idiopathic Ventricular Fibrillation

11.1.1 Introduction

Ventricular fibrillation (VF) is a chaotic electrical and mechanical activity of the cardiac ventricles and always results in severe hypotension leading to haemodynamic collapse. When not treated within minutes, the shortage of sufficient perfusion of the brain and other organs will result in permanent damage and ultimately death. Likewise, VF is still the most common cardiac rhythm documented at the time of sudden cardiac arrest and sudden cardiac death (Viskin and Belhassen 1990; Huikuri et al. 2001; Hulleman et al. 2015).

The cause of VF is not always known or uncovered afterwards, but in the majority of cases, atherosclerotic coronary artery disease is the underlying aetiology either due to acute myocardial ischemia or due to scar tissue that has developed after myocardial ischemia. This scenario is also considered to occur in the general population without previous symptoms as a sequence of previously silent coronary artery disease resulting at one time in an atherosclerotic plaque rupture, leading to obstruction in the coronary artery, myocardial ischemia, subsequent VF and cardiac arrest, and ultimately death when resuscitation is not directly started and/or not effective (Chugh et al. 2008). However, there are many more disease aetiologies that may proceed to VF, both cardiac and noncardiac in origin, such as intracranial haemorrhage, pulmonary embolism, myocarditis, cardiomyopathies, valvular heart disease, congenital cardiac anomalies and accessory pathways (Viskin and Belhassen 1990; Corrado et al. 2003; Puranik et al. 2005; Elliott et al. 2008; Maron et al. 2009; Postema et al. 2011a).

A specific group of causes of VF is inheritable cardiac disease, which may result in an excessively high VF occurrence in multiple family members. This is likely due to genetic mutations that share a propensity for the development of a substrate and/or trigger for VF. Such mutations may either disrupt the cardiac architecture, e.g. due to excessive hypertrophy, fibrosis or dilatation, or change the normal cardiac action potential morphology resulting in anomalous depolarization or repolarization characteristics (Postema et al. 2011a). Importantly, VF may not only occur at advanced age but also young adults, teenagers, children and even newborns may be at risk due to these underlying disease entities.

The scope of the problem of VF is large but not very easy to count because it takes only minutes before VF deteriorates in asystole and subsequent death, necessitating fast recognition and treatment. As a reference, in the United States, the estimations of cardiac arrest and subsequent sudden death range between 180,000 and 400,000 victims per year (Chugh et al. 2008; Zipes and Wellens 1998). In prospective studies, an incidence rate of 55 adult cardiac arrest cases per 100,000 person-years was documented (Berdowski et al. 2010). Because of the rising risk factors for cardiovascular disease such as obesity and diabetes, it is thought that these numbers will increase in the next decades (Chugh et al. 2008). There are several large efforts worldwide to investigate the occurrence, causes and modifiable factors of VF and cardiac arrest in the community. One such effort is the Amsterdam-based ARREST study. These investigators actually documented 15% *decline* in VF as the initial cardiac rhythm in cardiac arrest victims between 1995–1997 and 2006–2012 (Hulleman et al. 2015). This decline was explained by less initial VF (and thus more bradycardia, asystole or pulseless electrical activity) and an increase in unwitnessed cardiac arrest. The survival rates of out-of-hospital cardiac arrest victims are very low globally, around 7% (Berdowski et al. 2010). However, when the initial rhythm is VF, survival rates are better, around 17% (Berdowski et al. 2010). Likewise, the increased use of implantable cardioverter defibrillators (ICDs) to directly recognize and treat commencing VF was found to result in an estimated 33% decrease of cardiac arrest due to VF (Hulleman et al. 2012).

For the prevention of a first or recurrent episode of VF, it is very important to recognize the factors that precede the substrate and triggers that cause VF, as a means to treat these factors and prevent its occurrence. As mentioned, many disease entities may result in a higher propensity to VF. However, in some cases the cause of VF remains unknown, even after extensive evaluations. These cases can be labelled as idiopathic VF (IVF). It is thought that IVF accounts for approximately 5% of all VF cases (Joint Steering Committees of the Unexplained Cardiac Arrest Registry of Europe and of the Idiopathic Ventricular Fibrillation Registry of the United States 1997), although with the increasing knowledge of today, this percentage might currently be even lower.

Although IVF thus involves only a minority in the total burden of VF, it is a notoriously difficult disease entity to manage. This is because there are no signs to adequately identify the propensity to VF of these patients when they are still asymptomatic, and the first event is often lethal. Subsequently, modifiable factors are not recognizable nor treatable. In addition, IVF generally occurs in young patients, children and young adults, previously thought to be healthy. In addition, IVF may also tear through families because it might be inheritable, adding to its tragic character.

11.1.2 IVF Diagnosis

The diagnosis of IVF may well not be straight forward. There are two main obstacles that form the basis of the difficulties we have with an IVF diagnosis; (Viskin and Belhassen 1990) it is a very rare entity where only a minority survives the first manifestation of the disease and can proceed to a diagnostic evaluation, and (Huikuri et al. 2001) it is a *per exclusionem* diagnosis.

In those who are successfully resuscitated from a first VF episode, several evaluation schemes are available. These should include a detailed documentation of the event, medical history, family history, physical examination, electrocardiography (ECG), blood chemistry (including blood counts, electrolytes and cardiac enzymes) and other investigations aimed at diagnosing coronary artery disease (e.g. coronary angiography, also considering evaluation of coronary spasm) and/or structural heart disease (e.g. echocardiography, cardiac magnetic resonance imaging). The registration of the arrhythmia(s) can be very useful in this effort to delineate the course and origin of the event. When the diagnosis is not (fully) established after the previous investigation rounds, the further evaluations should be considered: exercise testing (in search of signs of ST-segment changes, abnormal conduction or repolarization characteristics or provocation of arrhythmias), Holter monitoring (also for abnormal conduction or repolarization or arrhythmias), toxicology screening, cardiac biopsies and/or drug provocation testing with sodium channel blocking drugs (e.g. ajmaline, flecainide or procainamide, in search of Brugada syndrome) or adrenaline (in search of long-QT syndrome or catecholaminergic polymorphic ventricular tachycardia) (Postema et al. 2009a, 2011a; Wolpert et al. 2005; Krahn et al. 2005). Clearly, depending on the particular case, there will be a differentiation in the necessity and order of these investigations. Coronary angiography in particular will be one of the first steps in the evaluation also because of the potential of therapeutic options during the procedure.

When no clues to the origin of the VF episode are found, an IVF diagnosis can be made. For the patient, this diagnosis may not have direct consequences as a secondary prevention ICD will generally be indicated (Priori et al. 2015; Zipes et al. 2006). However, choices for drug treatment to prevent IVF recurrences may well affected by an IVF diagnosis. For example, quinidine might be a life-saving treatment option in IVF (Dock 1929; Ten Sande et al. 2016) but can have devastating results in long-QT syndrome due to its QT prolonging effects (Jervell and Lange-Nielsen 1957; Selzer and Wray 1964). In addition, for the family members an IVF diagnosis in the index patient may have consequences. In about 5–20% of IVF victims, a family history of sudden unexpected deaths at young age is indeed present (Belhassen et al. 1999; Haissaguerre et al. 2002; Noda et al. 2005). When there is a possible or clear family history of inheritable IVF, the screening of the family members is particularly difficult because even the index patient does not have a phenotype that is indicative of the disease, apart from IVF. Primary prevention ICDs in the family members, based on the available data that underscores the age at which the risk is exceptionally high, might be the ultimate preventive treatment option in these families (Ten Sande et al. 2016; Alders et al. 2009; Postema et al. 2011b). However, implanting ICDs in these (often) young patients will surely result in ICD-related complications, which needs to be carefully weighed against the possible benefits (Olde Nordkamp et al. 2013, 2016).

Genetic analysis of the index patient might in some cases with a clear familial occurrence reveal the underlying genetic substrate (Alders et al. 2009; Marsman et al. 2014). This, however, often requires many years of dedicated research and particularly large families with multiple patients who survived the IVF event and who are able to undergo the necessary evaluations and comparisons in search of common genetic denominators.

When there is no VF documentation in a resuscitated or suddenly deceased patient, the sudden cardiac arrest or sudden death may well be due to another disease entity as opposed to IVF. Still, as previously mentioned, VF will only continue for minutes and will ultimately deteriorate into asystole. When asystole commences, the changes of survival decrease rapidly, and thus the possibilities to make an IVF diagnosis decrease simultaneously. When the patient does not survive the event, a detailed family history is still of paramount importance, as this may provide clues to an inheritable disease entity. When there is indeed a suspicion of inheritable disease, detailed evaluations of the family members will still be appropriate (Postema et al. 2011a; van der Werf et al. 2010).

11.1.3 IVF and Channelopathies

In the past centuries, sudden cardiac deaths in young and otherwise healthy individuals have challenged many families and doctors. Our ever-increasing knowledge on cardiac disorders and possibilities to delineate different phenotypes with an increased risk for malignant arrhythmias has impressively changed our understanding of sudden cardiac arrest. Importantly, what was considered idiopathic or without known cause in the past can at present be a clear and distinct disease entity which is recognizable and often also treatable. As such, many distinct disease entities have been described in the past 70 years (Table 11.1).

For nearly all of these disease entities, except for malignant early repolarization syndrome, a strong genetic background underlying these entities has been uncovered as well. The genetic mutations that are involved in these syndromes all impact on cardiac architecture (e.g. hypertrophic cardiomyopathy or arrhythmogenic (right) ventricular cardiomyopathy) or on cellular electrophysiology affecting either depolarization or repolarization characteristics (e.g. Brugada syndrome and long-QT syndrome). Interestingly, there are also overlap syndromes due to specific mutations which combine separate disease entities in individual patients (Bezzina et al. 1999; Postema et al. 2009b).

Although it is regarded that IVF is heritable in 5–20% of cases, there have until present only been two genetic substrates uncovered to explain the heritability of IVF (Alders et al. 2009; Marsman et al. 2014). In 2009, a haplotype involving the *DPP6* gene was found to be coinciding with IVF in several large and distantly related families involving over 600 family members (Alders et al. 2009). This gene is implicated in the transient outward current (I_{to}) in cardiac Purkinje tissue in particular and is believed to upregulate I_{to} (Alders et al. 2009; Xiao et al. 2013). This process may explain the propensity to short-coupled extrasystoles and VF in these patients although the exact pathophysiological mechanisms remain to be clarified (Fig. 11.1). In 2013/2014, a mutation in CALM1 was found to be associated with familial IVF (Marsman et al. 2014). This gene is implicated in calcium handling through calmodulin. Interestingly, mutations in this gene had earlier been implicated

Table 11.1 Milestone publications describing distinct disease entities with a propensity to inheritable malignant cardiac arrhythmias that were previously considered idiopathic

Year	Syndrome	Authors	Diagnosis by
1951	Long-QT syndrome	Jervell and Lange-Nielsen (1957)	ECG, exercise testing, Holter monitoring
1958	Hypertrophic cardiomyopathy	Teare (1958)	Post-mortem, echocardiography, ECG, biopsy
1978	Catecholaminergic polymorphic ventricular tachycardia	Coumel et al. (1978)	Exercise testing, Holter monitoring, adrenaline provocation
1982	Arrhythmogenic right ventricular cardiomyopathy/ dysplasia	Marcus et al. (1982)	Post-mortem, ECG, echocardiography, MRI, biopsy, Holter monitoring
1992	Brugada syndrome	Brugada and Brugada (1992)	ECG, sodium channel blocker provocation
2000	Short-QT syndrome	Gussak et al. (2000)	ECG
2008	VF associated with early repolarization	Haissaguerre et al. (2008)	ECG
2009	IVF associated with *DPP6*	Alders et al. (2009)	Genetic analysis
2013	IVF associated with *CALM1*	Marsman et al. (2014)	Genetic analysis

Table adapted from Postema et al. (2011a). IVF denotes idiopathic ventricular fibrillation. ECG denotes electrocardiogram

in cases of catecholaminergic polymorphic ventricular tachycardia (CPVT) and long-QT syndrome (LQTS) (Nyegaard et al. 2012; Crotti et al. 2013). Both *DPP6* and *CALM1* have not been replicated in other IVF families until date. Although this might partly be due to the rarity of the disease and the many years of intensive research that coined these discoveries, it also illustrates the probably diverse genetic background in the remainder of IVF families.

In this particular case, the patient went into VF storm without preceding symptoms, and after multiple ICD shocks on recurrent VF out of the hospital, atrial fibrillation developed and lasted until presentation in our clinic. On arrival at our cardiac care unit, VF initiated again from a short-coupled extrasystole from the RV free wall (putatively from Purkinje network) and was now captured on 12 L ECG. Further episodes of VF were prevented by administration of isoprenaline.

11.1.4 IVF Prognosis and Treatment

In contrast to the good long-term prognosis without VF recurrences in patients who survive their first episode of VF during acute myocardial ischemia (de Jong et al. 2009), the prognosis of patients with IVF is rather poor. It is reported that around 25–46% of the IVF patients have recurrent VF events, including recurrent VF storms

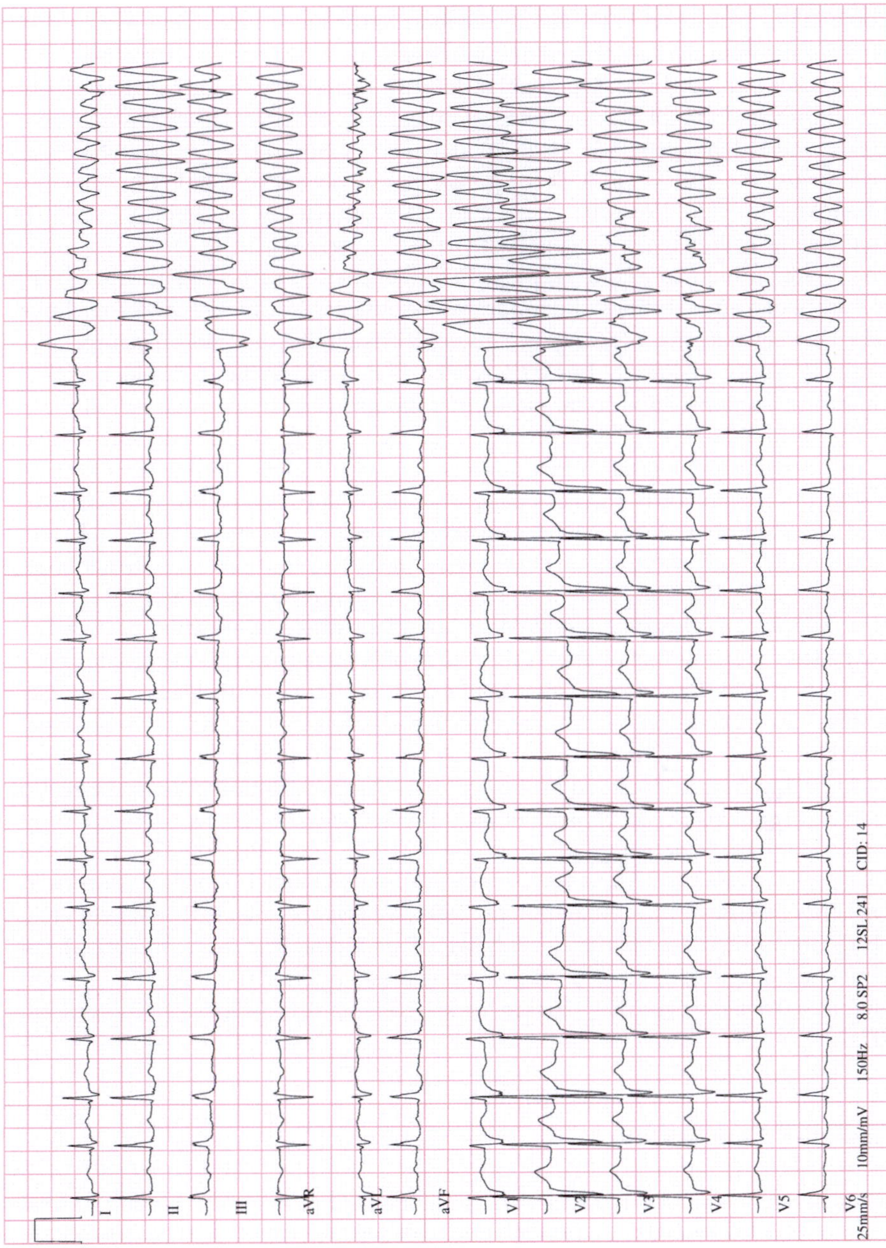

Fig. 11.1 Onset of ventricular fibrillation (VF) in a DPP6 patient

(Joint Steering Committees of the Unexplained Cardiac Arrest Registry of Europe and of the Idiopathic Ventricular Fibrillation Registry of the United States 1997; Ten Sande et al. 2016; Wever et al. 1993; Remme et al. 2001; Wever and Robles de Medina 2004). Although ICDs are designed to lower the risk of dying from VF, a risk of dying from cardiac arrhythmia even with an ICD remains (Brugada et al. 2009; Sacher et al. 2013a). Regrettably, this risk includes morbidity and mortality from ICD-related complications (Olde Nordkamp et al. 2013, 2016). In addition, patients with IVF are generally much younger and in much better general condition than patients who receive an ICD for primary or secondary prevention in ischemic cardiomyopathy. The exposure time of the IVF patients to possible ICD-related complications is thus much longer (Postema et al. 2011a).

As to antiarrhythmic drug therapy in IVF to decrease the recurrence rate of VF, there are therapeutic options. Importantly, the efficacy of the use of these drugs differs between patients, probably because of different, but currently unrecognized, underlying pathophysiology. In *DPP6*-related IVF, for example, beta blockers and amiodarone are considered to be ineffective. Instead, the drugs of choice in these patients are quinidine and isoproterenol (Ten Sande et al. 2016). This knowledge base is particularly important when there is need of treatment during VF storm. In VF storm isoproterenol is a quick-acting drug that can (only) be provided intravenously. In addition, sedation (either conscious sedation or deep sedation including intubation and respiratory support) might have antiarrhythmic effects. Whether high-frequency pacing also has beneficial effects is currently unknown in this specific patient population. However, addition of quinidine surely is effective in many of these patients either during or after VF storm to prevent or lower VF recurrence rates on the short and long term. Importantly and similarly to Brugada syndrome, low-dose quinidine might already be effective (Marquez et al. 2012). A potential problem of quinidine during VF storm is its oral administration route and hours of time to peak effect, necessitating treatment by other means during the acute phase (e.g. with isoproterenol). In general, in most IVF patients, quinidine will be the drug of choice for long-term therapy (Postema et al. 2011a; Joint Steering Committees of the Unexplained Cardiac Arrest Registry of Europe and of the Idiopathic Ventricular Fibrillation Registry of the United States 1997; Ten Sande et al. 2016; Wever et al. 1993). The difficulties on the availability of quinidine that have recently been encountered due to the threat of the manufacturers to terminate its production are therefore extremely alarming (Viskin et al. 2007, 2013; Wilde and Langendijk 2007).

11.1.5 IVF Conclusions

Idiopathic ventricular fibrillation (IVF) is a tragic and notoriously difficult to treat disease entity. Its hallmark is a presentation with sudden cardiac arrest due to VF, in individuals who were previously considered healthy and who do not show any other remarkable features upon rigorous evaluations. Its idiopathic nature and the low numbers of patients who survive its first manifestation trouble past and future research and development of effective treatment. In a considerable number of cases, IVF appears to be inheritable and threatens the lives of family members in a wide age range. In recent years, there has been progress in our understanding of IVF by the recognition of genetic variations that segregate with the phenotype. As such, presymptomatic family members can be identified and preventive treatment can be offered. Prevention of sudden cardiac death is now predominantly succeeded by implantation of implantable cardioverter defibrillators (ICDs). However, ICDs in this young and otherwise healthy population will coincide with ICD-related complications in the majority of patients during their lifetime. A better selection of patients who may benefit from preventive ICD therapy is therefore warranted. Drug therapy with quinidine in chronic and acute settings, and with isoproterenol in acute settings, can be life-saving options as well. Importantly, exclusion of non-IVF-related cardiac arrest is important as these drugs may aggravate the clinical condition of patients with other disease etiologies.

11.2 Early Repolarization

11.2.1 Introduction

Early repolarization is a morphological description of electrocardiographic variants in the terminal QRS complex or early ST segment. Although its definition is changing and debated throughout the past years and decades, it encompasses either ST or J-point elevation, and/or notching or slurring of the terminal QRS complex, and is particularly present in (relatively) healthy individuals without acute or chronic cardiac disease (e.g. acute myocardial infarction, chronic heart failure or bundle branch block, where ST-segment changes and abnormal terminal QRS morphology can be expected).

Since the early years of electrocardiography, such variants in the terminal QRS complex and ST segment have been noted. In 1936, in Cleveland, Ohio, Shipley and Hallaran described a prevalence of ST elevation and terminal QRS slurring or notching in 16–44% of cases in a series of young and apparently healthy adults (Shipley and Hallaran 1936). Such variations would today most probably be recognized as 'early repolarization'. In 1936, Shipley and Hallaran still used a string galvanometer (although already portable at that time), allowing analysis of three to four leads (lead I, II and III with or without one extra chest lead). However, in the years thereafter between 1947 and 1966, also with more modern electrocardiographs allowing analysis of 12 leads and also vectorcardiography, these phenomena were similarly noted. In particular, early repolarization patterns appeared to be present in young and athletic individuals and even more so in those from African descent (Myers and Klein 1947; Goldman 1953, 1960; Hiss et al. 1960; Seriki and Smith 1966).

In 1951, Grant, Estes and Doyle coined the term 'early repolarization' (Grant et al. 1951). In their paper on spatial vector electrocardiography combined with more traditional electrocardiography including (at least) 6 chest leads, they describe their experience in 3000 cases recorded in their hospital. Interestingly, they noted in the young and supposedly healthy individuals in their cohort that 'the repolarization forces are often unusually large in magnitude, producing large T waves and measurable ST-segment deviation in the limb as well as the precordial leads'. They continue their paper with some hypotheses on the cause of these repolarization forces and particularly on how a distinction can be made with pathology (e.g. in the presence of abnormal T-wave vectors). However, at the same time, they note that 'Occasionally, the S-T vector due to normal early repolarization forces is difficult to distinguish from the S-T vector due to acute pericarditis' (Grant et al. 1951). Importantly, longitudinal studies in these cases of normal ST elevation showed that these early repolarization changes can be present for years (Grant et al. 1951). Their notion that ST-segment deviations can be a normal variant with a benign prognosis lasted for over half a century thereafter, until a possible association was suggested with malignant ventricular arrhythmias (Haissaguerre et al. 2008; Rosso et al. 2008; Tikkanen et al. 2009).

11.2.2 Early Repolarization Definitions

From the 1950s, there have been different definitions of early repolarization, which partly complicates comparisons over the years. In 1961, Wasserburg, Alt and Lloyd considered early repolarization to consist of the following: (1) an elevated take-off of the ST segment and J-junction of the QRS complex, varying from 1 to 4 mm (0.1–0.4 mV), relative to succeeding T-P interval (isoelectric line), (2) a downward concavity of the ST segment and (3) symmetrically limbed T waves which are often of large amplitude (Wasserburger and Alt 1961). Interestingly, the authors also described the associated phenomenon of notching and slurring: 'the elevated S-T junction arose from a distinct notch on the downstroke of the R wave in most instances, occasionally it would be represented only as a well-defined slur. Thus, it superficially resembled a reversed Wolff-Parkinson-White pattern with notching at the distal segment of the QRS complex rather than at its inception' (Wasserburger and Alt 1961). The comparison with reversed Wolff-Parkinson-White patterns is especially memorable because current automatic algorithms to detect early repolarization are importantly coined on this feature (personal communication with Macfarlane 2016).

Between 2013 and 2016, there have been three international consensus reports on the definitions of early repolarization patterns, and unfortunately all three define different criteria. In 2013, the Heart Rhythm Society (HRS), European Heart Rhythm Association (EHRA) and the Asian Pacific Heart Rhythm Society (APHRS) consensus statement (Priori et al. 2013) first defines a distinction between an early repolarization pattern and early repolarization syndrome. This mirrors the previous distinction between a Brugada ECG pattern and Brugada syndrome. In early repolarization, this distinction makes even more sense as the pattern is extremely prevalent in young and healthy persons. The distinction between pattern and syndrome is made by the absence or presence of otherwise unexplained cardiac arrest or documented malignant ventricular arrhythmias. Moreover, the authors state the following electrocardiographic definition: the presence of J-point elevation ≥1 mm (0.1 mV) in ≥2 contiguous inferior and/or lateral leads of a standard 12-lead ECG. The authors do not touch upon the presence or absence, nor the value, of concomitant terminal QRS notching or slurring.

In contrast, in 2015 an international group of authors (Macfarlane et al. 2015) defined an early repolarization pattern to be *only* attributable to patients who display terminal QRS notching or slurring, where the notch should be entirely situated above the baseline, and the onset of a slur also be situated above the baseline. In addition, the peak of the notch or the onset of the slur should be ≥0.1 mV in ≥2 contiguous inferior and/or lateral leads and the QRS duration ≤120 ms. These authors specifically do not require J-point elevation (J-termination or 'Jt') as a criterion for early repolarization.

And lastly, in 2016 a consensus statement by the American Heart Association (AHA) includes for the definition of early repolarization, in contrast to the two earlier reports, either ECGs with or without terminal QRS notching or slurring and with or without ST elevation (Patton et al. 2016). In addition, the J-point elevation ≥0.1 mV in ≥2 contiguous inferior and/or lateral leads should in the presence of a notch or slur be measured at the peak of the notch or at the start of a slur, instead of the traditional definition of the J-point at the end of the notch or slur.

The differences between these criteria are depicted below in the figure.

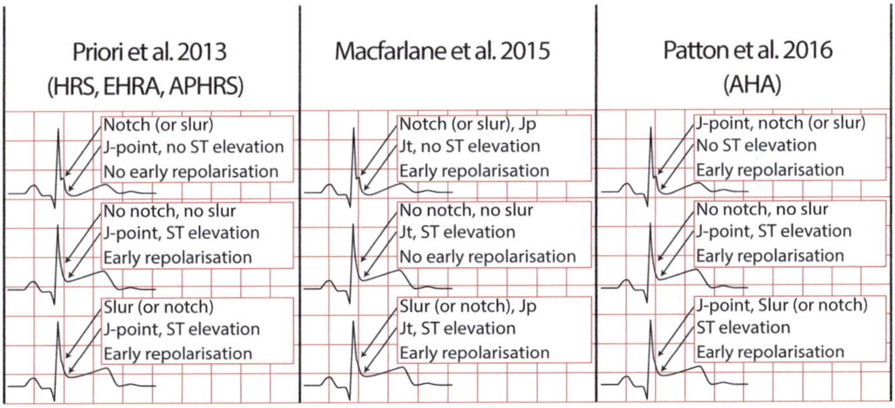

The differences between these three reports in the definition of electrocardiographic patterns of early repolarization and in clinical distinctions between those who only display the ECG features without any evidence of previous, or suspicion of future malignant arrhythmias, and those who did have malignant arrhythmias, clearly show a significant part of the difficulties we face with the past and future data on this subject. This not only impacts on the recognition of the phenomenon of early repolarization but will also impact on the therapeutic consequences that may be inflicted on these patients.

Only several years ahead of these reports, in 2009, early repolarization was labelled as ‘A statement that is used frequently to characterize a normal QRS-T variant with J-point elevation’ by the American Heart Association (AHA)/American College of Cardiology (ACC)/Heart Rhythm Association (HRS) (Rautaharju et al. 2009). However, in 2008, just before the AHA/ACC/HRS publication a possible association with malignant ventricular arrhythmias and sudden death was published in two reports (Haissaguerre et al. 2008; Rosso et al. 2008). Apparently, these reports could not be incorporated in the paper by the AHA/ACC/HRS committees.

11.2.3 Early Repolarization Associations with Malignant Arrhythmias

In 2008 the groups in Bordeaux, France, and Tel Aviv, Israel, published almost simultaneously on a possible association between early repolarization patterns and a history of sudden cardiac arrest (Haissaguerre et al. 2008; Rosso et al. 2008). In these papers, in cohorts of patients with a history of unexplained ventricular fibrillation, a subset of these patients appeared to have a far larger than expected prevalence of early repolarization patterns ($n = 64/142$ and $n = 19/45$, respectively). This association with sudden cardiac arrest was one year later verified in a larger general population cohort from Finland (Tikkanen et al. 2009). In addition, there has been further focus on the morphology of the subsequent ST segment on the association with ventricular arrhythmias (Viskin et al. 2013; Tikkanen et al. 2011). In these analyses, it appears that the association with a history of cardiac arrest of an early repolarization pattern with an accompanying horizontal or downsloping ST segment is particularly strong, while an upsloping ST segment appears to have a more benign evolution. However, about 30% of the patients with a history of cardiac arrest in the previous reports displays an upsloping ‘benign’ early repolarization pattern (Adler et al. 2013; Sacher et al. 2013b).

These data were truly pivotal for igniting medical and scientific interest in early repolarization patterns in patients and individuals from the general population. Particularly difficult in the association with malignant ventricular arrhythmias is the knowledge base that this electrocardiographic characteristic is very prevalent (as mentioned before, with a prevalence up to 44%) in healthy and young individuals.

11.2.4 Early Repolarization and Genetic Mutations

Although the prevalence of early repolarization patterns is large in the general population, its association with genetic mutations is only anecdotal and frustrated by a lack of segregation data for those with documentation of malignant arrhythmias. One year after their 2008 paper on the association between an early repolarization pattern and a history of cardiac arrest, the Bordeaux group published a paper on a genetic analysis of one of the cases (Haissaguerre et al. 2009a). In this study a severely affected young female patient who had suffered from over a hundred VF events was genetically evaluated with a panel of 21 candidate genes (*KCNQ1*, *KCNE1*, *KCNH2*, *KCNE2*, *KCNJ2*, *KCNJ8*, *KCNJ11*, *ABCC9*, *KCNJ5*, *KCNJ3*, *KCND3*, *IRX3*, *IRX5*, *SCN5A*, *SCN1B*, *NCX1*, *CACNA1C*, *CACNB2*, *CALR*, *CASQ2* and *ANK2*). This study revealed a missense variant in *KCNJ8*, p.S422 L. The *KCNJ8* gene encodes for the Kir6.1 subunit of the ATP-sensitive potassium (K-ATP) channel in heart and could thus be involved in aberrant action potential morphology and possibly also to an increased tendency towards the development of an early repolarization pattern and arrhythmias. This same mutation was not found in a, rather limited, control set of alleles ($n = 764$) and was also not uncovered in 156 other patients with a history of unexplained VF and early repolarization pattern. In addition, segregation data was not available as the mother of this patient did not have the mutation and the father denied testing.

In the following year, 2010, the *KCNJ8*-S422 L variant was replicated in a cohort of patients with Brugada syndrome or early repolarization syndrome (Medeiros-Domingo et al. 2010). In this cohort, the variant was uncovered in 1 out of 101 unrelated patients with Brugada syndrome and in 1 out of 14 patients with early repolarization syndrome. Moreover, the same variant was not documented in 1200 control alleles. Additionally, a gain-of-function effect was suggested by additional analyses using patch-clamping in heterologous co-expression with SUR2A in COS-1 cells.

Interestingly, the possible association of this variant with early repolarization and cardiac arrest was downplayed 4 years later. In 2014, it appeared that 4% of individuals in a cohort of Ashkenazi Jews carried the exact same *KCNJ8*-S422 L variant (Veeramah et al. 2014). Carriers of the variant did not display early repolarization patterns, and a possible association with possible malignant arrhythmias could also not be made.

Again in 2010, a possible association with cardiac calcium channel mutations and early repolarization syndrome was made (Burashnikov et al. 2010). In this study involving patients with Brugada syndrome, early repolarization syndrome and idiopathic ventricular fibrillation, variants were uncovered in the *CACNA1C*, *CACNB2* and *CACNA2D1* genes involved in cardiac calcium channels, in 4 out of 24 patients with early repolarization syndrome. These variants were not present in 400 tested reference alleles. However, segregation data was (again) not available.

In 2014, yet another gene, *ABCC9*, was implicated in early repolarization syndrome (Hu et al. 2014). In 5 out of 150 patients with either Brugada syndrome or early repolarization syndrome (number of patients with either syndrome was not reported), an *ABCC9* variant was uncovered. Three of these patients also carried a putatively pathogenic variant in other susceptibility genes (*SCN5A*, *CACNA1C* and *SCN10A*). The *ABCC9* gene encodes the sulfonylurea receptor 2 (SUR2) which interacts with Kir6.1 to form functional K-ATP channels. Functional analyses implicated that these *ABCC9* variants also affected the cardiac K-ATP channel with a gain-of-function effect.

The sparsity of genetic associations in patients with early repolarization syndrome led the European Society of Cardiology committee on ventricular arrhythmias to the conclusion that no clear familial evidence of familial transmission of early repolarization syndrome exists (Priori et al. 2015). However, there are two reports on a remarkable high prevalence of early repolarization patterns in families (Nunn et al. 2011; Gourraud et al. 2013). In these reports, families of early repolarization syndrome patients displayed a remarkable higher prevalence (23–61%) of early repolarization patterns as compared to their remaining relatives or unrelated controls. Despite these observations, more detailed underpinning of early repolarization patterns nor early repolarization syndrome has not yet surfaced.

11.2.5 Early Repolarization Syndrome Prognosis and Treatment

Like most patients who survived a cardiac arrest, patients with an early repolarization syndrome diagnosis after cardiac arrest have a high recurrence rate of malignant arrhythmias. A secondary prevention implantable cardioverter-defibrillator (ICD) is thus recommended in those with a reasonable life expectancy with a class I indication (Priori et al. 2013). In addition, in those patients with a suggested 'malignant' variant of an early repolarization pattern (downsloping) and symptoms (highly) suggestive of malignant ventricular arrhythmias, an ICD as primary prevention can be considered (Priori et al. 2013).

In contrast, in asymptomatic patients with suggested malignant variants of early repolarization pattern, and certainly not in patients with 'benign' early repolarization patterns (upsloping), preventive invasive or non-invasive therapies are not indicated (class III recommendation) (Priori et al. 2013). Importantly, other risk stratification modalities, such as electrophysiological studies with programmed stimulation to investigate VT/VF inducibility, do not seem to be of value (Mahida et al. 2015).

As for medical treatment, during VT/VF storm isoproterenol has been used effectively to suppress arrhythmias (Haissaguerre et al. 2009b), while chronic VT/VF suppression has been reached with quinidine therapy (Haissaguerre et al. 2009b). Importantly, many other tested drugs in the setting of this specific population with early repolarization syndrome and arrhythmias, did not seem to be of specific value (mexiletine, verapamil, flecainide, propafenone, pilsicainide and also amiodarone). However, amiodarone did seem to be effective in 1 out of 6 patients during VF storm.

11.2.6 Early Repolarization Conclusions

Early repolarization patterns, whether described as ST elevation and/or terminal QRS notching or slurring, have been described since the early years of electrocardiography. This phenomenon is very prevalent in young and otherwise healthy individuals, and even more so in athletes and individuals from African descent, and has until recently been evaluated as a benign condition. However, in the last decade an association with an increased propensity for malignant ventricular arrhythmias, sudden cardiac arrest and sudden death has been documented. With this sharp contrast between a prevalent and benign electrocardiographic phenomenon and an association with malignant ventricular arrhythmias and sudden death, a highly interesting but likewise difficult situation has emerged in the interpretation of early repolarization patterns. To complicate the matter further, international consensus has not even been reached on the definitions of early repolarization with three different definitions in three different international expert panels/guidelines. Still, consensus is rather clear on the treatment of individuals with early repolarization patterns; only those with proven or highly suspected malignant ventricular arrhythmias are currently candidates for invasive treatment (implantable cardioverter defibrillators) and drug treatment (isoproterenol during VT/VF storm and quinidine long-term therapy). Genetic associations have been anecdotal despite numerous efforts and are not supported by proof of genetic segregation with the phenotype. Future efforts to further delineate benign from malignant variants of early repolarization, aids in risk stratification and a better understanding underlying pathophysiology on cellular and genetic level, are imperative to further guide clinicians and patients in this matter.

Compliance with Ethical Standards

Sources of Funding None.

Conflict of Interest None.

Ethical Statement All procedures performed in studies involving human participants were in accordance with the ethical standards of the institutional and/or national research committee and with the 1964 Helsinki declaration and its later amendments or comparable ethical standards.

References

Adler A, Rosso R, Viskin D, Halkin A, Viskin S. What do we know about the "malignant form" of early repolarization? J Am Coll Cardiol. 2013;62:863–8.

Alders M, Koopmann TT, Christiaans I, et al. Haplotype-sharing analysis implicates chromosome 7q36 harboring DPP6 in familial idiopathic ventricular fibrillation. Am J Hum Genet. 2009;84:468–76.

Belhassen B, Viskin S, Fish R, Glick A, Setbon I, Eldar M. Effects of electrophysiologic-guided therapy with class IA antiarrhythmic drugs on the long-term outcome of patients with idiopathic ventricular fibrillation with or without the Brugada syndrome. J Cardiovasc. 1999;10:1301–12.

Berdowski J, Berg RA, Tijssen JGP, Koster RW. Global incidences of out-of-hospital cardiac arrest and survival rates: systematic review of 67 prospective studies. Resuscitation. 2010;81:1479–87.

Bezzina C, Veldkamp MW, van den Berg MP, et al. A single Na(+) channel mutation causing both long-QT and Brugada syndromes. Circ Res. 1999;85:1206–13.

Brugada P, Brugada J. Right bundle branch block, persistent ST segment elevation and sudden cardiac death: a distinct clinical and electrocardiographic syndrome. A multicenter report. J Am Coll Cardiol. 1992;20:1391–6.

Brugada P, Brugada J, Brugada R. When our best is not enough: the death of a teenager with Brugada syndrome. J Cardiovasc. 2009;20:108–9.

Burashnikov E, Pfeiffer R, Barajas-Martinez H, et al. Mutations in the cardiac L-type calcium channel associated with inherited J wave syndromes and sudden cardiac death. Heart Rhythm. 2010;7:1872–82.

Chugh SS, Reinier K, Teodorescu C, et al. Epidemiology of sudden cardiac death: clinical and research implications. Prog Cardiovasc Dis. 2008;51:213–28.

Corrado D, Basso C, Rizzoli G, Schiavon M, Thiene G. Does sports activity enhance the risk of sudden death in adolescents and young adults? J Am Coll Cardiol. 2003;42:1959–63.

Coumel P, Fidelle J, Lucet V, Attuel P, Bouvrain Y. Catecholamine-induced severe ventricular arrhythmias with Adams-stokes syndrome in children: report of four cases. Br Heart J. 1978;40:28–37.

Crotti L, Johnson CN, Graf E, et al. Calmodulin mutations associated with recurrent cardiac arrest in infants. Circulation. 2013;127:1009–17.

de Jong JS, Marsman RF, Henriques JP, et al. Prognosis among survivors of primary ventricular fibrillation in the percutaneous coronary intervention era. Am Heart J. 2009;158:467–72.

Dock W. Transitory ventricular fibrillation as a cause of syncope and its prevention by quinidine sulphate with case report and discussion of diagnostic criteria for ventricular fibrillation. Am Heart J. 1929;4:709–14.

Elliott P, Andersson B, Arbustini E, et al. Classification of the cardiomyopathies: a position statement from the European Society of Cardiology Working Group on myocardial and pericardial diseases. Eur Heart J. 2008;29:270–6.

Goldman MJ. RS-T segment elevation in mid- and left precordial leads as a normal variant. Am Heart J. 1953;46:817–20.

Goldman MJ. Normal variants in the electrocardiogram leading to cardiac invalidism. Am Heart J. 1960;59:71–7.

Gourraud J-B, Le Scouarnec S, Sacher F, et al. Identification of large families in early repolarization syndrome. J Am Coll Cardiol. 2013;61:164–72.

Grant RP, Estes EH, Doyle JT. Spatial vector electrocardiography; the clinical characteristics of S-T and T vectors. Circulation. 1951;3:182–97.

Gussak I, Brugada P, Brugada J, et al. Idiopathic short QT interval: a new clinical syndrome? Cardiology. 2000;94:99–102.

Haissaguerre M, Shoda M, Jais P, et al. Mapping and ablation of idiopathic ventricular fibrillation. Circulation. 2002;106:962–7.

Haissaguerre M, Derval N, Sacher F, et al. Sudden cardiac arrest associated with early repolarization. N Engl J Med. 2008;358:2016–23.

Haissaguerre M, Chatel S, Sacher F, et al. Ventricular fibrillation with prominent early repolarization associated with a rare variant of KCNJ8/K channel. J Cardiovasc. 2009a;20:93–8.

Haissaguerre M, Sacher F, Nogami A, et al. Characteristics of recurrent ventricular fibrillation associated with inferolateral early repolarization role of drug therapy. J Am Coll Cardiol. 2009b;53:612–9.

Hiss RG, Lamb LE, Allen MF. Electrocardiographic findings in 67,375 asymptomatic subjects. X. Normal values. Am J Cardiol. 1960;6:200–31.

Hu D, Barajas-Martínez H, Terzic A, et al. ABCC9 is a novel Brugada and early repolarization syndrome susceptibility gene. Int J Cardiol. 2014;171:431–42.

Huikuri HV, Castellanos A, Myerburg RJ. Sudden death due to cardiac arrhythmias. N Engl J Med. 2001;345:1473–82.

Hulleman M, Berdowski J, de Groot JR, et al. Implantable cardioverter-defibrillators have reduced the incidence of resuscitation for out-of-hospital cardiac arrest caused by lethal arrhythmias. Circulation. 2012;126:815–21.

Hulleman M, Zijlstra JA, Beesems SG, et al. Causes for the declining proportion of ventricular fibrillation in out-of-hospital cardiac arrest. Resuscitation. 2015;96:23–9.

Jervell A, Lange-Nielsen F. Congenital deaf-mutism, functional heart disease with prolongation of the Q-T interval and sudden death. Am Heart J. 1957;54:59–68.

Joint Steering Committees of the Unexplained Cardiac Arrest Registry of Europe and of the Idiopathic Ventricular Fibrillation Registry of the United States. Survivors of out-of-hospital cardiac arrest with apparently normal heart. Need for definition and standardized clinical evaluation. Consensus statement of the joint steering committees of the unexplained cardiac arrest registry of Europe and of the idiopathic ventricular fibrillation registry of the United States. Circulation. 1997;95:265–72.

Krahn AD, Gollob M, Yee R, et al. Diagnosis of unexplained cardiac arrest: role of adrenaline and procainamide infusion. Circulation. 2005;112:2228–34.

Macfarlane PW, Antzelevitch C, Haissaguerre M, et al. The early repolarization pattern: a consensus paper. J Am Coll Cardiol. 2015;66:470–7.

Mahida S, Derval N, Sacher F, et al. Role of electrophysiological studies in predicting risk of ventricular arrhythmia in early repolarization syndrome. J Am Coll Cardiol. 2015;65:151–9.

Marcus FI, Fontaine GH, Guiraudon G, et al. Right ventricular dysplasia: a report of 24 adult cases. Circulation. 1982;65:384–98.

Maron BJ, Doerer JJ, Haas TS, Tierney DM, Mueller FO. Sudden deaths in young competitive athletes: analysis of 1866 deaths in the United States, 1980-2006. Circulation. 2009;119:1085–92.

Marquez MF, Bonny A, Hernandez-Castillo E, et al. Long-term efficacy of low doses of quinidine on malignant arrhythmias in Brugada syndrome with an implantable cardioverter-defibrillator: a case series and literature review. Heart Rhythm. 2012;9:1995–2000.

Marsman RF, Barc J, Beekman L, et al. A mutation in CALM1 encoding calmodulin in familial idiopathic ventricular fibrillation in childhood and adolescence. J Am Coll Cardiol. 2014;63:259–66.

Medeiros-Domingo A, Tan B-H, Crotti L, et al. Gain-of-function mutation S422L in the KCNJ8-encoded cardiac K(ATP) channel Kir6.1 as a pathogenic substrate for J-wave syndromes. J Heart Rhythm Soc. 2010;7:1466–71.

Myers GB, Klein HA. Normal variations in multiple precordial leads. Am Heart J. 1947;34:785–808.

Noda T, Shimizu W, Taguchi A, et al. Malignant entity of idiopathic ventricular fibrillation and polymorphic ventricular tachycardia initiated by premature extrasystoles originating from the right ventricular outflow tract. J Am Coll Cardiol. 2005;46:1288–94.

Nunn LM, Bhar-Amato J, Lowe MD, et al. Prevalence of J-point elevation in sudden arrhythmic death syndrome families. J Am Coll Cardiol. 2011;58:286–90.

Nyegaard M, Overgaard MT, Søndergaard MT, et al. Mutations in calmodulin cause ventricular tachycardia and sudden cardiac death. Am J Hum Genet. 2012;91:703–12.

Olde Nordkamp LRA, Wilde AAM, Tijssen JGP, Knops RE, van Dessel PFHM, de Groot JR. The ICD for primary prevention in patients with inherited cardiac diseases: indications, utilization and outcome. A comparison with secondary prevention. Circ Arrhythm Electrophysiol. 2013;6:91–100.

Olde Nordkamp LRA, Postema PG, Knops RE, et al. Implantable cardioverter-defibrillator harm in young patients with inherited arrhythmia syndromes: a systematic review and meta-analysis of inappropriate shocks and complications. Heart Rhythm. 2016;13:443–54.

Patton KK, Ellinor PT, Ezekowitz M, et al. Electrocardiographic early repolarization: a scientific statement from the American Heart Association. Circulation. 2016;133:1520–9.

Postema PG, Wolpert C, Amin AS, et al. Drugs and Brugada syndrome patients: review of the literature, recommendations, and an up-to-date website (www.brugadadrugs.org). Heart Rhythm. 2009a;6:1335–41.

Postema PG, van den Berg MP, van Tintelen JP, et al. Founder mutations in the Netherlands. SCN5a 1795insD, the first described arrhythmia overlap syndrome and one of the largest and best described characterised families worldwide. Neth Heart J. 2009b;17:422–8.

Postema PG, van der Werf C, Wilde AA. Idiopathic ventricular fibrillation. In: Baars H, Doevendans P, van der Smagt J, editors. Clinical cardiogenetics. London: Springer; 2011a. p. 229–38.

Postema PG, Christiaans I, Hofman N, et al. Founder mutations in the Netherlands. Familial idiopathic ventricular fibrillation and DPP6. Neth Heart J. 2011b;19:290–6.
Priori SG, Wilde AA, Horie M, et al. HRS/EHRA/APHRS expert consensus statement on the diagnosis and management of patients with inherited primary arrhythmia syndromes expert consensus statement on inherited primary arrhythmia syndromes. Heart Rhythm. 2013;10:1932–63.
Priori SG, Blomström-Lundqvist C, Mazzanti A, et al. 2015 ESC guidelines for the management of patients with ventricular arrhythmias and the prevention of sudden cardiac death: the task force for the management of patients with ventricular arrhythmias and the prevention of sudden cardiac death of the European Society of Cardiology (ESC) endorsed by: Association for European Paediatric and Congenital Cardiology (AEPC). Europace. 2015;17:1601–87.
Puranik R, Chow CK, Duflou JA, Kilborn MJ, McGuire MA. Sudden death in the young. Heart Rhythm. 2005;2:1277–82.
Rautaharju PM, Surawicz B, Gettes LS, et al. AHA/ACCF/HRS recommendations for the standardization and interpretation of the electrocardiogram: part IV: the ST segment, T and U waves, and the QT interval: a scientific statement from the American Heart Association electrocardiography and arrhythmias committee, council on clinical cardiology; the American College of Cardiology Foundation; and the Heart Rhythm Society. Endorsed by the International Society for Computerized Electrocardiology. J Am Coll Cardiol. 2009;53:982–91.
Remme CA, Wever EF, Wilde AA, Derksen R, Hauer RN. Diagnosis and long-term follow-up of the Brugada syndrome in patients with idiopathic ventricular fibrillation. Eur Heart J. 2001;22:400–9.
Rosso R, Kogan E, Belhassen B, et al. J-point elevation in survivors of primary ventricular fibrillation and matched control subjects: incidence and clinical significance. J Am Coll Cardiol. 2008;52:1231–8.
Sacher F, Probst V, Maury P, et al. Outcome after implantation of cardioverter-defibrillator in patients with Brugada syndrome: a multicenter study—part 2. Circulation. 2013a;128:1739–47.
Sacher F, Lim HS, Haissaguerre M. Sudden cardiac death associated with J wave elevation in the inferolateral leads: insights from a multicenter registry. J Electrocardiol. 2013b;46:456–60.
Selzer A, Wray HW. Quinidine syncope. Paroxysmal ventricular fibrillation occurring during treatment of chronic atrial arrhythmias. Circulation. 1964;30:17–26.
Seriki O, Smith AJ. The electrocardiogram of young Nigerians. Am Heart J. 1966;72:153–7.
Shipley RA, Hallaran WR. The four-lead electrocardiogram in two hundred normal men and women. Am Heart J. 1936;11:325–245.
Teare D. Asymmetrical hypertrophy of the heart in young adults. Br Heart J. 1958;20:1–8.
Ten Sande JN, Postema PG, Boekholdt SM, et al. Detailed characterization of familial idiopathic ventricular fibrillation linked to the DPP6 locus. Heart Rhythm. 2016;13:905–12.
Tikkanen JT, Anttonen O, Junttila MJ, et al. Long-term outcome associated with early repolarization on electrocardiography. N Engl J Med. 2009;361:2529–37.
Tikkanen JT, Junttila MJ, Anttonen O, et al. Early repolarization: electrocardiographic phenotypes associated with favorable long-term outcome. Circulation. 2011;123:2666–73.
van der Werf C, van Langen IM, Wilde AA. Sudden death in the young: what do we know about it and how to prevent? Circ Arrhythm Electrophysiol. 2010;3:96–104.
Veeramah KR, Karafet TM, Wolf D, Samson RA, Hammer MF. The KCNJ8-S422L variant previously associated with J-wave syndromes is found at an increased frequency in Ashkenazi Jews. Eur J Hum Genet. 2014;22:94–8.
Viskin S, Belhassen B. Idiopathic ventricular fibrillation. Am Heart J. 1990;120:661–71.
Viskin S, Antzelevitch C, Marquez MF, Belhassen B. Quinidine: a valuable medication joins the list of "endangered species". Europace. 2007;9:1105–6.
Viskin S, M Wilde AA, Guevara-Valdivia ME, et al. Quinidine, a life-saving medication for Brugada syndrome, is inaccessible in many countries. J Am Coll Cardiol. 2013;61(23):2383–7.
Wasserburger RH, Alt WJ. The normal RS-T segment elevation variant. Am J Cardiol. 1961;8:184–92.

Wever EF, Robles de Medina EO. Sudden death in patients without structural heart disease. J Am Coll Cardiol. 2004;43:1137–44.

Wever EF, Hauer RN, Oomen A, Peters RH, Bakker PF, Robles de Medina EO. Unfavorable outcome in patients with primary electrical disease who survived an episode of ventricular fibrillation. Circulation. 1993;88:1021–9.

Wilde AA, Langendijk PN. Antiarrhythmic drugs, patients, and the pharmaceutical industry: value for patients, physicians, pharmacists or shareholders? Neth Heart J. 2007;15:127–8.

Wolpert C, Echternach C, Veltmann C, et al. Intravenous drug challenge using flecainide and ajmaline in patients with Brugada syndrome. Heart Rhythm. 2005;2:254–60.

Xiao L, Koopmann TT, Ordög B, et al. Unique cardiac purkinje fiber transient outward current β-subunit composition: a potential molecular link to idiopathic ventricular fibrillation. Circ Res. 2013;112:1310–22.

Zipes DP, Wellens HJ. Sudden cardiac death. Circulation. 1998;98:2334–51.

Zipes DP, Camm AJ, Borggrefe M, et al. ACC/AHA/ESC 2006 guidelines for management of patients with ventricular arrhythmias and the prevention of sudden cardiac death: a report of the American College of Cardiology/American Heart Association task force and the European Society of Cardiology Committee for practice guidelines (writing committee to develop guidelines for Management of Patients with Ventricular Arrhythmias and the prevention of sudden cardiac death): developed in collaboration with the European heart rhythm association and the Heart Rhythm Society. Circulation. 2006;114:e385–484.

Atrial Fibrillation 12

Ann-Kathrin Rahm, Hugo A. Katus, and Dierk Thomas

Abstract

Atrial fibrillation (AF) is the most common arrhythmia in clinical routine. AF is related to significant morbidity and mortality caused by thromboembolism, tachycardia-induced cardiomyopathy, and heart failure. The pathophysiological mechanisms leading to AF initiation and progression are complex.

Pedigrees of AF families enabled the identification of genetic factors predisposing to AF. In genotyped families, AF patients carry rare genetic variants in genes associated with ion channels, calcium handling protein, or genes related to fibrosis, conduction system disease, and inflammatory processes. Furthermore, common genetic variants have been linked directly to AF. However, in most cases the molecular mechanisms by which single nucleotide polymorphisms (SNPs) enhance AF susceptibility remain to be identified.

Optimized understanding of the molecular basis and genetics of AF will help to optimize risk stratification and therapeutic management of AF patients and lead to new therapeutic approaches in treating this epidemic disease.

A.-K. Rahm · H. A. Katus · D. Thomas (✉)
Department of Cardiology and HCR (Heidelberg Center for Heart Rhythm Disorders), University Hospital Heidelberg, Heidelberg, Germany

DZHK (German Centre for Cardiovascular Research), partner site Heidelberg/Mannheim, Heidelberg, Germany
e-mail: ann-kathrin.rahm@med.uni-heidelberg.de; sekretariat.katus@med.uni-heidelberg.de; dierk.thomas@med.uni-heidelberg.de

D. Thomas, C. A. Remme (eds.), *Channelopathies in Heart Disease*, Cardiac and Vascular Biology, https://doi.org/10.1007/978-3-319-77812-9_12

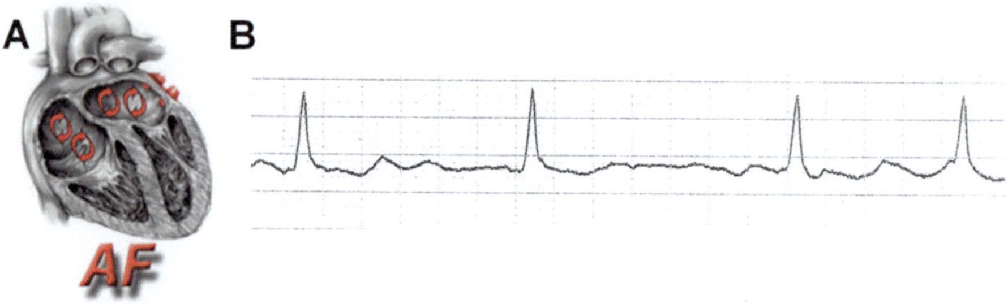

Fig. 12.1 Atrial fibrillation. Fibrillating electrical activity in the atria [**a**; reproduced with permission from Tucker and Ellinor (2014)] leads to the pathognomonic surface EGC (**b**) of irregular undulations around the isoelectric line and irregular conduction into the ventricles

12.1 Epidemiology and Clinical Significance of Atrial Fibrillation

12.1.1 Characteristics and Classification of Atrial Fibrillation

Atrial fibrillation (AF) is the most common arrhythmia in clinical routine and accounts for approximately one-third of hospitalizations for cardiac rhythm disorders (Go et al. 2001). AF is characterized by rapid and chaotic electrical activation of the atria, resulting in ineffective contraction and irregular ventricular heartbeat. The diagnosis of AF requires rhythm documentation using a surface ECG showing the typical pattern of AF, irregular RR intervals without distinct P waves (Fig. 12.1b).

AF may be classified by its temporal pattern, presentation, duration, and spontaneous termination into first diagnosed, paroxysmal, persistent, long-standing persistent, and permanent AF (Kirchhof et al. 2016). Approximately 35–40% of all patients exhibit paroxysmal AF, of which 30–50% convert into a chronic state depending on the underlying disease and risk factors (Vlachos et al. 2016).

12.1.2 Epidemiology and Health-Care Burden

The overall prevalence of AF in adults in the general population equals approximately around 1–3%, with higher rates in patients with cardiovascular comorbidities (Bjorck et al. 2013; Haim et al. 2015). Estimated counts of men and women with diagnosed AF in 2010 worldwide were 20.9 and 12.6 million, respectively (Chugh et al. 2014). Until the year 2030, 14–17 million patients are predicted to exhibit AF in the European Union (Fig. 12.2) with around 100–200 thousand new diagnoses every year (Collila et al. 2013; Krijthe et al. 2013; Zoni-Berisso et al. 2014). Improved detection methods of silent AF and the demographic development of AF-predisposing diseases contribute to increasing prevalence (Wang et al. 2003;

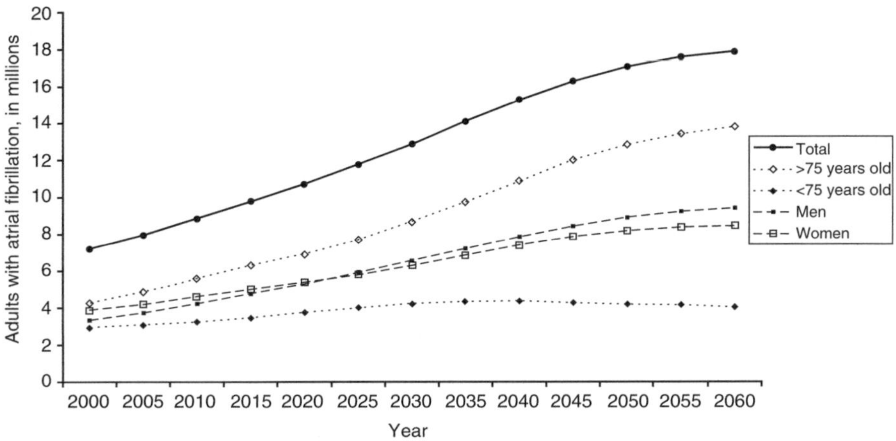

Fig. 12.2 Projected number of adults with AF in the European Union between 2000 and 2060 [reproduced with permission from Krijthe et al. (2013)]

Kishore et al. 2014; Sanna et al. 2014; Schnabel et al. 2015). Due to an aging population and rising AF prevalence, AF accounts for a considerable financial burden to health-care systems (Go et al. 2001), with more than 1% of health-care expenses being spent on AF-related hospitalization (Ball et al. 2013).

12.1.3 Risk Factors

One primary risk factor of AF is age: AF is present in 0.12–0.16% of patients younger than 49 years, in 3.7–4.2% aged 60–70 years, and in 10–17% of those aged 80 years or older (Zoni-Berisso et al. 2014). Second, patients' sex constitutes a risk factor. AF is more often encountered in males than females with a ratio of 1.2:1 (Zoni-Berisso et al. 2014). However, further studies are needed to clarify the unknown molecular basis for sex-specific differences. Regarding ethnicity notable differences have been observed, as patients with a genetically European background have a higher risk to develop AF compared to African-Americans despite a lower prevalence of risk factors. This difference may, at least in part, be attributed to genetics (Rodriguez et al. 2015; Roberts et al. 2016). Other clinical risk factors include hypertension, obesity, diabetes mellitus, and structural heart disease (Kannel and Benjamin 2008; McManus et al. 2012; Ball et al. 2013). Hyperthyroidism, alcohol consumption, and obstructive sleep disorders have been identified as additional risk factors. Obstructive sleep disorders interfere with the vegetative system and increase AF prevalence by producing repetitive arousal phases (Menezes et al. 2015).

12.1.4 Clinical Significance

Although AF itself is not a life-threatening arrhythmia, it may lead to significant morbidity and mortality. All-cause mortality is increased by twofold in women and by 1.5-fold in men (Benjamin et al. 1998; Stewart et al. 2002; Andersson et al. 2013). Thromboembolism is a potential cause of stroke, and tachycardia may induce cardiomyopathy and heart failure, both leading to significant morbidity and reduced quality of life (Wolf et al. 1991; Krahn et al. 1995; Stewart et al. 2002). Recent studies revealed that 20–30% of ischemic strokes are associated with AF (Henriksson et al. 2012; Grond et al. 2013). AF symptoms vary widely between patients. Their severity is categorized by the European Heart Rhythm Association (EHRA) classification (Wynn et al. 2014). Symptoms include palpitations, dizziness, shortness of breath, chest pain, and psychological distress. As current treatment options for AF remain suboptimal, the elucidation of genetic mechanisms may serve to enable individualized treatment of AF patients based on their underlying disease mechanism.

12.1.5 Family History

In some patients AF occurs without risk factors or structural heart disease and shows a strong heritable component (Fox et al. 2004; Arnar et al. 2006; Oyen et al. 2012). This subtype of AF is termed "lone AF" and accounts for 2–16% of all AF cases (Potpara and Lip 2015). Analyses among affected patients and their families indicated that genetic factors predispose to or even cause AF. A significant heritable component could be identified by analyzing pedigrees of AF patients. Family history of AF was shown to increase the risk for AF in the index person up to 40% (Lubitz et al. 2010), with a higher risk at younger age and when multiple relatives or first-degree relatives are affected (Ellinor et al. 2005; Oyen et al. 2012). Moreover it was shown that co-twins of AF patients belong to a high-risk group for developing AF (Christophersen et al. 2009). These lines of evidence indicate that genetic factors play a substantial role in the development of AF.

12.2 Ion Channel Remodeling in AF Pathophysiology

AF pathogenesis is complex. Mechanisms of AF pathophysiology will be presented briefly here to provide an overview over the molecular basis of AF with particular focus on ion channel remodeling. Further details have been summarized elsewhere (e.g., Schotten et al. 2011; Wakili et al. 2011).

In AF, pathophysiology factors leading to AF initiation may be differentiated from processes that promote perpetuation of the arrhythmia toward more chronic stages. Upon chronification, AF is associated with changes in atrial morphology, histology, and ion channel regulation and composition, leading to disease

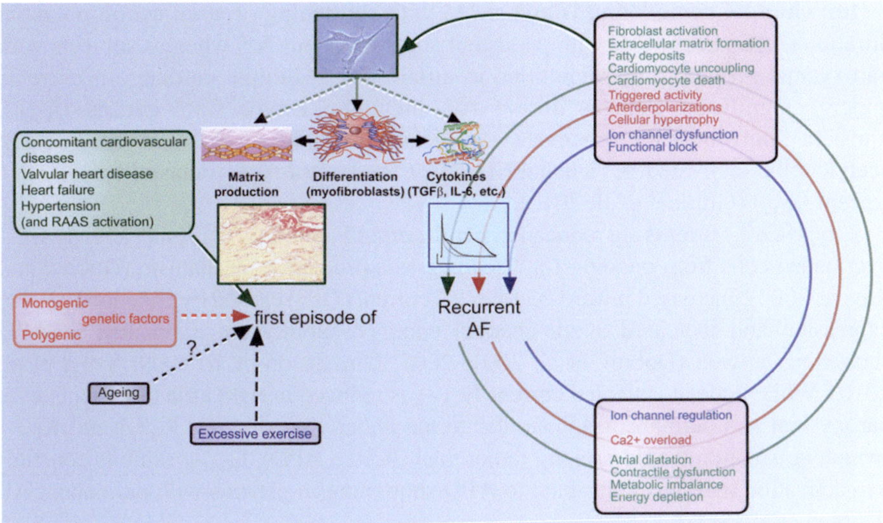

Fig. 12.3 Potential mechanisms underlying AF. Initial triggers (left part of the figure) and determinants of recurrent AF (right part) [reproduced with permission from Kirchhof and Fabritz (2014)]

progression and maintenance of AF known as "AF begets AF" (Fig. 12.3) (Wijffels et al. 1995).

While individual aspects of AF pathophysiology will be highlighted separately, it is important to note that in most cases a combination of multiple factors produces the phenotype of this complex disorder. Accelerated ventricular heart rates will cause Ca^{2+} overload and triggered activity by Ca^{2+} inhomogeneities. Delayed afterdepolarizations (**DADs**) caused by changes in intracellular Ca^{2+} levels and sodium calcium exchanger (NCX) activation and subsequent membrane depolarization may reach the threshold of action potential initiation and account for **focal arrhythmia** that may trigger AF. Focal arrhythmia may also arise from early afterdepolarizations (**EADs**) and spatial inhomogeneities in action potential duration, where prolonged action potentials give rise to increases in transmembrane Ca^{2+} currents and vice versa, as increased L-type Ca^{2+} current prolongs APD.

At the structural level, AF leads to **atrial remodelling**: fibroblasts are activated by profibrotic signaling through angiotensin II and TGFβ1 (Nattel et al. 2008; Tan and Zimetbaum 2011). In addition, platelet-derived and connective tissue growth factors cause extracellular matrix formation and cardiomyocyte uncoupling, contributing to changes in conduction velocity. Inhomogeneities in propagation of electrical activity form an electoanatomical basic for reentry. **Reentry** may be functional due to electrical remodeling and inhomogeneities in AP duration, or it may occur around anatomic obstacles such as scarring. Macroscopic consequences of histological and structural changes include atrial dilatation and contractile dysfunction of the atria (Heijman et al. 2014).

Ion channel remodeling is associated with shortening of atrial action potential duration (APD) in patients with persistent and permanent AF, whereas subjects with paroxysmal AF show no alterations in atrial APD. Multiple cardiac ion currents undergo significant changes during AF, including L-type Ca^{2+} current ($I_{Ca,L}$), transient outward K^+ channels (I_{to}), inward-rectifier K^+ channels (I_{K1}), acetylcholine-activated K^+ channel ($I_{K,ACh}$), and ultra-rapid delayed-rectifier K^+ current (I_{Kur}) (Grunnet et al. 2012).

L-type Ca^{2+} currents are reduced in permanent AF, possibly to protect already Ca^{2+} overloaded cells from ongoing Ca^{2+} loading as a protective mechanism (Dobrev and Ravens 2003). Increased inward-rectifier K^+ currents (**I_{K1}**) caused by enhanced protein expression and increased single-channel open probability are implicated in APD shortening as well (Dobrev et al. 2001, 2005; Girmatsion et al. 2009; Voigt et al. 2010). While agonist-activated current **$I_{K,ACh}$** is reduced in right atria of patients with paroxysmal and permanent AF, similar to the underlying subunits $K_{ir}3.1$ and $K_{ir}3.4$ (which in isolation would prolong rather than shorten APD), $I_{K,ACh}$ exhibits constitutive activation that may contribute to APD shortening in patients with permanent AF (Dobrev et al. 2005).

Small-conductance calcium-activated potassium currents (**I_{SK}**) appear to be affected as well. However, uniform up- or downregulation has not been established to date, as both increased SK currents and downregulation of SK3 subunits have been reported (Ling et al. 2013; Zhou et al. 2015).

Emerging evidence suggest a substantial role for **$K_{2P}3.1$** (TASK-1) two-pore-domain potassium channels in the regulation of atrial APD. $K_{2P}3.1$ channels are predominantly expressed in human atria vs. ventricles. $K_{2P}3.1$ mechanistically contributes to subtype-specific APD remodeling. In chronic AF patients, increased $K_{2P}3.1$ expression and function contribute to APD shortening (Schmidt et al. 2015). By contrast, atrial APD is prolonged in cases of severe heart failure, at least in part mediated by reduced abundance of $K_{2P}3.1$ channels in atrial cardiomyocytes of affected patients (Schmidt et al. 2017).

Despite the fact that APD at 90% repolarization is reduced in chronic forms of AF, the APD at 20% repolarization was reported to be prolonged due to diminished **I_{to}** currents in permanent AF (Workman et al. 2001). Similarly I_{Kur} and $K_v1.5$ subunits are reduced in permanent AF, and there is evidence that reduction in I_{Kur} may facilitate EADs in sympathetic stimulation (Wettwer et al. 2004; Olson et al. 2006).

Finally, abnormalities in intracellular Ca^{2+} handling facilitate arrhythmogenesis in the atria. Altered connexin-40 and connexin-43 mRNA and protein expression, as well as remodeling of their subcellular distribution, result in conduction abnormalities that promote AF perpetuation (Tribulova et al. 2015). Connexin remodeling functionally interacts with fibrogenesis in AF patients (Luo et al. 2007).

Taken together, atrial remodeling in AF involves multiple contributing factors (Fig. 12.4). Several of the ion channel proteins mentioned above are encoded by genes that have been associated with familial forms of AF.

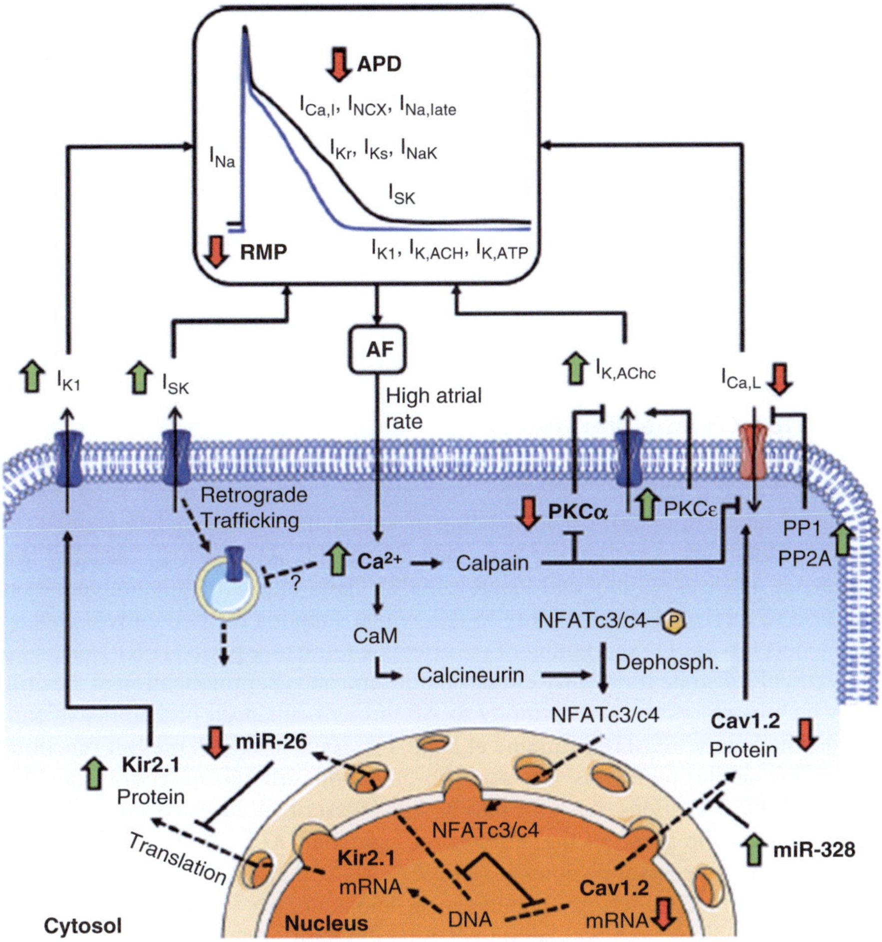

Fig. 12.4 Mechanisms of electrical remodeling in AF. Atrial tachycardia causes Ca^{2+} loading, which increases cellular signaling increasing K^+ currents and reduces L-type Ca^{2+} currents. APD indicates action potential duration; CaM, calmodulin; I_{K1}, basal inward-rectifier K^+ current; $I_{K,ACh}$, acetylcholine-dependent inward-rectifier K^+ current; I_{Kr}, rapid delayed-rectifier K^+ current; I_{Ks}, slow delayed-rectifier K^+ current; $I_{K,ATP}$, ATP-dependent K^+ current; I_{Na}, Na^+ current; I_{NaK}, Na^+-K^+-ATPase current; $I_{Na,late}$, persistent/late, Na^+ current; I_{NCX}, Na^+/Ca^{2+} exchanger current; I_{SK}, small-conductance Ca^{2+}-activated K^+ current; I_{to}, transient outward K^+ current; miR, microRNA; NFAT, nuclear factor of activated T cells; PKC, protein kinase C; PP1, protein phosphatase type, 1; PP2A, protein phosphatase type 2A; and RMP, resting membrane potential [reproduced with permission from Heijman et al. (2014)]

12.3 Variations of Single Genes in AF

Once heritability of AF was revealed, traditional genetic approaches were followed that led to the discovery of monogenetic traits in AF. Rare genetic variants with strong effects and mostly obvious phenotypes were identified, in contrast to common genetic variants (single nucleotide polymorphisms) with weaker effects and less obvious changes or a distinct phenotype (Campuzano and Brugada 2009). Variations of single genes involved in AF have been described in cardiac ion channels, in connexins, and in several non-ion channel genes which will be discussed here (Table 12.1). These variants and mutations are thought to predispose to AF either by modifying expression levels and/or by altering atrial conduction.

12.3.1 Ion Channel Variations

Mutations in ion channel genes are responsible for only a small subgroup of patients (Mahida et al. 2011). Sequencing of ion channel genes to date yielded few putative mutations, representing less than 2% of all AF cases if causality was assumed in all cases (Ellinor and MacRae 2008). Gain- or loss-of-function mutations may lead to increased susceptibility to AF, which supports the two current conceptual models for AF: (1) shortening of the effective refractory period (ERP) leading to an electrical substrate for reentry circuits in the atria and (2) prolongation of the atrial AP, which increases the susceptibility to AF through enhanced probability of early afterdepolarization (EAD) (Lemoine et al. 2011). Recent reports present functional effects of single mutations or describe mutations found in cohort analyses (Table 12.1). Mutations have primarily been described in potassium channels contributing to the shape of the atrial action potential (*KCNQ1*, *KCNA5*, *KCND3*, *KCNJ2*) and in accessory subunits (*KCNE1*, *KCNE2*, *KCNE3*, and *KCNE5*). In addition, calcium channels, sodium channels, and connexin 40 were affected in rare cases, each predicted to affect action potential duration or propagation of electrical conduction in the atria. Commonly involved ion channel gene mutations are elucidated in detail in the following text. Further mutations in other ion channel genes are presented in Table 12.1.

12.3.1.1 KCNQ1

The first mutation associated with familial AF was reported in 2003. In a large Chinese family with autosomal-dominant AF, a gain-of-function mutation in *KCNQ1*, encoding for the alpha subunit of I_{Ks}, was identified (Chen et al. 2003). This mutation led to enhanced I_{Ks}. Additional gain-of-function mutations were later detected, predicted to lead to atrial APD shortening (Chen et al. 2003; Hong et al. 2005; Lundby et al. 2007; Otway et al. 2007a, b; Das et al. 2009; Bartos et al. 2011, 2013; Hasegawa et al. 2014; Ki et al. 2014). In 2005 Hong et al. presented the *KCNQ1*-V141 M mutation in a newborn with AF and an abnormally short QT interval. The authors described pronounced instantaneous activation of I_{Ks} in the mutant channel underlying enhanced I_{Ks}. An independent heterozygous

Table 12.1 Genetic variants in ion channels identified in families and individuals with AF

Gene	Protein	Function	Mutation	Effect of mutation	Clinical phenotype	Reference
Genes with confirmed pathogenic mutation						
KCNQ1	$K_v7.1$	$\mathbf{I_{Ks}}$	**R14C**	Gain-of-function on stretch	Hypertension, late-onset AF	Otway et al. (2007a, b)
			A46T	Gain-of-function	Type of not reported	Olesen et al. (2014)
			S140G	Gain-of-function	Fam. Pers. AF	Chen et al. (2003)
			V141M	Coexpressed with KCNE1: Gain-of-function	Juvenile-onset AF	Hong et al. (2005)
			Q147R	Coexpressed with KCNE1: Loss-of-function	Perm. AF, LQTS	Lundby et al. (2007)
				Coexpressed with KCNE2: Gain-of-function		
			R195W	Gain-of-function	Parox. AF	Olesen et al. (2014)
			S209L	Gain-of-function	Lone AF	Das et al. (2009)
			G229D	Gain-of-function	Juvenile-onset AF	Hasegawa et al. (2014)
			R231C	Loss-of-function (LQT1) and gain-of-function (FAF) properties	LQTS and/or familial AF	Bartos et al. (2011)
			R231H	Gain-of-function	Familial AF, LQTS	Bartos et al. (2013)
			V241F	Gain-of-function	Age-dependent bradycardia and pers. AF	Ki et al. (2014)
			A302V	Loss-of-function	Perm. AF	Olesen et al. (2014)
			R670K	Gain-of-function	Type of not reported	Olesen et al. (2014)

(continued)

Table 12.1 (continued)

Gene	Protein	Function	Mutation	Effect of mutation	Clinical phenotype	Reference
KCNA5	$K_v1.5$	I_{Kur}	**E48G**	Gain-of-function	Early-onset lone AF	Christophersen et al. (2013)
			71–81del	Deletion of 11 amino acids in the N-terminus, reduced interaction with tyrosine kinase	Early-onset familial lone AF	Yang et al. (2010)
			Y155C	Loss-of-function, decreased membrane expression	Early-onset lone AF	Christophersen et al. (2013)
			A305T	Gain-of-function	Early-onset lone AF	Christophersen et al. (2013)
			D322H	Gain-of-function	Early-onset lone AF	Christophersen et al. (2013)
			E375X	Loss-of-function, dominant-negative effect on wild-type current	Type of not reported	Olson et al. (2006)
			H463A	Loss-of-function	Parox. AF	Hayashi et al. (2015)
			D469E	Loss-of-function: Decreased membrane expression	Early-onset lone AF	Christophersen et al. (2013)
			P488S	Loss-of-function: Decreased membrane expression	Early-onset lone AF	Christophersen et al. (2013)
			T527M	Loss-of-function	Parox. AF	Yang et al. (2009) and Hayashi et al. (2015)
			A576V	Loss-of-function	Type of not reported	Yang et al. (2009)
			E610K	Loss-of-function	Type of not reported	Yang et al. (2009)

KCND3	$K_v4.3$	I_{to}	**A545P**	Gain-of-function, increased peak current density	Early-onset persistent lone AF	Olesen et al. (2013)
KCNJ2	$K_{ir}2.1$	I_{Kur}	**V93I**	Gain-of-function	Type of not reported	Xia et al. (2005)
			M301K	Gain-of-function	Type of not reported	Hattori et al. (2012)
KCNK3	$K_{2P}3.1$/TASK-1	I_{K2P}	**V123L**	Loss-of-function	Lone AF	Liang et al. (2014)
KCNE1	Kv7.1/MinK	Reg. subunit of I_{Ks}	**G25V**	Gain-of-function for IKs	Parox. AF	Olesen et al. (2012a)
			G38S	Not reported	Type of not reported	Jiang et al. (2017)
			G60D	Gain-of-function for IKs	Lone AF	Olesen et al. (2012a)
KCNE2	MiRP1	Reg. subunit of I_{Ks}	**R27C**	Gain-of-function effect was observed upon coexpression with $K_V7.1$ and $K_V7.1$ + KCNE1	Predominantly parox. AF	Yang et al. (2004)
			M23L		Paroxysmal AF	Nielsen et al. (2014)
			I57T		Parox. AF	Nielsen et al. (2014)
KCNE3	MiRP2	Reg. subunit of I_{Ks}	**V17M**	Gain-of-function for the $K_v4.3$/KCNE3 and $K_v11.1$ channel complex	Early-onset lone AF	Lundby et al. (2008)
KCNE4	MiRP3	Reg. subunit of I_{Ks}	**E145D**	Gain-of-function for the $K_v7.1$/KCNE4 channel complex	Lone AF	Ma et al. (2007)
KCNH2	$K_v11.1$	I_{Kr}	**T436M**	Gain-of-function	Perm. AF	Hayashi et al. (2015)
			T895	Gain-of-function	Parox. AF	Hayashi et al. (2015)

(continued)

Table 12.1 (continued)

Gene	Protein	Function	Mutation	Effect of mutation	Clinical phenotype	Reference
SCN10A	$Na_v1.8$	I_{Na}	**R14L**	Not examined	Parox. AF, AVNRT	Jabbari et al. (2015)
			V94G	Loss-of-function	Parox./pers. AF	Jabbari et al. (2015)
			Y158D	Not examined	Pers. AF	Jabbari et al. (2015)
			R814H	Not examined	Pers. AF	Jabbari et al. (2015)
			E825D	Not examined	Parox.AF, AVNRT	Jabbari et al. (2015)
			I999L	Not examined, patient also has a *SCNB3* variant	Parox. AF	Jabbari et al. (2015)
			R1268Q	Not examined	Pers. AF	Jabbari et al. (2015)
			C1523Y	Not examined	Parox. AF	Jabbari et al. (2015)
			R1588Q	Loss-of-function	Parox. AF	Jabbari et al. (2015)
SCN1B		Sodium channel beta subunit	**R85H**	Loss-of-function	Parox.AF, aortic stenosis	Watanabe et al. (2009)
			D153N	Loss-of-function	Parox. Lone AF	Watanabe et al. (2009)
			T189M	Gain-of-function	Parox. AF	Hayashi et al. (2015)

SCN2B		Sodium channel beta subunit	**S18 N**	No data	Pers. AF	Olesen et al. (2014)
			R28Q	Loss-of-function	Parox. AF	Watanabe et al. (2009)
			R28W	Loss-of-function	Parox. Lone AF	Watanabe et al. (2009)
			Y69H	Loss-of-function	Perm. AF	Olesen et al. (2014)
SCN3B		Sodium channel beta subunit	**R6K**	Loss-of-function	Pers. AF	Olesen et al. (2014)
			L10P	Loss-of-function	Pers. AF	Olesen et al. (2014)
			M161T	Loss-of-function	Parox. AF	Olesen et al. (2014)
SCN4B		Sodium channel beta subunit	**V162G**	Not examined. Predicted to be disease causing	Familial AF	Li et al. (2013)
			I166L	Not examined. Predicted to be disease causing	Familial AF	Li et al. (2013)
SCN5A	$Na_v1.5$	I_{Na}	**T220I**	Loss-of-function	Parox. AF	Olesen et al. (2014)
			R340Q	Loss-of-function	Parox. AF	Olesen et al. (2014)
			R986Q	Loss-of-function	Parox. AF	Hayashi et al. (2015)
			T1304M	Gain-of-function	Parox. AF	Olesen et al. (2014)
			F1596I	No change	Pers. AF	Olesen et al. (2014)
			R1626H	Combined	Parox. AF	Olesen et al. (2014)

(continued)

Table 12.1 (continued)

Gene	Protein	Function	Mutation	Effect of mutation	Clinical phenotype	Reference
			D1819N	Gain-of-function	Parox. AF	Olesen et al. (2014)
			M1875 T	Gain-of-function	Familial lone AF	Makiyama et al. (2008)
			R1897W	Loss-of-function	Parox. AF	Olesen et al. (2014)
			V1951M	Gain-of-function	Pers. AF	Olesen et al. (2014)
HCN4	HCN4	I_f	**P257S**	Defective trafficking to the cell membrane, haploinsufficiency	Early-onset AF	Macri et al. (2014)
			K530N	Altered interaction between the adjacent wild-type and mutant C-linkers of HCN4, loss-of-function in heteromeric configuration	Familial age-dependent tachycardia-bradycardia syndrome and persistent AF	Duhme et al. (2013)
ABCC9	ATP-binding cassette subfamily 9, ATP-sensitive K^+ channel	Regulatory SUR2A channel subunit	**T1547I**	Compromised nucleotide-dependent K_{ATP} channel gating	Predisposition to adrenergic AF	Olson et al. (2007)
Genes with potential pathogenic mutation						
CACNB2i5	Forms protein complex coexpressed with $Ca_v1.2$		**Splice variant**	Not examined	Parox. AF	Weeke et al. (2014)
CACNA2D4	Regulatory subunit for voltage-gated Ca^{2+} channels ($Ca_v1.2$)		**S886R**	Not examined	Parox. AF	Weeke et al. (2014)
KCNN3	SK3	I_{SK}	**L66Q**	Not examined	Symptomatic AF	Tsai et al. (2015)

Genes with unclear linkage to AF						
KCNJ5	$K_{ir}4.3$	I_{KACh}	**C171T**	Not examined	Lone parox. AF	Zhang et al. (2009) and Jabbari et al. (2011)
			G810T	Not examined	Lone parox. AF	Zhang et al. (2009) and Jabbari et al. (2011)
KCNJ8	$K_{ir}6.1$	I_{KATP}	**S422L**	Gain-of-function	Lone AF	Delaney et al. (2012)
KCNE5	MiRP4	Beta subunit that interacts with $K_v7.1$	**L65F**	Gain-of-function	Type of not reported	Ravn et al. (2008)

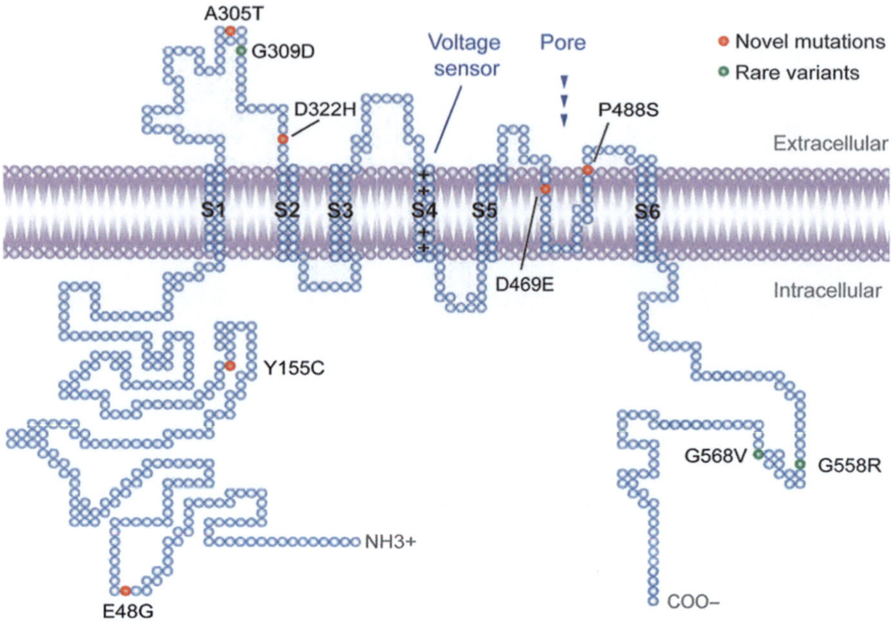

Fig. 12.5 Molecular model of $K_V1.5$ [modified from Yang et al. (2010)]. Red, mutations; green, rare variants [reproduced with permission from Christophersen et al. (2013)]

KCNQ1-S209 L mutation showed instantaneous opening on coexpression with WT *KCNQ1*, a negative shift of the half-maximal activation voltage, and slow current deactivation as biophysical mechanism of current increase (Das et al. 2009). More recently *KCNQ1*-R231C and R231H mutations in families with AF and mild QT prolongation were identified (Bartos et al. 2011, 2013). These mutations exhibited marked instantaneous activation and significant negative shifts in half- maximal activation voltages.

Recent mutagenesis experiments and structural modeling of the KCNQ1 protein revealed that residues S140, E160, R237, and R231 may associate in the closed state. Substitution of amino acids (S140, E160, R237, or R231) was shown to abolish KCNQ1 deactivation and to "lock" the I_{Ks} channel in the open state, indicating a role for normal channel gating (Restier et al. 2008).

12.3.1.2 KCNA5

KCNA5 encodes the potassium channel $K_v1.5$ (Fig. 12.5). This channel conducts the voltage-gated atrial-specific potassium current I_{Kur} that contributes to repolarization of human atrial cardiomyocytes.

After an initial description of AF-related mutations by the Olson laboratory (Olson et al. 2006), further mutations in *KCNA5* have been reported in the following years and were studied in expression systems (Christophersen et al. 2013; Yang et al. 2009, 2010). Both loss-of-function and gain-of-function have been reported, supporting the hypothesis that prolongation and shortening of atrial repolarization may promote AF. In addition, rare variants in *KCNA5* show significantly higher prevalence in AF patients compared with healthy controls. No other gene has been reported to have such a high frequency of rare variants associated with AF. This supports the hypothesis that *KCNA5* is substantially involved in early-onset lone AF (Christophersen et al. 2013).

12.3.1.3 KCND3

KCND3 encodes the voltage-sensitive potassium channel K_v4.3 underlying cardiac I_{to}, which is expressed in both atrial and ventricular tissue. I_{to} is involved in phase 1. A gain-of-function mutation was discovered by Olesen et al. (2013). KCND3-A545P was identified in a patient presenting with lone AF at the age of 22. The mutation results in increased current density when expressed in heterologous expression systems and shows a slower inactivation compared to wild type. In computer simulations the current alterations associated with AF lead to shortening of action potential duration. In contrast to this gain-of-function mutation, however, I_{to} currents have been reported to be reduced in chronic AF, possibly reflecting different stages of AF (Workman et al. 2008).

12.3.1.4 KCNK3

KCNK3 encodes TASK-1, a family of the two-pore-domain potassium family. TASK-1 contributes to background K^+ conductance in human atrial cardiomyocytes and has been proposed as a potential drug target for AF (Schmidt et al. 2015).

In a screening approach for *KCNK3* variants in lone AF patients, three mutations were found to be associated with AF. In patch-clamp experiments, the KCNK3-V123 L missense mutation depolarized the resting membrane potential in vitro and altered pH sensitivity. Structural modeling predicted channel instability at the TASK-1 pore (Liang et al. 2014). Two additional variants were found in the translation initiation sequence of *KCNK3*, leading to reduced expression levels in a constructed reporter compared to the wild-type Kozak sequence in vitro; however, in vivo significance of these findings remains to be established.

12.3.1.5 *SCN5A* and Beta Subunits

SCN5A encodes the predominant cardiac sodium channel alpha subunit Na_V1.5 responsible for I_{Na} current. Mutations in this gene cause a broad range of genetic arrhythmia disorders including LQTS, Brugada syndrome, sick sinus syndrome, and conduction diseases (Terrenoire et al. 2007).

SCN5A mutations as well as mutations in Na_V1.5-associated beta subunits *SCN1B*, *2B*, *3B*, *and 4B* have been associated with AF (Makiyama et al. 2008; Watanabe et al. 2009; Olesen et al. 2014). Both gain- and loss-of-function mutations have been observed in *SCN5A*, whereas interestingly, patients with beta subunit mutations are mostly presenting a cellular loss-of-function phenotype and show ST-segment elevation in the precordial leads. Possible arrhythmogenic mechanisms include enhanced late sodium current in gain-of-function mutations leading to prolonged repolarization and reduced peak sodium currents in loss-of-function mutations as basis for conduction slowing.

12.3.2 Mutations and Variants in Non-ion Channel Genes

Although most rare variants were identified in ion channel genes, variants in non-ion channel genes have been implicated in AF as well (Hucker et al. 2016), affecting more than 30 structural proteins involved in muscle function such as laminin, gap junction proteins (e.g., connexin 40), and transcription factors (GATA4, GATA5, GATA6, NKX2, Pitx2c, and TBX5) (Table 12.2). Mutations in transcription factors involved in cardiac development suggest that "lone" AF may at least partially share a common genetic origin with congenital structural malformations (Posch et al. 2010). In particular, defects in the pulmonary vein myocardium have been linked to the pathogenesis of lone AF (Mommersteeg et al. 2007a, 2009). The transcription factors NKX2–5 are expressed in the pulmonary vein myocardium and suppress the sinoatrial node lineage program. Therefore reduction of NKX expression may lead to ectopic formation of nodal-like cells with pacemaker activity (Mommersteeg et al. 2007a, b). Furthermore, genetic variants may disturb the functionality of transcription factors, resulting in abnormal ion channel composition, incorrect development of the cardiac conduction system, or the generation of fibrosis to support the development and maintenance of AF (Zhou et al. 2015).

12.3.3 AF in Channelopathies

Electrical dysfunction in cardiac channelopathies increases the risk of life-threatening ventricular arrhythmia. In addition to ventricular arrhythmogenesis, atrial electrical function may be affected as well by underlying mutations. Indeed, the association with AF is observed in several channelopathies and inherited cardiomyopathies (Enriquez et al. 2016) (Table 12.3), with prevalence ranging from 5 to 20% in patients with LQTS or Brugada syndrome and up to 70% in short QT syndrome (Gaita et al. 2003; Brugada et al. 2004; Giustetto et al. 2006; Antzelevitch et al. 2007). However, penetrance of the AF phenotype is variable. Both shortening and prolongation of the atrial action potential secondary to ion channel mutations have been suggested to increase the probability of AF occurrence.

Pharmacologic treatment of AF in channelopathies may be challenging as underlying ion channel defects may prevent the use of certain antiarrhythmic drugs that could aggravate the clinical phenotype.

12.3.3.1 Long QT Syndrome (LQTS)

Approximately 2% of LQTS patients present with AF before their fifth decade, compared to 0.1% in the general population. As episodes may be asymptomatic, the prevalence reported may be underestimated (Zellerhoff et al. 2009).

Table 12.2 Genetic variants in non-ion channel genes identified in families and individuals with AF

Gene	Protein	Function	Mutation	Type of mutation/effect	Clinical phenotype	Reference
Genes with confirmed pathogenic mutation						
GATA4	GATA4	Transcription factor	**G16C**	Loss-of-function: Significantly decreased transcriptional activity	Familial AF	Jiang et al. (2011)
			H28D		Familial AF	Jiang et al. (2011)
			Y38D		Familial AF	Wang et al. (2012)
			S70T		Aut.-dom. Familial AF	Yang et al. (2011)
			S160T		Aut.-dom. Familial AF	Yang et al. (2011)
			P103A		Familial AF	Wang et al. (2012)
			M247T	Affects a conserved domain adjacent to the first zinc finger domain	Familial lone AF and atrial septal aneurysm	Posch et al. (2010)
			A411V	Benign variation	Sporadic lone AF, mutation also detected in patients with cardiac septal defects	
GATA5	GATA5	Transcription factor	**G184V**	Not examined. Probably confers susceptibility to AF by reducing expression of target genes	Familial AF	Yang et al. (2012b)
			K218T		Familial AF	Yang et al. (2012b)
			A266P		Familial AF	Yang et al. (2012b)
			W200G	Loss-of-function: Significantly decreased transcriptional activity	Familial AF	Wang et al. (2013)
			C210G	Loss-of-function: Significantly decreased transcriptional activity	Familial parox. AF	Gu et al. (2012)
			Y138F	Loss-of-function: Significantly decreased transcriptional activity	Familial AF	Gu et al. (2012)

(continued)

Table 12.2 (continued)

Gene	Protein	Function	Mutation	Type of mutation/effect	Clinical phenotype	Reference
GATA6	GATA6	Transcription factor	**Q206P**	Not examined. Probably confers susceptibility to AF by reducing expression of target genes	Familial AF	Yang et al. (2012a)
			Y265X		Familial AF	Yang et al. (2012a)
			Y235S	Loss-of-function: Significantly decreased transcriptional activity	Lone AF	Yang et al. (2012c)
			G469V	Loss-of-function: Significantly decreased transcriptional activity	Familial AF	Li et al. (2012)
			R585L	Gain-of-function effect: Significantly increased transcriptional activity	Familial AF + ASD/VSD with incomplete penetrance	Tucker et al. (2017)
			A543G	Gain-of-function effect: Significantly increased transcriptional activity under coexpression with GATA4	Familial AF	Tucker et al. (2017)
			P91S	Gain-of-function effect: Significantly increased transcriptional activity	Familial AF	Tucker et al. (2017)
			A177T	Gain-of-function effect: Significantly increased transcriptional activity	Familial AF	Tucker et al. (2017)
NKX2.5		Transcription factor	**N19D**	Loss-of-function: Significantly decreased transcriptional activity	Familial AF	Xie et al. (2013)
			E21Q	Loss-of-function: Significantly decreased transcriptional activity	Familial AF	Yu et al. (2014)
			F145S	Loss-of-function: Significantly decreased transcriptional activity	Familial AF	Huang et al. (2013)
			F186S	Loss-of-function: Significantly decreased transcriptional activity	Familial AF	Xie et al. (2013)
			T180A	Loss-of-function: Significantly decreased transcriptional activity	Familial AF	Yu et al. (2014)

NKX2–6		Transcription factor	**Q175H**	Unknown transcriptional targets of NKX2–6, decreased transcriptional activity suspected	Lone AF	Wang et al. (2014)
TBX5		Transcription factor	**D118del**	Not examined	Perm. AF	Ma et al. (2016)
			R355C	Gain-of-function: Increased the expression of atrial natriuretic peptide (ANP) and connexin-40 (Cx40) in the primarily cultured rat atrial myocytes	Early-onset AF, perm. AF	Ma et al. (2016)
			S372L		Early-onset AF, parox. AF	Ma et al. (2016)
			Q376R		Early-onset AF, parox. AF	Ma et al. (2016)
			A428S		Early-onset AF, perm. AF, AV block	Ma et al. (2016)
NPPA	ANP		**D93E**	No data	Permanent AF	Olesen et al. (2014)
GJA5	Cx40		**A96S**	Loss-of-function	Parox. AF, Pers. AF	Gollob et al. 2006 and Olesen et al. (2014)
TNNI3	Troponin I3		**R186Q**	Not examined	Familial AF and HCM	Wang et al. (2016)
			E64G	Not examined	AF, hypertension, DM	Wang et al. (2016)
			M154L	Not examined	AF, hypertension, heart failure	Wang et al. (2016)
			E187G	Not examined	AF, coronary artery disease	Wang et al. (2016)
			D196G	Not examined	AF, coronary artery disease	Wang et al. (2016)
JPH2	Junctophilin 2	Stabilization of RyR2	**E169K**	Loss-of-function: Reduced binding of E169K-JPH2 to RyR2, abnormal SR Ca^{2+} release events	HCM, early-onset AF	Beavers et al. (2013)

(continued)

Table 12.2 (continued)

Gene	Protein	Function	Mutation	Type of mutation/effect	Clinical phenotype	Reference
EMD	Emerin	Stabilization of nuclear membrane	**K37del**	Loss-of-function	Progressive SND and AF (X-linked inheritance)	Karst et al. (2008)
NUP155	Nucleosporin 155	Nuclear pore complex protein	**R391H**	Loss-of- function: Altering mRNA and protein transport	AF and early sudden death	Zhang et al. (2008)
LMNA	Lamin A		**W498Ter**	Nonsense mutation: Effect not examined	Familial AF	Zhao et al. (2016)

Table 12.3 Inherited channelopathies associated with AF [adapted in parts from Kirchhof et al. (2016) with permission]

Syndrome	Gene	Functional alteration	Prevalence of AF	References
LQTS	***KCNQ1***	$I_{Kr}\downarrow$	5–10%	Moss et al. (1995), Schwartz et al. (2001), Kirchhof et al. (2003), Mohler et al. (2003), Johnson et al. (2008) and Zellerhoff et al. (2009)
	KCNH2	$I_{Kr}\downarrow$		
	SCN5A	$I_{Na}\uparrow$		
	ANK2	$I_{Na,K}\uparrow$		
Brugada syndrome	***SCN5A***	$I_{Na}\downarrow$	10–20%	Eckardt et al. (2001), Antzelevitch et al. (2007), London et al. (2007), Watanabe et al. (2008) and Kaufman (2009)
	GPD1L	$I_{Na}\downarrow$		
	SCN1B	$I_{Na}\downarrow$		
	CACNA1C	$I_{Ca}\downarrow$		
	CACNB2B	$I_{Ca}\downarrow$		
SQTS	***KCNQ1***	$I_{Kr}\uparrow$	Up to 70%	Gaita et al. (2003), Brugada et al. (2004), Giustetto et al. (2006) and Antzelevitch et al. (2007)
	KCNH2	$I_{Kr}\uparrow$		
	KCNJ2	$I_{K1}\uparrow$		
	CACNA1C	$I_{Ca}\downarrow$		
	CACNB2B	$I_{Ca}\downarrow$		
CPVT	***RYR2***	Abnormal Ca^{2+} release from SR	Variable	Bhuiyan et al. (2007), Napolitano and Priori (2007) and Mohamed et al. (2007)
	CASQ2			

Mutations in three ion channel genes *KCNQ1* (LQT1), *KCNH2* (LQT2), and *SCN5A* (LQT3) account for 75% of all LQTS cases (Ackerman et al. 2011; Schwartz et al. 2012). In each of these genes, mutations that predispose to AF have been reported (Benito et al. 2008; Bartos et al. 2011; Olesen et al. 2012a, b). The prevalence of AF appears to depend on the LQTS subtype. In the examined population, LQT1 had the highest prevalence (2.4%) compared with LQT2 (0%) and LQT3 (<0.1%) (Johnson et al. 2008). In addition to subtypes, sex affects the AF prevalence in LQTS patients. Male subjects are more often affected than females. Prolonged atrial action potentials and effective refractory periods were detected in LQTS, and 50% of affected patients developed polymorphic atrial tachycardias during electrophysiological examinations (Kirchhof et al. 2003). The length of the QT interval was identified as an additional AF risk factor. Of note, a J-shaped correlation highlights that both shortening and prolongation of the repolarization may put patients at risk of ventricular and atrial arrhythmias (Nielsen et al. 2013).

12.3.3.2 Brugada Syndrome

AF is the most common atrial arrhythmia in Brugada syndrome patients with a prevalence of 13.7% (Kusano et al. 2008). Both ventricular fibrillation and AF occur mainly (70%) at night, highlighting the role of the vagal tone in arrhythmogenesis (Kusano et al. 2008). AF has been associated with a more malignant clinical course and a higher incidence of ventricular arrhythmias in Brugada syndrome. AF has therefore been suggested as risk marker (Kusano et al. 2008; Giustetto et al. 2014). In turn, a spontaneous type I Brugada ECG is associated with a higher prevalence of AF compared to patients with drug induced type I ECG (Pappone et al. 2009).

Mutations in *SCN5A* account for the majority of Brugada syndrome cases with an identified gene mutation. Sodium channel dysfunction affects atrial electrophysiology, prolonging intra-atrial conduction and providing a substrate for AF initiation and maintenance (Amin et al. 2011).

12.3.3.3 Short QT Syndrome (SQTS)

SQTS displays the highest prevalence of AF (up to 70%) among cardiac channelopathies (Borggrefe et al. 2005), with higher prevalence being reported associated with *KCNQ1* mutations compared to other genes (63% vs. 21%) (Harrell et al. 2015). SQTS may be caused by gain-of-function mutations in potassium channel genes *KCNH2* (Brugada et al. 2004), *KCNQ1* (Bellocq et al. 2004), and *KCNJ2* (Priori et al. 2005) or loss-of-function mutations in the L-type calcium channel (Antzelevitch et al. 2007). These mutations shorten the ERP and increase dispersion of repolarization and susceptibility to reentry in the ventricles and the atria. In addition, AF is readily induced in SQTS patients upon electrophysiological study owing to short atrial ERPs (Gaita et al. 2004).

12.3.3.4 Catecholaminergic Polymorphic Ventricular Tachycardia (CPVT)

Atrial arrhythmias including AF are frequently observed in CPVT. CPVT-causing mutations encoding for the cardiac ryanodine receptor (*RYR2*) or for cardiac calsequestrin (*CASQ2*) are found in 60% of affected patients (Ackerman et al. 2011). Calcium leak from the sarcoplasmic reticulum is assumed to be involved in the pathophysiology of AF in CPVT patients. Pronounced calcium leak causes cytosolic calcium overload and triggers delayed afterdepolarizations that may increase AF susceptibility.

In genetically modified mouse models of CPVT with *RyR2* mutations, an increased diastolic Ca^{2+} release was reported (Cerrone et al. 2005; Kannankeril et al. 2006; Goddard et al. 2008; Lehnart et al. 2008; Suetomi et al. 2011), and in several models, similar to observations in patients, some mice showed atrial arrhythmias and evidence for diastolic SR Ca^{2+} leak (Chelu et al. 2009; Suetomi et al. 2011; Zhang et al. 2011; Neco et al. 2012; Shan et al. 2012).

12.4 Single Nucleotide Polymorphisms Associated with AF in Large Populations

Genome-wide association studies (GWAS) have been performed to link single nucleotide polymorphisms (SNPs) to increased risk of AF (Ellinor et al. 2012; Pérez-Serra et al. 2017). Multiple genome-wide loci enhancing susceptibility to AF have been identified, predominantly in noncoding regions. The adjacent genes include transcription factors, ion channels, and signaling molecules. In addition, a higher prevalence of rare variants was found associated with AF in a cohort of early-onset lone AF patients compared to a control cohort (Olesen et al. 2014).

Gudbjartsson et al. performed initial replication studies in populations of European descendants and in one Chinese AF population, identifying a strong association between two sequence variants on chromosome 4q25 and AF (rs2200733, rs10033464). Of European descendants, 35% carried at least one of the variants that increased AF risk by 1.72 and 1.39 per copy, respectively (Gudbjartsson et al. 2007). Both variants are close to *PITX2*, a gene implicated in heart development and left-right asymmetry (Faucourt et al. 2001; Franco and Campione 2003; Mommersteeg et al. 2007a, b). Downregulation of *PITX* predisposes to AF in a transgenic mouse model where a lack of Pitx2 in atrial myocardium impaired sodium channel and potassium channel expression (Chinchilla et al. 2011). A different SNP, rs2595104, close to *PITX2* showed a reduced enhancer activity within the risk allele. CRISPR-Cas9-mediated deletion of the rs2595104 region and editing of the rs2595104 risk allele in human stem-cell derived cardiomyocytes reduced PITX2c expression compared to the non-risk allele. TFAP2a, an enhancer binding protein, mediated the effect by being able to bind the normal allele, but unable to bind to the risk allele, giving rise to a possible pathway for AF susceptibility (Ye et al. 2016).

Subsequently, single nucleotide polymorphism rs2106261 was moderately correlated with AF (Benjamin et al. 2009), and rs719334 showed a significant relation to AF (Gudbjartsson et al. 2009). Both SNPs are located close to the gene *ZFHX3*, encoding a transcription factor. Ongoing GWAS analysis identified an intronic SNP (rs13376333) in the gene *KCNN3* correlating to lone AF (Ellinor et al. 2010). *KCNN3* encodes one member of the small-conductance calcium-activated potassium channels.

Moreover, the risk for postoperative AF after coronary artery bypass graft surgery (CABG) could be related to SNPs in *NEURL* (rs12415501) and *CAND2* (rs4642101) in a Chinese population (Wei et al. 2016). *NEURL* encodes an E3 ubiquitin ligase that altered atrial action potential duration when suppressed in zebrafish, without affecting cardiac contractile function or heart rate (Sinner et al. 2014). *CAND2* encodes a TATA-binding protein, TIP120b, with specific involvement in muscular tissue and myogenesis.

A summary of SNPs associated with AF is presented in Table 12.4. The latest update from a combined-ancestry GWAS meta-analysis in 2017 revealed 12 new loci associated with AF (Fig. 12.6) (Christophersen et al. 2017). Of note, significant differences regarding the association of each SNP with AF susceptibility may depend on specific allele frequencies in the respective population analyzed, indicating that the ethnic background needs to be considered when interpreting the results. The molecular mechanisms through which these SNPs affect AF are still unknown in most cases.

Despite the identification of multiple genetic factors including monogenic mutations and SNPs, no single change or combination of factors comprehensively explains or defines AF risk, suggesting that other genetic variants remain to be elucidated. In addition to SNPs, copy number variation (CNV) in the DNA sequence account for around 5% of human DNA (Conrad et al. 2010) and may affect

Table 12.4 Atrial fibrillation GWAS studies

Locus	Closest gene	SNP	Location	Reference
1p13.3	*PSRC1*	rs599839		Yamase et al. (2016)
1q21.3	*KCNN3*	rs6666258, rs13376333		Chang et al. (2012) and Yao et al. (2015)
1q24	*PRRX1*	rs3903239	Upstream gene variant	Ellinor et al. (2012)
1q24	*METTL11B-KIFAP3*	rs72700118	Intergenic variant	Christophersen et al. (2017)
2p13	*NXA4-GMCL1*	rs3771537	Intronic variant	Christophersen et al. (2017)
2p14	*CEP68*	rs2540949	Intronic variant	Christophersen et al. (2017)
2q31	*TTN-TTN-AS1*	rs2288327	Intronic variant	Christophersen et al. (2017)
3q25.1	*CAND2*	rs4642101	Intronic variant	Sinner et al. (2014) and Wei et al. (2016)
4q25	*PITX2*	rs2200733	Intergenic variant	Gudbjartsson et al. (2007)
4q25	*PITX2*	rs13143308	Upstream gene variant	Gudbjartsson et al. (2007)
4q25	*PITX2*	rs6817105, rs4400058, rs6838973, rs1448818, rs2200733, rs10033464, rs2595104		Gudbjartsson et al. (2007), Lubitz et al. (2014) and Ye et al. (2016)
5q22	*KCNN2*	rs337711	Intronic variant	Christophersen et al. (2017)
5q31	*WNT8A*	rs2040862		Lubitz et al. (2016)
5q31	*KLHL3*	rs2967791	Intronic variant	Christophersen et al. (2017)
5q35	*KCNIP1*	CN (4470 pb)	Intronic CNV	Tsai et al. (2016)
6q22	*GJA1*	rs13216675	Intergenic variant	Sinner et al. (2014)
6q22	*SLC3F1*	rs4946333	Intronic	Christophersen et al. (2017)
7q31.2	*CAV1*	rs3807989	Intronic variant	Ellinor et al. (2012), Liu et al. (2015) and Jia et al. (2016)
7q32	*ZC3HC1*	rs11556924		Yamase et al. (2016)
8p22	*ASAH1*	rs7508	3'-UTR	Christophersen et al. (2017)
9q22	*C9orf3*	rs10821415	Intronic variant	Ellinor et al. (2012)
10q22	*SYNPO2L*	rs10824026	Upstream gene variant	Ellinor et al. (2012) and Roberts et al. (2016)

(continued)

Table 12.4 (continued)

Locus	Closest gene	SNP	Location	Reference
10q24	*NEURL*	rs12415501, rs6584555	Intronic variants	Sinner et al. (2014) and Wei et al. (2016)
10q24	*SH3PD2A*	rs35176054	Intronic variants	Christophersen et al. (2017)
12q24	*CUX2*	rs6490029	Intronic variant	Sinner et al. (2014)
14q23	*SYNE2*	rs1152591	Intronic variant	Ellinor et al. (2012)
15q24	*HCN4*	rs7164883	Intronic variant	Ellinor et al. (2012) and Everett et al. (2013)
16q22	*ZFHX3*	rs2106261	Intronic variant	Benjamin et al. (2009) and Li et al. (2011)

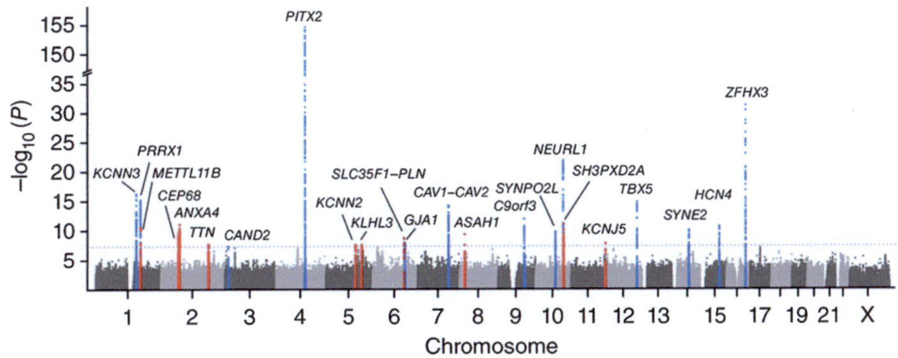

Fig. 12.6 Manhattan plot of a combined-ancestry GWAS meta-analysis. The plot shows genetic loci associated with AF in a combined-ancestry GWAS meta-analysis, in blue the replicated and in red the new genetic loci. The dashed line represents the threshold of statistical significance ($P = 5 \times 10^{-8}$). Gene names correspond to the gene in closest proximity to the most significant variant at each locus. There is a break in the *y* axis to increase the resolution of the genetic loci near the genome-wide significance threshold [reproduced with permission from Christophersen et al. (2017)]

expression of surrounding genes. SNPs and CNVs display little overlap (Stranger et al. 2007), revealing CNV as possible mechanism to induce complex phenotypes and diseases. The first CNV associated with AF has been identified in the first intron of *KCNIP1* (potassium interacting channel 1 gene) in a Taiwanese population (Tsai et al. 2016). Furthermore, epigenetic factors may contribute to the pathogenesis of AF. Recently Lin et al. (2017) performed genome-wide methylation profiling and identified seven methylation signatures associated with AF. These findings suggest

Table 12.5 DNA methylation patterns associated with AF

CpG site	Closest gene	SNP	Location	Reference
Cg13639451	WFIKKN2	–	Upstream gene	Lin et al. (2017)
Cg07191189	STRN	–	Upstream gene	Lin et al. (2017)

that DNA methylation might play a role in AF arrhythmogenesis (Table 12.5), but underlying epigenetic mechanisms remain to be identified in more detail.

12.5 The Impact of Genetics on AF Therapy

12.5.1 Current AF Therapy

AF treatment comprises anticoagulation in patients at risk for thromboembolic stroke, antiarrhythmic therapy to achieve rhythm control, reduction of AV-nodal conduction for rate control, and catheter ablation strategies. Additional targets include cardiovascular risk reduction by lifestyle changes, weight loss, treatment of obstructive sleep apnea, and physical training.

12.5.1.1 Anticoagulation

Oral anticoagulation (OAC) therapy prevents most ischemic strokes in AF patients and reduces morbidity. With the exception of patients at very low risk as assessed by the CHA_2DS_2-VASc-Score (Lip et al. 2010; Kirchhof et al. 2016), patients benefit from treatment with OAC. Anticoagulation may be achieved with non-vitamin K-anticoagulants (NOACs; apixaban, dabigatran, edoxaban, rivaroxaban) or "classical" vitamin K-antagonists (e.g., warfarin or phenprocoumon) in non-valvular forms of AF.

12.5.1.2 Rate Control Versus Rhythm Control

Pharmacological rate control can be achieved for acute or long-term rate control. The choice of the drug depends on patients' characteristics, concomitant disease, symptoms, and left ventricular function. If drugs fail to effectively reduce ventricular heart rates, AV-node ablation and pacemaker implantation may be considered. Rhythm control may be achieved by application of antiarrhythmic drugs and by catheter ablation. To date, no significant differences in overall mortality have been detected comparing rhythm vs. rate control.

12.5.1.3 Catheter Ablation of AF

Since the first discovery of triggers in the pulmonary veins initiating paroxysmal AF (Haïssaguerre et al. 1998), catheter ablation of AF has emerged as routine treatment to prevent AF recurrence. Catheter ablation is effective in restoring and maintaining sinus rhythm in patients with symptomatic paroxysmal AF and to reduced extent in persistent and long-standing persistent AF. In selected patients with paroxysmal AF, ablation is considered first-line therapy, whereas in the majority of patients catheter ablation constitutes second-line treatment after ineffective antiarrhythmic drug therapy.

12.5.1.4 Impact of Genetics on AF Management

Genetic discoveries highlight relevant interindividual heterogeneity in the pathophysiology of initiation and progression of AF. Thus, there is not "one" AF but rather multiple distinct sub-entities of the arrhythmia. The response to standardized therapy regimen will therefore differ among different patient groups, and therapeutic developments may be based on individual mechanistic subphenotypes.

Knowing the exact genetic background of an individual patient may guide clinicians toward a personalized therapy and management. The influence of SNPs on treatment outcome has been evaluated in the past. A recent study analyzed the outcome of pulmonary vein isolation depending of three common polymorphisms at chromosome loci 4q25, 1q21, and 16q22. In a multivariate analysis, rs2200733 was linked to AF recurrence, but none of the polymorphisms predicted AF recurrence in a long-term follow-up (Kiliszek et al. 2016). The same rs2200733 risk allele reduced clinical response to catheter ablation for AF in a Chinese population (Zhao et al. 2017), and it was suggested to use the rs2200733 polymorphism to sub-select patients for AF catheter ablation. The impact of the SNPs on procedural success and their high prevalence in the population show the potential for routine genotyping in the context of catheter ablation. However, additional studies are needed to further substantiate this concept of "ablatogenomics" (Roberts and Marcus 2015, 2016).

In addition to reduced effectiveness of catheter ablation, carriers of the SNP at chromosome 4q25 appear to be more refractory to antiarrhythmics (Parvez et al. 2012). Adaptation of AF management according to the patients' genetics with respect to pharmacology (pharmacogenetics) and interventional treatment (ablatogenomics) appears to be a promising clinical avenue for applied genetics (Roberts and Marcus 2016).

Genetics may guide antiarrhythmic therapy in channelopathy patients as well. Knowing the underlying mutation and its electrophysiological effects at cellular level may optimize the choice of antiarrhythmic drugs used to treat AF. Successful treatments of AF in LQT1 with mexiletine in one case report (El Yaman et al. 2008) or in LQT3 with flecainide (Benito et al. 2008; Chorin et al. 2018) have been reported. In addition, contraindications for class I antiarrhythmic drugs in Brugada syndrome and QT-prolonging drugs in congenital LQTS affect drug choice.

Nevertheless until now guidelines do not recommend genetic testing in the general AF population, as there is no clear link between specific mutations and detected outcomes or special therapeutic needs (Kirchhof et al. 2014).

12.5.1.5 Gene Therapy for AF

Gene therapy-based approaches for AF management have been successfully established in animal models, providing proof-of-concept for a novel antiarrhythmic paradigm (Arora 2017; Hucker et al. 2017).

Most preclinical gene-based approaches to treat AF employed a porcine model of right atrial tachypacing to induce AF. Most studies modulated expression levels of ion channels or gap junctions proteins that contribute to electrical and structural remodeling in the atria.

Specifically, overexpression of a dominant-negative mutant of the I_{Kr} channel decreased the occurrence of AF (Amit et al. 2010; Soucek et al. 2012) by prolonging atrial effective refractory periods. In addition, overexpression of connexin 40 or connexin 43 led to improved electrical conduction and reduced AF inducibility compared to control animals (Igarashi et al. 2012; Bikou et al. 2011). Overexpression of another atrial repolarizing channel, TREK-1 ($K_{2P}2.1$), shortened atrial refractory periods in pigs with heart failure and decreased AF prevalence (Lugenbiel et al. 2017).

In addition to ion channels and gap junctions, autonomic signaling in the atria has been explored as gene therapeutic target by interfering with sympathetic and vagal downstream signaling transduction pathways. Rate control was achieved by genetic upregulation of G_{ai} protein or by suppression of G_{as} protein in the porcine AV node (Lugenbiel et al. 2012; Donahue 2016). Moreover it was shown that vagally induced AF could be prevented by atrial injection of C-terminal G_{ai} and/or G_{ao} genes in dogs (Aistrup et al. 2011).

Other strategies have targeted molecular substrates for structural atrial remodeling including fibrosis, hypertrophy, and apoptotic cell loss. To reduce apoptosis in the atria, caspase 3 was suppressed by atrial Ad-siRNA-Cas3 gene transfer. The intervention delayed the onset of persistent AF associated with reduced apoptosis (Trappe et al. 2013). Furthermore, TGF-β contributes to the development of atrial fibrosis in AF. Dogs treated with a dominant-negative TGF-ß vector showed fewer conduction inhomogeneity and decreased duration of AF (Kunamalla et al. 2016).

In summary, several potential targets for gene therapeutic rhythm and rate control have been identified and tested. Long-term efficacy and safety remain to be demonstrated prior to transfer of gene therapy into humans (Lugenbiel et al. 2016).

12.6 Conclusions

AF is the most common arrhythmia. Despite the discoveries of several candidate genes following Mendelian genetics and potential targets in GWAS studies, the heritability of AF still remains to be fully understood.

Monogenetic causes for AF are rare. Genetics may account for increased susceptibility to AF or serve as modifiers of the disease. Genetic risk factors and their interplay with other nongenetic or environmental factors may influence the overall susceptibility to AF.

Therefore AF in general is a complex and heterogeneous disease in which multiple factors including genetic variants and environmental factors cause a predisposition to AF (Fig. 12.7). It remains incompletely understood how identified SNPs increase the susceptibility to AF. Further basic research and clinical studies are needed to identify underlying molecular pathways of AF pathophysiology to provide the basis for improved patient care and individualized therapy based on genetics.

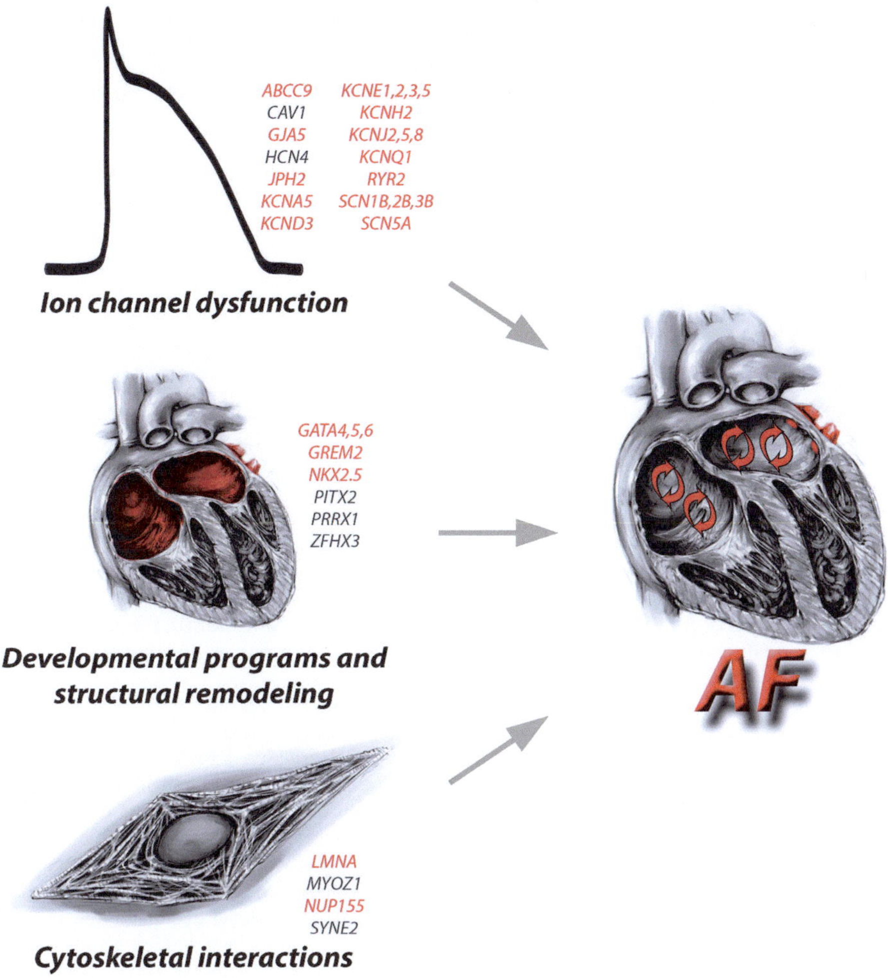

Fig. 12.7 Genetic pathways of atrial fibrillation (AF) pathogenesis. Schematic representation of AF-related genes. Genes listed in red indicate those identified by familial studies and candidate gene screens, whereas those listed in gray were identified by GWAS [reproduced with permission from Tucker and Ellinor (2014)]

Compliance with Ethical Standards

Funding This work was supported in part by grants from the University of Heidelberg, Faculty of Medicine (Physician Scientist Scholarship to A.K.R.) and the German Cardiac Society (DGK Research Scholarship to A.K.R,), from the German Cardiac Society and the Hengstberger Foundation (Klaus-Georg and Sigrid Hengstberger Scholarship to D.T.), from the German Heart Foundation/German Foundation of Heart Research (F/08/14 to D.T.), from the Joachim Siebenreicher Foundation (to D.T.), and from the Ministry of Science, Research, and the Arts, Baden-Wuerttemberg (Sonderlinie Medizin to D.T.).

Conflict of Interest D.T. reports receiving lecture fees/honoraria from Bayer Vital, Boehringer Ingelheim, Bristol-Myers Squibb, Daiichi Sankyo, Medtronic, Pfizer Pharma, Sanofi-Aventis, St. Jude Medical, and ZOLL CMS, and research grant support from Daiichi Sankyo. H.A.K. and D.T. filed a patent application for the use of K_{2P} potassium channels for altering cardiac electrophysiology.

Ethical Approval This article does not contain any studies with human participants or animals performed by any of the authors.

References

Ackerman MJ, Priori SG, Willems S, Berul C, Brugada R, Calkins H, Camm AJ, Ellinor PT, Gollob M, Hamilton R, Hershberger RE, Judge DP, Le Marec H, McKenna WJ, Schulze-Bahr E, Semsarian C, Towbin JA, Watkins H, Wilde A, Wolpert C, Zipes DP. HRS/EHRA expert consensus statement on the state of genetic testing for the channelopathies and cardiomyopathies. Europace. 2011;13(8):1077–109.

Aistrup GL, Cokic I, Ng J, Gordon D, Koduri H, Browne S, Arapi D, Segon Y, Goldstein J, Angulo A, Wasserstrom JA, Goldberger JJ, Kadish AH, Arora R. Targeted nonviral gene-based inhibition of Gα(i/o)-mediated vagal signaling in the posterior left atrium decreases vagal-induced atrial fibrillation. Heart Rhythm. 2011;8(11):1722–9.

Amin AS, Boink GJ, Atrafi F, Spanjaart AM, Asghari-Roodsari A, Molenaar RJ, Ruijter JM, Wilde AA, Tan HL. Facilitatory and inhibitory effects of SCN5A mutations on atrial fibrillation in Brugada syndrome. Europace. 2011;13(7):968–75.

Amit G, Kikuchi K, Greener ID, Yang L, Novack V, Donahue JK. Selective molecular potassium channel blockade prevents atrial fibrillation. Circulation. 2010;121(21):2263–70.

Andersson T, Magnuson A, Bryngelsson I-L, Frobert O, Henriksson KM, Edvardsson N, Poci D. All-cause mortality in 272,186 patients hospitalized with incident atrial fibrillation 1995-2008 a Swedish nationwide long-term case-control study. Eur Heart J. 2013;34(14):1061–7.

Antzelevitch C, Pollevick GD, Cordeiro JM, Casis O, Sanguinetti MC, Aizawa Y, Guerchicoff A, Pfeiffer R, Oliva A, Wollnik B, Gelber P, Bonaros EP, Burashnikov E, Wu Y, Sargent JD, Schickel S, Oberheiden R, Bhatia A, Hsu L-F, Haïssaguerre M, Schimpf R, Borggrefe M, Wolpert C. Loss-of-function mutations in the cardiac calcium channel underlie a new clinical entity characterized by ST-segment elevation, short QT intervals, and sudden cardiac death. Circulation. 2007;115(4):442–9.

Arnar DO, Thorvaldsson S, Manolio TA, Thorgeirsson G, Kristjansson K, Hakonarson H, Stefansson K. Familial aggregation of atrial fibrillation in Iceland. Eur Heart J. 2006;27 (6):708–12.

Arora R. Gene therapy for atrial fibrillation in heart failure. Clin Pharmacol Ther. 2017;102 (2):200–2.

Ball J, Carrington MJ, McMurray JJ, Stewart S. Atrial fibrillation profile and burden of an evolving epidemic in the 21st century. Int J Cardiol. 2013;167(5):1807–24.

Bartos DC, Duchatelet S, Burgess DE, Klug D, Denjoy I, Peat R, Lupoglazoff JM, Fressart V, Berthet M, Ackerman MJ, January CT, Guicheney P, Delisle BP. R231C mutation in KCNQ1 causes long QT syndrome type 1 and familial atrial fibrillation. Heart Rhythm. 2011;8(1):48–55.

Bartos DC, Anderson JB, Bastiaenen R, Johnson JN, Gollob MH, Tester DJ, Burgess DE, Homfray T, Behr ER, Ackerman MJ, Guicheney P, Delisle BP. A *KCNQ1* mutation causes a high penetrance for familial atrial fibrillation. J Cardiovasc Electrophysiol. 2013;24(5):562–9.

Beavers DL, Wang W, Ather S, Voigt N, Garbino A, Dixit SS, Landstrom AP, Li N, Wang Q, Olivotto I, Dobrev D, Ackerman MJ, Wehrens XH. Mutation E169K in junctophilin-2 causes atrial fibrillation due to impaired RyR2 stabilization. J Am Coll Cardiol. 2013;62(21):2010–9.

Bellocq C, van Ginneken AC, Bezzina CR, Alders M, Escande D, Mannens MM, Baró I, Wilde AA. Mutation in the *KCNQ1* gene leading to the short QT-interval syndrome. Circulation. 2004;109(20):2394–7.

Benito B, Brugada R, Perich RM, Lizotte E, Cinca J, Mont L, Berruezo A, Tolosana JM, Freixa X, Brugada P, Brugada J. A mutation in the sodium channel is responsible for the association of long QT syndrome and familial atrial fibrillation. Heart Rhythm. 2008;5(10):1434–40.

Benjamin EJ, Wolf PA, D'Agostino RB, Silbershatz H, Kannel WB, Levy D. Impact of atrial fibrillation on the risk of death. The Framingham heart study. Circulation. 1998;98(10):946–52.

Benjamin EJ, Rice KM, Arking DE, Pfeufer A, van Noord C, Smith AV, Schnabel RB, Bis JC, Boerwinkle E, Sinner MF, Dehghan A, Lubitz SA, D'Agostino RB, Lumley T, Ehret GB, Heeringa J, Aspelund T, Newton-Cheh C, Larson MG, Marciante KD, Soliman EZ, Rivadeneira F, Wang TJ, Eiríksdottir G, Levy D, Psaty BM, Li M, Chamberlain AM, Hofman A, Vasan RS, Harris TB, Rotter JI, Kao WH, Agarwal SK, Stricker BH, Wang K, Launer LJ, Smith NL, Chakravarti A, Uitterlinden AG, Wolf PA, Sotoodehnia N, Köttgen A, van Duijn CM, Meitinger T, Mueller M, Perz S, Steinbeck G, Wichmann H-E, Lunetta KL, Heckbert SR, Gudnason V, Alonso A, Kääb S, Ellinor PT, Witteman JC. Variants in *ZFHX3* are associated with atrial fibrillation in individuals of European ancestry. Nat Genet. 2009;41 (8):879–81.

Bhuiyan ZA, van den Berg MP, van Tintelen JP, Bink-Boelkens MT, Wiesfeld AC, Alders M, Postma AV, van Langen I, Mannens MM, Wilde AA. Expanding spectrum of human RYR2-related disease new electrocardiographic, structural, and genetic features. Circulation. 2007;116 (14):1569–76.

Bikou O, Thomas D, Trappe K, Lugenbiel P, Kelemen K, Koch M, Soucek R, Voss F, Becker R, Katus HA, Bauer A. Connexin 43 gene therapy prevents persistent atrial fibrillation in a porcine model. Cardiovasc Res. 2011;92(2):218–25.

Bjorck S, Palaszewski B, Friberg L, Bergfeldt L. Atrial fibrillation, stroke risk, and warfarin therapy revisited a population-based study. Stroke. 2013;44(11):3103–8.

Borggrefe M, Wolpert C, Antzelevitch C, Veltmann C, Giustetto C, Gaita F, Schimpf R. Short QT syndrome. Genotype-phenotype correlations. J Electrocardiol. 2005;38(4 Suppl):75–80.

Brugada R, Hong K, Dumaine R, Cordeiro J, Gaita F, Borggrefe M, Menendez TM, Brugada J, Pollevick GD, Wolpert C, Burashnikov E, Matsuo K, Wu YS, Guerchicoff A, Bianchi F, Giustetto C, Schimpf R, Brugada P, Antzelevitch C. Sudden death associated with short-QT syndrome linked to mutations in HERG. Circulation. 2004;109(1):30–5.

Campuzano O, Brugada R. Genetics of familial atrial fibrillation. Europace. 2009;11(10):1267–71.

Cerrone M, Colombi B, Santoro M, di Barletta MR, Scelsi M, Villani L, Napolitano C, Priori SG. Bidirectional ventricular tachycardia and fibrillation elicited in a knock-in mouse model carrier of a mutation in the cardiac ryanodine receptor. Circ Res. 2005;96:e77–82.

Chang SH, Chang SN, Hwang JJ, Chiang FT, Tseng CD, Lee JK, Lai LP, Lin JL, Wu CK, Tsai CT. Significant association of rs13376333 in *KCNN3* on chromosome 1q21 with atrial fibrillation in a Taiwanese population. Circ J. 2012;76(1):184–8.

Chelu MG, Sarma S, Sood S, Wang S, van Oort RJ, Skapura DG, Li N, Santonastasi M, Müller FU, Schmitz W, Schotten U, Anderson ME, Valderrábano M, Dobrev D, Wehrens XH. Calmodulin kinase II-mediated sarcoplasmic reticulum Ca^{2+} leak promotes atrial fibrillation in mice. J Clin Invest. 2009;119(7):1940–51.

Chen YH, Xu SJ, Bendahhou S, Wang XL, Wang Y, Xu WY, Jin HW, Sun H, Su XY, Zhuang QN, Yang YQ, Li YB, Liu Y, Xu HJ, Li XF, Ma N, Mou CP, Chen Z, Barhanin J, Huang W. KCNQ1 gain-of-function mutation in familial atrial fibrillation. Science. 2003;299(5604):251–4.

Chinchilla A, Daimi H, Lozano-Velasco E, Dominguez JN, Caballero R, Delpón E, Tamargo J, Cinca J, Hove-Madsen L, Aranega AE, Franco D. PITX2 insufficiency leads to atrial electrical and structural remodeling linked to arrhythmogenesis. Circ Cardiovasc Genet. 2011;4 (3):269–79.

Chorin E, Taub R, Medina A, Flint N, Viskin S, Benhorin J. Long-term flecainide therapy in type 3 long QT syndrome. Europace. 2018;20(2):370–6.

Christophersen IE, Ravn LS, Budtz-Joergensen E, Skytthe A, Haunsoe S, Svendsen JH, Christensen K. Familial aggregation of atrial fibrillation a study in Danish twins. Circ Arrhythm Electrophysiol. 2009;2(4):378–83.

Christophersen IE, Olesen MS, Liang B, Andersen MN, Larsen AP, Nielsen JB, Haunso S, Olesen S-P, Tveit A, Svendsen JH, Schmitt N. Genetic variation in *KCNA5*: impact on the atrial-specific potassium current I_{Kur} in patients with lone atrial fibrillation. Eur Heart J. 2013;34(20):1517–25.

Christophersen IE, Rienstra M, Roselli C, Yin X, Geelhoed B, Barnard J, Lin H, Arking DE, Smith AV, Albert CM, Chaffin M, Tucker NR, Li M, Klarin D, Bihlmeyer NA, Low S-K, Weeke PE, Müller-Nurasyid M, Smith JG, Brody JA, Niemeijer MN, Dörr M, Trompet S, Huffman J, Gustafsson S, Schurmann C, Kleber ME, Lyytikäinen L-P, Seppälä I, Malik R, Horimoto AR, Perez M, Sinisalo J, Aeschbacher S, Thériault S, Yao J, Radmanesh F, Weiss S, Teumer A, Choi SH, Weng L-C, Clauss S, Deo R, Rader DJ, Shah SH, Sun A, Hopewell JC, Debette S, Chauhan G, Yang Q, Worrall BB, Paré G, Kamatani Y, Hagemeijer YP, Verweij N, Siland JE, Kubo M, Smith JD, van Wagoner DR, Bis JC, Perz S, Psaty BM, Ridker PM, Magnani JW, Harris TB, Launer LJ, Shoemaker MB, Padmanabhan S, Haessler J, Bartz TM, Waldenberger M, Lichtner P, Arendt M, Krieger JE, Kähönen M, Risch L, Mansur AJ, Peters A, Smith BH, Lind L, Scott SA, Lu Y, Bottinger EB, Hernesniemi J, Lindgren CM, Wong JA, Huang J, Eskola M, Morris AP, Ford I, Reiner AP, Delgado G, Chen LY, Y-DI C, Sandhu RK, Li M, Boerwinkle E, Eisele L, Lannfelt L, Rost N, Anderson CD, Taylor KD, Campbell A, Magnusson PK, Porteous D, Hocking LJ, Vlachopoulou E, Pedersen NL, Nikus K, Orho-Melander M, Hamsten A, Heeringa J, Denny JC, Kriebel J, Darbar D, Newton-Cheh C, Shaffer C, Macfarlane PW, Heilmann-Heimbach S, Almgren P, Huang PL, Sotoodehnia N, Soliman EZ, Uitterlinden AG, Hofman A, Franco OH, Völker U, Jöckel K-H, Sinner MF, Lin HJ, Guo X, Dichgans M, Ingelsson E, Kooperberg C, Melander O, Loos RJ, Laurikka J, Conen D, Rosand J, van der Harst P, Lokki ML, Kathiresan S, Pereira A, Jukema JW, Hayward C, Rotter JI, März W, Lehtimäki T, Stricker BH, Chung MK, Felix SB, Gudnason V, Alonso A, Roden DM, Kääb S, Chasman DI, Heckbert SR, Benjamin EJ, Tanaka T, Lunetta KL, Lubitz SA, Ellinor PT. Large-scale analyses of common and rare variants identify 12 new loci associated with atrial fibrillation. Nat Genet. 2017;49(6):946–52.

Chugh SS, Havmoeller R, Narayanan K, Singh D, Rienstra M, Benjamin EJ, Gillum RF, Kim YH, McAnulty JH, JR ZZJ, Forouzanfar MH, Naghavi M, Mensah GA, Ezzati M, Murray CJ. Worldwide epidemiology of atrial fibrillation a global burden of disease 2010 study. Circulation. 2014;129(8):837–47.

Colilla S, Crow A, Petkun W, Singer DE, Simon T, Liu X. Estimates of current and future incidence and prevalence of atrial fibrillation in the U.S. adult population. Am J Cardiol. 2013;112 (8):1142–7.

Conrad DF, Pinto D, Redon R, Feuk L, Gokcumen O, Zhang Y, Aerts J, Andrews TD, Barnes C, Campbell P, Fitzgerald T, Hu M, Ihm CH, Kristiansson K, Macarthur DG, Macdonald JR, Onyiah I, Pang AW, Robson S, Stirrups K, Valsesia A, Walter K, Wei J, Tyler-Smith C, Carter NP, Lee C, Scherer SW, Hurles ME. Origins and functional impact of copy number variation in the human genome. Nature. 2010;464(7289):704–12.

Das S, Makino S, Melman YF, Shea MA, Goyal SB, Rosenzweig A, CA MR, Ellinor PT. Mutation in the S3 segment of KCNQ1 results in familial lone atrial fibrillation. Heart Rhythm. 2009;6 (8):1146–53.

Delaney JT, Muhammad R, Blair MA, Kor K, Fish FA, Roden DM, Darbar D. A *KCNJ8* mutation associated with early repolarization and atrial fibrillation. Europace. 2012;14(10):1428–32.

Dobrev D, Ravens U. Remodeling of cardiomyocyte ion channels in human atrial fibrillation. Basic Res Cardiol. 2003;98(3):137–48.

Dobrev D, Graf E, Wettwer E, Himmel HM, Hála O, Doerfel C, Christ T, Schüler S, Ravens U. Molecular basis of downregulation of G-protein-coupled inward rectifying K+ current ($I_{K,ACh}$) in chronic human atrial fibrillation. Circulation. 2001;104(21):2551–7.

Dobrev D, Friedrich A, Voigt N, Jost N, Wettwer E, Christ T, Knaut M, Ravens U. The G protein-gated potassium current $I_{K,ACh}$ is constitutively active in patients with chronic atrial fibrillation. Circulation. 2005;112(24):3697–706.

Donahue JK. Biological therapies for atrial fibrillation ready for prime time? J Cardiovasc Pharmacol. 2016;67(1):19–25.

Duhme N, Schweizer PA, Thomas D, Becker R, Schröter J, Barends TR, Schlichting I, Draguhn A, Bruehl C, Katus HA, Koenen M. Altered HCN4 channel C-linker interaction is associated with familial tachycardia-bradycardia syndrome and atrial fibrillation. Eur Heart J. 2013;34 (35):2768–75.

Eckardt L, Kirchhof P, Loh P, Schulze-Bahr E, Johna R, Wichter T, Breithardt G, Haverkamp W, Borggrefe M. Brugada syndrome and supraventricular tachyarrhythmias a novel association? J Cardiovasc Electrophysiol. 2001;12(6):680–5.

El Yaman M, Perry J, Makielski JC, Ackerman MJ. Suppression of atrial fibrillation with mexiletine pharmacotherapy in a young woman with type 1 long QT syndrome. Heart Rhythm. 2008;5 (3):472–4.

Ellinor PT, MacRae CA. Ion channel mutations in AF signal or noise? Heart Rhythm. 2008;5 (3):436–7.

Ellinor PT, Yoerger DM, Ruskin JN, MacRae CA. Familial aggregation in lone atrial fibrillation. Hum Genet. 2005;118(2):179–84.

Ellinor PT, Lunetta KL, Glazer NL, Pfeufer A, Alonso A, Chung MK, Sinner MF, de BPI, Mueller M, Lubitz SA, Fox E, Darbar D, Smith NL, Smith JD, Schnabel RB, Soliman EZ, Rice KM, van Wagoner DR, Beckmann B-M, van Noord C, Wang K, Ehret GB, Rotter JI, Hazen SL, Steinbeck G, Smith AV, Launer LJ, Harris TB, Makino S, Nelis M, Milan DJ, Perz S, Esko T, Köttgen A, Moebus S, Newton-Cheh C, Li M, Möhlenkamp S, Wang TJ, Kao WH, Vasan RS, Nöthen MM, MacRae CA, Stricker BH, Hofman A, Uitterlinden AG, Levy D, Boerwinkle E, Metspalu A, Topol EJ, Chakravarti A, Gudnason V, Psaty BM, Roden DM, Meitinger T, Wichmann H-E, Witteman JC, Barnard J, Arking DE, Benjamin EJ, Heckbert SR, Kääb S. Common variants in *KCNN3* are associated with lone atrial fibrillation. Nat Genet. 2010;42(3):240–4.

Ellinor PT, Lunetta KL, Albert CM, Glazer NL, Ritchie MD, Smith AV, Arking DE, Müller-Nurasyid M, Krijthe BP, Lubitz SA, Bis JC, Chung MK, Dörr M, Ozaki K, Roberts JD, Smith JG, Pfeufer A, Sinner MF, Lohman K, Ding J, Smith NL, Smith JD, Rienstra M, Rice KM, van Wagoner DR, Magnani JW, Wakili R, Clauss S, Rotter JI, Steinbeck G, Launer LJ, Davies RW, Borkovich M, Harris TB, Lin H, Völker U, Völzke H, Milan DJ, Hofman A, Boerwinkle E, Chen LY, Soliman EZ, Voight BF, Li G, Chakravarti A, Kubo M, Tedrow UB, Rose LM, Ridker PM, Conen D, Tsunoda T, Furukawa T, Sotoodehnia N, Xu S, Kamatani N, Levy D, Nakamura Y, Parvez B, Mahida S, Furie KL, Rosand J, Muhammad R, Psaty BM, Meitinger T, Perz S, Wichmann H-E, Witteman JC, Kao WH, Kathiresan S, Roden DM, Uitterlinden AG, Rivadeneira F, McKnight B, Sjögren M, Newman AB, Liu Y, Gollob MH, Melander O, Tanaka T, Stricker BH, Felix SB, Alonso A, Darbar D, Barnard J, Chasman DI, Heckbert SR, Benjamin EJ, Gudnason V, Kääb S. Meta-analysis identifies six new susceptibility loci for atrial fibrillation. Nat Genet. 2012;44(6):670–5.

Enriquez A, Antzelevitch C, Bismah V, Baranchuk A. Atrial fibrillation in inherited cardiac channelopathies from mechanisms to management. Heart Rhythm. 2016;13(9):1878–84.

Everett BM, Cook NR, Conen D, Chasman DI, Ridker PM, Albert CM. Novel genetic markers improve measures of atrial fibrillation risk prediction. Eur Heart J. 2013;34(29):2243–51.

Faucourt M, Houliston E, Besnardeau L, Kimelman D, Lepage T. The pitx2 homeobox protein is required early for endoderm formation and nodal signaling. Dev Biol. 2001;229(2):287–306.

Fox CS, Parise H, D'Agostino RB, SR L-JDM, Vasan RS, Wang TJ, Levy D, Wolf PA, Benjamin EJ. Parental atrial fibrillation as a risk factor for atrial fibrillation in offspring. JAMA. 2004;291 (23):2851–5.

Franco D, Campione M. The role of Pitx2 during cardiac development. Linking left-right signaling and congenital heart diseases. Trends Cardiovasc Med. 2003;13(4):157–63.

Gaita F, Giustetto C, Bianchi F, Wolpert C, Schimpf R, Riccardi R, Grossi S, Richiardi E, Borggrefe M. Short QT syndrome a familial cause of sudden death. Circulation. 2003;108 (8):965–70.

Gaita F, Giustetto C, Bianchi F, Schimpf R, Haissaguerre M, Calò L, Brugada R, Antzelevitch C, Borggrefe M, Wolpert C. Short QT syndrome pharmacological treatment. J Am Coll Cardiol. 2004;43(8):1494–9.

Girmatsion Z, Biliczki P, Bonauer A, Wimmer-Greinecker G, Scherer M, Moritz A, Bukowska A, Goette A, Nattel S, Hohnloser SH, Ehrlich JR. Changes in microRNA-1 expression and I_{K1} up-regulation in human atrial fibrillation. Heart Rhythm. 2009;6(12):1802–9.

Giustetto C, Di Monte F, Wolpert C, Borggrefe M, Schimpf R, Sbragia P, Leone G, Maury P, Anttonen O, Haissaguerre M, Gaita F. Short QT syndrome clinical findings and diagnostic-therapeutic implications. Eur Heart J. 2006;27(20):2440–7.

Giustetto C, Cerrato N, Gribaudo E, Scrocco C, Castagno D, Richiardi E, Giachino D, Bianchi F, Barbonaglia L, Ferraro A. Atrial fibrillation in a large population with Brugada electrocardiographic pattern prevalence, management, and correlation with prognosis. Heart Rhythm. 2014;11(2):259–65.

Go AS, Hylek EM, Phillips KA, Chang Y, Henault LE, Selby JV, Singer DE. Prevalence of diagnosed atrial fibrillation in adults: national implications for rhythm management and stroke prevention: the AnTicoagulation and risk factors in atrial fibrillation (ATRIA) study. JAMA. 2001;285(18):2370–5.

Goddard CA, Ghais NS, Zhang Y, Williams AJ, Colledge WH, Grace AA, Huang CL. Physiological consequences of the P2328S mutation in the ryanodine receptor (RyR2) gene in genetically modified murine hearts. Acta Physiol (Oxf). 2008;194(2):123–40.

Gollob MH, Jones DL, Krahn AD, Danis L, Gong XQ, Shao Q, Liu X, Veinot JP, Tang AS, Stewart AF, Tesson F, Klein GJ, Yee R, Skanes AC, Guiraudon GM, Ebihara L, Bai D. Somatic mutations in the connexin 40 gene (*GJA5*) in atrial fibrillation the greater cape floristic region. N Engl J Med. 2006;354(25):2677–88.

Grond M, Jauss M, Hamann G, Stark E, Veltkamp R, Nabavi D, Horn M, Weimar C, Kohrmann M, Wachter R, Rosin L, Kirchhof P. Improved detection of silent atrial fibrillation using 72-hour Holter ECG in patients with ischemic stroke a prospective multicenter cohort study. Stroke. 2013;44(12):3357–64.

Grunnet M, Bentzen BH, Sorensen US, Diness JG. Cardiac ion channels and mechanisms for protection against atrial fibrillation. Rev Physiol Biochem Pharmacol. 2012;162:1–58.

Gu JY, Xu JH, Yu H, Yang YQ. Novel *GATA5* loss-of-function mutations underlie familial atrial fibrillation. Clinics. 2012;67(12):1393–9.

Gudbjartsson DF, Arnar DO, Helgadottir A, Gretarsdottir S, Holm H, Sigurdsson A, Jonasdottir A, Baker A, Thorleifsson G, Kristjansson K, Palsson A, Blondal T, Sulem P, Backman VM, Hardarson GA, Palsdottir E, Helgason A, Sigurjonsdottir R, Sverrisson JT, Kostulas K, Ng MC, Baum L, So WY, Wong KS, Chan JC, Furie KL, Greenberg SM, Sale M, Kelly P, MacRae CA, Smith EE, Rosand J, Hillert J, Ma RC, Ellinor PT, Thorgeirsson G, Gulcher JR, Kong A, Thorsteinsdottir U, Stefansson K. Variants conferring risk of atrial fibrillation on chromosome 4q25. Nature. 2007;448(7151):353–7.

Gudbjartsson DF, Holm H, Gretarsdottir S, Thorleifsson G, Walters GB, Thorgeirsson G, Gulcher J, Mathiesen EB, Njølstad I, Nyrnes A, Wilsgaard T, Hald EM, Hveem K, Stoltenberg C, Kucera G, Stubblefield T, Carter S, Roden D, Ng MC, Baum L, So WY, Wong KS, Chan JC, Gieger C, Wichmann H-E, Gschwendtner A, Dichgans M, Kuhlenbäumer G, Berger K, Ringelstein EB, Bevan S, Markus HS, Kostulas K, Hillert J, Sveinbjörnsdóttir S, Valdimarsson EM, Løchen M-L, Ma RC, Darbar D, Kong A, Arnar DO, Thorsteinsdottir U, Stefansson K. A sequence variant in *ZFHX3* on 16q22 associates with atrial fibrillation and ischemic stroke. Nat Genet. 2009;41(8):876–8.

Haim M, Hoshen M, Reges O, Rabi Y, Balicer R, Leibowitz M. Prospective national study of the prevalence, incidence, management and outcome of a large contemporary cohort of patients with incident non-valvular atrial fibrillation. J Am Heart Assoc. 2015;4(1):e001486. https://doi.org/10.1161/JAHA.114.001486.

Haïssaguerre M, Jaïs P, Shah DC, Takahashi A, Hocini M, Quiniou G, Garrigue S, Le Mouroux A, Le Métayer P, Clémenty J. Spontaneous initiation of atrial fibrillation by ectopic beats originating in the pulmonary veins. N Engl J Med. 1998;339(10):659–66.

Harrell DT, Ashihara T, Ishikawa T, Tominaga I, Mazzanti A, Takahashi K, Oginosawa Y, Abe H, Maemura K, Sumitomo N, Uno K, Takano M, Priori SG, Makita N. Genotype-dependent differences in age of manifestation and arrhythmia complications in short QT syndrome. Int J Cardiol. 2015;190:393–402.

Hasegawa K, Ohno S, Ashihara T, Itoh H, Ding WG, Toyoda F, Makiyama T, Aoki H, Nakamura Y, Delisle BP, Matsuura H, Horie M. A novel KCNQ1 missense mutation identified in a patient with juvenile-onset atrial fibrillation causes constitutively open IKs channels. Heart Rhythm. 2014;11(1):67–75.

Hattori T, Makiyama T, Akao M, Ehara E, Ohno S, Iguchi M, Nishio Y, Sasaki K, Itoh H, Yokode M, Kita T, Horie M, Kimura T. A novel gain-of-function KCNJ2 mutation associated with short-QT syndrome impairs inward rectification of $K_{ir}2.1$ currents. Cardiovasc Res. 2012;93(4):666–73.

Hayashi K, Konno T, Tada H, Tani S, Liu L, Fujino N, Nohara A, Hodatsu A, Tsuda T, Tanaka Y, Kawashiri MA, Ino H, Makita N, Yamagishi M. Functional characterization of rare variants implicated in susceptibility to lone atrial fibrillation study of thyroid dysfunction in diabetic patients. Circ Arrhythm Electrophysiol. 2015;8(5):1095–104.

Heijman J, Voigt N, Nattel S, Dobrev D. Cellular and molecular electrophysiology of atrial fibrillation initiation, maintenance, and progression. Circ Res. 2014;114(9):1483–99.

Henriksson KM, Farahmand B, Asberg S, Edvardsson N, Terent A. Comparison of cardiovascular risk factors and survival in patients with ischemic or hemorrhagic stroke. Int J Stroke. 2012;7(4):276–81.

Hong K, Piper DR, Diaz-Valdecantos A, Brugada J, Oliva A, Burashnikov E, Santos-de-Soto J, Grueso-Montero J, Diaz-Enfante E, Brugada P, Sachse F, Sanguinetti MC, Brugada R. De novo *KCNQ1* mutation responsible for atrial fibrillation and short QT syndrome in utero. Cardiovasc Res. 2005;68(3):433–40.

Huang RT, Xue S, Xu YJ, Zhou M, Yang YQ. A novel NKX2.5 loss-of-function mutation responsible for familial atrial fibrillation. Int J Mol Med. 2013;31(5):1119–26.

Hucker WJ, Saini H, Lubitz SA, Ellinor PT. Atrial fibrillation genetics is there a practical clinical value now or in the future? Can J Cardiol. 2016;32(11):1300–5.

Hucker WJ, Hanley A, Ellinor PT. Improving atrial fibrillation therapy is there a gene for that? J Am Coll Cardiol. 2017;69(16):2088–95.

Igarashi T, Finet JE, Takeuchi A, Fujino Y, Strom M, Greener ID, Rosenbaum DS, Donahue JK. Connexin gene transfer preserves conduction velocity and prevents atrial fibrillation. Circulation. 2012;125(2):216–25.

Jabbari J, Olesen MS, Holst AG, Nielsen JB, Haunso S, Svendsen JH. Common polymorphisms in *KCNJ5* [corrected] are associated with early-onset lone atrial fibrillation in Caucasians. Cardiology. 2011;118(2):116–20.

Jabbari J, Olesen MS, Yuan L, Nielsen JB, Liang B, Macri V, Christophersen IE, Nielsen N, Sajadieh A, Ellinor PT, Grunnet M, Haunsø S, Holst AG, Svendsen JH, Jespersen T. Common and rare variants in *SCN10A* modulate the risk of atrial fibrillation. Circ Cardiovasc Genet. 2015;8(1):64–73.

Jia W, Qi X, Li Q. Association between Rs3807989 polymorphism in Caveolin-1 (*CAV1*) gene and Atrial fibrillation a meta-analysis. Med Sci Monit. 2016;22:3961–6.

Jiang JQ, Shen FF, Fang WY, Liu X, Yang YQ. Novel *GATA4* mutations in lone atrial fibrillation. Int J Mol Med. 2011;28(6):1025–32.

Jiang YF, Chen M, Zhang NN, Yang HJ, Xu LB, Rui Q, Sun SJ, Yao JL, Zhou YF. Association between KCNE1 G38S gene polymorphism and risk of atrial fibrillation a PRISMA-compliant meta-analysis. Medicine. 2017;96(25):e7253. https://doi.org/10.1097/MD.0000000000007253.

Johnson JN, Tester DJ, Perry J, Salisbury BA, Reed CR, Ackerman MJ. Prevalence of early-onset atrial fibrillation in congenital long QT syndrome. Heart Rhythm. 2008;5(5):704–9.

Kannel WB, Benjamin EJ. Status of the epidemiology of atrial fibrillation. Med Clin North Am. 2008;92(1):17–40, ix.

Kannankeril PJ, Mitchell BM, Goonasekera SA, Chelu MG, Zhang W, Sood S, Kearney DL, Danila CI, De Biasi M, Wehrens XH, Pautler RG, Roden DM, Taffet GE, Dirksen RT, Anderson ME, Hamilton SL. Mice with the R176Q cardiac ryanodine receptor mutation exhibit catecholamine-induced ventricular tachycardia and cardiomyopathy. Proc Natl Acad Sci U S A. 2006;103:12179–84.

Karst ML, Herron KJ, Olson TM. X-linked nonsyndromic sinus node dysfunction and atrial fibrillation caused by emerin mutation. J Cardiovasc Electrophysiol. 2008;19(5):510–5.

Kaufman ES. Mechanisms and clinical management of inherited channelopathies long QT syndrome, Brugada syndrome, catecholaminergic polymorphic ventricular tachycardia, and short QT syndrome. Heart Rhythm. 2009;6(8 Suppl):51–5.

Ki CS, Jung CL, Kim HJ, Baek KH, Park SJ, On YK, Kim KS, Noh SJ, Youm JB, Kim JS, Cho H. A KCNQ1 mutation causes age-dependant bradycardia and persistent atrial fibrillation. Pflugers Archiv: European Journal of Physiology. 2014;466(3):529–40.

Kiliszek M, Kozluk E, Franaszczyk M, Lodzinski P, Piatkowska A, Ploski R, Opolski G. The 4q25, 1q21, and 16q22 polymorphisms and recurrence of atrial fibrillation after pulmonary vein isolation. Arch Med Sci. 2016;12(1):38–44.

Kirchhof P, Fabritz L. Of hammers and screws renin-angiotensin-aldosterone system inhibition to prevent atrial fibrillation in patients with hypertension. Eur Heart J. 2014;35(18):1169–71.

Kirchhof P, Eckardt L, Franz MR, Mönnig G, Loh P, Wedekind H, Schulze-Bahr E, Breithardt G, Haverkamp W. Prolonged atrial action potential durations and polymorphic atrial Tachyarrhythmias in patients with long QT syndrome. J Cardiovasc Electrophysiol. 2003;14 (10):1027–33.

Kirchhof P, Sipido KR, Cowie MR, Eschenhagen T, Fox KA, Katus H, Schroeder S, Schunkert H, Priori S. The continuum of personalized cardiovascular medicine a position paper of the European Society of Cardiology. Eur Heart J. 2014;35(46):3250–7.

Kirchhof P, Benussi S, Kotecha D, Ahlsson A, Atar D, Casadei B, Castella M, Diener HC, Heidbuchel H, Hendriks J, Hindricks G, Manolis AS, Oldgren J, Popescu BA, Schotten U, van Putte B, Vardas P, Agewall S, Camm J, Baron Esquivias G, Budts W, Carerj S, Casselman F, Coca A, de CR, Deftereos S, Dobrev D, Ferro JM, Filippatos G, Fitzsimons D, Gorenek B, Guenoun M, Hohnloser SH, Kolh P, Lip GY, Manolis A, McMurray J, Ponikowski P, Rosenhek R, Ruschitzka F, Savelieva I, Sharma S, Suwalski P, Tamargo JL, Taylor CJ, van Gelder IC, Voors AA, Windecker S, Zamorano JL, Zeppenfeld K. 2016 ESC guidelines for the management of atrial fibrillation developed in collaboration with EACTS. Eur Heart J. 2016;37(38):2893–962.

Kishore A, Vail A, Majid A, Dawson J, Lees KR, Tyrrell PJ, Smith CJ. Detection of atrial fibrillation after ischemic stroke or transient ischemic attack a systematic review and meta-analysis. Stroke. 2014;45(2):520–6.

Krahn AD, Manfreda J, Tate RB, Mathewson FA, Cuddy TE. The natural history of atrial fibrillation incidence, risk factors, and prognosis in the Manitoba follow-up study. Am J Med. 1995;98(5):476–84.

Krijthe BP, Kunst A, Benjamin EJ, Lip GY, Franco OH, Hofman A, Witteman JC, Stricker BH, Heeringa J. Projections on the number of individuals with atrial fibrillation in the European Union, from 2000 to 2060. Eur Heart J. 2013;34(35):2746–51.

Kunamalla A, Ng J, Parini V, Yoo S, McGee KA, Tomson TT, Gordon D, Thorp EB, Lomasney J, Zhang Q, Shah S, Browne S, Knight BP, Passman R, Goldberger JJ, Aistrup G, Arora R. Constitutive expression of a dominant-negative TGF-β type II receptor in the posterior left atrium leads to beneficial remodeling of atrial fibrillation substrate. Circ Res. 2016;119 (1):69–82.

Kusano KF, Taniyama M, Nakamura K, Miura D, Banba K, Nagase S, Morita H, Nishii N, Watanabe A, Tada T, Murakami M, Miyaji K, Hiramatsu S, Nakagawa K, Tanaka M, Miura A, Kimura H, Fuke S, Sumita W, Sakuragi S, Urakawa S, Iwasaki J, Ohe T. Atrial fibrillation in patients with Brugada syndrome relationships of gene mutation, electrophysiology, and clinical backgrounds. J Am Coll Cardiol. 2008;51(12):1169–75.

Lehnart SE, Mongillo M, Bellinger A, Lindegger N, Chen BX, Hsueh W, Reiken S, Wronska A, Drew LJ, Ward CW, Lederer WJ, Kass RS, Morley G, Marks AR. Leaky Ca^{2+} release channel/ ryanodine receptor 2 causes seizures and sudden cardiac death in mice. J Clin Invest. 2008;118 (6):2230–45.

Lemoine MD, Duverger JE, Naud P, Chartier D, Qi XY, Comtois P, Fabritz L, Kirchhof P, Nattel S. Arrhythmogenic left atrial cellular electrophysiology in a murine genetic long QT syndrome model. Cardiovasc Res. 2011;92(1):67–74.

Li C, Wang F, Yang Y, Fu F, Xu C, Shi L, Li S, Xia Y, Wu G, Cheng X, Liu H, Wang C, Wang P, Hao J, Ke Y, Zhao Y, Liu M, Zhang R, Gao L, Yu B, Zeng Q, Liao Y, Yang B, Tu X, Wang QK. Significant association of SNP rs2106261 in the ZFHX3 gene with atrial fibrillation in a Chinese Han GeneID population. Hum Genet. 2011;129(3):239–46.

Li J, Liu WD, Yang ZL, Yang YQ. Novel GATA6 loss-of-function mutation responsible for familial atrial fibrillation. Int J Mol Med. 2012;30(4):783–90.

Li RG, Wang Q, Xu YJ, Zhang M, Qu XK, Liu X, Fang WY, Yang YQ. Mutations of the *SCN4B*-encoded sodium channel β4 subunit in familial atrial fibrillation. Int J Mol Med. 2013;32 (1):144–50.

Liang B, Soka M, Christensen AH, Olesen MS, Larsen AP, Knop FK, Wang F, Nielsen JB, Andersen MN, Humphreys D, Mann SA, Huttner IG, Vandenberg JI, Svendsen JH, Haunsø S, Preiss T, Seebohm G, Olesen SP, Schmitt N, Fatkin D. Genetic variation in the two-pore domain potassium channel, TASK-1, may contribute to an atrial substrate for arrhythmogenesis. J Mol Cell Cardiol. 2014;67:69–76.

Lin H, Yin X, Xie Z, Lunetta KL, Lubitz SA, Larson MG, Ko D, Magnani JW, Mendelson MM, Liu C, McManus DD, Levy D, Ellinor PT, Benjamin EJ. Methylome-wide association study of atrial fibrillation in Framingham heart study. Sci Rep. 2017;7:40377.

Ling TY, Wang XL, Chai Q, Lau TW, Koestler CM, Park SJ, Daly RC, Greason KL, Jen J, Wu LQ, Shen WF, Shen WK, Cha YM, Lee HC. Regulation of the SK3 channel by MicroRNA-499 - potential role in atrial fibrillation. Heart Rhythm. 2013;10(7):1001–9.

Lip GY, Nieuwlaat R, Pisters R, Lane DA, Crijns HJ. Refining clinical risk stratification for predicting stroke and thromboembolism in atrial fibrillation using a novel risk factor-based approach the euro heart survey on atrial fibrillation. Chest. 2010;137(2):263–72.

Liu Y, Ni B, Lin Y, Chen XG, Chen M, Hu Z, Zhang F. The rs3807989 G/a polymorphism in *CAV1* is associated with the risk of atrial fibrillation in Chinese Han populations. Pacing Clin Electrophysiol. 2015;38(2):164–70.

London B, Michalec M, Mehdi H, Zhu X, Kerchner L, Sanyal S, Viswanathan PC, Pfahnl AE, Shang LL, Madhusudanan M, Baty CJ, Lagana S, Aleong R, Gutmann R, Ackerman MJ, McNamara DM, Weiss R, Dudley SC. Mutation in glycerol-3-phosphate dehydrogenase 1 like gene (*GPD1-L*) decreases cardiac Na^{+} current and causes inherited arrhythmias. Circulation. 2007;116(20):2260–8.

Lubitz SA, Yin X, Fontes JD, Magnani JW, Rienstra M, Pai M, Villalon ML, Vasan RS, Pencina MJ, Levy D, Larson MG, Ellinor PT, Benjamin EJ. Association between familial atrial fibrillation and risk of new-onset atrial fibrillation. JAMA. 2010;304(20):2263–9.

Lubitz SA, Lunetta KL, Lin H, Arking DE, Trompet S, Li G, Krijthe BP, Chasman DI, Barnard J, Kleber ME, Dörr M, Ozaki K, Smith AV, Müller-Nurasyid M, Walter S, Agarwal SK, Bis JC, Brody JA, Chen LY, Everett BM, Ford I, Franco OH, Harris TB, Hofman A, Kääb S, Mahida S, Kathiresan S, Kubo M, Launer LJ, Macfarlane PW, Magnani JW, McKnight B, McManus DD, Peters A, Psaty BM, Rose LM, Rotter JI, Silbernagel G, Smith JD, Sotoodehnia N, Stott DJ, Taylor KD, Tomaschitz A, Tsunoda T, Uitterlinden AG, van Wagoner DR, Völker U, Völzke H, Murabito JM, Sinner MF, Gudnason V, Felix SB, März W, Chung M, Albert CM, Stricker BH, Tanaka T, Heckbert SR, Jukema JW, Alonso A, Benjamin EJ, Ellinor PT. Novel genetic markers associate with atrial fibrillation risk in Europeans and Japanese. J Am Coll Cardiol. 2014;63(12):1200–10.

Lubitz SA, Brody JA, Bihlmeyer NA, Roselli C, Weng L-C, Christophersen IE, Alonso A, Boerwinkle E, Gibbs RA, Bis JC, Cupples LA, Mohler PJ, Nickerson DA, Muzny D, Perez MV, Psaty BM, Soliman EZ, Sotoodehnia N, Lunetta KL, Benjamin EJ, Heckbert SR, Arking

DE, Ellinor PT, Lin H. Whole exome sequencing in atrial fibrillation. PLoS Genet. 2016;12(9): e1006284.

Lugenbiel P, Thomas D, Kelemen K, Trappe K, Bikou O, Schweizer PA, Voss F, Becker R, Katus HA, Bauer A. Genetic suppression of Gαs protein provides rate control in atrial fibrillation. Basic Res Cardiol. 2012;107(3):265.

Lugenbiel P, Schweizer PA, Katus HA, Thomas D. Antiarrhythmic gene therapy - will biologics replace catheters, drugs and devices? Eur J Pharmacol. 2016;791:264–73.

Lugenbiel P, Wenz F, Syren P, Geschwill P, Govorov K, Seyler C, Frank D, Schweizer PA, Franke J, Weis T, Bruehl C, Schmack B, Ruhparwar A, Karck M, Frey N, Katus HA, Thomas D. TREK-1 ($K_{2P}2.1$) K+ channels are suppressed in patients with atrial fibrillation and heart failure and provide therapeutic targets for rhythm control. Basic Res Cardiol. 2017;112:8.

Lundby A, Ravn LS, Svendsen JH, Olesen SP, Schmitt N. KCNQ1 mutation Q147R is associated with atrial fibrillation and prolonged QT interval. Heart Rhythm. 2007;4(12):1532–41.

Lundby A, Ravn LS, Svendsen JH, Hauns S, Olesen SP, Schmitt N. KCNE3 mutation V17M identified in a patient with lone atrial fibrillation. Cell Physiol Biochem. 2008;21(1–3):47–54.

Luo MH, Li YS, Yang KP. Fibrosis of collagen I and remodeling of Connexin 43 in atrial myocardium of patients with atrial fibrillation. Cardiology. 2007;107(4):248–53.

Ma KJ, Li N, Teng SY, Zhang YH, Sun Q, Gu DF, Pu JL. Modulation of KCNQ1 current by atrial fibrillation-associated KCNE4 (145E/D) gene polymorphism. Chin Med J. 2007;120(2):150–4.

Ma JF, Yang F, Mahida SN, Zhao L, Chen X, Zhang ML, Sun Z, Yao Y, Zhang YX, Zheng GY, Dong J, Feng MJ, Zhang R, Sun J, Li S, Wang QS, Cao H, Benjamin EJ, Ellinor PT, Li YG, Tian XL. *TBX5* mutations contribute to early-onset atrial fibrillation in Chinese and Caucasians. Cardiovasc Res. 2016;109(3):442–50.

Macri V, Mahida SN, Zhang ML, Sinner MF, Dolmatova EV, Tucker NR, McLellan M, Shea MA, Milan DJ, Lunetta KL, Benjamin EJ, Ellinor PT. A novel trafficking-defective HCN4 mutation is associated with early-onset atrial fibrillation. Heart Rhythm. 2014;11(6):1055–62.

Mahida S, Lubitz SA, Rienstra M, Milan DJ, Ellinor PT. Monogenic atrial fibrillation as pathophysiological paradigms. Cardiovasc Res. 2011;89(4):692–700.

Makiyama T, Akao M, Shizuta S, Doi T, Nishiyama K, Oka Y, Ohno S, Nishio Y, Tsuji K, Itoh H, Kimura T, Kita T, Horie M. A novel SCN5A gain-of-function mutation M1875T associated with familial atrial fibrillation. J Am Coll Cardiol. 2008;52(16):1326–34.

McManus DD, Rienstra M, Benjamin EJ. An update on the prognosis of patients with atrial fibrillation. Circulation. 2012;126(10):e143–6. https://doi.org/10.1161/CIRCULATIONAHA.112.129759.

Menezes AR, Lavie CJ, de SA, Milani RV, O'Keefe J, DiNicolantonio JJ, Morin DP, Abi-Samra FM. Lifestyle modification in the prevention and treatment of atrial fibrillation. Prog Cardiovasc Dis. 2015;58(2):117–25.

Mohamed U, Napolitano C, Priori SG. Molecular and electrophysiological bases of catecholaminergic polymorphic ventricular tachycardia. J Cardiovasc Electrophysiol. 2007;18(7):791–7.

Mohler PJ, Schott JJ, Gramolini AO, Dilly KW, Guatimosim S, duBell WH, Song LS, Haurogné K, Kyndt F, Ali ME, Rogers TB, Lederer WJ, Escande D, Le Marec H, Bennett V. Ankyrin-B mutation causes type 4 long-QT cardiac arrhythmia and sudden cardiac death. Nature. 2003;421 (6923):634–9.

Mommersteeg MT, Brown NA, Prall OW, de G-d VC, Harvey RP, Moorman AF, Christoffels VM. Pitx2c and Nkx2-5 are required for the formation and identity of the pulmonary myocardium. Circ Res. 2007a;101(9):902–9.

Mommersteeg MT, Hoogaars WM, Prall OW, de G-d VC, Wiese C, Clout DE, Papaioannou VE, Brown NA, Harvey RP, Moorman AF, Christoffels VM. Molecular pathway for the localized formation of the sinoatrial node. Circ Res. 2007b;100(3):354–62.

Mommersteeg MT, Christoffels VM, Anderson RH, Moorman AF. Atrial fibrillation a developmental point of view. Heart Rhythm. 2009;6(12):1818–24.

Moss AJ, Zareba W, Benhorin J, Locati EH, Hall WJ, Robinson JL, Schwartz PJ, Towbin JA, Vincent GM, Lehmann MH. ECG T-wave patterns in genetically distinct forms of the hereditary long QT syndrome. Circulation. 1995;92(10):2929–34.

Napolitano C, Priori SG. Diagnosis and treatment of catecholaminergic polymorphic ventricular tachycardia. Heart Rhythm. 2007;4(5):675–8.

Nattel S, Burstein B, Dobrev D. Atrial remodeling and atrial fibrillation mechanisms and implications. Circ Arrhythm Electrophysiol. 2008;1(1):62–73.

Neco P, Torrente AG, Mesirca P, Zorio E, Liu N, Priori SG, Napolitano C, Richard S, Benitah JP, Mangoni ME, Gómez AM. Paradoxical effect of increased diastolic Ca^{2+} release and decreased sinoatrial node activity in a mouse model of catecholaminergic polymorphic ventricular tachycardia. Circulation. 2012;126(4):392–401.

Nielsen JB, Graff C, Pietersen A, Lind B, Struijk JJ, Olesen MS, Haunsø S, Gerds TA, Svendsen JH, Køber L, Holst AG. J-shaped association between QTc interval duration and the risk of atrial fibrillation results from the Copenhagen ECG study. J Am Coll Cardiol. 2013;61(25):2557–64.

Nielsen JB, Bentzen BH, Olesen MS, David JP, Olesen SP, Haunsø S, Svendsen JH, Schmitt N. Gain-of-function mutations in potassium channel subunit KCNE2 associated with early-onset lone atrial fibrillation. Biomark Med. 2014;8(4):557–70.

Olesen MS, Bentzen BH, Nielsen JB, Steffensen AB, David JP, Jabbari J, Jensen HK, Haunso S, Svendsen JH, Schmitt N. Mutations in the potassium channel subunit KCNE1 are associated with early-onset familial atrial fibrillation. BMC Med Genet. 2012a;13:24.

Olesen MS, Yuan L, Liang B, Holst AG, Nielsen N, Nielsen JB, Hedley PL, Christiansen M, Olesen S-P, Haunsø S, Schmitt N, Jespersen T, Svendsen JH. High prevalence of long QT syndrome-associated SCN5A variants in patients with early-onset lone atrial fibrillation. Circ Cardiovasc Genet. 2012b;5(4):450–9.

Olesen MS, Refsgaard L, Holst AG, Larsen AP, Grubb S, Haunso S, Svendsen JH, Olesen S-P, Schmitt N, Calloe K. A novel KCND3 gain-of-function mutation associated with early-onset of persistent lone atrial fibrillation. Cardiovasc Res. 2013;98(3):488–95.

Olesen MS, Andreasen L, Jabbari J, Refsgaard L, Haunsø S, Olesen S-P, Nielsen JB, Schmitt N, Svendsen JH. Very early-onset lone atrial fibrillation patients have a high prevalence of rare variants in genes previously associated with atrial fibrillation. Heart Rhythm. 2014;11(2):246–51.

Olson TM, Alekseev AE, Liu XK, Park S, Zingman LV, Bienengraeber M, Sattiraju S, Ballew JD, Jahangir A, Terzic A. $K_V1.5$ channelopathy due to KCNA5 loss-of-function mutation causes human atrial fibrillation. Hum Mol Genet. 2006;15(14):2185–91.

Olson TM, Alekseev AE, Moreau C, Liu XK, Zingman LV, Miki T, Seino S, Asirvatham SJ, Jahangir A, Terzic A. K_{ATP} channel mutation confers risk for vein of Marshall adrenergic atrial fibrillation. Nat Clin Pract Cardiovasc Med. 2007;4(2):110–6.

Otway R, Vandenberg JI, Fatkin D. Atrial fibrillation - a new cardiac channelopathy. Heart Lung Circ. 2007a;16(5):356–60.

Otway R, Vandenberg JI, Guo G, Varghese A, Castro ML, Liu J, Zhao J, Bursill JA, Wyse KR, Crotty H, Baddeley O, Walker B, Kuchar D, Thorburn C, Fatkin D. Stretch-sensitive KCNQ1 mutation a link between genetic and environmental factors in the pathogenesis of atrial fibrillation? J Am Coll Cardiol. 2007b;49(5):578–86.

Oyen N, Ranthe MF, Carstensen L, Boyd HA, Olesen MS, Olesen S-P, Wohlfahrt J, Melbye M. Familial aggregation of lone atrial fibrillation in young persons. J Am Coll Cardiol. 2012;60(10):917–21.

Pappone C, Radinovic A, Manguso F, Vicedomini G, Sala S, Sacco FM, Ciconte G, Saviano M, Ferrari M, Sommariva E, Sacchi S, Ciaccio C, Kallergis EM, Santinelli V. New-onset atrial fibrillation as first clinical manifestation of latent Brugada syndrome prevalence and clinical significance. Eur Heart J. 2009;30(24):2985–92.

Parvez B, Vaglio J, Rowan S, Muhammad R, Kucera G, Stubblefield T, Carter S, Roden D, Darbar D. Symptomatic response to antiarrhythmic drug therapy is modulated by a common single nucleotide polymorphism in atrial fibrillation. J Am Coll Cardiol. 2012;60(6):539–45.

Pérez-Serra A, Campuzano O, Brugada R. Update about atrial fibrillation genetics. Curr Opin Cardiol. 2017. https://doi.org/10.1097/HCO.0000000000000387.

Posch MG, Boldt LH, Polotzki M, Richter S, Rolf S, Perrot A, Dietz R, Ozcelik C, Haverkamp W. Mutations in the cardiac transcription factor GATA4 in patients with lone atrial fibrillation. Eur J Med Genet. 2010;53(4):201–3.

Potpara TS, Lip GY. A brief history of 'lone' atrial fibrillation from 'a peculiar pulse irregularity' to a modern public health concern. Curr Pharm Des. 2015;21(5):679–96.

Priori SG, Pandit SV, Rivolta I, Berenfeld O, Ronchetti E, Dhamoon A, Napolitano C, Anumonwo J, Di Barletta MR, Gudapakkam S, Bosi G, Stramba-Badiale M, Jalife J. A novel form of short QT syndrome (SQT3) is caused by a mutation in the *KCNJ2* gene. Circ Res. 2005;96(7):800–7.

Ravn LS, Aizawa Y, Pollevick GD, Hofman-Bang J, Cordeiro JM, Dixen U, Jensen G, Wu Y, Burashnikov E, Haunso S, Guerchicoff A, Hu D, Svendsen JH, Christiansen M, Antzelevitch C. Gain of function in I_{Ks} secondary to a mutation in *KCNE5* associated with atrial fibrillation. Heart Rhythm. 2008;5(3):427–35.

Restier L, Cheng L, Sanguinetti MC. Mechanisms by which atrial fibrillation-associated mutations in the S1 domain of KCNQ1 slow deactivation of I_{Ks} channels. J Physiol. 2008;586 (17):4179–91.

Roberts JD, Marcus GM. The burgeoning field of ablatogenomics. Circ Arrhythm Electrophysiol. 2015;8(2):258–60.

Roberts JD, Marcus GM. Ablatogenomics can genotype guide catheter ablation for cardiac arrhythmias? Pharmacogenomics. 2016. https://doi.org/10.2217/pgs-2016-0114.

Roberts JD, Hu D, Heckbert SR, Alonso A, Dewland TA, Vittinghoff E, Liu Y, Psaty BM, Olgin JE, Magnani JW, Huntsman S, Burchard EG, Arking DE, Bibbins-Domingo K, Harris TB, Perez MV, Ziv E, Marcus GM. Genetic investigation into the differential risk of atrial fibrillation among black and white individuals. JAMA Cardiol. 2016;1(4):442–50.

Rodriguez CJ, Soliman EZ, Alonso A, Swett K, Okin PM, Goff DC, Heckbert SR. Atrial fibrillation incidence and risk factors in relation to race-ethnicity and the population attributable fraction of atrial fibrillation risk factors the multi-ethnic study of atherosclerosis. Ann Epidemiol. 2015;25 (2):71–6, 76.e1. https://doi.org/10.1016/j.annepidem.2014.11.024.

Sanna T, Diener H-C, Passman RS, Di Lazzaro V, Bernstein RA, Morillo CA, Rymer MM, Thijs V, Rogers T, Beckers F, Lindborg K, Brachmann J. Cryptogenic stroke and underlying atrial fibrillation. N Engl J Med. 2014;370(26):2478–86.

Schmidt C, Wiedmann F, Voigt N, Zhou XB, Heijman J, Lang S, Albert V, Kallenberger S, Ruhparwar A, Szabó G, Kallenbach K, Karck M, Borggrefe M, Biliczki P, Ehrlich JR, Baczkó I, Lugenbiel P, Schweizer PA, Donner BC, Katus HA, Dobrev D, Thomas D. Upregulation of $K_{2P}3.1$ K^{+} current causes action potential shortening in patients with chronic atrial fibrillation. Circulation. 2015;132(2):82–92.

Schmidt C, Wiedmann F, Kallenberger SM, Ratte A, Schulte JS, Scholz B, Müller FU, Voigt N, Zafeiriou M-P, Ehrlich JR, Tochtermann U, Veres G, Ruhparwar A, Karck M, Katus HA, Thomas D. Stretch-activated two-pore-domain (K2P) potassium channels in the heart. Focus on atrial fibrillation and heart failure. Prog Biophys Mol Biol. 2017;130(Pt B):233–43. https://doi.org/10.1016/j.pbiomolbio.2017.05.004.

Schnabel RB, Yin X, Gona P, Larson MG, Beiser AS, McManus DD, Newton-Cheh C, Lubitz SA, Magnani JW, Ellinor PT, Seshadri S, Wolf PA, Vasan RS, Benjamin EJ, Levy D. 50 year trends in atrial fibrillation prevalence, incidence, risk factors, and mortality in the Framingham heart study a cohort study. Lancet. 2015;386(9989):154–62.

Schotten U, Verheule S, Kirchhof P, Goette A. Pathophysiological mechanisms of atrial fibrillation a translational appraisal. Physiol Rev. 2011;91(1):265–325.

Schwartz P, Priori S, Spazzolini G. Genotype-phenotype correlation in the long-QT syndrome. ACC Curr J Rev. 2001;10(3):74.

Schwartz PJ, Crotti L, Insolia R. Long-QT syndrome from genetics to management. Circ Arrhythm Electrophysiol. 2012;5(4):868–77.

Shan J, Xie W, Betzenhauser M, Reiken S, Chen BX, Wronska A, Marks AR. Calcium leak through ryanodine receptors leads to atrial fibrillation in 3 mouse models of catecholaminergic polymorphic ventricular tachycardia. Circ Res. 2012;111(6):708–17.

Sinner MF, Tucker NR, Lunetta KL, Ozaki K, Smith JG, Trompet S, Bis JC, Lin H, Chung MK, Nielsen JB, Lubitz SA, Krijthe BP, Magnani JW, Ye J, Gollob MH, Tsunoda T, Müller-Nurasyid M, Lichtner P, Peters A, Dolmatova E, Kubo M, Smith JD, Psaty BM, Smith NL, Jukema JW, Chasman DI, Albert CM, Ebana Y, Furukawa T, Macfarlane PW, Harris TB, Darbar D, Dörr M, Holst AG, Svendsen JH, Hofman A, Uitterlinden AG, Gudnason V, Isobe M, Malik R, Dichgans M, Rosand J, van Wagoner DR, Benjamin EJ, Milan DJ, Melander O, Heckbert SR, Ford I, Liu Y, Barnard J, Olesen MS, Stricker BH, Tanaka T, Kääb S, Ellinor PT. Integrating genetic, transcriptional, and functional analyses to identify 5 novel genes for atrial fibrillation. Circulation. 2014;130(15):1225–35.

Soucek R, Thomas D, Kelemen K, Bikou O, Seyler C, Voss F, Becker R, Koenen M, Katus HA, Bauer A. Genetic suppression of atrial fibrillation using a dominant-negative ether-a-go-go-related gene mutant. Heart Rhythm. 2012;9(2):265–72.

Stewart S, Hart CL, Hole DJ, McMurray JJ. A population-based study of the long-term risks associated with atrial fibrillation 20-year follow-up of the Renfrew/Paisley study. Am J Med. 2002;113(5):359–64.

Stranger BE, Forrest MS, Dunning M, Ingle CE, Beazley C, Thorne N, Redon R, Bird CP, Grassi A, de Lee C, Tyler-Smith C, Carter N, Scherer SW, Tavaré S, Deloukas P, Hurles ME, Dermitzakis ET. Relative impact of nucleotide and copy number variation on gene expression phenotypes. Science. 2007;315(5813):848–53.

Suetomi T, Yano M, Uchinoumi H, Fukuda M, Hino A, Ono M, Xu X, Tateishi H, Okuda S, Doi M, Kobayashi S, Ikeda Y, Yamamoto T, Ikemoto N, Matsuzaki M. Mutation-linked defective interdomain interactions within ryanodine receptor cause aberrant Ca^{2+}release leading to catecholaminergic polymorphic ventricular tachycardia. Circulation. 2011;124(6):682–94.

Tan AY, Zimetbaum P. Atrial fibrillation and atrial fibrosis. J Cardiovasc Pharmacol. 2011;57 (6):625–9.

Terrenoire C, Simhaee D, Kass RS. Role of sodium channels in propagation in heart muscle how subtle genetic alterations result in major arrhythmic disorders. J Cardiovasc Electrophysiol. 2007;18(8):900–5.

Trappe K, Thomas D, Bikou O, Kelemen K, Lugenbiel P, Voss F, Becker R, Katus HA, Bauer A. Suppression of persistent atrial fibrillation by genetic knockdown of caspase 3A pre-clinical pilot study. Eur Heart J. 2013;34(2):147–57.

Tribulova N, Egan Benova T, Szeiffova Bacova B, Viczenczova C, Barancik M. New aspects of pathogenesis of atrial fibrillation: remodelling of intercalated discs. J Physiol Pharmacol. 2015;66(5):625–34.

Tsai CT, Hsieh CS, Chang SN, Chuang EY, Juang JMJ, Lin LY, Lai LP, Hwang JJ, Chiang FT, Lin JL. Next-generation sequencing of nine atrial fibrillation candidate genes identified novel de novo mutations in patients with extreme trait of atrial fibrillation. J Med Genet. 2015;52 (1):28–36.

Tsai CT, Hsieh CS, Chang SN, Chuang EY, Ueng KC, Tsai CF, Lin TH, Wu CK, Lee JK, Lin LY, Wang YC, Yu CC, Lai LP, Tseng CD, Hwang JJ, Chiang FT, Lin JL. Genome-wide screening identifies a *KCNIP1* copy number variant as a genetic predictor for atrial fibrillation. Nat Commun. 2016;7:10190.

Tucker NR, Ellinor PT. Emerging directions in the genetics of atrial fibrillation. Circ Res. 2014;114 (9):1469–82.

Tucker NR, Mahida S, Ye J, Abraham EJ, Mina JA, Parsons VA, McLellan MA, Shea MA, Hanley A, Benjamin EJ, Milan DJ, Lin H, Ellinor PT. Gain-of-function mutations in GATA6 lead to atrial fibrillation. Heart Rhythm. 2017;14(2):284–91.

Vlachos K, Letsas KP, Korantzopoulos P, Liu T, Georgopoulos S, Bakalakos A, Karamichalakis N, Xydonas S, Efremidis M, Sideris A. Prediction of atrial fibrillation development and progression current perspectives. World J Cardiol. 2016;8(3):267–76.

Voigt N, Trausch A, Knaut M, Matschke K, Varró A, van Wagoner DR, Nattel S, Ravens U, Dobrev D. Left-to-right atrial inward rectifier potassium current gradients in patients with paroxysmal versus chronic atrial fibrillation. Circ Arrhythm Electrophysiol. 2010;3(5):472–80.

Wakili R, Voigt N, Kääb S, Dobrev D, Nattel S. Recent advances in the molecular pathophysiology of atrial fibrillation. J Clin Invest. 2011;121(8):2955–68.

Wang TJ, Larson MG, Levy D, Vasan RS, Leip EP, Wolf PA, D'Agostino RB, Murabito JM, Kannel WB, Benjamin EJ. Temporal relations of atrial fibrillation and congestive heart failure and their joint influence on mortality the Framingham heart study. Circulation. 2003;107 (23):2920–5.

Wang J, Sun YM, Yang YQ. Mutation spectrum of the GATA4 gene in patients with idiopathic atrial fibrillation. Mol Biol Rep. 2012;39(8):8127–35.

Wang XH, Huang CX, Wang Q, Li RG, Xu YJ, Liu X, Fang WY, Yang YQ. A novel GATA5 loss-of-function mutation underlies lone atrial fibrillation. Int J Mol Med. 2013;31(1):43–50.

Wang J, Zhang DF, Sun YM, Li RG, Qiu XB, Qu XK, Liu X, Fang WY, Yang YQ. NKX2-6 mutation predisposes to familial atrial fibrillation. Int J Mol Med. 2014;34(6):1581–90.

Wang C, Wu M, Qian J, Li B, Tu X, Xu C, Li S, Chen S, Zhao Y, Huang Y, Shi L, Cheng X, Liao Y, Chen Q, Xia Y, Yao W, Wu G, Cheng M, Wang QK. Identification of rare variants in *TNNI3* with atrial fibrillation in a Chinese GeneID population. Mol Gen Genomics. 2016;291 (1):79–92.

Watanabe H, Koopmann TT, Le Scouarnec S, Yang T, Ingram CR, Schott J-J, Demolombe S, Probst V, Anselme F, Escande D, Wiesfeld AC, Pfeufer A, Kääb S, Wichmann HE, Hasdemir C, Aizawa Y, Wilde AA, Roden DM, Bezzina CR. Sodium channel β1 subunit mutations associated with Brugada syndrome and cardiac conduction disease in humans. J Clin Invest. 2008;118(6):2260–8.

Watanabe H, Darbar D, Kaiser DW, Jiramongkolchai K, Chopra S, Donahue BS, Kannankeril PJ, Roden DM. Mutations in sodium channel β1- and β2-subunits associated with atrial fibrillation. Circ Arrhythm Electrophysiol. 2009;2(3):268–75.

Weeke P, Muhammad R, Delaney JT, Shaffer C, Mosley JD, Blair M, Short L, Stubblefield T, Roden DM, Darbar D. Whole-exome sequencing in familial atrial fibrillation. Eur Heart J. 2014;35(36):2477–83.

Wei T, Song J, Xu M, Lv L, Liu C, Shen J, Huang Y. *NEURL* rs6584555 and *CAND2* rs4642101 contribute to postoperative atrial fibrillation a prospective study among Chinese population. Oncotarget. 2016;7(27):42617–24.

Wettwer E, Hála O, Christ T, Heubach JF, Dobrev D, Knaut M, Varró A, Ravens U. Role of I_{Kur} in controlling action potential shape and contractility in the human atrium. Circulation. 2004;110 (16):2299–306.

Wijffels MC, Kirchhof CJ, Dorland R, Allessie MA. Atrial fibrillation begets atrial fibrillation. A study in awake chronically instrumented goats. Circulation. 1995;92(7):1954–68.

Wolf PA, Abbott RD, Kannel WB. Atrial fibrillation as an independent risk factor for stroke the Framingham study. Stroke. 1991;22(8):983–8.

Workman AJ, Kane KA, Rankin AC. The contribution of ionic currents to changes in refractoriness of human atrial myocytes associated with chronic atrial fibrillation. Cardiovasc Res. 2001;52 (2):226–35.

Workman AJ, Kane KA, Rankin AC. Cellular bases for human atrial fibrillation. Heart Rhythm. 2008;5(6 Suppl):1–6.

Wynn GJ, Todd DM, Webber M, Bonnett L, McShane J, Kirchhof P, Gupta D. The European heart rhythm association symptom classification for atrial fibrillation validation and improvement through a simple modification. Europace. 2014;16(7):965–72.

Xia M, Jin Q, Bendahhou S, He Y, Larroque MM, Chen Y, Zhou Q, Yang Y, Liu Y, Liu B, Zhu Q, Zhou Y, Lin J, Liang B, Li L, Dong X, Pan Z, Wang R, Wan H, Qiu W, Xu W, Eurlings P, Barhanin J, Chen Y. A $K_{ir}2.1$ gain-of-function mutation underlies familial atrial fibrillation. Biochem Biophys Res Commun. 2005;332(4):1012–9.

Xie WH, Chang C, Xu YJ, Li RG, Qu XK, Fang WY, Liu X, Yang YQ. Prevalence and spectrum of Nkx2.5 mutations associated with idiopathic atrial fibrillation. Clinics (Sao Paulo, Brazil). 2013;68(6):777–84.

Yamase Y, Kato K, Horibe H, Ueyama C, Fujimaki T, Oguri M, Arai M, Watanabe S, Murohara T, Yamada Y. Association of genetic variants with atrial fibrillation. Biomedical Reports. 2016;4 (2):178–82.

Yang Y, Xia M, Jin Q, Bendahhou S, Shi J, Chen Y, Liang B, Lin J, Liu Y, Liu B, Zhou Q, Zhang D, Wang R, Ma N, Su X, Niu K, Pei Y, Xu W, Chen Z, Wan H, Cui J, Barhanin J, Chen Y. Identification of a KCNE2 gain-of-function mutation in patients with familial atrial fibrillation. Am J Hum Genet. 2004;75(5):899–905.

Yang Y, Li J, Lin X, Yang Y, Hong K, Wang L, Liu J, Li L, Yan D, Liang D, Xiao J, Jin H, Wu J, Zhang Y, Chen YH. Novel KCNA5 loss-of-function mutations responsible for atrial fibrillation. J Hum Genet. 2009;54(5):277–83.

Yang T, Yang P, Roden DM, Darbar D. Novel KCNA5 mutation implicates tyrosine kinase signaling in human atrial fibrillation. Heart Rhythm. 2010;7(9):1246–52.

Yang YQ, Wang MY, Zhang XL, Tan HW, Shi HF, Jiang WF, Wang XH, Fang WY, Liu X. GATA4 loss-of-function mutations in familial atrial fibrillation. Clin Chim Acta. 2011;412 (19–20):1825–30.

Yang YQ, Li L, Wang J, Zhang XL, Li RG, Xu YJ, Tan HW, Wang XH, Jiang JQ, Fang WY, Liu X. GATA6 loss-of-function mutation in atrial fibrillation. Eur J Med Genet. 2012a;55 (10):520–6.

Yang YQ, Wang J, Wang XH, Wang Q, Tan HW, Zhang M, Shen FF, Jiang JQ, Fang WY, Liu X. Mutational spectrum of the GATA5 gene associated with familial atrial fibrillation. Int J Cardiol. 2012b;157(2):305–7.

Yang YQ, Wang XH, Tan HW, Jiang WF, Fang WY, Liu X. Prevalence and spectrum of GATA6 mutations associated with familial atrial fibrillation. Int J Cardiol. 2012c;155(3):494–6.

Yao JL, Zhou YF, Yang XJ, Qian XD, Jiang WP. *KCNN3* SNP rs13376333 on chromosome 1q21 confers increased risk of atrial fibrillation. Int Heart J. 2015;56(5):511–5.

Ye J, Tucker NR, Weng LC, Clauss S, Lubitz SA, Ellinor PT. A functional variant associated with atrial fibrillation regulates PITX2c expression through TFAP2a. Am J Hum Genet. 2016;99 (6):1281–91.

Yu H, Xu JH, Song HM, Zhao L, Xu WJ, Wang J, Li RG, Xu L, Jiang WF, Qiu XB, Jiang JQ, Qu XK, Liu X, Fang WY, Jiang JF, Yang YQ. Mutational spectrum of the NKX2-5 gene in patients with lone atrial fibrillation. Int J Med Sci. 2014;11(6):554–63.

Zellerhoff S, Pistulli R, Mönnig G, Hinterseer M, Beckmann BM, Köbe J, Steinbeck G, Kääb S, Haverkamp W, Fabritz L, Gradaus R, Breithardt G, Schulze-Bahr E, Böcker D, Kirchhof P. Atrial arrhythmias in long-QT syndrome under daily life conditions a nested case control study. J Cardiovasc Electrophysiol. 2009;20(4):401–7.

Zhang X, Chen S, Yoo S, Chakrabarti S, Zhang T, Ke T, Oberti C, Yong SL, Fang F, Li L, de La Fuente R, Wang L, Chen Q, Wang QK. Mutation in nuclear pore component *NUP155* leads to atrial fibrillation and early sudden cardiac death. Cell. 2008;135(6):1017–27.

Zhang C, Yuan GH, Cheng ZF, Xu MW, Hou LF, Wei FP. The single nucleotide polymorphisms of $K_{ir}3.4$ gene and their correlation with lone paroxysmal atrial fibrillation in Chinese Han population. Heart Lung Circul. 2009;18(4):257–61.

Zhang Y, Fraser JA, Jeevaratnam K, Hao X, Hothi SS, Grace AA, Lei M, Huang CL. Acute atrial arrhythmogenicity and altered Ca^{2+} homeostasis in murine RyR2-P2328S hearts. Cardiovasc Res. 2011;89(4):794–804.

Zhao J, Yao H, Li Z, Wang L, Liu G, Wang DW, Wang DW, Liang Z. A novel nonsense mutation in *LMNA* gene identified by exome sequencing in an atrial fibrillation family. Eur J Med Genet. 2016;59(8):396–400.

Zhao LQ, Zhang GB, Wen ZJ, Huang CK, Wu HQ, Xu J, Qi BZ, Wang ZM, Shi YY, Liu SW. Common variants predict recurrence after nonfamilial atrial fibrillation ablation in Chinese Han population. Int J Cardiol. 2017;227:360–6.

Zhou M, Liao Y, Tu X. The role of transcription factors in atrial fibrillation. J Thorac Dis. 2015;7 (2):152–8.

Zoni-Berisso M, Lercari F, Carazza T, Domenicucci S. Epidemiology of atrial fibrillation European perspective. Clin Epidemiol. 2014;6:213–20.

Genetic Testing for Inheritable Cardiac Channelopathies

13

Florence Kyndt, Jean-Baptiste Gourraud, and Julien Barc

Abstract

Inheritable cardiac channelopathies (ICC) are defined as primary electrical disorders without identifiable cardiac structural abnormalities and are mostly encountered in young adults (under 40 years). Diagnosis of ICC is often established after the first symptoms such as recurrent palpitations and syncope or more dramatically after unexplained sudden cardiac death (SCD). In this context, familial clinical screening coupled with genetic testing are required to prevent additional (fatal) arrhythmia events in relatives. This review presents an update of the ICC-associated genes and proposes a screening hierarchy according to the phenotype. The impact of the new sequencing technologies on the genetic testing as well as on the patient management will be also discussed.

F. Kyndt (✉)
l'institut du thorax, INSERM, CNRS, UNIV Nantes, Nantes, France

CHU Nantes, Service de Génétique Médicale, Nantes, France
e-mail: florence.kyndt@chu-nantes.fr

J.-B. Gourraud
l'institut du thorax, INSERM, CNRS, UNIV Nantes, Nantes, France

L'institut du thorax, CHU Nantes, Service de Cardiologie, Nantes, France
e-mail: jeanbaptiste.gourraud@chu-nantes.fr

J. Barc
l'institut du thorax, INSERM, CNRS, UNIV Nantes, Nantes, France
e-mail: julien.barc@inserm.fr

D. Thomas, C. A. Remme (eds.), *Channelopathies in Heart Disease*, Cardiac and Vascular Biology, https://doi.org/10.1007/978-3-319-77812-9_13

13.1 Genetic Testing and Cardiac Channelopathies

Genetic diagnostics were proposed 20 years ago when the first associated channelopathy genes were discovered (Wilde and Behr 2013). Today, genetic information is used to confirm a clinical diagnosis and as a powerful preventive tool to identify family members at risk of developing electrical disorders despite having a normal ECG. Genetic diagnostics are currently performed on six highly phenotypically characterized channelopathies such as Brugada syndrome (BrS), cardiac conduction defects (CCD), long QT syndrome (LQTS), short QT syndrome (SQTS), catecholaminergic polymorphic ventricular tachycardia (CPVT) and arrhythmogenic right ventricular cardiomyopathy (ARVC). These channelopathies present specific ECG indices at baseline or are unmasked by provocative tests that are often coupled with a singular circumstance of an arrhythmic event outcome. Such clinical information guides genetic testing towards candidate genes harbouring the highest rate of causal mutations. Therefore, the added value of molecular diagnostics could be inherent in the preventive screening of relatives who are either negative or borderline after clinical evaluation. Furthermore, according to the gene identified as being associated with the channelopathy, therapy and care can be adapted accordingly (Imbrici et al. 2016). However, in certain cases, such as young patients presenting idiopathic ventricular fibrillation or (resuscitated) sudden cardiac death, neither ECG indices nor relatives can be identified. In this context, genetic testing could uncover variant(s)/gene(s) and guide clinicians to a diagnosis. However, extreme caution must be taken vis-à-vis our ability to interpret rare genetic variants without supporting clinical and functional or familial segregation data (Wijeyeratne and Behr 2017).

13.2 Present Context of Genetic Testing

In the last few years, the number of genes associated with arrhythmias has considerably increased. However, the major genes that account for a large fraction of the cases remain limited. “Minor” genes are usually poorly characterized in terms of function and pathophysiological role. As a consequence, the identification of variations in these genes often leads to results that are difficult to interpret. Genetic testing interpretation is also particularly challenging owing to the overlapping, variable and incompletely penetrant nature of the clinical presentations of the channelopathies.

Next-generation sequencing (NGS) technologies rapidly developed in diagnostic laboratories and have become a routine approach. However, the sequencing scale is a matter of debate. Indeed, the technologies allow to focus the screening on panels of 20–200 genes, explore the whole coding region of the genome (whole exome sequencing (WES)) or even give access to the whole genome of a patient. Target resequencing panels (TRP) is commonly employed for diagnostic purposes given the reasonable cost, high sequencing coverage and limited interpretation and ethical issues. Different TRP strategies can be applied from a cardiac panel with 60–180 genes covering "all" (channelopathies and cardiomyopathies) known genes or an arrhythmia panel focused on 20–60 genes directly associated with genetic electrical defects or even a restricted panel containing only the "key" genes corresponding to the most prevalent cardiac arrhythmia genes. The use of small panel focusing on well-characterized genes aims to reduce the number of genes tested and then limit the identification of variant with uncertain interpretation. The pros and cons related to the different approaches using NGS technologies are discussed further.

13.3 Genetic Testing for Inheritable Cardiac Channelopathies: Daily Practice

13.3.1 Brugada Syndrome (BrS)

13.3.1.1 Clinical Description

Brugada syndrome (BrS) was first described in 1992 (Brugada and Brugada 1992) and was based on a familial form of ventricular fibrillation associated with a singular ECG pattern. The latest guidelines define a diagnosis of BrS diagnostic as follows: "patients with ST-segment elevation with type 1 morphology $\geq$2 mm in one or more leads among the right precordial leads V1 and/or V2 positioned in the second, third, or fourth intercostal space, occurring either spontaneously or after provocative drug test with intravenous administration of sodium channel blockers" (Priori et al. 2015). The prevalence of BrS is estimated at approximately 0.05% in Europe and is higher in Asia (0.12%)—particularly in Thailand, the Philippines and Japan. The mean age on diagnosis is 40. BrS predominantly affects (80%) males after puberty (Andorin et al. 2016) and is associated with sudden cardiac death occurring mostly at rest (Tomaselli and Barth 2016).

13.3.1.2 Genetic Testing

The first gene identified was the *SCN5A* gene encoding for the major cardiac sodium channel NaV1.5 (Chen et al. 1998). The prevalence of mutations in *SCN5A* is from 20 to 25% of the cases and remains the major gene in BrS (Crotti et al. 2012; Le Scouarnec et al. 2015). So far, 24 genes have been described as harbouring rare genetic variations in BrS patients (Table 13.1). They can be categorized in three main groups according to whether they affect the sodium (I_{Na}; *SCN5A, GPD1L, SCN1B, SCN3B, RANGRF, SCN2B, PKP2, SLMAP, SCN10A* and *FGF12*), potassium (I_K; *KCNJ8, KCNH2, KCNE3, KCND3, KCNE5, KCND2, SEMA3A, ABCC9* and *KCNAB2*) or L-type calcium ($I_{Ca\text{-}L}$; *CACNA1C, CACNB2B* and *CACNA2D1*) currents. The last two associated genes affect non-selective channels (*HCN4* and *TRPM4*). It should be noted that among the BrS-associated genes, only two (*GPD1L* and *KCNAB2*) have been uncovered by the powerful and hypothesis-free familial approach. Moreover, apart from the *SCN5A* gene, the contribution of other genes remains extremely low or uncertain for some of them (Fig. 13.1) (Le Scouarnec et al. 2015).

Table 13.1 Genes identified in Brugada syndrome (BrS)

Type	Locus	Gene	Protéine, fonction/courant	Effet des mutations	Transmission	OMIM
BrS1	3p21	*SCN5A*	$Na_v1.5$ (alpha subunit sodium channel), depolarization/I_{Na}	$\downarrow I_{Na}$	AD	600163
BrS2	3p22.3	*GPD1L*	Glycerol-3-phosphate deshydrogenase 1-like	$\downarrow I_{Na}$	AD	611778
BrS3	12p13.3	*CACNA1C*	$Ca_v1.2$ (alpha1C subunit calcium channel), depolarization/$I_{Ca\text{-}L}$	$\downarrow I_{Ca\text{-}L}$	AD	114205
BrS4	10p12	*CACNB2*	Ca_{vB2b} (beta2b subunit calcium channel), depolarization/$I_{Ca\text{-}L}$	$\downarrow I_{Ca\text{-}L}$	AD	600003
BrS5	19q13.1	*SCN1B*	$Na_v\beta1$ (beta1 subunit sodium channel), depolarization/I_{Na}	$\downarrow I_{Na}$	AD	600235
BrS6	11q13-q14	*KCNE3*	MiRP2 (beta subunit potassium channel), repolarization/I_{to}	$I_{to/IKs}$	AD	604433
BrS7	11q23.3	*SCN3B*	$Na_v\beta3$ (beta3 subunit sodium channel), depolarization/I_{Na}	$\downarrow I_{Na}$	AD	608214
BrS8	15q24-q25	*HCN4*	Hyperpolarization-activated cyclic nucleotide-gated potassium channel 4/if current	Courant de fuite	AD	605206
BrS9	1p13.3	*KCND3*	Kv4.3 (alpha subunit potassium channel), repolarization/I_{to}	I_{to}	AD	605411
BrS10	12p11.23	*KCNJ8*	Kir6.1 (ATP potassium channel)/I_{KATP}	I_{KATP}	AD	600935
BrS11	7q21-q22	*CACNA2D1*	$Ca_{v2\delta1}$ (alpha2d1 subunit calcium channel), depolarization/$I_{Ca\text{-}L}$	$\downarrow I_{Ca\text{-}L}$	AD	114204
BrS12	Xq22.3	*KCNE5*	KCNE1-L (beta subunit potassium channel), repolarization/I_{to}	I_{to}/I_{Ks}	lié à l'X	300328
BrS13	17p13.1	*RANGRF/MOG1*	Ran guanine nucleotide release factor/multicopy suppressor of Gsp1	$\downarrow I_{Na}$	AD	607954
BrS14	3p14.3	*SLMAP*	Sarcolemma-associated protein	$\downarrow I_{Na}$	AD	602701
BrS15	19q13.33	*TRPM4*	Transient receptor potential cation channel subfamily M member 4	↓ et NSCCa	AD	606936
BrS16	11q23	*SCN2B*	$Na_v\beta2$ (beta2 subunit sodium channel), depolarization/I_{Na}	$\downarrow I_{Na}$	AD	601327
BrS17	3q28-q29	*FGF12*	Fibroblast growth factor 12	$\downarrow I_{Na}$	AD	601513
BrS18	12p11	*PKP2*	Plakophilin-2	$\downarrow I_{Na}$	AD	602861

BrS19	12p12.1	*ABCC9*	ATP-sensitive potassium channels	I_{KATP}	AD	601439
BrS20	7q31.31	*KCND2*	Alpha subunit of the Kv4.2 potassium channel	I_{to}	AD	605410
BrS21	7q36.1	*KCNH2*	Kv11.1 (Alpha subunit potassium channel), repolarization/IKr	I_{Kr}	AD	152427
BrS22	7q21.11	*SEMA3A*	Semaphorin family protein	I_{to}	AD	603961
BrS23	3p22.2	*SCN10A*	$Na_v1.8$ (alpha subunit sodium channel)	$\downarrow I_{Na}$	AD	604427
BrS24	1p36.31	*KCNAB2*	Potassium voltage-gated channel subfamily A regulatory beta subunit 2	I_{to}	AD	601142

AD autosomal dominant, *AR* autosomal recessive

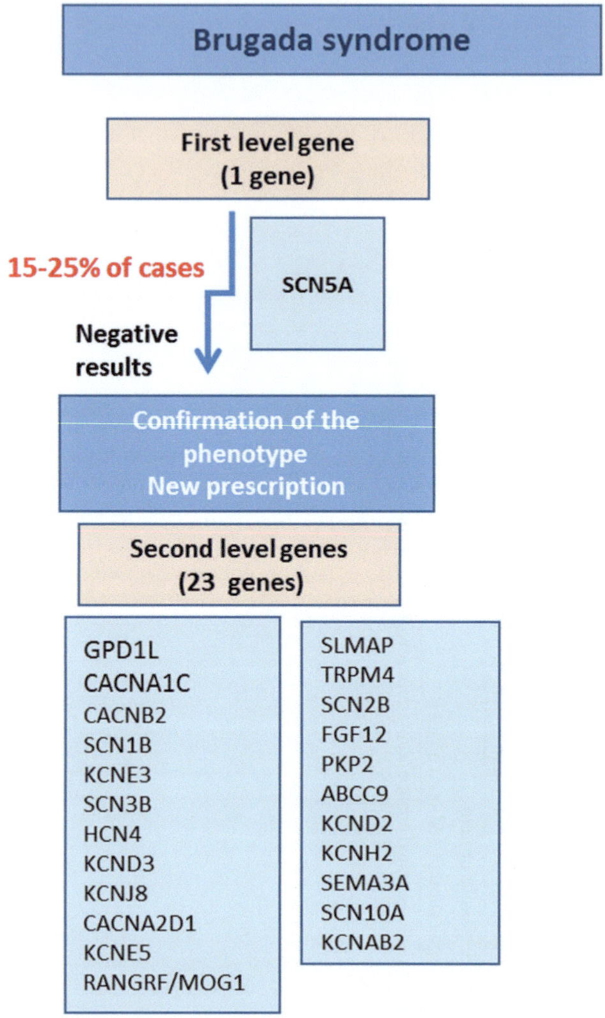

Fig. 13.1 Decision tree model for genetic testing in Brugada syndrome

The autosomal dominant model was the first genetic model proposed. However, the segregation study of *SCN5A* mutations among BrS pedigrees revealed a low penetrance and a variable expressivity of the pathology suggesting modulating factors (Probst et al. 2009). Despite the low predictive value of *SCN5A* status, the genotyping of this sodium channel remains the gold standard for BrS genetic testing since loss of function of the $Na_V1.5$ channel has been clearly implicated in BrS pathophysiology (Tan et al. 2003). Furthermore, clear evidence exists describing the genotype/phenotype relationship between *SCN5A* carriers and conduction defects increasing the risk of arrhythmia (Probst et al. 2010). All in all, approximately 300 distinct *SCN5A* mutations have been described in patients with the disorder.

Of these, two-thirds are missense variations and one-third are nonsense mutations, splice-site mutations and small insertions/deletions that lead to a truncated channel protein (Kapplinger et al. 2010). Other genes can be screened secondarily if *SCN5A* screening remains negative (see decision tree in Fig. 13.1), but extreme caution must be taken for variant interpretation since little evidence has been noted until now (Le Scouarnec et al. 2015).

Genetic testing plays a major role in the presymptomatic screening of BrS relatives. In the context of BrS families presenting a *SCN5A* mutation, genetic screening turns out to be of interest especially when ECG (even after provocative test) fails to diagnose the patient (Probst et al. 2009). Presymptomatic genetic testing appears to be even more pertinent in the context of early childhood when provocative tests using sodium blockers are not systematically performed for questionable relevance. Moreover, episodes of fever are frequent in the young, increase the risk of arrhythmia and emphasize the importance of identifying such individuals. Of note, a recent study on young BrS cases shows that genetic testing can uncover a higher proportion of *SCN5A* mutations (47%) than in adults—supporting the important role of genetic testing.

About 70% of BrS cases remain negative after genetic testing. The first explanation could be the fact that variants within the candidate genes cannot be detected with the sequencing technologies. Indeed, large rearrangements such as large deletion or insertion [also called copy number variation (CNV)] require additional methods for detection such as MLPA (multiplex ligation-dependent probe amplification). This investigation appears to be more and more performed (Selga et al. 2015; García-Molina et al. 2013; Koopmann et al. 2007), but few CNV have been reported so far (Eastaugh et al. 2011). Another hypothesis would be that additional genes could remain uncovered. Furthermore, a recent genome-wide association study suggested that BrS could follow a more complex genetic model than the Mendelian model commonly applied with the combination of common and rare variants of different size effects influencing the risk of developing BrS (Bezzina et al. 2013). Another hypothesis to explain the two-thirds of BrS cases with missing molecular diagnosis could be the presence of acquired Type I BrS ECG in the general population. Drugs have been identified as capable of inducing Type I BrS ECG—especially psychotropic and analgesic-anaesthetic drugs (Konigstein et al. 2016). An up-to-date list of drugs is accessible online to warn BrS patients about increasing their risk of arrhythmia with such drugs (http://www.brugadadrugs.org/) (Postema et al. 2009).

13.3.2 Long QT Syndrome (LQTS)

13.3.2.1 Clinical Description

Congenital long QT syndrome (LQTS) is a group of cardiac "channelopathies" characterized by delayed ventricular repolarization manifesting as QT interval prolongation on the ECG in the setting of a structurally normal heart (Morita et al. 2008). The prevalence of LQTS varies from 1/2000 to 1/5000 (Goldenberg and Moss 2008; Schwartz et al. 2009), with a female predominance (2/3 of the patients)

(Imboden et al. 2006). LQTS also shows variable expressivity and incomplete penetrance (Roden 2008). The first descriptions of the disease were provided by Jervell and Lange-Nielsen in 1957 (Jervell and Lange-Nielsen 1957) and by Romano and Ward (Ward 1964; Romano et al. 1963). Electrocardiographic findings are characterized by the presence of a prolonged heart rate-corrected QT interval on ECG (QTc). LQTS can be diagnosed in the presence of a QTc > 480 ms (Priori et al. 2015). The length of the QT interval is associated with the risk of syncope and sudden death. There is a high risk when QTc > 500 ms and an extremely high risk when QTc > 600 ms (Goldenberg and Moss 2008). The presence of T-wave alternans despite proper therapy is a sign of electrical instability and requires preventive measures. Patients with syncope or cardiac arrest before the age of 7, especially in their first year of life, have a higher risk of arrhythmias and sudden death (Priori et al. 2004; Spazzolini et al. 2009).

13.3.2.2 Genetic Testing

Genetic testing is part of the diagnostic criteria of LQTS (Priori et al. 2013). LQTS is most often inherited in an autosomal dominant manner (Romano-Ward syndrome) (Schwartz et al. 1993) and rarely in an autosomal recessive manner associated with sensorineural deafness (Jervell and Lange-Nielsen syndrome).

To date, more than 1200 pathogenic variations have been identified in 17 different genes (*KCNQ1, KCNH2, SCN5A, ANK2, KCNE1, KCNE2, KCNJ2, CACNA1C, CAV3, SCN4B, AKAP9, SNTA1, KCNJ5, CALM1, CALM2, TRDN and TECRL*) (Table 13.2) (Tester and Ackerman 2014). A causal mutation is found in these genes in 80–85% of LQTS patients. However, LQTS1 (*KCNQ1*; 35% of the cases), LQTS2 (*KCNH2*; 30% of the cases) and LQTS3 (*SCN5A*; 10% of the cases) comprise 90% of the LQTS mutations (Fig. 13.2) (Tester and Ackerman 2014, 2008). The majority of LQTS-causing mutations are coding-region single-nucleotide substitutions or small insertions/deletions. However, a few large gene rearrangements involving single or multiple exons deletions/duplications have been described (Barc et al. 2011).

LQTS is probably the heritable arrhythmia syndrome for which the genotype-phenotype relationship has been the most understood in terms of clinical manifestations, risk stratification and response to therapy. The phenotype-genotype association could facilitate phenotype-directed genetic testing. Indeed, genotype can be inferred by thorough clinical evaluation: swimming, physical exertion or emotional stress cardiac events strongly indicate mutations in *KCNQ1*. Cardiac events triggered by auditoria stimulation strongly indicate mutations in *KCNH2*, and symptoms at rest or during sleep are generally observed in LQT3 (Schwartz et al. 2001). T-wave morphology could also help to distinguish different LQTS subtypes: LQT1 exhibits a broad-based T-wave, LQT2 exhibits a low-amplitude notched or biphasic T-wave and LQT3 exhibits a late-appearing T-wave. These gene-suggestive ECG patterns could be helpful in guiding genetic testing, but exceptions to these relatively gene-specific T-wave patterns exist. Of note, from 20% to 25% of the patients with a long QT syndrome confirmed by the presence of a mutation could have a normal QTc (Priori et al. 2003; Goldenberg et al. 2011).

Table 13.2 Genes identified in long QT syndrome (LQTS)

Type (OMIM)	Locus	Gene	Protein, function/current	Effect of mutations	Transmission	Frequency	OMIM
LQTS1	11p15.5-p15.4	*KCNQ1*	Kv7.1 (alpha subunit potassium channel), repolarization/I_{Ks}	↓ I_{ks}	Ad+AR	30–35%	607542
LQTS2	7q36.1	*KCNH2*	Kv11.1 (alpha subunit potassium channel), repolarization/IKr	↓ I_{kr}	AD	25–30%	152427
LQTS3	3p21	*SCN5A*	Na_v1.5 (alpha subunit sodium channel), depolarization/I_{Na}	I_{Na}	AD	5–10%	600163
LQTS4	4q25-q26	*ANK2*	Ankyrin B (membrane anchoring/adapter protein)	Loss of function	AD	<1%	106410
LQTS5	21q22.12	*KCNE1*	MinK (beta subunit potassium channel), repolarization/IKs	↓ I_{Ks}	Ad+AR	<1%	176261
LQTS6	21q22.11	*KCNE2*	MiRP1 (beta subunit potassium channel), repolarization/I_{Kr}	↓ I_{Kr}	AD	<1%	603796
LQTS7 (ATS1)	17q24.3	KCNJ2	Kir2.1 (alpha subunit potassium channel), repolarization /IK1	↓ IK1	AD	<1%	600681
LQTS8	12p13.33	*CACNA1C*	Ca_v1.2 (alpha1C subunit calcium channel), depolarization/$I_{Ca\text{-}L}$	I_{CaL}	AD	<1%	114205
LQTS9	3p25.3	*CAV3*	Caveolin 3 (caveolae coat protein)	I_{Na}	AD	Rare	601253
LQTS10	11q23.3	SCN4B	NaVβ4 (beta1 subunit sodium channel), depolarization/INa	I_{Na}	AD	Rare	608256
LQTS11	7q21.2	AKAP9	A-kinase anchoring protein 9 (adapter protein)	↓ I_{Ks}	AD	Rare	604001
LQTS12	20q11.21	SNTA1	Syntrophin alpha 1 (membrane scaffold)	I_{Na}	AD	Rare	601017
LQTS13	11q24.3	KCNJ5	KIR3.4 (alpha subunit potassium channel)	↓ IK_{Ach}	AD	Rare	600734
LQTS14	14q32.11	*CALM1*	Calmodulin 1	↓ signalization Ca	AD	<1%	114180

(continued)

Table 13.2 (continued)

Type (OMIM)	Locus	Gene	Protein, function/current	Effect of mutations	Transmission	Frequency	OMIM
LQTS15	2p21	*CALM2*	Calmodulin 2	↓ signalization Ca	AD	<1%	114182
	19q13.32	*CALM3*	Calmodulin 3	↓ signalization Ca	AD	<1%	114183
	6q22.31	*TRDN*	Triadin		AR	<1%	603283
	4q13.1	*TECRL*	Trans-2.3-enoyl-CoA reductase-like protein	↓ signalization Ca	AR		617242

AD autosomal dominant, *AR* autosomal recessive

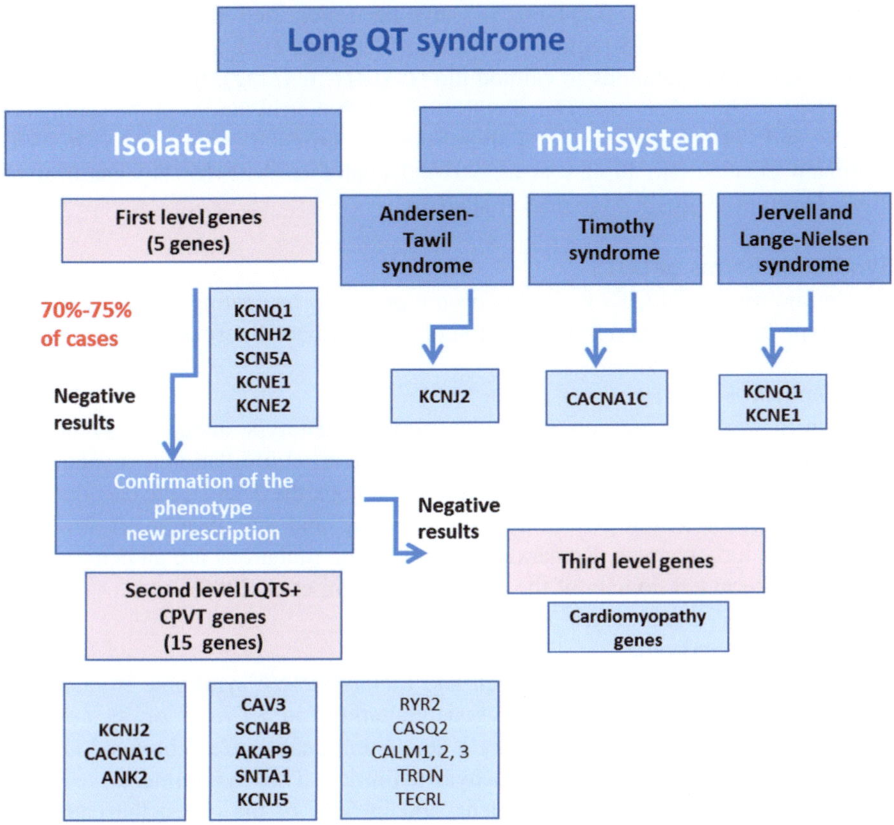

Fig. 13.2 Decision tree model for genetic testing in long QT syndrome

Moreover, the risk of arrhythmia is higher in LQT1 and LQT2 patients with a QTc > 500 ms and LQT3 in men whatever the interval QT duration. Mutation location can also be interpreted: mutations in the cytoplasmic loops of LQT1, mutations with dominant negative ion current effects and mutations in the pore-loop region of LQT2 have been associated with a different risk of cardiac events (Barsheshet et al. 2012; Moss et al. 2002; Shimizu et al. 2009; Migdalovich et al. 2011). In contrast, mutations in the C-terminal region tend to be associated with a milder phenotype (Donger et al. 1997; Crotti et al. 2007).

Genetic testing for the major forms of LQTS (*KCNQ1, KCNH2 and SCN5A)* has a major role in the diagnosis of probands, risk stratification, familial screening and treatment and could be proposed soon after birth. This is applicable only for pathogenic variations. However, many variations are classified as variants of unknown significance (VUS) and remain uninterpretable and therefore have no clinical impact (Steffensen et al. 2015). In contrast, 15–20% of patients have negative genetic tests, but that does not rule out the presence of the disease.

Mutations in other genes (LQT9 to 13) have been identified in a few patients, and their prevalence in LQTS requires evaluation (Schwartz et al. 2012).

Very recently, mutations in calmodulin (*CALM1*, *CALM2*), triadin (*TRDN*) and *TECRL* genes have been associated with highly lethal arrhythmias in the settings of LQTS and catecholaminergic polymorphic ventricular tachycardia (CPVT) (Altmann et al. 2015; Pipilas et al. 2016; Devalla et al. 2016; Nyegaard et al. 2012; Crotti et al. 2013; Marsman et al. 2014).

Syndromic Forms of LQTS

Syndromic forms of LQTS have also been reported as presenting a broader cardiac phenotype than QT interval prolongation or extracardiac abnormalities.

LQT4: Ankyrin-B Syndrome

Ankyrin-B syndrome presents a complex cardiac phenotype including QT interval prolongation, bradycardia, sinus node dysfunction, atrial fibrillation and ventricular arrhythmia (Schott et al. 1995; Le et al. 2008). Of note, the *ANK2* gene encodes for a non-channel protein but is essential for targeting and stabilization of structural proteins and ion channels. Remarkably, most of the mutations are gathered within the death/C-terminal domain of the protein (Hashemi et al. 2009).

Lange-Nielsen Syndrome

A recessive form of LQTS, the Jervell and Lange-Nielsen syndrome is caused by homozygous or compound heterozygous mutations in *KCNQ1* or its auxiliary subunit *KCNE1*. It is characterized by a congenital deafness, a prolonged QTc on ECG (>500 ms) and ventricular tachyarrhythmias. This syndrome is very rare (1/200,000 to 1/1,000,000) but virulent, and in 50% of the cases, the pathology manifests before the age of 3 (Pagon et al. 1993).

Andersen-Tawil Syndrome (ATS)

The Andersen-Tawil syndrome is characterized by periodic paralysis, a prominent U wave, a long QT interval with ventricular arrhythmias predisposing to sudden cardiac death and dysmorphic features: short stature, scoliosis, low-set ears, widely spaced eyes, small mandible, clinodactyly, brachydactyly and syndactyly. ATS is inherited in an autosomal dominant manner with variable penetrance. Mutations in the *KCNJ2* gene, encoding the alpha subunit of the Kir2.1 potassium channel, are implicated in 60% of the cases. A mutation in the *KCNJ5* gene, encoding the alpha subunit of the Kir3.4 potassium channel, has also been described (Kokunai et al. 2014).

Timothy Syndrome (TS)

The Timothy syndrome is caused by mutations in the *CACNA1C* gene (LQT8) encoding the alpha subunit of the CaV1.2 calcium channel. The syndrome often results from de novo mutation or from germinal mosaicism in one of the parents. Genetic testing of this syndrome is singular owing to its genetic homogeneity since the cases described so far have been carrying the p.G406R or p.G402R gain-of-function mutations (Tester and Ackerman 2014). This is a multisystem disease characterized by cardiac, hand/foot, facial and neurodevelopmental features. Typical cardiac findings include a rate-corrected QT interval >480 ms, functional 2:1 AV block with bradycardia, tachyarrhythmias and congenital heart defects. The diagnosis of Timothy syndrome is generally made within the first few days of life although it may be suspected prenatally by identifying 2:1 AV block or bradycardia in the foetus (Pagon et al. 1993).

Thus, based on the LQT clinical presentation and potential associated cardiac or not cardiac symptoms, genetic testing is proposed to be performed according to the decision tree presented in Fig. 13.2.

13.3.3 Short QT Syndrome (SQTS)

13.3.3.1 Clinical Description

SQTS was first described in 2000 and remains one of the rarest channelopathies (Gussak et al. 2000). Prevalence is difficult to estimate because of the limited number of patients (Gollob et al. 2011). SQTS leads to palpitation, syncope and sudden cardiac death, typically during childhood. It is characterized by a short QT interval on the ECG with peaked T-wave and a high risk of fatal arrhythmias (Patel et al. 2010).

A complete consensus of the cut-off value of QT interval does not exist yet. Diagnostic criteria have been proposed: for QTc <330 ms or <360 ms (350 ms in men and 365 ms in women) if associated with a pathogenic mutation, a family history of SQTS, VT/VF with normal structural heart or of sudden death before the age of 40 (Gollob et al. 2011; Bjerregaard 2011; Veltmann and Borggrefe 2011). Of note, patients with SQTS present a fairly similar QT interval whatever the heart rate variations requiring no or very little heart rate correction. Patients with the shortest QT duration present the highest risk of rhythmic events (Giustetto et al. 2011).

An implantable defibrillator is indicated in high-risk SQTS patients who have experienced VT/VF or resuscitated sudden death and in patients with a familial story of cardiac sudden death. Treatment with quinidine could be an alternative because of its prolongation of QT duration (Priori et al. 2015; Giustetto et al. 2011; Gaita et al. 2004).

13.3.3.2 Genetic Testing

Pathogenic variations have been described in 6 genes (*KCNQ1, KCNH2, KCNJ2, CACNA1C, CACNB2B and CACNA2D1*) harbouring mutations that mirror the functional effect of those encountered in LQTS (Brugada et al. 2004; Bellocq et al. 2004; Priori et al. 2005; Antzelevitch et al. 2007). Mutations in potassium channel genes produce a gain of function, and in contrast, mutations in the calcium genes produce a loss of function resulting in shorter AP durations. Mutations in these genes account for 50% of SQTS cases according to the literature, but in clinical practice, the diagnostic yield of genetic testing is very modest in these genes, which supposes that there are other causal genes and that a negative genetic cannot exclude the diagnosis.

13.3.4 Catecholaminergic Polymorphic Ventricular Tachycardia (CPVT)

13.3.4.1 Clinical Description

Catecholaminergic polymorphic ventricular tachycardia (CPVT) is a hereditary arrhythmogenic disease that is responsible for familial forms of sudden death in infants and young adults. The prevalence for CPVT in the general population is estimated to be 1/10,000 (Van der Werf and Wilde 2013). Arrhythmias occur as recurrent syncope triggered by adrenergic stimulation such as emotional or exercise stress, while baseline ECG is often normal and cardiac imaging reveals no structural abnormality (George et al. 2007). Approximately, 30% of patients with CPVT develop symptoms before 10 years of age and 60% before 40 years (Liu et al. 2008; Ackerman et al. 2011). A family history of exercise-related syncope, seizure or sudden death is reported in 30% of the patients. As with LQTS1, swimming is a potentially lethal arrhythmia precipitating CPVT, and the symptoms are often attributed to alternative diagnoses, most commonly epilepsy. Drug challenge with epinephrine or isoproterenol may elicit arrhythmias. In the absence of beta-blocker treatment, VT/VF and syncope are associated with a high rate of recurrence with 30–50% mortality before the age of 40 years (Hayashi et al. 2009). Beta blockers, combined with restriction of exercise and avoidance of stressful situations, are the first-line therapy for CPVT. All patients and family members who harbour a disease-causing mutation, irrespective of symptoms, should be treated with beta blockers. Left cardiac sympathetic denervation or ICD implantation could be effective to prevent recurrent VT/VF and SCD in patients intolerant to beta blockers or for those with persistent symptoms or arrhythmias despite treatment (Wilde et al. 2008). However, CPVT patients appear to be very sensitive to catecholamines, and the subsequent ICD shock has a high probability of initiating a vicious cycle with multiple shocks caused by recurrent ventricular tachycardia or VF (Wilde et al. 2008).

13.3.4.2 Genetic Testing

The presence of a pathogenic mutation is of diagnostic value. Two genetic models of CPVT have been identified: an autosomal dominant form due to mutations in the cardiac ryanodine receptor 2 (*RYR2*) gene and less common autosomal recessive forms due to mutations in the *CASQ2*, *TRDN* and more recently *TECRL* genes (Table 13.3) (Devalla et al. 2016). Mutations in *RYR2* are found in 50%–60% of CPVT cases, whereas mutations in *CASQ2* and *TRDN* are found in less than 10% of CPVT patients (Fig. 13.3) (Roux-Buisson et al. 2012; Rooryck et al. 2015). Rarely, mutations are found in calmodulin genes (*CALM1*, *CALM2* and *CALM3*), the *KCNJ2* gene (Andersen-Tawil syndrome also known as LQT7) and the *ANK2* gene (also known as LQT4) (Nyegaard et al. 2012; Mohler et al. 2004; Tully et al.

Table 13.3 Genes identified in catecholaminergic polymorphic ventricular tachycardia (CPVT)

Type (OMIM)	Locus	Gene	Protein	Transmission	Frequency	OMIM
CPVT1	1q43	*RYR2*	Ryanodine receptor 2	AD	40–60%	180902
CPVT2	1p13.1	*CASQ2*	Calsequestrin 2	AR	1–2%	114251
	17q24.3	*KCNJ2*	Potassium voltage-gated channel subfamily J member 2	AD	Rare	600681
CPVT5	6q22.31	*TRDN*	Triadin	AR	Rare	603283
	4q25-q26	*ANK2*	Ankyrin 2	AD	Rare	106410
CPVT4	14q32.11	*CALM1*	Calmodulin 1	AD	Rare	114180
	2p21	*CALM2*	Calmodulin 2	AD	Rare	114182
	19q13.32	*CALM3*	Calmodulin 3	AD	Rare	114183
	4q13.1	*TECRL*	Trans-2.3-enoyl-CoA reductase-like protein	AR	Rare	617242

AD autosomal dominant, *AR* autosomal recessive

2015). A locus associated with an early-onset lethal form of recessive CPVT was found on chromosome 7p14-p22, but so far no gene has been identified (Bhuiyan et al. 2007).

Interestingly, *RYR2* mutations cluster in both N- and C-terminal domains of the protein and within transmembrane domains (Priori and Napolitano 2005). Forty percent of all *RYR2* mutations are estimated to be de novo mutations. Only rare cases of rearrangements have been described and remarkably all concern *RYR2* exon3 (Bhuiyan et al. 2007; Campbell et al. 2015). Since SCD can be the first manifestation of CPVT, genetic testing is clinically relevant, especially for the care management of family members.

At present, there is no genotype-based risk stratification or therapeutic approach in CPVT. Nearly one-third of possible or atypical LQTS cases with exertion-induced syncope have also been identified as positive for an RYR2 mutation, and accordingly, a clinical presentation of exercise-induced syncope and a QTc <460 ms should always first point to the consideration of CPVT rather than the so-called concealed or normal QT interval LQT1 (Medeiros-Domingo et al. 2009).

Since clinical overlap have been described between CPVT and cardiomyopathy cases (OMIM#604772), cardiomyopathy-related genes may be considered in case of negative diagnostic from the screening of the first and second gene panels (Fig. 13.3).

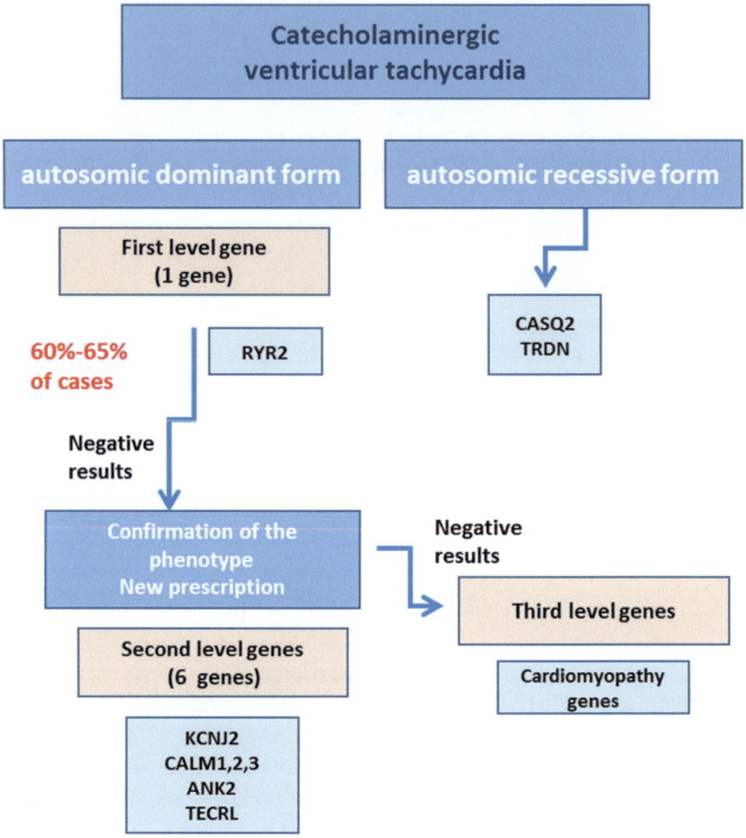

Fig. 13.3 Decision tree model for genetic testing in catecholaminergic polymorphic ventricular tachycardia

13.3.5 Cardiac Conduction Defect (CCD)

13.3.5.1 Clinical Description

Cardiac conduction defect (CCD) is a rare disorder associated with a risk of sudden death in the absence of pacing. Progressive cardiac conduction defect (PCCD), also called Lenegre-Lev disease, is the most frequent form, affecting the His-Purkinje pathway, and may progress to complete atrioventricular block (AVB) (Lev 1964; Lenegre 1964). Diagnosis relies on echocardiogram (ECG) findings showing an advanced conduction defect. Congenital structural CCD can be part of a syndrome and be associated with abnormalities in other organ systems or cardiac malformations including atrial septal defects, ventricular septal defects and tetralogy of Fallot (Baruteau et al. 2015).

Patients who receive pacemaker implantation appear to present an excellent prognosis except in those with *LMNA* mutations (see also below) which can lead to ventricular tachycardia and sudden cardiac death. In this population, ICD implantation is recommended in case of severe cardiac conduction defect (Anselme et al. 2013).

13.3.5.2 Genetic Testing

Transmission is usually autosomal dominant with incomplete penetrance and variable expressivity. Inherited forms of CCD are attributed to mutations in a wide variety of genes impacting the process of action potential generation and propagation. These electrical disturbances concern mainly isolated form of conduction defects and involve cardiac ion channel genes (*SCN5A, HCN4, TRPM4, SCN1B*) and cytoskeletal and inner nuclear membrane genes (*LMNA*). Mutations appear more frequently in *SCN5A* (Schott et al. 1999; Probst et al. 2003; Watanabe et al. 2008; Liu et al. 2010; Stallmeyer et al. 2012; Daumy et al. 2016). Point mutations and more rarely rearrangement in the gene encoding lamin A/C (LMNA) can lead to dilated cardiomyopathy associated with cardiac conduction system disease and arrhythmias and also to isolated CCD (Marsman et al. 2011). Genetic testing in suspected *LMNA* patients is of great interest since *LMNA* carriers present a high risk of SCD and a poor prognosis (Wolf and Berul 2006; Wolf et al. 2008). A mutation in the *GJA5* gene encoding the gap junction Cx40 gene was identified in a family with progressive cardiac conduction defect (PCCD) characterized by AV block and wide QRS bundle branch block (Makita et al. 2012). Mutations in PRKAG2, which encodes for a regulatory subunit (γ-2) of adenosine monophosphate-activated protein kinase (AMPK), were found in patients with the Wolff-Parkinson-White (WPW) syndrome, a disease characterized by ventricular pre-excitation, atrial fibrillation and conduction defects such as sinoatrial and atrioventricular block and cardiac hypertrophy (Fig. 13.4; Table 13.4) (Wolf and Berul 2006). In the absence of mutations among the above candidate genes, the screening of the cardiomyopathy-associated genes might be considered (Sisakian 2014). Furthermore, in case of atrial septal defect (ASD), abnormalities in transcription-factor *NKX2–5*, *GATA4* and *TBX5* have been reported in congenital atrioventricular block (Schott et al. 1998; Garg et al. 2003; Basson et al. 1994).

13.3.6 Arrhythmogenic Right Ventricular Cardiomyopathy/ Dysplasia (ARVC/D)

13.3.6.1 Clinical Description

Arrhythmogenic right ventricular cardiomyopathy/dysplasia is a heritable cardiomyopathy characterized by life-threatening ventricular arrhythmias and progressive dystrophy of the ventricular myocardium with fibrofatty replacement. Generally referred to as a right ventricular disease, left or bi-ventricular forms prompted the use of the broader term arrhythmogenic cardiomyopathy (Bhonsale et al. 2015). The estimated prevalence in the general population is 1/2000–1/5000 with an age-related penetrance (usually adolescence-young adulthood) (Corrado et al. 2017). Moreover, ARVC/D more frequently affects males than females (up to 3:1) despite a similar

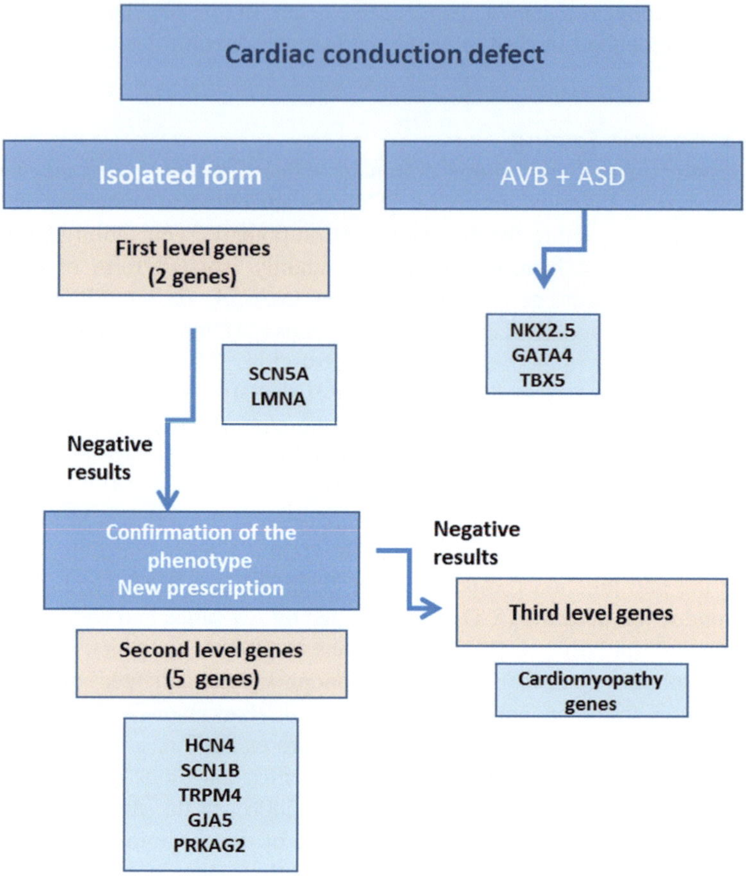

Fig. 13.4 Decision tree model for genetic testing in cardiac conduction defect

prevalence of carrier status between sexes (Bhonsale et al. 2015). ARVC/D could account for up to 20% of cases of aborted cardiac arrest and SCD in the young, especially in athletes (Corrado et al. 2017).

Multiple criteria are required for diagnosis combining repolarization abnormalities, morpho-functional alterations, histopathological features on endomyocardial biopsy, family history and genetics, but diagnosis of ARVC/D is often complicated by its evolution over time (Marcus et al. 2010). Patients typically present with symptomatic ventricular arrhythmias, characterized by premature ventricular contraction (PVC) or VT with LBBB morphology and T-wave inversion in V1–V3 leads on basal electrocardiogram, leading to syncope or cardiac arrest. The first symptoms manifest before the age of 40 and are encountered in the large majority of patients (Orgeron and Calkins 2016).

Table 13.4 Genes identified in cardiac conduction defect (CCD)

Type (omim)	Locus (HGNC)	Gene	Protein	Transmission	OMIM
PFHB-1A	3p22.2	SCN5A	$Na_v1.5$	AD	600163
	1q22	LMNA	Lamin A/C	AD/AR	150330
	19q13.11	SCN1B	Sodium channel, voltage-gated, type i, beta subunit	AD	600235
PFHB-1B	19q13.33	TRPM4	Transient receptor potential cation channel, subfamily M, member 4	AD	606936
		HCN4		AD	
	5q35.1	Nkx2.5	nk2, drosophila, homolog of, e; nkx2e	AD	600584
			Cardiac-specific homeobox 1; csx1		
	1q21.2	GJA5	Gap junction protein, alpha-5	AD	121013
	7q36.1	PRKAG2	Protein kinase, amp-activated, noncatalytic, gamma-2	AD	602743

AD autosomal dominant, *AR,* autosomal recessive, *PFHB* progressive familial heart block

13.3.6.2 Genetic Testing

ARVC/D is inherited predominantly as an autosomal dominant trait with incomplete penetrance and with a variable age-dependent expressivity. Inter- and intra-familial variation in disease severity and expressivity is frequently observed with coexistence of both classic RV and dominant LV pattern in the same family and/or life-threatening ventricular arrhythmias in probands vs. a more favourable prognosis in relatives (Sen-Chowdhry et al. 2007). In a few cases, ARVC/D inheritance follows an autosomal recessive trait (Naxos disease or Carvajal syndrome). These ARVC/D patients present extracardiac symptoms such as a cutaneous phenotype (palmoplantar keratoderma) and woolly hair (Corrado et al. 2017). ARVC/D is widely considered to be a desmosomal disease. Desmosomes are one of the junctional complexes at the intercalated disc that are essential for the structural and functional integrity of cardiac tissue. Desmosomal gene mutations lead to defective desmosomal components and to the development of structural abnormalities. Mutations in five desmosomal genes have been identified: plakophilin-2 (*PKP2*) desmoplakin (*DSP*), desmocollin-2 (*DSC2*), desmoglein-2 (*DSG2*), plakoglobin (*JUP*) and more recently the cadherin 2 (*CDH2*) (Mayosi et al. 2017). Non-desmosomal genes, transmembrane protein 43 (*TMEM43*), desmin (DES), lamin A/C (*LMNA*), titin (*TTN*), phospholamban (*PLN*) and alpha T-catenin (*CTNNA3*), have been associated with atypical forms of ARVC/D. Mutations in the regulatory region of transforming growth factor beta-3 (*TGFB3*) have also been reported, but their pathogenicity is still controversial. Rarely, mutations in the *RYR2* gene and the *SCN5A* gene have been described (Table 13.5) (Te Riele et al. 2017). ARVC/D gene mutations are found in approximately 60% of ARVC/D index cases, of which the majority are desmosomal gene mutations (*PKP2* 10–45%, *DSP*

Table 13.5 Genes identified in arrhythmogenic right ventricular dysplasia (ARVD)

Type (OMIM)	Locus	Gene	Protein	Transmission	Frequency (%)	OMIM
ARVD9	12p11	*PKP2*	Plakophilin-2	AD/AR	30–40%	602861
ARVD8	6p24	DSP	Desmoplakin	AD/AR	10–15%	125647
ARVD10	18q12.1	*DSG2*	Desmoglein-2	AD/AR	3–8%	125671
ARVD11	18q21	*DSC2*	Desmocollin-2	AD/AR	1–5%	125645
ARVD12	17q21	*JUP*	Junction plakoglobin	AD/AR	<1%	173325
ARVD2	1q43	*RYR2*	Ryanodine receptor-2	AD	<1%	180902
ARVD1	14q24	*TGFB3*	Transforming growth factor b3	AD	<1%	190230
ARVD5	3p25.1	*TMEM43*	Transmembrane protein 43	AD	<1%	612048
ARVD13	10q22.3	CTNNA3	Alpha T-catenin	AD	Rare	607667
	2q31	*TTN*	Titin	AD	Rare	188840
	6q22.1	*PLN*	Phospholamban	AD	Rare	172405
	1q22	LMNA	Lamin A/C	AD	Rare	150330
	2q35	DES	Desmin	AD	Rare	125660
	3p22.2	SCN5A	Sodium channel, voltage-gated, type v, alpha subunit	AD	Rare	600163
	18q12.1	CDH2	Cadherin 2	AD	Rare	114020

AD autosomal dominant, *AR* autosomal recessive

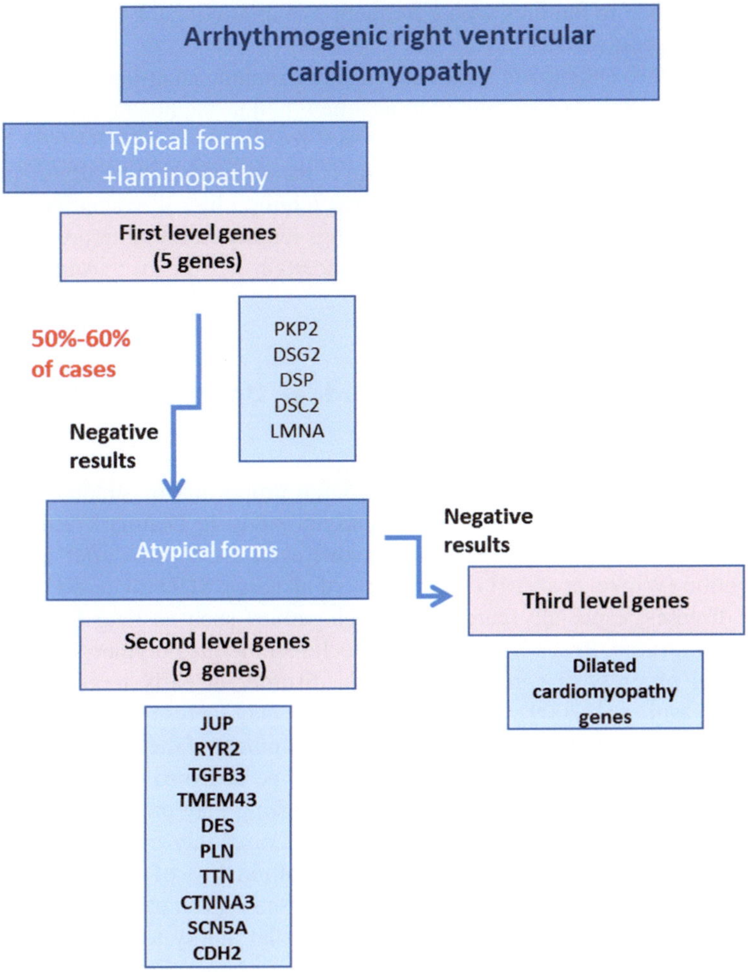

Fig. 13.5 Decision tree model for genetic testing in arrhythmogenic right ventricular cardiomyopathy/dysplasia

10–15%, *DSG2* 7–10%, *DSC2* 1–2%, *JUP* 1–2%) (Fig. 13.5), but the proportion of causal genes differs according to cohort location and ethnicity (Ohno 2016). Compound/digenic mutations may be found in 10–25% of patients and lead to earlier onset of symptoms, and carriers of mutations in *DSP* are more likely to present with left ventricular dysfunction, heart failure and SCD than carriers with mutations in *PKP2* (Bhonsale et al. 2015; Rigato et al. 2013). Among single mutation carriers, premature truncating, splice-site and missense mutations were identified in, respectively, 60%, 23% and 14% of cases, but the outcomes did not differ significantly (Bhonsale et al. 2015). Entire *PKP2* exons or even whole gene deletions have been recently described in families with a frequency of approximately 2% (Roberts et al. 2013; Li Mura et al. 2013).

While the 2010 Diagnostic Task Force Criteria include identification of a putative ARVC/D susceptibility gene mutation as a new major diagnostic criterion, genetic testing should be used as a confirmatory tool for diagnosis of ARVC/D in index cases (Fig. 13.5). Presymptomatic screening is recommended in family members. If they carry the same mutation, they will then require clinical evaluation and long-term observation. Nevertheless, genetic findings require careful interpretation owing to the large number of genetic variants of uncertain significance (Andreasen et al. 2013).

13.3.7 Role of Genetic Testing in Sudden Unexplained Death (SUD)

13.3.7.1 Clinical Description

Sudden cardiac death is responsible for a large proportion of sudden unexpected deaths (SUD) in young individuals. A national study in Denmark estimated that SUD represents 29% of the deaths in persons aged 1–35 years after medicolegal investigation (Winkel et al. 2011). In 5–10% of all cases, SUD is caused by inherited cardiac diseases, especially cardiac channelopathies such as long QT syndrome, short QT syndrome, Brugada syndrome, catecholaminergic polymorphic ventricular tachycardia, progressive cardiac conduction disorder or early repolarization syndrome (Hayashi et al. 2009). In these cases, SUD may represent an early malignant arrhythmogenic presentation, preceding the development of the ECG indices. Therefore, when no identifiable ECG abnormalities can be observed or when only a post-mortem blood sample can be collected, genetic testing may provide insight into the causal mechanism leading to (fatal) arrhythmia. Comprehensive clinical investigation of SCD families would identify causes of death in 40% of cases, demonstrating that SCD were probably due to inherited heart disease (Tan et al. 2005; Behr et al. 2008). Next-generation sequencing (NGS), with the ability to screen hundreds of genes simultaneously, could provide confirmation of the clinical investigation or become the only possibility to point clinicians to a clinical diagnosis and identify the relatives at risk of similar severe arrhythmia. Furthermore, when patients present a borderline phenotype, genetic testing could encourage clinicians to further look for specific ECG abnormalities and employ adapted provocative tests.

13.3.7.2 Genetic Testing

The identification of the genes associated with cardiomyopathies and channelopathies has provided insights into disease mechanisms of cardiac arrest. Most genetic studies have focused on cardiac channelopathy-associated genes. Mutation analysis in 5 LQTS genes (*KCNQ1, KCNH2, SCN5A, KCNE1, KCNE2*) identified pathogenic variations in 11–17% of SCD cases (Chugh et al. 2003; Skinner et al. 2011; Winkel et al. 2012). More recently, post-mortem genetic testing for a large cohort of cases of autopsy-negative sudden unexplained deaths identified putative pathogenic variations in LQTS and CPVT genes in 20% of the cases (Tester et al. 2012). Another study on 34 genes of channelopathies identified 3 likely pathogenic variants in *KCNH2, SNTA1* and *RYR2* in 3 out of 15 (20%) SUD cases (Hertz et al. 2015). A prevalence of 5–10% of variants in the CPVT-associated gene *RYR2* has been found in sudden infant death syndrome cases (Larsen et al. 2013). Exome analysis of 28 cases of autopsy-negative sudden unexplained death identified 3 rare variants in *KCNQ1* and *KCNH2* genes in 3 SUD cases (10%) and 4 rare variations in cardiomyopathy genes (*DSP, MYBPC3* and *TTN*) (Bagnall et al. 2014). A recent screening of 100 genes in 61 cases of autopsy-negative sudden unexplained death (<50 years) identified likely pathogenic variants in 34% of the SCD cases, 40% associated with inherited cardiomyopathies and 60% associated with channelopathies (Christiansen et al. 2016).

However, the use of genetics as a clinical diagnostic tool must always be considered the context of the clinical case. Therefore, particularly in the context of SCD, genetic testing should complement or guide but never supplant the clinical investigation and segregation analyses. Without a clear clinical picture and a clear correlation between the disease and the genetic variants, causality may be nonconclusive, since several genes have been linked to several phenotypes or SCD could be caused by other unknown genetic causes or factors.

13.4 Genetic Testing and Technological Evolution: Progress and Side Effects

13.4.1 The Next-Generation Sequencing: High Throughput and Low Cost

As described above, the number of channelopathy-associated genes has increased dramatically in the last decades. With capillary sequencing, genetic testing was restricted to the major genes since this molecular analysis was expensive and had limited throughput. Since 2011, next-generation sequencing has revolutionized genetic diagnostics by offering the possibility to screen thousands of genes simultaneously coupled with a dramatic reduction (10,000-fold; www.genome.gov/sequencing costs) in sequencing cost (Kingsmore and Saunders 2011). However, knowing the number of genes to explore is crucial since one must maintain a good ratio in order to guarantee a reasonable cost (number of genes to be sequenced), data that are of sufficient quality to interpret the results (depth of sequencing) and a throughput that is compatible with an analysis of numerous samples. A minimum mean depth of 30X has been considered reasonable to achieve confident variant calls. In the absence of consensus gene design for channelopathies, a rationalized approach could consist in defining a set of genes comprising the channel subunits and partners predominantly implicated in rhythm disorders (see above). In case of negative results, a second panel that includes all genes associated with any primary rhythm disorders or even cardiomyopathies could be employed (Novelli et al. 2016). However, our broad sequencing capability could be moderated by interpretation, ethical and storage issues.

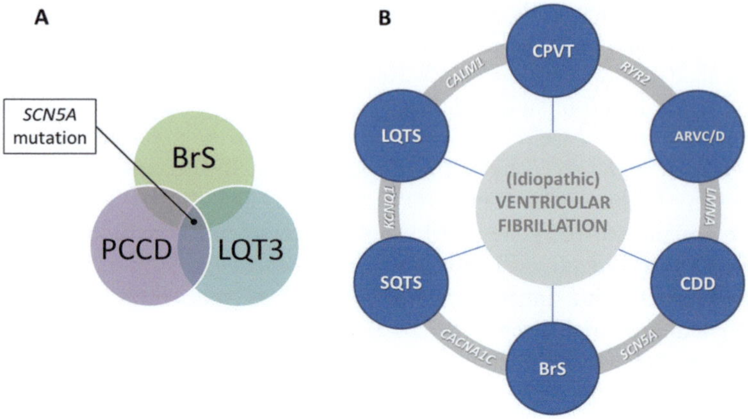

Fig. 13.6 (**a**) *SCN5A* overlap syndromes and (**b**) genetic continuum among the cardiac channelopathies

13.4.2 The More You Sequence, the More Complex Will Be the Interpretation

As illustrated by the genetics of the channelopathies, we observed a large genetic heterogeneity among each clinical entity coupled with gene overlap between them (Fig. 13.6). This gene overlap could also reveal a potential continuum among these channelopathies (Makita 2009). Moreover, several studies have reported channelopathies due to double mutations harboured by different genes (Kapplinger et al. 2010, 2009; Bauce et al. 2010). These observations suggest the need to screen large panels of genes. These panels are now widely employed within the molecular diagnostic centres covering between 50 and 200 candidate genes according technologies and diseases targeted. The continuous reduction of the sequencing cost tends to switch to whole exome (all coding regions of the genome) sequencing (WES), or even whole genome sequencing (WGS), to overcome the limitation of the panel(s). However, reagent cost for capturing approximately 100 genes remains 4 times lower than for WES and 10 times cheaper than for WGS. Aside from the financial point of view, drawbacks need to be anticipated before producing and analysing a large/whole gene scan. First of all, besides the promise that all coding regions and then the "Mendelianome" will be covered by a WES, daily practice reveals that WES actually provides a poorer coverage (% of targeted bases covered at a defined mean depth) than the targeted gene panels (Pua et al. 2016). Coverage for genetic diagnostic is crucial since all candidate DNA base pairs have to be confidently explored. Whole genome sequencing presents the advantage of uniformly covering the genome and facilitating genomic rearrangement identification (Sudmant et al. 2015) but at twice the WES price.

Today, genetic diagnostics face the challenge of variant interpretation uncovered by the large-scale sequencing technologies offering the advantage of simultaneously screening hundreds of candidate genes but at the same time revealing multiple rare genetic variations of unknown significance (Novelli et al. 2016). For example, rare genetic variations in associated channelopathy genes are also commonly found in the general population (Risgaard et al. 2013; Ghouse et al. 2017; Paludan-Müller et al. 2017). Large databases such as gnomad.broadinstitute.org consisting of >120,000 exomes and >15,000 whole genomes currently constitute a powerful tool to estimate the minor allele frequency (MAF) of a candidate variant. In addition, such databases furnish MAF in different broad categories of ethnicities which could help in the possible absence of demographically matched controls to prevent false-positive mutation. For this reason, numerous countries have developed their own national databases to reflect the specific genetic architecture of their population (Zawistowski et al. 2014). However, the rarity of a variant does not make it necessarily causal, and it can even turn out to be a variant of uncertain/unknown significance (VUS). Indeed, within an exome, between 300 and 600 rare variants will be identified which obviously do not at all cause a pathologic phenotype (Tennessen et al. 2012). Then, considering that 12,958 (64%) out of 20,344 genes in the genome (http://www.proteinatlas.org/) are expressed in the cardiac tissue, we can estimate that between 200 and 400 of them could be of potential interest in a context of a patient presenting a cardiac disease. This estimation puts into perspective the challenge of the number of variants generated by the NGS and our ability to provide a pertinent molecular diagnosis. Misinterpretation of a VUS can lead to inappropriate treatment such as an implantable cardioverter-defibrillator which could have a dramatic impact on the patient's quality of life with possible severe complications and potentially serious psychologic consequences (Ackerman 2015).

This led to re-evaluation of the criteria defining the pathogenicity of a variant and then to modification of genetic test interpretations in order to avoid false positives (Ackerman 2015). Segregation studies among family members and/or functional experiments constitute the best approaches for causal mutation validation but remain rare owing to the time and cost investment required and are not adapted to a large number of variants. This sequencing technology revolution could therefore change the skill profiles required for genetic testing. High-throughput functional screening (in vitro) technologies to cope with massive VUS identification could emerge to characterize the variants isolated by data analysis and interpretations performed by bioinformatics. Indeed, after filtering out common variants by checking their MAF which is available in large public databases, additional tools could guide a geneticist to isolate the causal variant. The type of mutation can be considered (missense, nonsense, frameshift or splice-site variants) especially when loss of function of the protein can be predicted. Another argument of causality can reside into a previous observation of this variation in patients presenting the same phenotype and listed in dedicated databases (HGMD http://www.hgmd.cf.ac.uk/ac/index.php, CLINVAR www.ncbi.nlm.nih.gov/clinvar). When a variant is rare but never found in patients or in dedicated databases and no family members could have been tested, multiple algorithms can be used to indicate the likelihood of variant pathogenicity. These tools can combine information from nucleotide or/and amino acid conservation among species or include parameters such as modification of amino acid polarity or/and 3D protein structure. Others combine several algorithms to provide a consensus score (Table 13.6). Guidelines for interpretation and classification of genetic

Table 13.6 Non-exhaustive list of tools for variant consequence prediction

Name	Methods	REF/PMID
CADD	Combined annotation-dependent depletion (CADD) is a framework that integrates multiple annotations into one metric by contrasting variants that survived natural selection with simulated mutations	24487276
Condel	**Consensus deleteriousness score of** MutationAssessor and FATHMM	21457909
ENTPRISE	A boosted tree regression **machine-learning** approach to predict human disease-associated amino acid variations by utilizing a comprehensive combination of protein sequence and structure features	26982818
FunSeq2	The pipeline has a weighted scoring system combining: Inter- and intraspecies conservation, loss- and gain-of-function events for transcription-factor binding, enhancer-gene linkages and network centrality, and per-element recurrence across samples. We further highlight putative drivers with information specific to a particular sample, such as differential expression	25273974
GWAVA	Based on a wide range of annotations of non-coding elements (largely from ENCODE/ GENCODE), along with genome-wide properties such as evolutionary conservation and GC-content	https://www.sanger.ac.uk/sanger/StatGen_Gwava
iFish	**Gene-specific and family-specific customized classifiers. Bayesian model**	27527004
LoFtool	LoFtool provides a rank of genic intolerance and consequent susceptibility to disease based on the **ratio of loss of function (LoF)** to synonymous mutations for each gene in 60,706 individuals from ExAC, adjusting for the gene de novo mutation rate and evolutionary protein conservation	27563026
Mechismo	Enables simultaneous consideration of thousands of **3D structures** and biomolecular interactions to predict rapidly mechanistic consequences for mutations and modifications. Compares to protein close in structure but not necessarily in sequence	25392414
MutationAssessor	Functional impact score (FIS) for amino acid residue changes using **evolutionary conservation patterns**	21727090
MutationTaster2	**Comprises evolutionary conservation, splice-site changes, loss of proteinfeatures and changes that might affect the amount of mRNA. Bayesian model**	20676075

(continued)

Table 13.6 (continued)

Name	Methods	REF/PMID
PaPi	Classify and score human coding variants by estimating the probability to damage their protein-related function. Consists in using **pseudo amino acid composition (PseAAC)** through which wild and mutated protein sequences are represented in a discrete model. Then **machine learning (random forest)**. Uses the scores of **SIFT** and **PolyPhen** for training	25928477
PolyPhen-2	Compare to homologous protein sequences, then prediction of 3D structure. Compare of structures and merging proteins in cluster. Score calculation on the cluster with the interest protein	12202775
PON-P2	**Random forest algorithm and machine learning.** PON-P2 is a novel tool that employs features of evolutionary sequence conservation, properties of amino acids, GO annotations and functional annotations	25647319
PredictSNP2	**Consensus** on eight established prediction tools: MAPP, nsSNPAnalyzer, PANTHER, PhD-SNP, PolyPhen-1, PolyPhen-2, SIFT and SNAP	27224906
PROVEAN	This **alignment-based score** measures the change in sequence similarity of a query sequence to a protein sequence homolog before and after the introduction of an amino acid variation to the query sequence	25851949
REVEL	Make a **consensus** between: MutPred, FATHMM, VEST, PolyPhen, SIFT, PROVEAN, MutationAssessor, MutationTaster, LRT, GERP, SiPhy, phyloP and phastCons	27666373
SIFT	Compare to homologous protein sequences: The more the position is conserved in evolution, the more mutation is supposed to be damaging	11337480

variations are proposed by the American College of Medical Genetics and Genomics to evaluate the pathogenicity of a variant (Richards et al. 2015).

This unexpected number of rare variants that harbours a genome and the difficulty of their interpretation could reinforce, especially for rare diseases such as channelopathies, the benefit of gathering clinical and genotype information in common databases to improve the knowledge acquired by the number of cases explored.

13.4.3 Genetic Testing and the Impact of Identifying a Mutation for Relatives and Predictive Testing

Because of the complexities involved, genetic testing should only be performed in dedicated specialized centres such as a cardiac genetics clinic or clinical genetics service where appropriate family management and genetic counselling before and after testing can be offered. This should be performed by cardiologists, nurses, clinical geneticists and genetic counsellors with specialized training in cardiovascular genetics.

Identification of a mutation in a family can often explain why the disease has occurred. It also allows for cascade testing of other affected and unaffected family members. Testing of affected family members is performed as a confirmation of their disease status and to exclude the possibility of a "phenocopy" (i.e. an individual who has an acquired rather than genetic cause of the same condition as other members of a family). This enables an accurate risk assessment for their offspring. Asymptomatic family members can be offered a predictive genetic test to clarify whether they are at risk of developing clinical disease and to determine the inheritance risk to their children. There may also be implications for participation in sports and employment.

13.4.4 Genetic Testing and Psychological Impact

The quality of life and psychological stress of cardiac variant carriers are most often not evaluated. A few studies have provided a quantification of the level of anxiety and depression in cardiac mutation carriers and their relatives (Hendriks et al. 2008; Christiaans et al. 2009). Genetic testing can help with anxiety regarding the disease especially when the clinical diagnostic is uncertain. Stress should also be considered in partners since the degree of anxiety is higher in carrier partners than in those of noncarriers. In this context, genetic counselling is essential to answer the questions generated by the clinical and genetic diagnosis to prevent anxiety. The uncertainty of the functional effect of VUS and, more generally, the large number of rare variants present in all individuals could also give rise to anxiety in genetic counsellors (Spoonamore and Ware 2016).

13.4.5 Large-Scale Sequencing and Incidental Findings

Aside from the complex VUS interpretation, known causal mutation may be uncovered and sometimes aside from the initial indication of cardiac diseases. These incidental findings are a matter of clinical and ethical debate since it is difficult to assess a clear benefit for a patient and/or a clinician to be informed of such mutations. The American College of Medical Genetics and Genomics (ACMG) recommends reporting variants belonging to a defined list of genes to an appropriate clinician for re-evaluation and surveillance of the patient and his family. The ACMG proposes a list of 57 genes associated with 24 syndromes, cancer, endocrinal or cardiac diseases.

Of note, this recommendation concerns 9 channelopathy-associated genes and 11 cardiomyopathy genes (Green et al. 2013; Kalia et al. 2017). This risk should be well explained to the patient before performing a whole exome scan.

13.4.6 Large-Scale Sequencing Data, Storage and Privacy

The rapid development of sequencing throughput has encouraged us to screen increasingly larger numbers of genes and the entire genome in the very short term. Thus, there have been problems in bioinformatics to handle and characterize the great number of rare variants contained within the genome. Moreover, large-scale sequencing, in particular whole genome screening, also implies anticipating storage capability since data associated with the sequencing of a whole genome represents approximately 120GB. A 3-day run of a last-generation sequencer (i.e. Illumina X Ten solution) produces 16 whole genomes and 1.8 TB of data. For example, a centre such as UPPMAX accumulated 2200 TB of data from 2011 to the end of 2015 (Spjuth et al. 2016). In addition, whole genome data requires secure storage since they contain sensitive and private information. Aside from storage, privacy must be maintained during the phases of alignment, calling or comparison with public databases when they are not performed on a local level (Akgün et al. 2015). These large genome scans shake our habits and require thought before rushing into data production.

13.5 Conclusion

The last decade saw the new high-throughput sequencing technologies offering the opportunity to screen first a large panel of candidate genes, then all coding region and now the whole genome for a continuously decreasing cost. This allowed improving the efficiency of the genetic testing and consequently the care management for the patient and their family members. This opportunity to always screen more genes revealed also the high prevalence of rare variants present in the genome and therefore brings us to a new challenge consisting of isolate one(s) with an impact on the cardiac phenotype.

Compliance with Ethical Standards

Sources of Funding Julien Barc was supported by the H2020-MSCA-IF-2014 Program of the European Commission (RISTRAD-661617).

Conflict of Interest Florence Kyndt declares that she has no conflict of interest. Jean-Baptiste Gourraud declares that he has no conflict of interest. Julien Barc declares that he has no conflict of interest.

This article does not contain any studies with human participants or animals performed by any of the authors.

References

Ackerman MJ. Genetic purgatory and the cardiac channelopathies: exposing the variants of uncertain/unknown significance issue. Heart Rhythm. 2015;12:2325–31.

Ackerman MJ, Priori SG, Willems S, et al. HRS/EHRA expert consensus statement on the state of genetic testing for the channelopathies and cardiomyopathies: this document was developed as a partnership between the Heart Rhythm Society (HRS) and the European heart rhythm association (EHRA). Europace. 2011;13:1077–109.

Akgün M, Bayrak AO, Ozer B, Sağıroğlu MŞ. Privacy preserving processing of genomic data: a survey. J Biomed Inform. 2015;56:103–11.

Altmann HM, Tester DJ, Will ML, et al. Homozygous/compound heterozygous triadin mutations associated with autosomal-recessive long-QT syndrome and pediatric sudden cardiac arrest: elucidation of the triadin knockout syndrome. Circulation. 2015;131:2051–60.

Andorin A, Behr ER, Denjoy I, et al. Impact of clinical and genetic findings on the management of young patients with Brugada syndrome. Heart Rhythm. 2016;13:1274–82.

Andreasen C, Nielsen JB, Refsgaard L, et al. New population-based exome data are questioning the pathogenicity of previously cardiomyopathy-associated genetic variants. Eur J Hum Genet. 2013;21:918–28.

Anselme F, Moubarak G, Savouré A, et al. Implantable cardioverter-defibrillators in Lamin a/C mutation carriers with cardiac conduction disorders. Heart Rhythm. 2013;10:1492–8.

Antzelevitch C, Pollevick GD, Cordeiro JM, et al. Loss-of-function mutations in the cardiac calcium channel underlie a new clinical entity characterized by ST-segment elevation, short QT intervals, and sudden cardiac death. Circulation. 2007;115:442–9.

Bagnall RD, Das KJ, Duflou J, Semsarian C. Exome analysis-based molecular autopsy in cases of sudden unexplained death in the young. Heart Rhythm. 2014;11:655–62.

Barc J, Briec F, Schmitt S, et al. Screening for copy number variation in genes associated with the long QT syndrome: clinical relevance. J Am Coll Cardiol. 2011;57:40–7.

Barsheshet A, Goldenberg I, O-Uchi J, et al. Mutations in cytoplasmic loops of the KCNQ1 channel and the risk of life-threatening events: implications for mutation-specific response to β-blocker therapy in type 1 long-QT syndrome. Circulation. 2012;125:1988–96.

Baruteau A-EE, Probst V, Abriel H. Inherited progressive cardiac conduction disorders. Curr Opin Cardiol. 2015;30:33–9.

Basson CT, Cowley GS, Solomon SD, et al. The clinical and genetic spectrum of the Holt-Oram syndrome (heart-hand syndrome). N Engl J Med. 1994;330:885–91.

Bauce B, Nava A, Beffagna G, et al. Multiple mutations in desmosomal proteins encoding genes in arrhythmogenic right ventricular cardiomyopathy/dysplasia. Heart Rhythm. 2010;7:22–9.

Behr ER, Dalageorgou C, Christiansen M, et al. Sudden arrhythmic death syndrome: familial evaluation identifies inheritable heart disease in the majority of families. Eur Heart J. 2008;29:1670–80.

Bellocq C, van Ginneken AC, Bezzina CR, et al. Mutation in the KCNQ1 gene leading to the short QT-interval syndrome. Circulation. 2004;109:2394–7.

Bezzina CR, Barc J, Mizusawa Y, et al. Common variants at SCN5A-SCN10A and HEY2 are associated with Brugada syndrome, a rare disease with high risk of sudden cardiac death. Nat Genet. 2013;45:1044–9.

Bhonsale A, Groeneweg JA, James CA, et al. Impact of genotype on clinical course in arrhythmogenic right ventricular dysplasia/cardiomyopathy-associated mutation carriers. Eur Heart J. 2015;36:847–55.

Bhuiyan ZA, van den Berg MP, van Tintelen JP, et al. Expanding spectrum of human RYR2-related disease: new electrocardiographic, structural, and genetic features. Circulation. 2007;116:1569–76.

Bjerregaard P. Proposed diagnostic criteria for short QT syndrome are badly founded. J Am Coll Cardiol. 2011;58:549–50; author reply 550–1.

Brugada P, Brugada J. Right bundle branch block, persistent ST segment elevation and sudden cardiac death: a distinct clinical and electrocardiographic syndrome. A multicenter report. J Am Coll Cardiol. 1992;20:1391. Available at: PM:1309182

Brugada R, Hong K, Dumaine R, et al. Sudden death associated with short-QT syndrome linked to mutations in HERG. Circulation. 2004;109:30–5.

Campbell MJ, Czosek RJ, Hinton RB, Miller EM. Exon 3 deletion of ryanodine receptor causes left ventricular noncompaction, worsening catecholaminergic polymorphic ventricular tachycardia, and sudden cardiac arrest. Am J Med Genet A. 2015;167A:2197–200.

Chen Q, Kirsch GE, Zhang D, et al. Genetic basis and molecular mechanism for idiopathic ventricular fibrillation. Nature. 1998;392:293. Available at: PM:9521325

Christiaans I, van Langen IM, Birnie E, Bonsel GJ, Wilde AA, Smets EM. Quality of life and psychological distress in hypertrophic cardiomyopathy mutation carriers: a cross-sectional cohort study. Am J Med Genet A. 2009;149A:602–12.

Christiansen SL, Hertz CL, Ferrero-Miliani L, et al. Genetic investigation of 100 heart genes in sudden unexplained death victims in a forensic setting. Eur J Hum Genet. 2016;24:1797–802.

Chugh SS, Chung K, Zheng Z-JJ, John B, Titus JL. Cardiac pathologic findings reveal a high rate of sudden cardiac death of undetermined etiology in younger women. Am Heart J. 2003;146:635–9.

Corrado D, Link MS, Calkins H. Arrhythmogenic right ventricular cardiomyopathy. N Engl J Med. 2017;376:61–72.

Crotti L, Spazzolini C, Schwartz PJ, et al. The common long-QT syndrome mutation KCNQ1/A341V causes unusually severe clinical manifestations in patients with different ethnic backgrounds: toward a mutation-specific risk stratification. Circulation. 2007;116:2366–75.

Crotti L, Marcou CA, Tester DJ, et al. Spectrum and prevalence of mutations involving BrS1- through BrS12-susceptibility genes in a cohort of unrelated patients referred for Brugada syndrome genetic testing: implications for genetic testing. J Am Coll Cardiol. 2012;60:1410. Available at: PM:22840528

Crotti L, Johnson CN, Graf E, et al. Calmodulin mutations associated with recurrent cardiac arrest in infants. Circulation. 2013;127:1009–17.

Daumy X, Amarouch M-YY, Lindenbaum P, et al. Targeted resequencing identifies TRPM4 as a major gene predisposing to progressive familial heart block type I. Int J Cardiol. 2016;207:349–58.

Devalla HD, Gélinas R, Aburawi EH, et al. TECRL, a new life-threatening inherited arrhythmia gene associated with overlapping clinical features of both LQTS and CPVT. EMBO Mol Med. 2016;8:1390–408.

Donger C, Denjoy I, Berthet M, et al. KVLQT1 C-terminal missense mutation causes a forme fruste long-QT syndrome. Circulation. 1997;96:2778–81.

Eastaugh LJ, James PA, Phelan DG, Davis AM. Brugada syndrome caused by a large deletion in SCN5A only detected by multiplex ligation-dependent probe amplification. J Cardiovasc Electrophysiol. 2011;22:1073–6.

Gaita F, Giustetto C, Bianchi F, et al. Short QT syndrome: pharmacological treatment. J Am Coll Cardiol. 2004;43:1494–9.

García-Molina E, Lacunza J, Ruiz-Espejo F, et al. A study of the SCN5A gene in a cohort of 76 patients with Brugada syndrome. Clin Genet. 2013;83:530–8.

Garg V, Kathiriya IS, Barnes R, et al. GATA4 mutations cause human congenital heart defects and reveal an interaction with TBX5. Nature. 2003;424:443–7.

George CH, Jundi H, Thomas NL, Fry DL, Lai FA. Ryanodine receptors and ventricular arrhythmias: emerging trends in mutations, mechanisms and therapies. J Mol Cell Cardiol. 2007;42:34–50.

Ghouse J, Have CT, Skov MW, et al. Numerous Brugada syndrome-associated genetic variants have no effect on J-point elevation, syncope susceptibility, malignant cardiac arrhythmia, and all-cause mortality. Genet Med. 2017;19:521–8.

Giustetto C, Schimpf R, Mazzanti A, et al. Long-term follow-up of patients with short QT syndrome. J Am Coll Cardiol. 2011;58:587–95.

Goldenberg I, Moss AJ. Long QT syndrome. J Am Coll Cardiol. 2008;51:2291–300. Available at: PM:18549912

Goldenberg I, Horr S, Moss AJ, et al. Risk for life-threatening cardiac events in patients with genotype-confirmed long-QT syndrome and normal-range corrected QT intervals. J Am Coll Cardiol. 2011;57:51–9.

Gollob MH, Redpath CJ, Roberts JD. The short QT syndrome: proposed diagnostic criteria. J Am Coll Cardiol. 2011;57:802–12.

Green RC, Berg JS, Grody WW, et al. ACMG recommendations for reporting of incidental findings in clinical exome and genome sequencing. Genet Med. 2013;15:565–74.

Gussak I, Brugada P, Brugada J, et al. Idiopathic short QT interval: a new clinical syndrome? Cardiology. 2000;94:99–102.

Hashemi SM, Hund TJ, Mohler PJ. Cardiac ankyrins in health and disease. J Mol Cell Cardiol. 2009;47:203–9.

Hayashi M, Denjoy I, Extramiana F, et al. Incidence and risk factors of arrhythmic events in catecholaminergic polymorphic ventricular tachycardia. Circulation. 2009;119:2426–34.

Hendriks KS, Hendriks MM, Birnie E, et al. Familial disease with a risk of sudden death: a longitudinal study of the psychological consequences of predictive testing for long QT syndrome. Heart Rhythm. 2008;5:719. Available at: PM:18452877

Hertz CL, Christiansen SL, Ferrero-Miliani L, et al. Next-generation sequencing of 34 genes in sudden unexplained death victims in forensics and in patients with channelopathic cardiac diseases. Int J Legal Med. 2015;129:793–800.

Imboden M, Swan H, Denjoy I, et al. Female predominance and transmission distortion in the long-QT syndrome. N Engl J Med. 2006;355:2744. Available at: PM:17192539

Imbrici P, Liantonio A, Camerino GM, et al. Therapeutic approaches to genetic ion channelopathies and perspectives in drug discovery. Front Pharmacol. 2016;7:121.

Jervell A, Lange-Nielsen F. Congenital deaf-mutism, functional heart disease with prolongation of the Q-T interval and sudden death. Am Heart J. 1957;54:59–68.

Kalia SS, Adelman K, Bale SJ, et al. Recommendations for reporting of secondary findings in clinical exome and genome sequencing, 2016 update (ACMG SF v2.0): a policy statement of the American College of Medical Genetics and Genomics. Genet Med. 2017;19(2):249–55.

Kapplinger JD, Tester DJ, Salisbury BA, et al. Spectrum and prevalence of mutations from the first 2,500 consecutive unrelated patients referred for the FAMILION long QT syndrome genetic test. Heart Rhythm. 2009;6:1297. Available at: PM:19716085

Kapplinger JD, Tester DJ, Alders M, et al. An international compendium of mutations in the SCN5A-encoded cardiac sodium channel in patients referred for Brugada syndrome genetic testing. Heart Rhythm. 2010;7:33–46.

Kingsmore SF, Saunders CJ. Deep sequencing of patient genomes for disease diagnosis: when will it become routine? Sci Transl Med. 2011;3:87ps23.

Kokunai Y, Nakata T, Furuta M, et al. A Kir3.4 mutation causes Andersen-Tawil syndrome by an inhibitory effect on Kir2.1. Neurology. 2014;82:1058–64.

Konigstein M, Rosso R, Topaz G, et al. Drug-induced Brugada syndrome: clinical characteristics and risk factors. Heart Rhythm. 2016;13:1083–7.

Koopmann TT, Beekman L, Alders M, et al. Exclusion of multiple candidate genes and large genomic rearrangements in SCN5A in a Dutch Brugada syndrome cohort. Heart Rhythm. 2007;4:752–5.

Larsen MK, Berge KE, Leren TP, et al. Postmortem genetic testing of the ryanodine receptor 2 (RYR2) gene in a cohort of sudden unexplained death cases. Int J Legal Med. 2013;127:139–44.

Le SS, Bhasin N, Vieyres C, et al. Dysfunction in ankyrin-B-dependent ion channel and transporter targeting causes human sinus node disease. Proc Natl Acad Sci U S A. 2008;105:15617. Available at: PM:18832177

Le Scouarnec S, Karakachoff M, J-BB G, et al. Testing the burden of rare variation in arrhythmia-susceptibility genes provides new insights into molecular diagnosis for Brugada syndrome. Hum Mol Genet. 2015;24:2757–63.

Lenegre J. Etiology and pathology of bilateral bundle branch block in relation to complete heart block. Prog Cardiovasc Dis. 1964;6:409–44.

Lev M. The pathology of complete atrioventricular block. Prog Cardiovasc Dis. 1964;6:317–26.

Li Mura IE, Bauce B, Nava A, et al. Identification of a PKP2 gene deletion in a family with arrhythmogenic right ventricular cardiomyopathy. Eur J Hum Genet. 2013;21:1226–31.

Liu N, Ruan Y, Priori SG. Catecholaminergic polymorphic ventricular tachycardia. Prog Cardiovasc Dis. 2008;51:23–30.

Liu H, El Zein L, Kruse M, et al. Gain-of-function mutations in TRPM4 cause autosomal dominant isolated cardiac conduction disease. Circ Cardiovasc Genet. 2010;3:374–85.

Makita N. Phenotypic overlap of cardiac sodium channelopathies. individual-specific or mutation-specific? Circ J. 2009;73:810–7.

Makita N, Seki A, Sumitomo N, et al. A connexin40 mutation associated with a malignant variant of progressive familial heart block type I. Circ Arrhythm Electrophysiol. 2012;5:163–72.

Marcus FI, McKenna WJ, Sherrill D, et al. Diagnosis of arrhythmogenic right ventricular cardiomyopathy/dysplasia: proposed modification of the task force criteria. Eur Heart J. 2010;31:806–14.

Marsman RF, Bardai A, Postma AV, et al. A complex double deletion in LMNA underlies progressive cardiac conduction disease, atrial arrhythmias, and sudden death. Circ Cardiovasc Genet. 2011;4:280–7.

Marsman RF, Barc J, Beekman L, et al. A mutation in CALM1 encoding calmodulin in familial idiopathic ventricular fibrillation in childhood and adolescence. J Am Coll Cardiol. 2014;63:259–66.

Mayosi BM, Fish M, Shaboodien G, et al. Identification of cadherin 2 (CDH2) mutations in arrhythmogenic right ventricular cardiomyopathy. Circ Cardiovasc Genet. 2017;10

Medeiros-Domingo A, Bhuiyan ZA, Tester DJ, et al. The RYR2-encoded ryanodine receptor/calcium release channel in patients diagnosed previously with either catecholaminergic polymorphic ventricular tachycardia or genotype negative, exercise-induced long QT syndrome: a comprehensive open reading frame mutational analysis. J Am Coll Cardiol. 2009;54:2065–74.

Migdalovich D, Moss AJ, Lopes CM, et al. Mutation and gender-specific risk in type 2 long QT syndrome: implications for risk stratification for life-threatening cardiac events in patients with long QT syndrome. Heart Rhythm. 2011;8:1537–43.

Mohler PJ, Splawski I, Napolitano C, et al. A cardiac arrhythmia syndrome caused by loss of ankyrin-B function. Proc Natl Acad Sci U S A. 2004;101:9137–42.

Morita H, Wu J, Zipes DP. The QT syndromes: long and short. Lancet. 2008;372:750. Available at: PM:18761222

Moss AJ, Zareba W, Kaufman ES, et al. Increased risk of arrhythmic events in long-QT syndrome with mutations in the pore region of the human ether-a-go-go-related gene potassium channel. Circulation. 2002;105:794–9.

Novelli V, Gambelli P, Memmi M, Napolitano C. Challenges in molecular diagnostics of channelopathies in the next-generation sequencing era: less is more? Front Cardiovasc Med. 2016;3:29.

Nyegaard M, Overgaard MT, Sondergaard MT, et al. Mutations in calmodulin cause ventricular tachycardia and sudden cardiac death. Am J HumGenet. 2012;91:703. Available at: PM:23040497

Ohno S. The genetic background of arrhythmogenic right ventricular cardiomyopathy. J Arrhythm. 2016;32:398–403.

Orgeron GM, Calkins H. Advances in the diagnosis and management of arrhythmogenic right ventricular dysplasia/cardiomyopathy. Curr Cardiol Rep. 2016;18:53.

Pagon RA, Adam MP, Ardinger HH, et al. GeneReviews(®). Seattle: University of Washington; 1993.

Paludan-Müller C, Ahlberg G, Ghouse J, et al. Integration of 60,000 exomes and ACMG guidelines question the role of Catecholaminergic polymorphic ventricular tachycardia-associated variants. Clin Genet. 2017;91:63–72.

Patel C, Yan G-XX, Antzelevitch C. Short QT syndrome: from bench to bedside. Circ Arrhythm Electrophysiol. 2010;3:401–8.

Pipilas DC, Johnson CN, Webster G, et al. Novel calmodulin mutations associated with congenital long QT syndrome affect calcium current in human cardiomyocytes. Heart Rhythm. 2016;13:2012–9.

Postema PG, Wolpert C, Amin AS, et al. Drugs and Brugada syndrome patients: review of the literature, recommendations, and an up-to-date website (www.brugadadrugs.org). Heart Rhythm. 2009;6:1335–41.

Priori SG, Napolitano C. Cardiac and skeletal muscle disorders caused by mutations in the intracellular Ca2+ release channels. J Clin Invest. 2005;115(8):2033.

Priori SG, Schwartz PJ, Napolitano C, et al. Risk stratification in the long-QT syndrome. N Engl J Med. 2003;348:1866–74.

Priori SG, Napolitano C, Schwartz PJ, et al. Association of long QT syndrome loci and cardiac events among patients treated with beta-blockers. JAMA. 2004;292:1341–4.

Priori SG, Pandit SV, Rivolta I, et al. A novel form of short QT syndrome (SQT3) is caused by a mutation in the KCNJ2 gene. Circ Res. 2005;96(7):800.

Priori SG, Wilde AA, Horie M, et al. HRS/EHRA/APHRS expert consensus statement on the diagnosis and Management of Patients with inherited primary arrhythmia syndromes expert consensus statement on inherited primary arrhythmia syndromes: document endorsed by HRS, EHRA, and APHRS in may 2013 and by ACCF, AHA, PACES, and AEPC in June 2013. Heart Rhythm. 2013;19:e75. Available at: PM:24011539

Priori SG, Blomström-Lundqvist C, Mazzanti A, et al. 2015 ESC guidelines for the management of patients with ventricular arrhythmias and the prevention of sudden cardiac death: the task force for the management of patients with ventricular arrhythmias and the prevention of sudden cardiac death of the European Society of Cardiology (ESC). Endorsed by: Association for European Paediatric and Congenital Cardiology (AEPC). Eur Heart J. 2015;36:2793–867.

Probst V, Kyndt F, Potet F, et al. Haploinsufficiency in combination with aging causes SCN5A-linked hereditary Lenègre disease. J Am Coll Cardiol. 2003;41:643–52.

Probst V, Wilde AA, Barc J, et al. SCN5A mutations and the role of genetic background in the pathophysiology of Brugada syndrome. Circ Cardiovasc Genet. 2009;2:552. Available at: PM:20031634

Probst V, Veltmann C, Eckardt L, et al. Long-term prognosis of patients diagnosed with Brugada syndrome: results from the FINGER Brugada syndrome registry. Circulation. 2010;121:635. Available at: PM:20100972

Pua CJ, Bhalshankar J, Miao K, et al. Development of a comprehensive sequencing assay for inherited cardiac condition genes. J Cardiovasc Transl Res. 2016;9:3–11.

Richards S, Aziz N, Bale S, et al. Standards and guidelines for the interpretation of sequence variants: a joint consensus recommendation of the American College of Medical Genetics and Genomics and the Association for Molecular Pathology. Genet Med. 2015;17:405–24.

Rigato I, Bauce B, Rampazzo A, et al. Compound and digenic heterozygosity predicts lifetime arrhythmic outcome and sudden cardiac death in desmosomal gene-related arrhythmogenic right ventricular cardiomyopathy. Circ Cardiovasc Genet. 2013;6:533–42.

Risgaard B, Jabbari R, Refsgaard L, et al. High prevalence of genetic variants previously associated with Brugada syndrome in new exome data. Clin Genet. 2013;84:489–95.

Roberts JD, Herkert JC, Rutberg J, et al. Detection of genomic deletions of PKP2 in arrhythmogenic right ventricular cardiomyopathy. Clin Genet. 2013;83:452–6.

Roden DM. Clinical practice. Long-QT syndrome. N Engl J Med. 2008;358:169–76.

Romano C, Gemme G, Pongiglione R. Rare cardiac arrythmias of the pediatric age. II. Syncopal attacks due to paroxysmal ventricular fibrillation (presentation of 1st case in Italian pediatric literature). Clin Pediatr (Bologna). 1963;45:656–83.

Rooryck C, Kyndt F, Bozon D, et al. New family with catecholaminergic polymorphic ventricular tachycardia linked to the triadin gene. J Cardiovasc Electrophysiol. 2015;26:1146–50.

Roux-Buisson N, Cacheux M, Fourest-Lieuvin A, et al. Absence of triadin, a protein of the calcium release complex, is responsible for cardiac arrhythmia with sudden death in human. Hum Mol Genet. 2012;21:2759–67.

Schott JJ, Charpentier F, Peltier S, et al. Mapping of a gene for long QT syndrome to chromosome 4q25-27. Am J Hum Genet. 1995;57:1114–22.

Schott JJ, Benson DW, Basson CT, et al. Congenital heart disease caused by mutations in the transcription factor NKX2-5. Science. 1998;281:108–11.

Schott JJ, Alshinawi C, Kyndt F, et al. Cardiac conduction defects associate with mutations in SCN5A. Nat Genet. 1999;23:20–1.

Schwartz PJ, Moss AJ, Vincent GM, Crampton RS. Diagnostic criteria for the long QT syndrome. An update. Circulation. 1993;88:782–4.

Schwartz PJ, Priori SG, Spazzolini C, et al. Genotype-phenotype correlation in the long-QT syndrome: gene-specific triggers for life-threatening arrhythmias. Circulation. 2001;103:89–95.

Schwartz PJ, Stramba-Badiale M, Crotti L, et al. Prevalence of the congenital long-QT syndrome. Circulation. 2009;120:1761–7.

Schwartz PJ, Crotti L, Insolia R. Long-QT syndrome: from genetics to management. Circ Arrhythm Electrophysiol. 2012;5:868–77.

Selga E, Campuzano O, Pinsach-Abuin ML, et al. Comprehensive genetic characterization of a Spanish Brugada syndrome cohort. PLoS One. 2015;10:e0132888.

Sen-Chowdhry S, Syrris P, Ward D, Asimaki A, Sevdalis E, McKenna WJ. Clinical and genetic characterization of families with arrhythmogenic right ventricular dysplasia/cardiomyopathy provides novel insights into patterns of disease expression. Circulation. 2007;115:1710–20.

Shimizu W, Moss AJ, Wilde AA, et al. Genotype-phenotype aspects of type 2 long QT syndrome. J Am Coll Cardiol. 2009;54:2052–62.

Sisakian H. Cardiomyopathies: evolution of pathogenesis concepts and potential for new therapies. World J Cardiol. 2014;6:478–94.

Skinner JR, Crawford J, Smith W, et al. Prospective, population-based long QT molecular autopsy study of postmortem negative sudden death in 1 to 40 year olds. Heart Rhythm. 2011;8:412–9.

Spazzolini C, Mullally J, Moss AJ, et al. Clinical implications for patients with long QT syndrome who experience a cardiac event during infancy. J Am Coll Cardiol. 2009;54:832–7.

Spjuth O, Bongcam-Rudloff E, Dahlberg J, et al. Recommendations on e-infrastructures for next-generation sequencing. Gigascience. 2016;5:26.

Spoonamore KG, Ware SM. Genetic testing and genetic counseling in patients with sudden death risk due to heritable arrhythmias. Heart Rhythm. 2016;13:789–97.

Stallmeyer B, Koopmann M, Schulze-Bahr E. Identification of novel mutations in LMNA associated with familial forms of dilated cardiomyopathy. Genet Test Mol Biomarkers. 2012;16:543–9.

Steffensen AB, Refaat MM, J-PP D, et al. High incidence of functional ion-channel abnormalities in a consecutive long QT cohort with novel missense genetic variants of unknown significance. Sci Rep. 2015;5:10009.

Sudmant PH, Rausch T, Gardner EJ, et al. An integrated map of structural variation in 2,504 human genomes. Nature. 2015;526:75–81.

Tan HL, Bezzina CR, Smits JP, Verkerk AO, Wilde AA. Genetic control of sodium channel function. Cardiovasc Res. 2003;57:961. Available at: PM:12650874

Tan HL, Hofman N, van Langen IM, van der Wal AC, Wilde AA. Sudden unexplained death: heritability and diagnostic yield of cardiological and genetic examination in surviving relatives. Circulation. 2005;112:207–13.

Te Riele AS, Agullo-Pascual E, James CA, et al. Multilevel analyses of SCN5A mutations in arrhythmogenic right ventricular dysplasia/cardiomyopathy suggest non-canonical mechanisms for disease pathogenesis. Cardiovasc Res. 2017;113:102–11.

Tennessen JA, Bigham AW, O'Connor TD, et al. Evolution and functional impact of rare coding variation from deep sequencing of human exomes. Science. 2012;337:64–9.

Tester DJ, Ackerman MJ. Novel gene and mutation discovery in congenital long QT syndrome: let's keep looking where the street lamp standeth. Heart Rhythm. 2008;5:1282–4.

Tester DJ, Ackerman MJ. Genetics of long QT syndrome. Methodist Debakey Cardiovasc J. 2014;10:29–33.

Tester DJ, Medeiros-Domingo A, Will ML, Haglund CM, Ackerman MJ. Cardiac channel molecular autopsy: insights from 173 consecutive cases of autopsy-negative sudden unexplained death referred for postmortem genetic testing. Mayo Clin Proc. 2012;87:524–39.

Tomaselli GF, Barth AS. Ion channel diseases: an update for 2016. Curr Treat Options Cardiovasc Med. 2016;18:21.

Tully I, Atherton J, Hunt L, Ingles J, Semsarian C, McGaughran J. Rarity and phenotypic heterogeneity provide challenges in the diagnosis of Andersen-Tawil syndrome: two cases presenting with ECGs mimicking catecholaminergic polymorphic ventricular tachycardia (CPVT). Int J Cardiol. 2015;201:473–5.

Van der Werf C, Wilde AA. Catecholaminergic polymorphic ventricular tachycardia: from bench to bedside. Heart. 2013;99:497–504.

Veltmann C, Borggrefe M. Arrhythmias: a "Schwartz score" for short QT syndrome. Nat Rev Cardiol. 2011;8:251–2.

Ward OC. A new familial cardiac syndrome in children. J Irish Med Assoc. 1964;54:103–6.

Watanabe H, Koopmann TT, Le Scouarnec S, et al. Sodium channel β1 subunit mutations associated with Brugada syndrome and cardiac conduction disease in humans. J Clin Invest. 2008;118:2260–8.

Wijeyeratne YD, Behr ER. Sudden death and cardiac arrest without phenotype: the utility of genetic testing. Trends Cardiovasc Med. 2017;27(3):207–13.

Wilde AA, Behr ER. Genetic testing for inherited cardiac disease. Nat Rev Cardiol. 2013;10:571–83.

Wilde AA, Bhuiyan ZA, Crotti L, et al. Left cardiac sympathetic denervation for catecholaminergic polymorphic ventricular tachycardia. N Engl J Med. 2008;358:2024–9.

Winkel BG, Holst AG, Theilade J, et al. Nationwide study of sudden cardiac death in persons aged 1-35 years. Eur Heart J. 2011;32:983–90.

Winkel BG, Larsen MK, Berge KE, et al. The prevalence of mutations in KCNQ1, KCNH2, and SCN5A in an unselected national cohort of young sudden unexplained death cases. J Cardiovasc Electrophysiol. 2012;23:1092–8.

Wolf CM, Berul CI. Inherited conduction system abnormalities--one group of diseases, many genes. J Cardiovasc Electrophysiol. 2006;17:446–55.

Wolf CM, Wang L, Alcalai R, et al. Lamin a/C haploinsufficiency causes dilated cardiomyopathy and apoptosis-triggered cardiac conduction system disease. J Mol Cell Cardiol. 2008;44:293–303.

Zawistowski M, Reppell M, Wegmann D, et al. Analysis of rare variant population structure in Europeans explains differential stratification of gene-based tests. Eur J Hum Genet. 2014;22:1137–44.

Part III

Research into Cardiac Channelopathies: New Avenues

Novel Imaging Techniques in Cardiac Ion Channel Research

14

Esperanza Agullo-Pascual, Alejandra Leo-Macias, Donna R. Whelan, Mario Delmar, and Eli Rothenberg

Abstract

Light microscopy has long been at the forefront of biological research, perhaps most significantly in the form of fluorescence microscopy. This technique, paired with the ongoing discovery and synthesis of increasingly brilliant fluorophores, allows for visualization of the internal machinations of cells with molecular specificity. However, until recently, a persistent limitation of fluorescence microscopy—the diffraction of visible light—has restricted elucidation of the subcellular organization and localization of molecules to spatial resolutions of 200–300 nanometers. The invention and implementation of several super-resolution fluorescence microscopies (SRFMs) over the last 10 years have circumvented this diffraction limit and allowed up to tenfold improvements in resolution. Applications of SRFM in cardiology research have already illuminated aspects of the cardiac nanoscale architecture which were previously unobservable, opening the door for new avenues of research. These discoveries include the sub-diffraction structure of the intercalated disk, the t-tubular network, and excitation-contraction coupling. In this chapter we will review SRFM methodologies, present some examples of their successful application in cardiac research, and discuss the techniques' advantages, ongoing challenges, and future potential.

E. Agullo-Pascual · A. Leo-Macias · M. Delmar
The Leon H Charney Division of Cardiology, New York University School of Medicine, New York, NY, USA
e-mail: esperanza.agullo-pascual@nyumc.org; alejandra.leo-macias@nyumc.org; mario.delmar@nyumc.org

D. R. Whelan · E. Rothenberg (✉)
Department of Biochemistry and Molecular Pharmacology, New York University School of Medicine, New York, NY, USA
e-mail: donna.whelan@nyumc.org; eli.rothenberg@nyumc.org

D. Thomas, C. A. Remme (eds.), *Channelopathies in Heart Disease*, Cardiac and Vascular Biology, https://doi.org/10.1007/978-3-319-77812-9_14

Abbreviations

CLEM	Correlative light-electron microscopy
FP	Fluorescent protein
LM	Light microscopy
PALM	Photoactivation localization microscopy
SIM	Structured illumination microscopy
SMLM	Single-molecule localization microscopy
SRFM	Super-resolution fluorescence microscopy
STED	Stimulated emission depletion microscopy
STORM	Stochastic optical reconstruction microscopy

14.1 Introduction

Although the inventor of the compound microscope is a lingering point of contention, simple optical microscopes became available sometime in the seventeenth century, and in 1665 Robert Hooke coined the word "cell" and published what many consider to be the first important work on microscopy, *Micrographia*, a book which included illustrations of different microscopically examined specimens. Following this, single high-magnification lenses and compound microscopes were popularized as a technique for biologists by other scientists including Antonie van Leeuwenhoek and Marcello Malpighi. This soon allowed for the discovery and characterization of numerous previously undetectable organisms and cells such as bacteria, muscle fibers, and erythrocytes.

However, the technical difficulties of improving seventeenth-century microscopes, in particular due to problems with lens development and configuration, severely hindered further applications. Then, during the second half of the nineteenth century, Ernst Abbe described the principles of image formation in light microscopy and developed various new and improved lenses, and August Köhler invented Köhler illumination allowing for even sample illumination, adjustable contrast, and exclusion of the illumination source. This led to a breakthrough in the design and manufacturing of commercially available microscopes and reinstated optical microscopy as the key tool of biologists.

Following this, fluorescence microscopy was developed, with fluorophore discovery and synthesis, alongside labeling approaches, taking center stage. The first major advancement was the production and application of fluorescently labeled antibodies by Albert Coons in 1941 that enabled labeling of various cellular structures with bright synthetic fluorophores. This was followed in the 1990s by the utilization of the green fluorescence protein (GFP) as a fusion tag for live-cell imaging, which in turn catalyzed the development of a plethora of other FPs spanning the visible spectrum. Together, these advances allow for the fluorescent visualization of the subcellular landscape with, importantly, multiplexed molecular specificity. However, despite these improvements in labeling, as well as technological improvements to the microscopes themselves, fluorescence imaging remained limited by the diffraction of visible light, a physical reality of the wave nature of photons described by Ernst Abbe in the nineteenth century. Diffraction limited the best possible spatial resolution of light microscopy to no better than a few hundred nanometers, causing blurriness in images and dictating hundreds of nanometers of uncertainty to molecular localizations and structures.

Recently, however, a revolution in fluorescence microscopy occurred with the development of techniques, collectively known as super-resolution fluorescence microscopy (SRFM), which circumvent the diffraction limit. This is achieved in one of two ways: either deterministically by structuring the illumination light or by stochastically imaging, and thus localizing, fluorophores individually. The importance of these novel technologies was recently recognized with the Nobel Prize in Chemistry in 2014.

SRFM has instigated a new era in fluorescence microscopy in which molecular specificity can now be paired with molecular localization. This allows for visualization of nanoscale architecture providing insight into the organization, orientation, and interactions of biomolecules in complex structures such as the synaptic button in neurons (Tang et al. 2016; Wilhelm et al. 2014; Dani et al. 2010), focal adhesions (Case et al. 2015; Kanchanawong et al. 2010), and nuclear pores (Loschberger et al. 2014; Szymborska et al. 2013).

Here we describe the different SRFM techniques available and review some of the applications of these methods in cardiac research. We also discuss SRFM limitations, other applications, and future potential.

14.2 Methods

SRFM encompasses several techniques which can broadly be classified as either relying on a deterministic approach which uses spatially patterned illumination light or on a stochastic approach which detects single fluorophores. Both (1) structured illumination microscopy (SIM) and (2) stimulated emission depletion microscopy (STED) are popular examples of the former, whereas (3) single-molecule localization microscopy (SMLM) describes several variants of the latter.

14.2.1 Structured Illumination Microscopy

SIM is a widefield approach which relies on the structuring of light using a grid pattern which generates a sinusoidal excitation pattern of known spatial frequency, orientation, and phase. The grid is rotated in steps to acquire different images that are processed to extract high-frequency information and produce a reconstructed image with a twofold improvement in lateral resolution (~100 nm) (Fig. 14.1) (Langhorst et al. 2009).

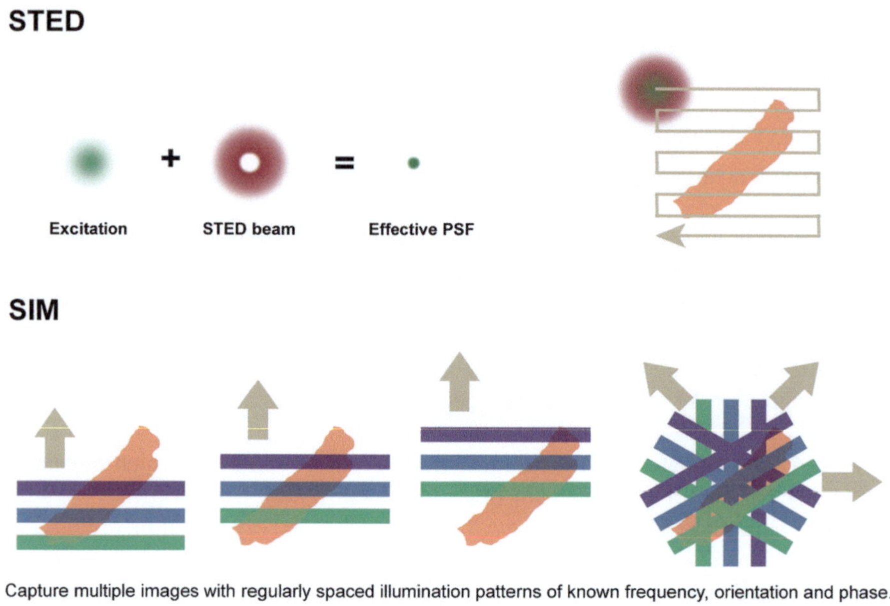

Fig. 14.1 An overview of the STED and SIM approaches to SRFM

While SIM setups have been available for longer than both STED and SMLM, there are relatively few published studies that use SIM to study cardiac architecture. In one of these studies, Granzier et al. labeled different regions of the protein Titin to study the role of the I-band/A-band junction on the strain on the molecular spring elements, uncovering biomechanical sensing associated with cardiac hypertrophy (Granzier et al. 2014). In another study, Lukyanenko et al. used SIM to localize Ca^{2+} /calmodulin-activated phosphodiesterase type 1A (PDE1) in sinoatrial nodal cells (Lukyanenko et al. 2016). This study shows that PDE1 localizes beneath the cell surface in sinoatrial nodal cells, an observation which provides insight into mechanisms for maintaining Ca^{2+} signaling and pacemaker function.

14.2.2 Stimulated Emission Depletion Microscopy

STED microscopy also relies on introducing patterns into the illumination light, but instead of a grid arrangement as in SIM, STED couples a conventional confocal illumination pattern with an overlapping torus-shaped "depletion" beam. Fluorophores illuminated by this depletion beam, i.e., those located away from the center of the focused confocal excitation beam, undergo stimulated emission and do not fluoresce. Consequently, any detected photons are known to originate in the sub-diffraction area where the torus beam does not overlap the excitation beam (Hell and Wichmann 1994). By raster scanning a sample with these two overlapping lasers using discrete acquisition steps of sub-diffraction size (usually 10–20 nm), a super-resolution image can be obtained without any further processing (Fig. 14.1). The lateral resolution that can be achieved with this technique ranges from 20 to 70 nm and is dependent on the intensity and size of the depletion beam, as well as the scanning step size.

Because of its commercial availability, live-cell applicability, and compatibility with many conventional fluorophores, STED microscopy has been used in several investigations into cardiac cell architecture and molecular organization. In a study published by Wagner and colleagues, STED microscopy was used to image t-tubular organization in live cardiomyocytes and to follow remodeling after myocardial infarction (Wagner et al. 2012). The improved image resolution allowed detection of enlargement of the t-tubule cross sections as well as network-wide remodeling. These changes in the t-tubule morphology could explain the early excitation-contraction uncoupling observed during heart failure development after myocardial infarction.

STED was also used to unravel the molecular mechanisms of atrial fibrillation (AF) in research by Macquaide et al. which visualized the organization of the ryanodine receptor (RyR) in atrial myocytes (Macquaide et al. 2015). Analysis of these images revealed that although there was no change in RyR cluster size, the distance between clusters was reduced in AF. It was also found that RyR clusters grouped into Ca^{2+} release units (CRUs) were bigger and contained more RyR, although CRU organization (ratio RyR: total area per CRU) was more fragmented in AF myocytes. Moreover, the frequency of CRU along the z-line, as well as between z-lines, was increased in AF. Together these observations suggest that the probability of CRU firing and propagated Ca^{2+} release in AF could be increased, which would correlate with a higher Ca^{2+} spark frequency and duration.

The organization of ion channels at the intercalated disk has also been investigated using a modified version of STED microscopy called gated STED (gSTED). Veeraraghavan et al. described in two papers the distribution of sodium channels ($Na_V1.5$) (Veeraraghavan et al. 2015) and potassium channels ($K_{ir}2.1$) (Veeraraghavan et al. 2016) in the connexin43 (Cx43) gap junction perinexus area. gSTED differs from conventional STED microscopy in that it takes into consideration the fluorescence lifetime of detected emitters. Although the STED depletion beam cannot deplete all fluorophores outside the center of the excitation ring, it does affect the lifetimes of these fluorophores, thus providing a useful means of excluding erroneous emissions and allowing improved spatial resolutions. gSTED of guinea pig tissue sections and neonatal rat ventricular myocytes demonstrated that approximately one third of $Na_V1.5$ and $K_{ir}2.1$ clusters localize within 200 nm of a Cx43 cluster (perinexus area). These results, combined with some functional data and mathematical modeling, suggest that the perinexus is a specialized nanodomain surrounding the gap junction plaque, which plays a role in the propagation of electrical excitation via ephaptic transmission.

14.2.3 Single-Molecule Localization Microscopy

SMLM encompasses dozens of methods which all make use of stochastic single-molecule fluorescence emission and localization to construct super-resolution images. In 2006 three groups independently published the first SMLM methods: stochastic optical reconstruction microscopy (STORM) (Rust et al. 2006), photoactivation localization microscopy (PALM) (Betzig et al. 2006) and fluorescence PALM (FPALM) (Hess et al. 2006). These methods, along with the variations that followed, take advantage of the intrinsic photophysics of fluorophores which allow switching between emissive and dark states or between different absorption/emission wavelengths. In taking control of this switching, single fluorophores in densely labeled areas can be imaged individually by switching all surrounding fluorophores to a dark state or different color. This can be achieved with small synthetic fluorophores by inducing a reduced state via the semi-stable triplet state, or in fluorescent proteins via bleaching, or by several other photophysical/photochemical means (Reid and Rothenberg 2015). This on/off blinking behavior (rate of transition and "off" state duration) is different for each fluorophore and nanoscale cellular environment, and so optimization of fluorophore blinking must be achieved empirically through a combination of irradiation intensity, fluorophore concentration, and buffer conditions (e.g., oxidizing agents, reducing agents, and aqueous oxygen).

Once good fluorophore blinking is achieved, to obtain a super-resolved image, many thousands of frames are acquired with a different subset of fluorophores stochastically switched "on" in each frame. These movies are then processed using single-molecule emission finding and fitting algorithms which can determine, to within a few nanometers precision, the locations of each individual fluorophore (Fig. 14.2).

In particular, the direct STORM (*d*STORM) (Heilemann et al. 2008) imaging approach which makes use of many commercially available antibody-conjugated fluorophores has found much use in biological research including applications to elucidate macromolecular complexes in cardiac cells. Soeller and colleagues have focused on the organization of the ryanodine receptor in several papers (Baddeley et al. 2009; Hou et al. 2015; Wong et al. 2013). In their SMLM imaging, they demonstrated for the first time that RyR clusters vary widely in both shape and size. Interestingly, edge-to-edge distances showed that most of the clusters were within a distance of less than 100 nm with neighboring clusters suggesting the potential formation of "superclusters" that could facilitate the coupling of these receptors and trigger Ca^{2+} release cascades. Further studies have looked at the interaction of junctophilin-2 with RyR (Jayasinghe et al. 2012; Munro et al. 2016) and the Na^{+}/Ca^{2+} exchanger (NCX) (Wang et al. 2014). De La Fuente et al. also performed super-resolution imaging of the mitochondrial calcium uniporter in cardiac mitochondria, demonstrating its association to the sarcoplasmic reticulum and RyR clusters (De La Fuente et al. 2016).

Our group has further taken advantage of the improved resolution of SRFM to probe the molecular architecture of the intercalated disk in cardiomyocytes. In our

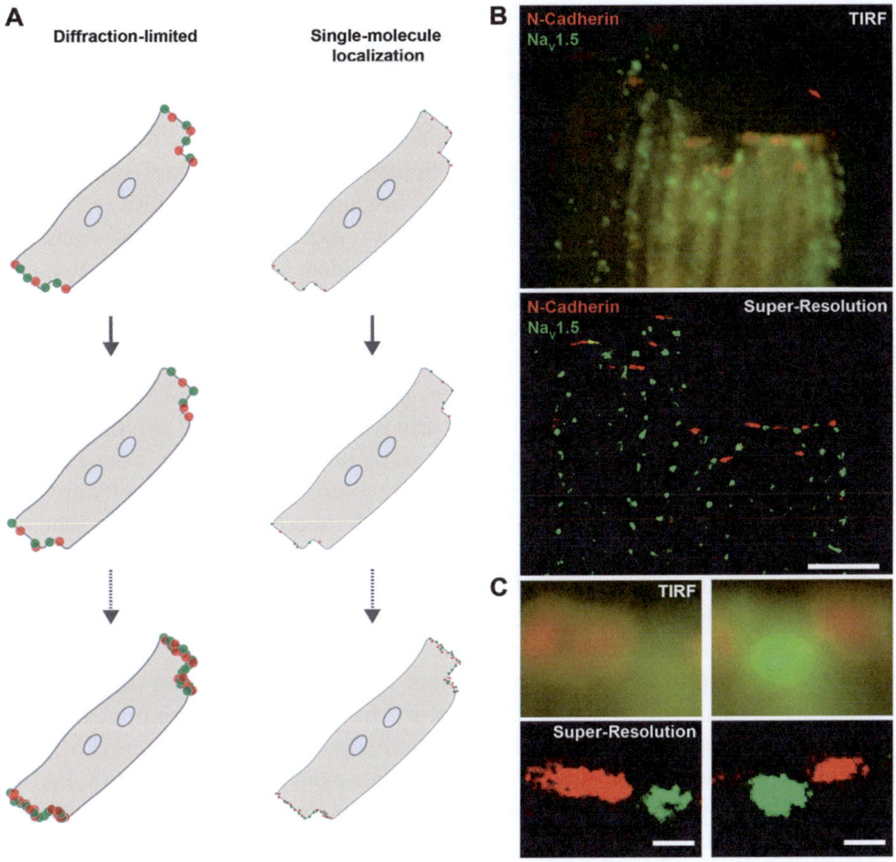

Fig. 14.2 Single-molecule localization microscopy. (**a**) Illustration of data acquisition for SMLM. Only a subset of fluorophores are in the "on" state in each frame allowing localization of each single molecule with nanometer accuracy. (**b**) An isolated adult mouse cardiomyocyte stained for N-cadherin (red) and Na$_V$1.5 (green). The top panel shows a TIRF image (diffraction-limited), while the bottom panel shows the super-resolved image of the same cell with improved resolution. (**c**) Zoom regions of B show the enhanced spatial resolution achieved by SMLM. Scale bars: 5 μm (**b**), 400 nm (**c**)

first study, we demonstrated that Cx43 and plakophilin-2 (PKP2), molecules that pertain to gap junctions and desmosomes, respectively, also interact at the membrane in neonatal cardiomyocytes (Agullo-Pascual et al. 2013). Moreover, when silencing the anchoring protein ankyrin-G, this interaction was reduced, and Cx43 organization was altered. These observations were further supported by Monte-Carlo simulations that demonstrated that the detected cluster overlaps were not random but a functional event. Follow-up studies continued to investigate the interactions between the different complexes that localize at the intercalated disk such as Cx43/gap junctions with sodium channels (Na$_V$1.5) (Agullo-Pascual et al. 2014) and PKP2 with Na$_V$1.5 (Cerrone et al. 2014). The nanoscale resolution obtained with *d*STORM images allowed us to look at the organization and interaction between proteins as well as trafficking to the intercalated disk. Furthermore, the decreased sodium current measured by patch clamp in cardiomyocytes from these mice (Cx43D378stop (Agullo-Pascual et al. 2014) and PKP2-Hz[30]) correlated with a decrease in both the number of $Na_V1.5$ clusters and the number of microtubule plus-end protein EB1 clusters at the intercalated disk.

So far, we have only discussed the highlighted SRFM methodologies in their 2D super-resolution capabilities and applications because in the vast majority of SRFM research, only lateral imaging is undertaken or required. 3D modifications are available; however, they are often more complicated to implement and use and inevitably suffer from decreased temporal resolution and increased data handling and storage difficulties, two problems which are already limiting to 2D SRFM. Nonetheless, visualization of 3D organization can be key to an experiment, and so, in some cases, the trade-off is deemed necessary. In these cases, 3D-SIM is usually the most approachable because it is inherently a three-dimensional approach, needing only for the grated illumination pattern to be oriented axially as well. Furthermore, most commercial SIM microscopes are outfitted with both software and hardware for 3D imaging. Similarly, because of its confocal excitation scheme, STED can be used to produce 3D data simply by producing z-slices, albeit with diffraction limited z-resolutions. Alternatively, sub-diffraction axial resolutions can be realized by applying a second depletion beam which depletes fluorophores away from the center of the illumination beam in xz space.

In SMLM methodologies, many different approaches have been described for generating 3D data; in particular, these efforts have focused on manipulation and examination of single-molecule emission patterns [point spread functions (PSFs)] to encode and extract axial position information. For example, interferometric PALM uses two objectives and imaging paths allowing for the same photon to travel two paths and then recombine and self-interfere. The difference in distance traveled by the photon along the two paths is dependent on the fluorophore's axial position which can thus be calculated from the detected difference in phase amplitude of the self-interfering photon and the unaffected photon (Shtengel et al. 2009). Structures like focal adhesions and microtubules have been resolved in three dimensions at the nanometer scale using this approach (Case et al. 2015; Kanchanawong et al. 2010; Shtengel et al. 2009).

Biplane 3D also employs two imaging paths, introducing a slight difference in the distance between the objective and two cameras so that PSFs appear with different intensities and sizes (Proppert et al. 2014). More recently, algorithms have been put forward to extract z-information from 2D SMLM data, with the caveat that the distance from the imaging plane deduced is not identifiable as above or below (Franke et al. 2017). Huang et al. proposed an astigmatic approach to achieving 3D-SMLM in which a weak cylindrical lens is introduced into the emission path to distort the PSF in a predictable fashion based on axial position. Fluorophores above the imaging plane appear elongated along one lateral axis, while those below the imaging plane appear elongated in the other lateral axis (Huang et al. 2008). The ratio of the PSF size in the two axes can then be used with a calibration curve of known z-position PSF xy ratios to construct a 3D image (Fig. 14.3).

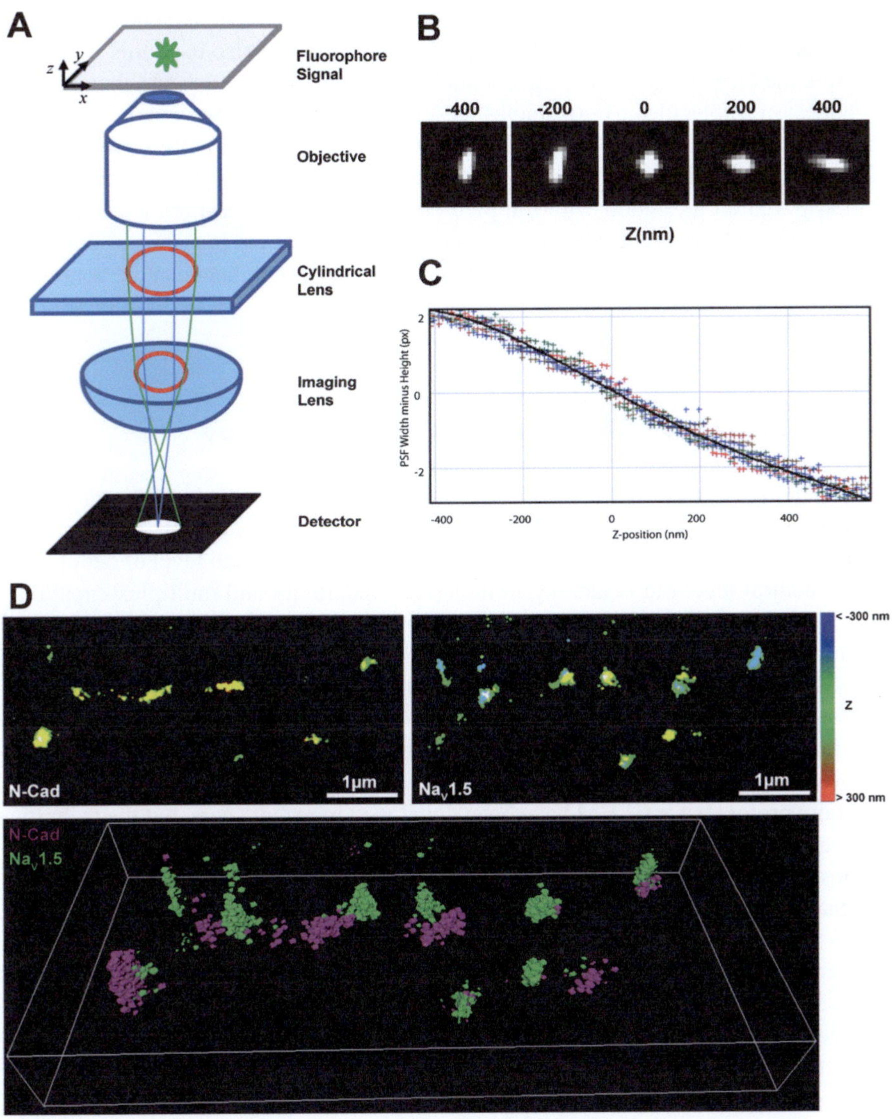

Fig. 14.3 Three-dimensional SMLM by astigmatism. (**a**) Diagram of 3D-SMLM system. (**b**) Examples of astigmatic images of a 100 nm diameter fluorescent bead in different axial positions. (**c**) A calibration curve used to generate 3D-SMLM images. The z-position is determined by the width minus height of the point spread function (PSF). (**d**) 3D-SMLM image of the cell end of an isolated adult mouse cardiomyocyte stained for N-cadherin (magenta) and $Na_V1.5$ (green). Top images show a 2D z-color coded image of N-cadherin (left) and of $Na_V1.5$ (right). Bottom image shows the 3D view of the same region

Using astigmatic 3D-SMLM, the periodic actin/spectrin cytoskeletal structure in axons has been discovered (Xu et al. 2013). Our group has also used this method to describe the distribution of sodium channels at the intercalated disk and their relation to adhesion molecules (Leo-Macias et al. 2016). Using two-color 3D super-resolution, we found that 35% of $Na_V1.5$ clusters co-localized with, or were within 100 nm of, N-cadherin clusters. Moreover, analysis of the cluster dimensions demonstrated a correlation with the predicted dimensions measured by angle view patch clamp. These observations lead us to speculate that the presence of adhesion/ excitability nodes at the intercalated disk facilitates cross talk between the contractile and electrical apparatus.

14.3 Correlative SMLM and Electron Microscopy

Conventional fluorescence microscopy allows protein imaging in living cells with high molecular specificity but without significant insight into their ultrastructural context. Electron microscopy, on the other hand, provides a high level of detail at the ultrastructural level but is lacking in molecular specificity and multiplex capability. Because the strengths of these techniques are exceptionally complementary, their combination in correlative light-electron microscopy (CLEM) has the potential to fill many gaps in biological imaging research. Indeed, successful CLEM studies, especially those that make use of SRFM, have demonstrated the ability to precisely localize specific proteins within the highly detailed ultrastructural landscapes of their native cellular contexts.

Many protocols for CLEM have been developed in recent years, but the technique remains challenging due to differences in sample preparation requirements for fluorescence microscopy and EM (Johnson et al. 2015; Paez-Segala et al. 2015).

One particular example of CLEM is the visualization of $Na_V1.5$ ion channels. These were imaged in relation to the adhesion protein N-cadherin within the ultrastructural context of thin sections of ventricular tissue (Fig. 14.4) (Leo-Macias et al. 2016). After SMLM imaging of the ion channel and adhesion proteins, the sample was processed for, and imaged using, transmission electron microscopy (TEM). The SMLM and EM images were overlaid using fiducial markers visible in both images, allowing specific selection and further study of those $Na_V1.5$ clusters located at the membrane of the end-to-end contact of the cardiac cells, the so-called intercalated disc. Further analysis of these regions revealed that $Na_V1.5$ preferentially aggregates with the adhesion molecule N-cadherin. The combination of this result with electrophysiology and adhesion strength experiments demonstrated that these clusters are major contributors to cardiac sodium current and that loss of $Na_V1.5$ expression reduces intercellular adhesion. These adhesion/excitability nodes are proposed to be key sites for cross talk of the contractile and electrical molecular apparatus and may represent the structural substrate of cardiomyopathies in patients with mutations in molecules of the VGSC complex, as was the case in this report (Te Riele et al. 2017).

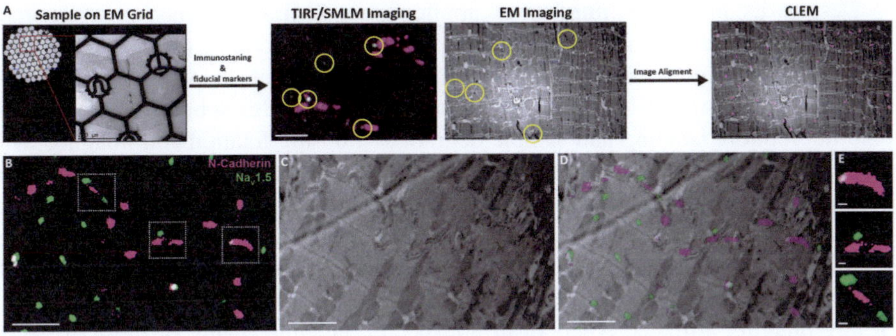

Fig. 14.4 Correlative light—electron microscopy. (**a**) Protocol for CLEM. Sample is mounted on an EM finder grid. After immunolabeling and addition of fiducial markers, the sample is imaged first for SMLM and then further processed for EM imaging. The same imaged region can be localized using the features on the grid, and the images are aligned using the fiducial markers that can be detected on both images. (**b–d**) Detection of N-cadherin (magenta) and Na$_V$1.5 (green) in a mouse heart tissue section by CLEM. (**e**) Zoom regions at the intercalated disk show localization of adhesion molecules and sodium channels in close apposition. Scale bar: 5 μm (**a**), 2 μm (**b–d**), 200 nm (**e**)

14.4 Advantages of SMLM and Ongoing Challenges

The SRFM approaches described here offer imaging capabilities that surpass the diffraction limit of light, but each comes with its own advantages and limitations. While SIM provides only a twofold improvement in resolution (lateral resolution of 120 nm) and is technologically complex, the availability of commercial instruments and its ease of use makes the technique approachable. This is particularly true when one takes into account the interchangeability of conventional fluorescence microcopy samples with SIM samples: any conventional fluorophore can be used, and live-cell imaging is constrained only by temporal resolution. Given the resolution provided in SIM, it is therefore recommended only when studying big cellular complexes like sarcomeres (Granzier et al. 2014). In comparison, STED and SMLM offer resolutions well below 100 nm that allow the study of molecular interactions and complexes, but despite commercial instruments being available and homebuilt instrumentation relatively straightforward (especially for SMLM), both methods suffer from more complicated sample preparations. STED relies on extremely resilient fluorophores which can be depleted but do not easily bleach or blink, while SMLM requires dyes which can be manipulated to blink. In both cases, the dyes must fulfill the requirements of the imaging methodology while also integrating into the sample with high specificity and, especially in the case of live cells, minimal perturbation to the sample. Of the available SRFMs, SMLM is arguably the most widely used (Huang et al. 2009), and so below we will primarily focus on the specifics of SMLM.

Despite its increasing number of applications, SMLM remains an emerging, highly specialized method, and several considerations need to be taken into account in order to perform a successful experiment. Optimization of sample preparation and imaging conditions are among the most important steps. This is because all nonspecific fluorophore labeling and sample degradation during fixation and imaging will result in false-positive localizations in the final SMLM image, as will low signal to noise and overlapping single molecule blinks. Furthermore, while the localization precision of SMLM data relies predominantly on the number of photons detected in a single-fluorophore signal, the spatial resolution relies not only on the precision but also the labeling type and degree of success and is strongly impacted by the presence of any imaging artifacts. Apart from nonspecific labeling, many of these artifacts come from suboptimal fluorophore blinking, which can be affected by several parameters:

- Imaging buffer: The blinking properties of fluorophores arise from electronic and structural changes in dye molecules. In many respects it is a stochastic process that arises from the recurring transition of the emitter between a non-emissive state (off) and an emissive state (on). A table of the most prevalent SMLM fluorophores is provided in Table 1 in Reid and Rothenberg (2015). For many conventional organic fluorophores, the off state is achieved by reduction of the semi-stable triplet state by an added reductant such a mercaptoethylamine. Return to the emissive on state is often caused by collision of dark reduced fluorophores with aqueous oxygen, and so controlling its concentration can also control blinking kinetics. This can be achieved by addition of an oxygen scavenging system, the most common of which comprises of glucose, glucose oxidase, and catalase. An added advantage of oxygen scavenging is that aqueous oxygen is responsible for some irreversible photobleaching events which are also undesired in SMLM experiments.
- Laser intensity: The number of cycles per second of fluorophores, between the ground and excited electronic states as achieved by photon absorption, vibrational relaxation, and photon emission, is largely dependent on photon flux. The probability of transition of an excited fluorophore to the semi-stable triplet state from which it can be reduced and stably switched off is stochastic, and so the probability of such a transition occurring in a given time also correlates directly with laser intensity. In contrast, laser intensity does not largely affect the probability of an off to on transition over a given time.
- Signal to noise ratio: In SMLM it is important to detect as many photons as possible from a single fluorophore in order to localize the molecule with nanometer precision. Therefore, overlapping PSFs, out-of-plane fluorescence, and autofluorescence can all interfere with the successful detection and localization of single molecules. In order to improve signal to noise, a total internal reflection fluorescence (TIRF) or highly inclined and laminated optical sheet (HiLo) configuration can be implemented. Both techniques angle the incident light so that only a fraction of the sample interacts with the incident light, thereby removing noise from out-of-plane fluorescence. In TIRF, an evanescent wave is used to

excite fluorophores within only a few hundred nanometers of the coverslip, while HiLo can be adjusted to excite fluorophores across variable depths of several microns.

– Labeling density: To achieve ultrastructural insight using SMLM, very high labeling densities must be used as described by the Nyquist sampling theory. Simply put, to achieve a true spatial resolution of x nm, a fluorophore must be localized every $x/2$ nm in the target structure. If SMLM is being used to quantify single-molecule distributions or interactions, it is similarly very important to achieve a high degree of labeling so as to not underestimate interactions. By altering acquisition parameters, namely, laser intensity and the concentrations of reductant and molecular oxygen, very densely labeled samples can be manipulated to optimize the ratio of fluorophores in the dark state to those in the emissive state. Over-labeling of samples for SMLM therefore only occurs when increased antibody aggregation and nonspecific labeling are observed.

Ideal SRFM experimental conditions vary widely depending on the target of interest, the fluorophores used, and the sample itself; therefore, extensive and careful optimization must be undertaken for each new experiment (Whelan and Bell 2015).

Analysis and rendering of SMLM data present a further challenge despite the availability of dozens of free software suites, ImageJ plug-ins, and open access codes [for a summary and comparison of algorithms, see Sage et al. (2015)]. Each of these analytical tools attempts to detect and localize each single-molecule emission within a data set, thus generating a list of coordinates which can be rendered into an "image." Most of these approaches can be distilled into three steps: first, detection of a potential single-molecule emission, usually by searching for local maxima; second, determination of the precise localization of the single-molecule emitter by fitting a Gaussian or arbitrary PSF; and finally rendering of the localized coordinates in 2D or 3D space. Various degrees of complexity and flexibility exist within and between the algorithms. Denoising (e.g., band/low-pass filtering and wavelet transformation) and thresholding can be used to process the raw data for detection of single-molecule emissions, while localization can be achieved using least squares, maximum likelihood, center of mass, or center of symmetry methods, among others. Each method generates different SMLM data with quantifiable detection rates, including false negatives and positives, localization accuracies, and spatial resolutions. The usability (computational costs and interface) and speed of different algorithms should also be taken into account. Because of this variability, multiple algorithms should be considered and trialed. Ongoing awareness of the degree, to which SMLM renderings are often affected by artifacts, are not "real" images, and are not comparable to other fluorescence images, is also important.

Partly because of the non-real nature of SMLM images, analysis often poses new challenges with many established approaches, such as those used for confocal fluorescence images, frequently proving inadequate. Inherent to the tenfold improvement in lateral resolution, SMLM provides nanometer-precise dimensions of ultrastructures and clusters and elucidates sub-diffraction distributions and intermolecular/intercluster distances. Because SMLM provides images of single-molecule locations, it also depicts molecular-level colocalizations and interactions, even allowing for quantification of these relations as well as molecular density. This is a key advantage over conventional fluorescence imaging which lacks single-molecule sensitivity and routinely relies on detection of clustered molecules, disregarding the diffuse "background" level of fluorophores. While many image parameters can be extracted from SR images manually, in order to harness the full capabilities of SMLM in biomedical research, new standardized tools for automated big data analysis are needed. Several such tools are already available for analysis of protein clusters (Andronov et al. 2016), intermolecular interactions and the formation of higher-order molecular complexes (Caetano et al. 2015), microtubule networks (Zhang et al. 2017), three-color molecular correlation (Yin and Rothenberg 2016), and general SMLM data (Malkusch and Heilemann 2016). However, with any new research endeavor, a robust analytical approach must be devised; thus, most successful applications of SMLM will rely on collaboration with computational science and bioinformatics specialists.

14.5 Future Potential

Taking into account the many different aspects and complexities of SMLM discussed, its use remains highly specialized and challenging. In particular the highly interdisciplinary demands of SMLM require that research groups making use of the method possess a breadth of skills including optics, photophysics, photochemistry, coding, and mathematics, and the biology required to not only prepare and handle samples but to pose relevant questions. Increased interest from the biomedical research community has encouraged various microscopy companies to develop systems with integrated single-molecule capabilities, but while these commercial microscopy systems provide an impressive array of imaging modalities, they are extremely costly, especially as compared to in-house customized or standard commercial microscopy systems (Holm et al. 2014). Moreover, having a commercial SMLM setup does not assist with the more difficult aspects of devising and carrying out experiments, including optimization of sample preparation and imaging acquisition parameters and data analysis. Because these challenges are ongoing, an ability to carry out these optimizations and troubleshoot SMLM experiments is integral.

Of the various potential uses of SMLM in biomedical research, perhaps the most sought after is live-cell imaging. It is particularly difficult to generate live-cell SMLM data because typically a single SMLM image takes longer than a minute to acquire; this temporal resolution is not useful for imaging many cellular processes, and so methods which make use of far fewer frames or rolling averages are often implemented. Live-cell imaging is also limited by the fluorophores available, the majority of which are less bright, less capable of "blinking," and more likely to photobleach than the synthetic dyes used for fixed-cell imaging. This has led to continuing development of better fluorescent proteins for live-cell SMLM applications, as well as novel methods for delivery of synthetic fluorophores (Chang et al. 2012; Teng et al. 2016; Hennig et al. 2015; Kube et al. 2017).

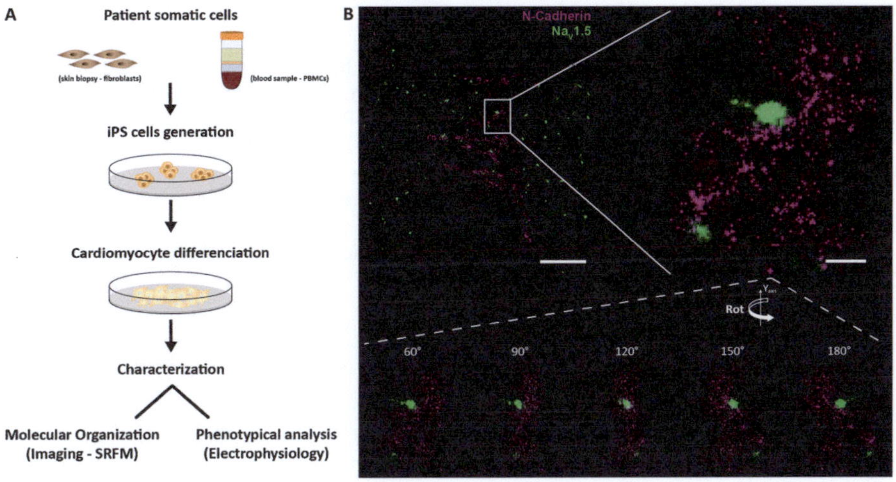

Fig. 14.5 Application of SRFM for patient-specific diagnosis. (**a**) Patient cells can be reprogrammed to pluripotent stem cells (iPS cells) and then differentiated to cardiomyocytes. These cells can then be characterized using different techniques like imaging or electrophysiology. (**b**) Human iPSC-derived cardiomyocytes analyzed by SMLM. Staining for N-cadherin (magenta) and $Na_V1.5$ (green) shows the localization of N-cadherin at regions of cell–cell contact and $Na_V1.5$ in close apposition. Scale bar: 4 μm (**b** left), 400 nm (**b** right)

Despite the difficulties associated with establishing and applying SRFM methods within the lab, the potential insights afforded by these new and exciting techniques unquestionably make their uptake worth the effort. Aside from the specific applications of SMLM to cardiac research that we have outlined in this chapter, many other successful applications to biomedical research can be found in the literature. Moreover, SMLM has potential clinical applications and has already been used in personalized disease modeling. Te Riele et al. used SMLM to assess the molecular basis for arrhythmogenic cardiomyopathy in induced pluripotent stem cell-derived cardiomyocytes generated from peripheral blood mononuclear cells from a patient carrying a *SCN5A* mutation (Fig. 14.5) (Te Riele et al. 2017). Using SMLM, they demonstrated a reduction in the number of sodium channel clusters at junctional sites, which correlated with a reduced sodium current. A reduction in the number of N-cadherin clusters was also observed at the junctional sites potentially revealing a noncanonical mechanism of $Na_V1.5$ to alter intercellular adhesion that can lead to an AC phenotype. This study, along with the others presented here, highlights the vast future potential of SMLM, spanning fundamental and biomedical research and even opening new avenues of investigation in translational and clinical medicine.

Acknowledgments The authors acknowledge the help and critical discussions with members of the Rothenberg and Delmar labs.

Compliance with Ethical Standards

Sources of Funding Work in the Rothenberg lab is funded by the NIH grants R01-GM057691 and R21-CA187612 and the American Cancer Society grant (ACS 130304-RSG-16-241-01-DMC). Research in the Delmar lab is supported by NIH grants RO1-GM57691, RO1-HL134328, and RO1-HL136179.

Conflict of Interest Authors declare that they have no conflict of interest.

Ethical Approval This article does not contain any studies with human participants or animals performed by any of the authors.

References

Agullo-Pascual E, Reid DA, Keegan S, Sidhu M, Fenyö D, Rothenberg E, et al. Super-resolution fluorescence microscopy of the cardiac connexome reveals plakophilin-2 inside the connexin43 plaque. Cardiovasc Res. 2013;100:231–40.

Agullo-Pascual E, Lin X, Leo-Macias A, Zhang M, Liang FX, Li Z, et al. Super-resolution imaging reveals that loss of the C-terminus of connexin43 limits microtubule plus-end capture and NaV1.5 localization at the intercalated disc. Cardiovasc Res. 2014;104:371–81.

Andronov L, Lutz Y, Vonesch JL, Klaholz BP. SharpViSu: integrated analysis and segmentation of super-resolution microscopy data. Bioinformatics. 2016;32:2239–41.

Baddeley D, Jayasinghe ID, Lam L, Rossberger S, Cannell MB, Soeller C. Optical single-channel resolution imaging of the ryanodine receptor distribution in rat cardiac myocytes. Proc Natl Acad Sci U S A. 2009;106:22275–80.

Betzig E, Patterson GH, Sougrat R, Lindwasser OW, Olenych S, Bonifacino JS, et al. Imaging intracellular fluorescent proteins at nanometer resolution. Science. 2006;313:1642–5.

Caetano FA, Dirk BS, Tam JH, Cavanagh PC, Goiko M, Ferguson SS, et al. MIiSR: molecular interactions in super-resolution imaging enables the analysis of protein interactions, dynamics and formation of multi-protein structures. PLoS Comput Biol. 2015;11:e1004634.

Case LB, Baird MA, Shtengel G, Campbell SL, Hess HF, Davidson MW, et al. Molecular mechanism of vinculin activation and nanoscale spatial organization in focal adhesions. Nat Cell Biol. 2015;17:880–92.

Cerrone M, Lin X, Zhang M, Agullo-Pascual E, Pfenniger A, Chkourko Gusky H, et al. Missense mutations in plakophilin-2 cause sodium current deficit and associate with a Brugada syndrome phenotype. Circulation 2014;129:1092–1103.

Chang H, Zhang M, Ji W, Chen J, Zhang Y, Liu B, et al. A unique series of reversibly switchable fluorescent proteins with beneficial properties for various applications. Proc Natl Acad Sci U S A. 2012;109:4455–60.

Dani A, Huang B, Bergan J, Dulac C, Zhuang X. Superresolution imaging of chemical synapses in the brain. Neuron. 2010;68:843–56.

De La Fuente S, Fernandez-Sanz C, Vail C, Agra EJ, Holmstrom K, Sun J, et al. Strategic positioning and biased activity of the mitochondrial calcium Uniporter in cardiac muscle. J Biol Chem. 2016;291:23343–62.

Franke C, Sauer M, van de Linde S. Photometry unlocks 3D information from 2D localization microscopy data. Nat Methods. 2017;14:41–4.

Granzier HL, Hutchinson KR, Tonino P, Methawasin M, Li FW, Slater RE, et al. Deleting titin's I-band/A-band junction reveals critical roles for titin in biomechanical sensing and cardiac function. Proc Natl Acad Sci U S A. 2014;111:14589–94.

Heilemann M, van de Linde S, Schuttpelz M, Kasper R, Seefeldt B, Mukherjee A, et al. Subdiffraction-resolution fluorescence imaging with conventional fluorescent probes. Angew Chem Int Ed Engl. 2008;47:6172–6.

Hell SW, Wichmann J. Breaking the diffraction resolution limit by stimulated emission: stimulated-emission-depletion fluorescence microscopy. Opt Lett. 1994;19:780–2.

Hennig S, van de Linde S, Lummer M, Simonis M, Huser T, Sauer M. Instant live-cell super-resolution imaging of cellular structures by nanoinjection of fluorescent probes. Nano Lett. 2015;15:1374–81.

Hess ST, Girirajan TP, Mason MD. Ultra-high resolution imaging by fluorescence photoactivation localization microscopy. Biophys J. 2006;91:4258–72.

Holm T, Klein T, Loschberger A, Klamp T, Wiebusch G, van de Linde S, et al. A blueprint for cost-efficient localization microscopy. Chemphyschem. 2014;15:651–4.

Hou Y, Jayasinghe I, Crossman DJ, Baddeley D, Soeller C. Nanoscale analysis of ryanodine receptor clusters in dyadic couplings of rat cardiac myocytes. J Mol Cell Cardiol. 2015;80: 45–55.

Huang B, Wang W, Bates M, Zhuang X. Three-dimensional super-resolution imaging by stochastic optical reconstruction microscopy. Science. 2008;319:810–3.

Huang B, Bates M, Zhuang X. Super-resolution fluorescence microscopy. Annu Rev Biochem. 2009;78:993–1016.

Jayasinghe ID, Baddeley D, Kong CH, Wehrens XH, Cannell MB, Soeller C. Nanoscale organization of junctophilin-2 and ryanodine receptors within peripheral couplings of rat ventricular cardiomyocytes. Biophys J. 2012;102:L19–21.

Johnson E, Seiradake E, Jones EY, Davis I, Grunewald K, Kaufmann R. Correlative in-resin super-resolution and electron microscopy using standard fluorescent proteins. Sci Rep. 2015;5:9583.

Kanchanawong P, Shtengel G, Pasapera AM, Ramko EB, Davidson MW, Hess HF, et al. Nanoscale architecture of integrin-based cell adhesions. Nature. 2010;468:580–4.

Kube S, Hersch N, Naumovska E, Gensch T, Hendriks J, Franzen A, et al. Fusogenic liposomes as nanocarriers for the delivery of intracellular proteins. Langmuir. 2017;33(4):1051–9.

Langhorst MF, Schaffer J, Goetze B. Structure brings clarity: structured illumination microscopy in cell biology. Biotechnol J. 2009;4:858–65.

Leo-Macias A, Agullo-Pascual E, Sanchez-Alonso JL, Keegan S, Lin X, Arcos T, et al. Nanoscale visualization of functional adhesion/excitability nodes at the intercalated disc. Nat Commun. 2016;7:10342.

Loschberger A, Franke C, Krohne G, van de Linde S, Sauer M. Correlative super-resolution fluorescence and electron microscopy of the nuclear pore complex with molecular resolution. J Cell Sci. 2014;127:4351–5.

Lukyanenko YO, Younes A, Lyashkov AE, Tarasov KV, Riordon DR, Lee J, et al. Ca(2+)/calmodulin-activated phosphodiesterase 1A is highly expressed in rabbit cardiac sinoatrial nodal cells and regulates pacemaker function. J Mol Cell Cardiol. 2016;98:73–82.

Macquaide N, Tuan HT, Hotta J, Sempels W, Lenaerts I, Holemans P, et al. Ryanodine receptor cluster fragmentation and redistribution in persistent atrial fibrillation enhance calcium release. Cardiovasc Res. 2015;108:387–98.

Malkusch S, Heilemann M. Extracting quantitative information from single-molecule super-resolution imaging data with LAMA—LocAlization microscopy Analyzer. Sci Rep. 2016;6:34486.

Munro ML, Jayasinghe ID, Wang Q, Quick A, Wang W, Baddeley D, et al. Junctophilin-2 in the nanoscale organisation and functional signalling of ryanodine receptor clusters in cardiomyocytes. J Cell Sci. 2016;129:4388–98.

Paez-Segala MG, Sun MG, Shtengel G, Viswanathan S, Baird MA, Macklin JJ, et al. Fixation-resistant photoactivatable fluorescent proteins for CLEM. Nat Methods 2015;12:215–218, 214 p following 218

Proppert S, Wolter S, Holm T, Klein T, van de Linde S, Sauer M. Cubic B-spline calibration for 3D super-resolution measurements using astigmatic imaging. Opt Express. 2014;22:10304–16.

Reid DA, Rothenberg E (2015) Single-molecule fluorescence imaging techniques. In: Encyclopedia of analytical chemistry, pp 1–20.

Rust MJ, Bates M, Zhuang X. Sub-diffraction-limit imaging by stochastic optical reconstruction microscopy (STORM). Nat Methods. 2006;3:793–5.

Sage D, Kirshner H, Pengo T, Stuurman N, Min J, Manley S, et al. Quantitative evaluation of software packages for single-molecule localization microscopy. Nat Methods. 2015;12:717–24.

Shtengel G, Galbraith JA, Galbraith CG, Lippincott-Schwartz J, Gillette JM, Manley S, et al. Interferometric fluorescent super-resolution microscopy resolves 3D cellular ultrastructure. Proc Natl Acad Sci U S A. 2009;106:3125–30.

Szymborska A, de Marco A, Daigle N, Cordes VC, Briggs JA, Ellenberg J. Nuclear pore scaffold structure analyzed by super-resolution microscopy and particle averaging. Science. 2013;341: 655–8.

Tang AH, Chen H, Li TP, Metzbower SR, MacGillavry HD, Blanpied TA. A trans-synaptic nanocolumn aligns neurotransmitter release to receptors. Nature. 2016;536:210–4.

Te Riele AS, Agullo-Pascual E, James CA, Leo-Macias A, Cerrone M, Zhang M, et al. Multilevel analyses of SCN5A mutations in arrhythmogenic right ventricular dysplasia/cardiomyopathy suggest non-canonical mechanisms for disease pathogenesis. Cardiovasc Res. 2017;113: 102–11.

Teng KW, Ishitsuka Y, Ren P, Youn Y, Deng X, Ge P, et al. Labeling proteins inside living cells using external fluorophores for microscopy. elife. 2016;5

Veeraraghavan R, Lin J, Hoeker GS, Keener JP, Gourdie RG, Poelzing S. Sodium channels in the Cx43 gap junction perinexus may constitute a cardiac ephapse: an experimental and modeling study. Pflugers Archiv Eur J Physiol. 2015;467:2093–105.

Veeraraghavan R, Lin J, Keener JP, Gourdie R, Poelzing S. Potassium channels in the Cx43 gap junction perinexus modulate ephaptic coupling: an experimental and modeling study. Pflugers Archiv Eur J Physiol. 2016;468:1651–61.

Wagner E, Lauterbach MA, Kohl T, Westphal V, Williams GS, Steinbrecher JH, et al. Stimulated emission depletion live-cell super-resolution imaging shows proliferative remodeling of T-tubule membrane structures after myocardial infarction. Circ Res. 2012;111:402–14.

Wang W, Landstrom AP, Wang Q, Munro ML, Beavers D, Ackerman MJ, et al. Reduced junctional Na+/Ca2+-exchanger activity contributes to sarcoplasmic reticulum Ca2+ leak in junctophilin-2-deficient mice. Am J Phys Heart Circ Phys. 2014;307:H1317–26.

Whelan DR, Bell TD. Super-resolution single-molecule localization microscopy: tricks of the trade. J Phys Chem Lett. 2015;6:374–82.

Wilhelm BG, Mandad S, Truckenbrodt S, Krohnert K, Schafer C, Rammner B, et al. Composition of isolated synaptic boutons reveals the amounts of vesicle trafficking proteins. Science. 2014;344:1023–8.

Wong J, Baddeley D, Bushong EA, Yu Z, Ellisman MH, Hoshijima M, et al. Nanoscale distribution of ryanodine receptors and caveolin-3 in mouse ventricular myocytes: dilation of T-tubules near junctions. Biophys J. 2013;104:L22–4.

Xu K, Zhong G, Zhuang X. Actin, spectrin, and associated proteins form a periodic cytoskeletal structure in axons. Science. 2013;339:452–6.

Yin Y, Rothenberg E. Probing the spatial organization of molecular complexes using triple-pair-correlation. Sci Rep. 2016;6:30819.

Zhang Z, Nishimura Y, Kanchanawong P. Extracting microtubule networks from superresolution single-molecule localization microscopy data. Mol Biol Cell. 2017;28:333–45.

Transgenic Animal Models of Cardiac Channelopathies: Benefits and Limitations

15

Katja E. Odening and David Ziupa

Abstract

Ideally, studies investigating pathophysiological mechanisms of human arrhythmia disorders should be performed in human subjects, their hearts, tissue, and cells. Human cardiac tissues, however, are not easily accessible to experimental electrophysiologists. Therefore, transgenic animal models (mouse, rabbit, and pig) mimicking (several aspects of) the human disease phenotype have been generated and utilized to gather mechanistic insight into cardiac channelopathies.

In this overview, we summarize advantages, limitations, and translational value of the different available genetic animal models (mouse, rabbit, and pig) for potassium channelopathies (long QT syndromes), sodium channelopathies (LQT3, Brugada syndrome, cardiac conduction disease, and overlap syndrome), and catecholaminergic polymorphic ventricular tachycardia (CPVT).

K. E. Odening (✉)
Department of Cardiology and Angiology I, Heart Center University of Freiburg, Freiburg, Germany

Faculty of Medicine, University of Freiburg, Freiburg, Germany

Institute for Experimental Cardiovascular Medicine, Heart Center University of Freiburg, Freiburg, Germany
e-mail: katja.odening@universitaets-herzzentrum.de

D. Ziupa
Department of Cardiology and Angiology I, Heart Center University of Freiburg, Freiburg, Germany

Faculty of Medicine, University of Freiburg, Freiburg, Germany
e-mail: david.ziupa@universitaets-herzzentrum.de

D. Thomas, C. A. Remme (eds.), *Channelopathies in Heart Disease*, Cardiac and Vascular Biology, https://doi.org/10.1007/978-3-319-77812-9_15

15.1 Introduction

To increase our understanding of the pathophysiology of human diseases, ideally, human subjects, their organs, tissues, and cells should be studied. However, in-depth mechanistic studies on alterations of cardiac electrophysiology and arrhythmogenesis in channelopathies can only be performed to a very limited extent in human patients. Particularly, mechanisms of arrhythmogenesis can only be assessed on certain levels, e.g., in vivo and exceptionally—if available—on cellular levels in isolated or induced pluripotent stem cell-derived cardiomyocytes (iPSC-CM) (Hoekstra et al. 2012), while assessment of whole human hearts remains an exception (Opthof et al. 2017). Similarly, although the expression and biophysical characterization of mutated cardiac ion channels in heterologous cellular expression systems have increased our understanding of the electrophysiological basis of cardiac channelopathies (Nerbonne et al. 2001; Nerbonne and Kass 2005), these cells cannot reflect the endogenous cardiomyocyte environment and any potential electrical remodeling that may occur in it due to the disease-specific ion channel mutations.

The advantage of using animals over human patients (or cellular systems) for mechanistic studies is that animal models allow (1) to identify pathophysiological mechanisms on multiple levels and (2) to conduct longitudinal studies in subjects with a defined genetic background and without confounding comorbidities for the assessment of factors that may alter the arrhythmic disease phenotype, allowing not only observations but also defined quantitative pro- or anti-arrhythmic interventions. This comes at the cost, however, of partially limited clinical transferability due to some species differences in features of cardiac electrical function (Nerbonne et al. 2001; Salama and London 2007; Baczkó et al. 2016) that are responsible for incomplete recapitulation of different aspects of the human disease. Therefore, animal models that capture disease-specific essential aspects of human cardiac pathophysiology (e.g., repolarizing ion channel function for long QT syndromes, sodium channel function for Brugada syndrome and long QT type 3, and Ca^{2+} handling/ryanodine receptor properties for CPVT) are required for improving our understanding of the complex, multidimensional alteration of physiological cardiac function in channelopathies with the goal of "bench-to-bedside" translation (Odening and Kohl 2016).

Both drug-induced and genetically modified animal models of various species have been generated and utilized to investigate arrhythmic mechanisms in different channelopathies [reviewed in Nerbonne et al. (2001), Salama and London (2007), Nattel et al. (2008), Derangeon et al. (2012), Choy et al. (2016) and Lang et al. (2016b)]. One main shortcoming of drug-induced animal models, however, is the fact that most ion channel-blocking or ion channel-activating drugs are not channel-selective, thus causing a "mixed" disease phenotype. In addition, drugs have to be administered continuously to sustain the drug-induced "channelopathy," thus impeding detailed investigation of arrhythmic mechanisms and triggering free-moving, non-anesthetized animals. Therefore, transgenic or genetically modified animal models for channelopathies are generated aiming at (1) mimicking the human disease genotype and phenotype and (2) gathering insights into disease-specific electrophysiological cardiac function and mechanisms of arrhythmogenesis on cellular, tissue, organ, and in vivo levels and (3) for "bench-to-bedside" translation to improve diagnostic and therapeutic strategies in patients with channelopathy.

Small animals such as mice and rabbits are among the most commonly used animal models in cardiac research, since they have relatively short generation times and their handling is rather cost-effective. Most importantly, they have the added advantage that they can be more easily subjected to genetic manipulation than larger animals. Therefore, despite apparent differences between human and mouse cardiac electrophysiology [reviewed in Nerbonne et al. (2001), Salama and London (2007) and Baczkó et al. (2016)], the first (and most) genetic channelopathy models were mouse models [reviewed in Nerbonne et al. (2001), Salama and London (2007), Derangeon et al. (2012) and Choy et al. (2016)]. Thanks to novel developments in animal transgenesis (Bősze et al. 2016), rabbits—representing a species that mimics human cardiac electrophysiology surprisingly well (Nerbonne 2000; Valentin et al. 2004; Hondeghem 2016)—have also entered the range of species in whom genetic manipulation can successfully replicate certain human cardiac diseases (Sanbe et al. 2005; Brunner et al. 2008; Major et al. 2016). Last but not least, a much larger species, the pig—which even more closely resembles humans—has recently been successfully modified genetically to represent a channelopathy model (Park et al. 2015).

In the following, we will summarize advantages, limitations, and translational value of the different currently available genetic (mouse, rabbit, and pig) animal models for cardiac channelopathies: potassium channelopathies (long QT syndromes; Tables 15.1, 15.2 and 15.3), sodium channelopathies (LQT3, Brugada syndrome, cardiac conduction disease, and overlap syndrome; Table 15.4), and catecholaminergic polymorphic ventricular tachycardia (CPVT; Table 15.5). The different diseases and their clinical manifestation and treatment options, however, are not reviewed again in detail; here the interested readers are encouraged to refer to the respective preceding chapters in this volume.

15.2 Available Transgenic Animal Models of Cardiac Channelopathies

15.2.1 Transgenic Animal Models for Long QT Syndrome Based on Potassium Channel Mutations

Long QT syndrome (LQTS) is an inherited channelopathy characterized by prolonged QT duration as manifestation of a prolonged cardiac repolarization (Priori et al. 2001a). The disease is predominantly caused by autosomal dominant mutations in genes encoding for repolarizing potassium channels (90%, *KCNQ1*: LQT1, *KCNH2*: LQT2) and depolarizing sodium channels (5%, *SCN5A*: LQT3) (Priori et al. 2001a). Patients are prone to develop polymorphic *torsade de pointes* ventricular tachycardia (VT) and sudden cardiac death (SCD).

Table 15.1 LQTS—mouse models with alterations of mouse potassium channels

Channel subunit	Modification	Current	APD prolongation	QT prolongation	Arrhythmia	Reference
Kv1.1	Kv1.1 DN	↓IK,slow	Yes	Yes	Yes, spontaneous and inducible	London et al. (1998a)
Kv1.4	Kv1.4$^{-/-}$	↓Ito,s	No	No	No	London et al. (1998b)
Kv1.5	Kv1.5 replaced with Kv1.1 (SWAP)	↓ IK, slow1 ↑ IK, slow2	No	No	No	London et al. (2001)
	Kv1.5 DN	↓ Ito,s ↓ IK, slow1	Yes	Yes	No	Li et al. (2004)
Kv2.1	Kv2.1 DN	↓ IK, slow2	Yes	Yes	Yes, spontaneous and inducible	Xu et al. (1999)
Kv4.2	Kv4.2 DN	↓ Ito,f	Yes	n.d.	n.d.	Wickenden et al. (1999)
	Kv4.2 DN	↓ Ito,f ↑ Ito,s	Yes	Yes	No	Barry et al. (1998)
	Kv4.2$^{-/-}$	↓ Ito,f ↑ Ito,s	No	No	n.d.	Guo et al. (2005)
KChIP2 (subunit to Kv4.2 and Kv4.3)	KChIP2$^{-/-}$	↓ Ito,f	Yes	No	Yes, inducible	Kuo et al. (2001)
Multiple channels (cross)						
Kv1.1 + Kv2.1	Kv1.1 DN × Kv2.1 DN	↓ Ito,s ↓ IK, slow1 ↓ IK, slow2	Yes	Yes	Yes, spontaneous and inducible	Kodirov et al. (2004)

Kv1.1 + Kv4.2	Kv1.1 DN × Kv4.2 DN	↓ Ito,f	Yes	Yes	No	Brunner et al. (2001)
		↓ Ito,s				
		↓ IK, slow1				
Kv1.4 + Kv4.2	Kv1.4$^{-/-}$ × Kv4.2 DN	↓Ito,f	Yes	Yes	Yes, spontaneous	Guo et al. (2000)
		↓Ito,s				

Abbreviations: −/−, homozygous knockout; *DN* dominant-negative, *n.d* not done

Table 15.2 LQTS—mouse models with alterations of "human" potassium channels or mouse-equivalent genes

Human disease and gene	Modification	Current	APD prolongation	QT prolongation	Arrhythmia	Reference
LQT1: Kcnq1/ KCNQ1	Kcnq1$^{-/-}$ (exon 1)	n.d.	n.d.	No	n.d.	Lee et al. (2000)
	Kcnq1$^{-/-}$ (exon 2)	n.d.	No	Yes	n.d.	Casimiro et al. (2001)
	KCNQ1 DN (TG)	↓ IKs	Yes	Yes[a]	No	Demolombe et al. (2001) and Lande et al. (2001)
LQT2: Kcnh1/ Merg/KCNH2/ HERG	Merg$^{+/-}$	n.d.	Yes	n.d.	Yes, induced VT	Salama et al. (2009)
	Merg1b$^{-/-}$	↓ IKr	n.d.	No	No	Lees-Miller et al. (2003)
	HERG1-G628S DN (TG)	↓ IKr	Yes/No[b]	No	No	Babij et al. (1998)
LQT5: Kcne1/ KCNE1	Kcne1/ minK$^{-/-}$	↓ IKs	n.d.	Yes[c]	n.d.	Drici et al. (1998)
	Kcne1/ minK$^{-/-}$	n.d.	No	n.d.	n.d.	Charpentier et al. (1998)
	Kcne1/ minK$^{-/-}$	n.d.	Yes + APD alternans/ dispersion	n.d.	Pacing- and isoprenaline-induced arrhythmia (but also in WT), anti-arrhythmic effect of nifedipine	Balasubramaniam et al. (2003) and Thomas et al. (2007)
	Kcne1/ minK$^{-/-}$	n.d.	No	n.d.	No	Salama et al. (2009)
	minK replaced with lacZ	↓ IKs	No	No	n.d.	Kupershmidt et al. (1999)

LQT7: Kcnj2/ KCNJ2	Kir2.1$^{-/-}$	↓ IK1	Yes	Yes[c]	Spontaneous AP in vitro, but no PVC/VT in vivo	Zaritsky et al. (2000)
	Kir2.1 DN (TG)	↓ IK1	Yes	Yes	No	McLerie and Lopatin (2003)

Abbreviation: –/– homozygous knockout, –/+ heterozygous knockout, *DN* dominant-negative; *TG* transgenic; *AP(D)* action potential (duration), *PVC* premature ventricular contraction, *n.d* not done, *WT* wild type

[a]No QT prolongation induced by dofetilide, E 4031, haloperidol, sultopride, astemizole, cisapride; QT shortening: lidocaine, nicardipine

[b]Yes in single cardiomyocytes, No in ventricular strips

[c]Only during bradycardia

Table 15.3 LQTS—rabbit models based on potassium channel mutations

Human disease and gene	Modification	Curr.	APD prolong.	QT prolong.	Arrhythmia	Clinical implication: alteration by drugs or hormones	Clinical implication: electromechanical insight	Reference
LQT1: KCNQ1	KCNQ1/KvLQT1-Y315S DN (TG), pore region	↓ IKs (loss)	Yes	Yes	No	*Parameters analyzed*: Assessment of QT interval, EAD, conduction velocities, VERP and VERP dispersion, APD and APD dispersion/triangulation/alternans *Drugs*: Isoflurane, thiopental, midazolam, propofol, ketamine, isoproterenol, dofetilide, nicorandil (+/− isoproterenol), NS-1643, E-4031, erythromycin, ryanodine, tetrodotoxin, ranolazine [as reviewed in Lang et al. (2016b)]	Electrical: – APD dispersion not increased – VT/VF inducible in tachymyopathy (continuous tachypacing) – Focal excitations arising from the RV initiates arrhythmia – EAD formation with continuous adrenergic stimulation	Brunner et al. (2008), Odening et al. (2008, 2013), Liu et al. (2012), Ziupa et al. (2014), Lau et al. (2015), Kim et al. (2015) and Lang et al. (2016b)
LQT2: KCNH2	KCNH2/HERG-G628S DN (TG), pore region	↓ IKr (loss)	Yes, even more at slow	Yes, even more at slow heart rate	Yes: Spontaneous TdP resulting in SCD,	*Parameters analyzed*: Assessment of arrhythmia (TdP), QT interval, T wave alternans, EAD,	Electrical: – Increased APD dispersion, leading to unidirectional functional block,	Brunner et al. (2008), Odening et al. (2008, 2010, 2012, 2013), Ziv et al. (2009), Liu

			heart rate		inducible VT/VF	conduction velocities, VERP, VERP dispersion	reentry formation and VF – "discordant alternans" preceded VT/VF – EAD formation with sudden adrenergic surge	et al. (2012) and Lang et al. 2016a, b)
						Drugs: Isoflurane, thiopental, midazolam, propofol, ketamine, isoproterenol, dofetilide [as reviewed in Lang et al. (2016b)]	Mechanical: – Echo/MRI: Normal global function – TPM-MRI: Regional diastolic dysfunction at baseline; particularly prolonged contraction duration in symptomatic animals	
						Hormones: Pro-arrhythmic effect of estradiol (APD dispersion, EAD formation, lethal pVT), anti-arrhythmic effect of progesterone		
LQT5: KCNE1	KCNE1-G52R DN (TG)	↓ IKs	No	Slightly prolonged, increased STVQT	No	Dofetilide provoked TdP and further increase in STVQT	n.d.	Major et al. (2016)

Abbreviations: *DN* dominant-negative, *TG* transgenic, *STVQT* short-term variability of the QT interval, *TPM-MRI* tissue phase mapping MRI

Table 15.4 Animal models for sodium channelopathies (*Scn5a*—LQT3, Brugada syndrome, CCD, and overlap syndrome)

Human disease	Modification	Curr.	APD prolong.	QT prolong.	Conduction disease	Arrhythmia	Clinical implication: evaluation of drug effects	Reference
Features of LQT3 *(gain-of-function)*								
LQT3 (+CCD)	$Scn5a^{+/\Delta\ KPQ}$ (KI)	↑ INa ↑ INa, L	Yes, EAD in vitro after AV block	Yes	Lower heart rate, sinus pauses, AV block	Yes, spontaneous and induced	– Adrenergic agonists suppressed induced arrhythmias – No arrhythmias were provoked by physical stress, isoproterenol, or atropine – Carbachol induced bigemini and TdP – Propranolol and esmolol did not prevent arrhythmias (except in[a]) – Propranolol prevented carbachol-induced VT/VF[a] – Mexiletine and flecainide suppressed arrhythmias	Nuyens et al. (2001), Fabritz et al. (2003, 2010) and Calvillo et al. (2014)[a]
LQT3 (+CCD)	$Scn5a^{+/\Delta\ KPQ}$ (KI)	↑ INa ↑ INa, L	Yes, EAD in vitro	n.d.	n.d.	Yes, induced	– Propranolol did not suppress isoproterenol-induced arrhythmia – Mexiletine suppressed arrhythmia	Head et al. (2005)

LQT3 (+CCD)	SCN5A-N1325S (TG)	↑ INa, L	Yes, EAD in vitro	Yes	RR and PR shortened, QRS similar PR shortened	Yes, spontaneous	– Mexiletine shortened APD and suppressed arrhythmia	Tian et al. (2004)
LQT3 (+atrial fibrillation + structural changes)	SCN5A-F1759A (TG)	↑ INa, L	Yes	Yes	PR shortened	Yes, spontaneous PVC, pVT (and AF)	– Diminished use-dependent lidocaine block of INa – Persistent INa resistant to ranolazine – NCX inhibitor SEA-0400 reduced burden to PVC and atrial fibrillation	Wan et al. (2016)
Features of BrS and CCD *(loss-of-function)*								
CCD (+ BrS)	$Scn5a^{+/-}$ (KO)	↓ INa	No	No	P wave, PR, and QRS prolongation	Triggered VT (spontaneous in old)	n.d.	Papadatos et al. (2002)
CCD	Scn5a $\Delta SIV/\Delta SIV$ (KI)	↓ INa[a]	No, but reduced AP upstroke	No	P wave, PR, and QRS prolongation	No	– Flecainid aggravated conduction disease	Shy et al. (2014)
Pig model[b] of BrS (+CCD)	$Scn5a^{E558X/+}$ (KI)	↓ INa	No	No	P wave, PR, and QRS prolonged, atrial-his and his-ventricular conduction delay	– No, in vivo (no SCD, no arrhythmia)– Yes, ex vivo (spontaneous and pacing-induced, VF at 39 °C, stable at 35 °C)	– Combination of propranolol and atropine did not provoke arrhythmia– Flecainide induced PR and QRS prolongation, but not BrS-type ECG changes	Park et al. (2015)

(continued)

Table 15.4 (continued)

Human disease	Modification	Curr.	APD prolong.	QT prolong.	Conduction disease	Arrhythmia	Clinical implication: evaluation of drug effects	Reference
Features of overlap syndromes (and biophysical overlap)								
Overlap: −LQT3—BrS—CCD	Scn5a $^{1798insD/+}$ (KI)	↓ INa ↑ INa, L	Yes	Yes	RR, PR, and QRS prolonged, RV conduction slowing	Sinus pauses in vivo, EADs in vitro	– Flecainide induced sinus bradycardia and/or sinus arrest	Remme et al. (2006)
Overlap: LQT3—CCD	SCN5A-D1275N(TG)	↓ INa ↑ INa, L	Yes	Yes	P wave, PR, and QRS prolongation	Yes, spontaneous mVT/pVT	n.d.	Watanabe et al. (2011)

Abbreviations: *KI* knock-in, *KO* knockout, *TG* transgenic, *CCD* cardiac conduction disease, *BrS* Brugada syndrome, *INa* peak sodium current, *INa,L* late sodium current, *PVC* premature ventricular contraction

[a]Decreased INa most likely due to defects in cell surface expression of sodium channel

[b]With the exception of this pig model, all other models are mouse models

Table 15.5 Mouse models for catecholaminergic polymorphic VT

Gene	Modification	Normal ECG at baseline	Arrhythmia (trigger)	Clinical implication: evaluation of drug effects	Reference
FKBP12.6	FKBP12.6$^{-/-}$ (KO)	Yes	Yes (exercise)	n.d.	Wehrens (2003)
RyR2	RyR2$^{R4496C/+}$ (KI)	Yes	Yes (catecholamines)	Mice pretreated with propranolol developed VT	Cerrone et al. (2005, 2007) and Fernández-Velasco (2009)
	RyR2$^{R176Q/+}$ (KI)	Yes	Yes (programmed stimulation, catecholamines, caffeine)	n.d.	Kannankeril (2006)
	RyR2$^{P2328S/+}$ RyR2$^{P2328S/P2328S}$ (KI)	Yes	Yes (programmed stimulation, catecholamines, caffeine)	Catecholamines and caffeine reduced myocardial conduction velocity	Goddard (2008) and Zhang (2013)
	RyR2$^{R2474S/+}$ (KI)	Yes	Yes (exercise, catecholamines)	Dantrolene prevented inducible VT	Kobayashi et al. (2010)
	RyR2-$^{S2246L/+}$ (KI)	Yes	Yes (exercise)	Dantrolene stopped the exercise-induced ventricular tachycardia	Suetomi et al. (2011)
	RyR2$^{+/Ex3\text{-}del}$ (KO)	No (bradycardia)	No	n.d.	Liu et al. (2014)
	RyR2$^{A4860G/+}$ (KI)	n.d.	Yes (catecholamines)	n.d.	Zhao (2015)
CASQ2	CASQ2$^{\Delta E9/\Delta E9}$ (KO)	Yes	Yes (spontaneous, exercise, catecholamines)	– Flecainide (but not lidocaine) suppressed VT (conflicting data in and) – Propafenone prevented exercise-induced CPVT (but not procainamide or lidocaine) – Propranolol and sotalol had little anti-arrhythmic effect during exercise	Knollmann (2006), Song et al. (2007), Watanabe et al. (2009), Katz et al. (2010), Hwang et al. (2011) and Kurtzwald-Josefson et al. (2014)

(continued)

Table 15.5 (continued)

Gene	Modification	Normal ECG at baseline	Arrhythmia (trigger)	Clinical implication: evaluation of drug effects	Reference
				but not after epinephrine – Verapamil significantly lowered VT prevalence – Phentolamine (alpha-antagonist) or labetalol (alpha/beta-antagonist) abolished exercise- and epinephrine-induced arrhythmia – Mg2+ significantly lowered the incidence of catecholamine-induced sustained VT	
	CASQ2 $^{D307H/D307H}$ (KO)	Yes	Yes (spontaneous, exercise, catecholamines)	– Flecainide, procainamide, and lidocaine were ineffective in controlling arrhythmia – Propranolol and sotalol had little anti-arrhythmic effect during exercise but not after epinephrine – Verapamil completely abolished arrhythmia in D307H mice – Mg2+ significantly lowered the incidence of catecholamine-induced sustained VT	Song et al. (2007), Katz et al. (2010) and Kurtzwald-Josefson et al. (2014)

Abbreviations: *KO* knockout, *KI* knock-in, n.d. not done

15.2.1.1 Genetically Modified Mouse Models for Long QT Syndromes with Potassium Channel Mutations

Since genetic manipulation has for a long time nearly exclusively been feasible in mice and not in other larger mammals, the first transgenic (and knockout) animal models for LQTS based on potassium channel mutations were mouse models (London et al. 1998a, b). However, it has to be noted that pronounced species differences in the expression and functional relevance of ion channel proteins and currents conveying cardiac repolarizing, action potential shape and duration, and consecutive surface ECG morphology have been identified between mice and humans [reviewed in Nerbonne et al. (2001), Salama and London (2007) and Baczkó et al. (2016)]. Briefly (and simplified), in mouse cardiomyocytes, the major repolarizing currents are the fast and slow components of the transient outward K^+ current $I_{to,f}$ and $I_{to,s}$ and the rapidly activating, slowly inactivating delayed rectifier currents $I_{K,slow1}$ and $I_{K,slow2}$ (Nerbonne 2000; Nerbonne et al. 2001). In human cardiomyocytes, in contrast, repolarization is driven by the transient outward K^+ current I_{to}, the slow delayed rectifier K^+ current I_{Ks}, the rapid delayed rectifier K^+ current I_{Kr}, and the inward rectifier K^+ current I_{K1}—with I_{Kr} and I_{Ks} being by far the most important repolarizing currents (Nerbonne 2000; Nerbonne et al. 2001).

Since the 1990s a variety of different models (1) targeting mouse repolarizing potassium currents or (2) introducing mutated human repolarizing potassium channels have been generated and investigated in detail (Nerbonne et al. 2001; Salama and London 2007) (see sections "Genetically Modified Mouse Models for Long QT Syndrome with Alterations of Mouse Potassium Channels" and "Genetically Modified Mouse Models for Long QT Syndrome with Alterations of "Human" Potassium Channels"). In these models, the so-called "dominant-negative" (DN) transgenic strategy, e.g., the fact that the co-assembly of mutated and normal channel subunits completely disrupts ion channel function, was utilized to decrease the expression of functionally normal repolarizing potassium channel proteins.

Genetically Modified Mouse Models for Long QT Syndrome with Alterations of Mouse Potassium Channels

(Partial) Imitation of Long QT Phenotype: APD, QT, and Arrhythmia

Mouse models with altered expression of mouse potassium channel α-subunits have been crucial for our understanding of the functional relevance of molecularly distinct subunits of repolarizing potassium channels (Nerbonne et al. 2001). Moreover, these models either overexpressing dominant-negative loss-of-function mutations in mouse repolarizing potassium channel genes or harboring targeted deletions of these genes were able to partially mimic the human LQTS disease phenotype. Dominant-negative transgenic mice expressing (1) an N-terminal fragment of Kv1.1 (lack of the 4-aminopyridine (4-AP)-sensitive current $I_{K,slow1}$), (2) a Kv1.5 pore mutation (similar lack of the 4-AP-sensitive current $I_{K,slow1}$), (3) a Kv2.1 mutation (lack of $I_{K,slow2}$), or (4) a truncated Kv4.2 as well as a Kv4.2 pore mutation (lack of $I_{to,f}$) all demonstrated prolongation of the action potential duration (APD) and/or the QT interval (Table 15.1; London et al. 1998a; Barry et al. 1998; Xu et al.

1999; Wickenden et al. 1999; Li et al. 2004). Similarly, the targeted disruption of KChIP2—an auxiliary subunit of the Kv4.x family necessary for regular trafficking of channel proteins to the cell membrane—that results in a lack of I_{to} led to a prolongation of APD and QT (Kuo et al. 2001). Other mice with targeted deletions of mouse channel subunits (Kv1.4$^{-/-}$ lack of $I_{to,s}$; Kv4.2$^{-/-}$ lack of $I_{to,f}$), in contrast, had no APD or QT prolongation at all (London et al. 1998b; Guo et al. 2005).

However, only some of these models exhibited short spontaneous and/or inducible ventricular arrhythmia (London et al. 1998a; Xu et al. 1999; Kuo et al. 2001; Kodirov et al. 2004), particularly when dominant-negative mutations in several ion channel genes were combined (Guo et al. 2000; Kodirov et al. 2004)—while others seemed to be protected from arrhythmia despite their prolonged cardiac repolarization (Li et al. 2004; Barry et al. 1998; Brunner et al. 2001). Of note, interestingly, in Kv1.1DN mice, the major arrhythmia was monomorphic and not polymorphic *torsade de pointes* tachycardia that typically develops in human LQTS (London et al. 1998a), indicating potentially different mechanisms of arrhythmogenesis in murine and human hearts. SCD due to VT/VF, however, was not observed in any of these LQTS mouse models.

Electrical Remodeling

A compensatory upregulation of repolarizing currents not affected by the mutation was observed in many of these mouse models and may be (partially) responsible for a lack of APD/QT prolongation or lack of arrhythmia: Kv4.2 DN transgenic mice lacked $I_{to,f}$ but demonstrated a compensatory upregulation of Kv1.4/$I_{to,s}$ leading to APD and QT prolongation but without arrhythmia (Barry et al. 1998). Similarly, mice with a targeted deletion of Kv4.2 completely lacked $I_{to,f}$ and showed compensatory upregulation of $I_{to,s}$ and downregulation of the accessory subunit KChIP2 and consequently lacked APD or QT prolongation (Guo et al. 2005). Gene-targeted mice in which Kv1.5 was replaced by Kv1.1 (SWAP) lacked $I_{K,slow1}$ and demonstrated upregulation of Kv2.1/$I_{K,slow2}$ resulting in a lack of APD prolongation (London et al. 2001). Whether a likewise electrical remodeling with compensatory upregulation of other repolarizing ion currents also occurs in human cardiomyocytes is unclear and remains to be elucidated.

Mechanisms of Long QT-Related Arrhythmia

To gather insights into mechanisms responsible for long QT-related arrhythmia, not only the abovementioned "monogenic" LQTS mouse models were generated, but experiments intercrossing two specific models were also performed to investigate the impact of the combined lack of several repolarizing ion currents on arrhythmogenesis. Cross-breeding Kv4.2 DN mice with Kv1.4$^{-/-}$ mice yielded mice that completely lacked I_{to} ($I_{to,s}$ and $I_{to,f}$) and had very pronounced APD/QT prolongation, with increased early afterdepolarization (EAD) formation and spontaneous ventricular arrhythmia (Guo et al. 2000). Similarly, crossing Kv1.1 DN and Kv2.1 DN mice resulted in mice lacking both $I_{K,slow1}$ and $I_{K,slow2}$ with pronounced APD/QT prolongation and spontaneous and inducible arrhythmia (Kodirov et al. 2004). Cross-breeding Kv1.1 DN with Kv4.2 DN mice (lack of $I_{K,slow1}$ and both components of I_{to}), in contrast, let to mice with pronounced prolongation of cardiac repolarization but lack of arrhythmia (Brunner et al. 2001). These models revealed the importance of a heterogeneously prolonged cardiac repolarization for LQTS-related arrhythmia formation and an anti-arrhythmic effect of a regionally more homogenous AP prolongation (Baker et al. 2000; Brunner et al. 2001; Kodirov et al. 2004; London et al. 2007).

However, due to the above indicated differences in cardiac electrophysiology, the clinical relevance of mechanistic findings gathered in these potassium channel-targeting mouse models to human LQT1 and LQT2 remains elusive. Moreover, in transgenic or gene-targeted mouse models of LQTS in general, electrical remodeling (compensatory upregulation of other ion currents) and structural remodeling (fibrosis) are common, which may itself affect arrhythmogenesis, thus limiting their transferability to human pathophysiology (Koren 2004; Salama and London 2007).

Genetically Modified Mouse Models for Long QT Syndrome with Alterations of "Human" Potassium Channels

Another approach to investigate LQTS in mouse models is to modify human potassium channel genes or their mouse equivalents. Several groups have generated mouse models with dominant-negative loss-of-function mutations of human voltage-gated potassium channels KvLQT1/*KCNQ1* (LQT1), HERG/*KCNH2* (LQT2), minK/*KCNE1* (LQT5), or Kir2.1/*KCNJ2* (LQT7) or knockouts of the mouse-equivalent genes (Table 15.2), aiming at a better imitation of the human LQTS phenotype [reviewed in Nerbonne et al. (2001) and Salama and London (2007)]. As different repolarizing voltage-gated potassium currents determine cardiac repolarization in murine and human cardiomyocytes (Nerbonne 2000), however, these mouse models representing LQT types 1, 2, 5, or 7 have failed to completely mimic the human disease phenotype. Only some of the models demonstrated APD/QT prolongation, and all failed to show any spontaneous ventricular arrhythmia (Table 15.2).

(Partial) Imitation of Long QT Phenotype—APD, QT, and Arrhythmia: KCNQ1—LQT1

Consequences of a reduction or elimination of I_{Ks} have been investigated in *KCNQ1* knockout and dominant-negative mouse models. Similarly as human patients with Jervell and Lange-Nielsen syndrome—an autosomal recessive form of LQT1—homozygote mice with targeted disruption of *Kcnq1* (*Kcnq1*$^{-/-}$) had bilateral sensorineural deafness (Lee et al. 2000; Casimiro et al. 2001). The cardiac phenotype, however, was less clear. While some *Kcnq1*$^{-/-}$ mice [deletion of exon 1 (Lee et al. 2000)] had normal QT, other *Kcnq1*$^{-/-}$ mice [deletion of exon 2 (Casimiro et al. 2001)] demonstrated QT prolongation but no corresponding changes in endocardial or epicardial monophasic action potentials (Table 15.2), indicating a less predominant role for KCNQ1/I_{Ks} in ventricular repolarization in mice than in human. In dominant-negative *KCNQ1* DN mice (Demolombe et al. 2001), in contrast, prolonged QT and APD were observed and were associated with sinus node dysfunction and alterations of AV nodal conduction, suggesting that KCNQ1/I_{Ks} may play a role in sinus node automaticity and impulse propagation through the AV node in murine hearts. *KCNQ1* DN mice were further utilized to investigate differential effects of I_{Kr} and I_{to} blocking drugs (Lande et al. 2001), demonstrating a QT prolonging effect of I_{to} blockers (but—contrasting with effects observed in humans—not of I_{Kr} blockers) and slowing of sinus automaticity with I_{Kr} blockade, thus further dissecting the functional role of various repolarizing ion currents in different parts of the heart.

(Partial) Imitation of Long QT Phenotype—APD, QT, and Arrhythmia: KCNH2—LQT2

Similarly as suggested by pharmacological experiments, the selective knockout of mouse Merg1b potassium channel (*Kcnh2/Merg*$^{-/-}$) (Lees-Miller et al. 2003) or the elimination of I_{Kr} due to a cardiac-specific overexpression of the dominant-negative pore mutation HERG-G628S (Babij et al. 1998) [which confers a pronounced phenotype in human patients (Sanguinetti et al. 1996)] both did not prolong QT intervals but caused sinus bradycardia (Table 15.2). These findings indicate a very limited role of I_{Kr} in murine ventricular repolarization but some importance in sinus node electrophysiology. Similarly as in models targeting mouse potassium channels, the lack of a cardiac phenotype may also be partly due to compensatory upregulation of other ion currents such as I_{Ks} (Lees-Miller et al. 2003). In contrast to these observations, a more recent study (Salama et al. 2009) demonstrated that a partial reduction of Merg1a protein/I_{Kr} current in heterozygous *Merg1*$^{+/-}$ mice may cause APD prolongation and increased base-to-apex dispersion of repolarization thus predisposing the heart to arrhythmia as demonstrated by increased VT inducibility.

(Partial) Imitation of Long QT Phenotype—APD, QT, and Arrhythmia: KCNE1—LQT5

Consequences of *Kcne1/minK*$^{-/-}$ knockout on murine cardiac electrophysiology are conflicting: Several groups reported that *minK*$^{-/-}$ mouse models have reduced I_{Ks} current densities but lack changes in QT (Kupershmidt et al. 1999) or in (right and left ventricular) APD (Charpentier et al. 1998; Salama et al. 2009)—while others have described QT abnormalities in response to heart rate changes with longer QT at slow heart rates and a paradoxical shorter QT at fast heart rates (Drici et al. 1998) (Table 15.2). In line with these observations, a recent study demonstrated prolonged epicardial APD, resulting in increased transmural dispersion of repolarization (Thomas et al. 2007). Earlier studies by the same group had additionally identified slowing and increased dispersion of conduction velocities (Balasubramaniam et al. 2003). However, whether these are due to potential fibrotic remodeling or stem directly from the knockout remains unclear. In *Kcne1/minK*$^{-/-}$ mice demonstrating prolongation and increased dispersion of APD and conduction velocities, an increased incidence of EADs and of inducible and spontaneous VT (particularly upon complete cessation of ventricular pacing and hence during bradycardia) was observed in Langendorff-perfused hearts ex vivo (Balasubramaniam et al. 2003; Thomas et al. 2007). Of note, these ventricular tachycardias, however, were monomorphic [similarly as observed in Kv1.1DN mice (London et al. 1998a)] and not polymorphic thus not corresponding to *torsade de pointes* tachycardia typically observed in human LQTS patients. Interestingly, in these models the Ca^{2+} channel blocker nifedipine exerted a pronounced anti-arrhythmic effect (Balasubramaniam et al. 2003; Thomas et al. 2007). In another study, in which optical mapping experiments were performed in *Kcne1/minK*$^{-/-}$ mouse hearts that lacked APD prolongation, in contrast, no arrhythmia was observed neither spontaneously nor after ventricular stimulation (Salama et al. 2009).

(Partial) Imitation of Long QT Phenotype—APD, QT, and Arrhythmia: KCNJ2—LQT7
The effects of a reduction or elimination of Kir2.1/I_{K1} have been investigated using gene-targeted knockout and dominant-negative transgenesis (Table 15.2): Ventricular cardiomyocytes from *Kcnj2/Kir2.1*$^{-/-}$ knockout mice lacked I_{K1} and demonstrated a pronounced APD prolongation and increased rates of spontaneous APs. No spontaneous ventricular arrhythmia, however, was observed in *Kcnj2/Kir2.1*$^{-/-}$ mice in vivo (Zaritsky et al. 2000). Similarly, cardio-selective overexpression of dominant-negative Kir2.1 channel subunits (Kir2.1 DN mice) led to a nearly absent I_{K1} and prolonged APD and QT (McLerie and Lopatin 2003). In addition, a prolongation of PR intervals and QRS duration was observed, indicating a role of I_{K1} in controlling repolarization and impulse propagation. Despite the prolonged cardiac repolarization in this model, however, no ectopic electrical activity and no VT were observed (McLerie and Lopatin 2003), further stressing the electrophysiological differences between mice and human in terms of arrhythmogenesis.

Mechanisms of Long QT-Related Arrhythmia
Despite their lack of complete imitation of the human disease phenotype, insights into mechanisms underlying long QT-related arrhythmia were also gathered with these mouse models: the importance of an increased dispersion of repolarization for arrhythmia formation (Thomas et al. 2007; Salama et al. 2009) was similarly highlighted as with mouse models with alterations in mouse potassium channels.

15.2.1.2 Genetically Modified Rabbit Models for Long QT Syndromes Based on Potassium Channel Mutations

Although the above described LQTS mouse models were able to partly mimic the human LQTS phenotype, these models failed to show spontaneous sustained ventricular arrhythmia or SCD thus limiting their value for the investigation of arrhythmic mechanisms or anti-arrhythmic therapeutic approaches. This calls for a species that more closely resembles humans such as the rabbit, which demonstrates pronounced similarities to humans in terms of ion currents determining cardiac repolarization, intracellular ion concentrations, and responses to electrophysiologically relevant pharmacological interventions (Nerbonne 2000; Valentin et al. 2004; Hondeghem 2016). In rabbit cardiomyocytes, cardiac repolarization is mainly driven by HERG/I_{Kr} and (to a lesser degree) KvLQT1/I_{Ks}—similar as in human cardiomyocytes (Salata et al. 1996; Nerbonne et al. 2001; Baczkó et al. 2016). In addition, regional contractile and diastolic behavior of rabbit hearts resembles that of human (Jung et al. 2012), and similar cardiac mechano-electrical coupling mechanisms have been described (Quinn and Kohl 2016).

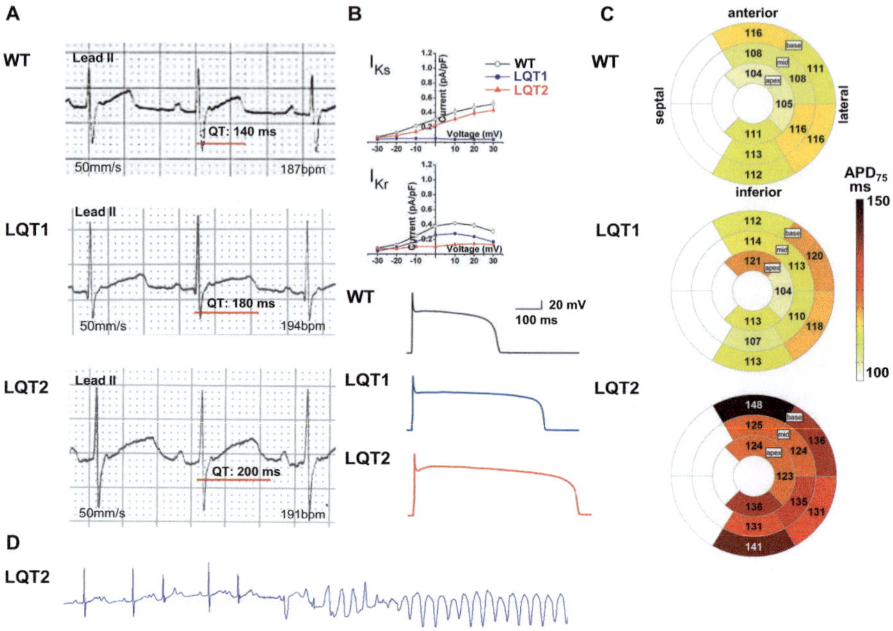

Fig. 15.1 Electrical phenotype of transgenic LQT1 and LQT2 rabbits. (**a**) Representative ECGs from sedated wild-type (WT), transgenic LQT1 and LQT2 rabbits depicting prolonged QT intervals in LQT1 and LQT2 rabbits. (**b**) Current-voltage diagrams of I_{Kr} (top) and I_{Ks} (mid) current densities in cardiomyocytes from WT, LQT1, and LQT2 rabbits demonstrating loss of I_{Kr} in LQT2 and loss of I_{Ks} in LQT1. Representative cellular action potentials (bottom) from WT, transgenic LQT1 and LQT2 rabbits with prolonged action potential durations in LQT1 and LQT2. (**c**) Bull's-eye plots from average monophasic APD at 75% of repolarization (APD_{75} in ms) in different segments of the left ventricular base, mid, and apex derived from Langendorff-perfused hearts of wild-type ($n = 10$), LQT1 ($n = 12$), and LQT2 ($n = 9$) rabbits. (**d**) Exemplary ECG episode of ventricular fibrillation in a free-moving LQT2 rabbit (telemetric ECG monitoring). Modified and adapted from Brunner et al. (2008), Odening et al. (2012, 2013), Ziupa et al. (2014) and Lang et al. (2016a, b)

Transgenic Rabbit Models for LQT1, LQT2, and LQT5—Imitation of Long QT Phenotype

All available transgenic LQTS rabbit models have been engineered by cardio-selective overexpression of dominant-negative mutated human genes encoding for voltage-gated K^+ channels *KCNQ1*/KvLQT1 (KvLQT1-Y315S, LQT1), *KCNH2*/HERG (HERG-G628S, LQT2), or *KCNE1*/minK (KCNE1-G52R) driven by β-myosin heavy chain promoters (Brunner et al. 2008; Major et al. 2016) [reviewed in detail in Lang et al. (2016b)] (Table 15.3). In LQT1 or LQT2, rabbit cardiomyocytes I_{Ks} (LQT1) or I_{Kr} (LQT2), respectively, were completely eliminated resulting in prolongation of APD on the cellular and whole heart levels and prolongation of ventricular refractoriness and QT duration in vivo (Brunner et al. 2008; Odening et al. 2010) (Table 15.3, Fig. 15.1a–c). In LQT2 rabbit hearts, an increased spatial dispersion of APD was observed (Brunner et al. 2008; Odening et al. 2013), and VT/VF were easily inducible with left ventricular (LV) epicardial stimulation (Brunner et al. 2008). Importantly, LQT2 rabbits even developed spontaneous polymorphic VT and SCD (Brunner et al. 2008; Odening et al. 2012) (Fig. 15.1d), thus representing the first transgenic animal models mimicking the complete electrical phenotype of LQT2. Transgenic LQT1 rabbits with a more homogeneously prolonged APD without dispersion of repolarization, in contrast, developed no spontaneous VT or SCD (Brunner et al. 2008). In transgenic LQT5 rabbits (Major et al. 2016), biophysical properties of I_{Ks} were altered with accelerated deactivation kinetics. These rabbits exhibited only a very slightly prolonged QT but an increased short-term beat-to-beat variability of the QT (Major et al. 2016). Due to their reduced repolarization reserve, the phenotype was augmented by I_{Kr}-blocking drug dofetilide, which further increased short-term variability of QT and promoted drug-induced VT (Major et al. 2016) (Table 15.3).

In the following, we highlight mechanistic findings on the arrhythmic substrate, pro-arrhythmic triggering factors, anti-arrhythmic agents, and electromechanical dysfunction in transgenic LQTS rabbit models and their potential translational application in the clinical management of LQTS patients in more detail.

Arrhythmic Substrate: Role of Spatial and Temporal Dispersion of Repolarization

Studies in transgenic LQT1 and LQT2 rabbits highlight the major role of an enhanced dispersion of repolarization in LQTS-related arrhythmogenesis: In LQT2 rabbit hearts, a pronounced dispersion of repolarization was identified in left and right ventricles (Brunner et al. 2008; Odening et al. 2010, 2013), and optical mapping visualizing VF initiation demonstrated that this enhanced regional dispersion of repolarization may cause unidirectional functional block and reentry formation (Brunner et al. 2008). Dispersion of repolarization can also occur in a dynamic spatiotemporal fashion with pronounced beat-to-beat alternations and "out-of-phase" heterogeneities between adjacent regions, the so-called discordant alternans. In transgenic LQT2 rabbit hearts, this discordant alternans developed at physiological heart rates and preceded VT/VF formation (Ziv et al. 2009). In contrast, in LQT1 hearts lacking regional or temporal dispersion of repolarization (Brunner et al. 2008; Odening et al. 2010, 2013; Ziupa et al. 2014), no VT/VF could be induced, suggesting that a regionally more homogeneous APD prolongation may exert a protective effect. When LQT1 hearts were further stressed, however, by continuous tachypacing to induce cardiac tachymyopathy, APD dispersion increased, spatially discordant alternans developed, and VT/VF was easily inducible (Lau et al. 2015).

Triggers: Role and Mechanisms of Early Afterdepolarization

Clinical registry data suggest genotype-specific arrhythmic triggers in LQTS patients: The constantly elevated adrenergic tone during physical exercise (particularly during swimming) has been determined to promote arrhythmia in LQT1, while a sudden sympathetic surge in episodes of rest through emotional stress and (auditory) startle may trigger arrhythmia in LQT2 (Schwartz et al. 2001; Morita et al. 2008). In line with these observations, genotype-specific differences in the mechanisms of EAD formation and arrhythmia initiation were demonstrated in

LQT1 and LQT2 rabbits. In LQT2 cardiomyocytes, EADs developed during sudden sympathetic surge, while continuous perfusion with isoproterenol prevented EAD formation. In LQT1 cardiomyocytes, in contrast, continuous adrenergic stimulation facilitated the occurrence of EADs (Liu et al. 2012). Different time courses in sympathetic activation of cardiac ion currents may explain why different sympathetic modes are associated with arrhythmia formation in different genotypes of LQTS: Upon sympathetic stimulation, activation of $I_{Ca,L}$ that may elicit EADs is faster than the activation of I_{Ks} that shortens APD in LQT2 and acts as antiarrhythmic mechanism upon continuous adrenergic stimulation in LQT2. In addition, different modes of arrhythmia initiation and maintenance in different LQTS genotypes were identified. While in LQT2 reentry formation played an important role (Brunner et al. 2008), in LQT1 hearts, a novel mechanistic concept of LQTS-related arrhythmogenesis was identified: Arrhythmia was initiated by focal excitations arising particularly from the RV and was maintained by multiple shifting excitation foci and bi-excitability (Kim et al. 2015).

Role of Electrical Remodeling

In transgenic LQTS rabbit models, a distinct interaction between I_{Kr} and I_{Ks} was identified. In contrast to transgenic mouse models, in which electrical remodeling with a partially compensatory upregulation of reciprocal repolarizing ion currents has been observed (Koren 2004), in transgenic LQT1 and LQT2 rabbits, the impaired repolarization reserve was further limited and the phenotype aggravated by downregulation of the reciprocal repolarizing currents (Brunner et al. 2008; Ren et al. 2010) due to direct KCNQ1 and HERG protein interactions (Organ-Darling et al. 2013). Likewise, in transgenic LQT5 rabbits, the faster deactivation of I_{Ks} was paralleled by a faster deactivation of I_{Kr} (Major et al. 2016). Whether electrical remodeling mechanisms occurring in human LQTS cardiomyocytes more closely resemble the changes in transgenic rabbit or mouse models still warrants detailed investigation.

Investigation of Pro- and Anti-arrhythmic Effects of Drugs and Hormones in LQTS

As transgenic LQTS rabbit models demonstrate a reduced repolarization reserve, they may serve as particularly sensitive tools for in vivo and ex vivo drug testing to identify potential pro-arrhythmic ion channel-blocking drugs and may be similarly used to investigate genotype-specific efficacy of ion channel-activating drugs and their potential application as genotype-specific therapies.

LQT1 rabbits lacking I_{Ks} and LQT5 rabbits with impaired I_{Ks} were demonstrated to be particularly sensitive in identifying I_{Kr}-blocking properties of drugs showing prolonged APD/QT, increased spatial dispersion of APD, increased short-term variability of QT, and increased arrhythmia formation (Odening et al. 2008, 2010; Ziupa et al. 2014; Major et al. 2016). Similarly, transgenic LQT2 rabbits demonstrated a particularly high sensitivity to I_{Ks}- or I_{K1}-blocking anesthetic agents (Odening et al. 2008).

Pronounced sex differences in arrhythmic risk have been identified in LQTS patients with an increased risk for cardiac arrhythmic events in women after puberty and a particularly high risk during the postpartum (particularly in LQT2 patients) (Sauer et al. 2007), strongly suggesting that changing sex hormone levels may affect LQTS-related arrhythmogenesis. Consequently, identifying pro- and anti-arrhythmic effects of sex hormones has become an emerging field of interest in LQTS research, with the goal of revealing underlying molecular mechanisms and identifying novel potential therapeutic targets (Odening and Koren 2014). As in transgenic LQT2 rabbits ventricular arrhythmia and SCD also often occurred postpartum-related (Brunner et al. 2008; Odening et al. 2012)—suggesting the existence of similar arrhythmia-triggering mechanisms as in human LQTS patients—these models were further utilized to explore sex hormone effects on arrhythmic triggers and substrate (Odening et al. 2012): Estradiol exerted a pro-arrhythmic effect with an increased incidence of lethal pVT by changing the pattern of APD dispersion and increasing EAD formation upon sympathetic stimuli, while progesterone had an anti-arrhythmic, protective effect that was based on a shortening of cardiac refractoriness, a reduced formation of EAD, and stabilizing Ca^{2+} effects (decreased $I_{Ca,L}$ density, increased SERCA expression) (Odening et al. 2012). Further studies revealed that progesterone increased SERCA by slowing its degradation, thereby shortening the decay and duration of Ca^{2+} transients (Moshal et al. 2014). These studies suggest that progesterone-based therapies may be considered as novel anti-arrhythmic approaches in female LQTS patients.

Insights into Electro-mechanical Dysfunction in LQTS

Because electrical and mechanical cardiac functions are closely coupled, it stands to reason that LQTS also causes mechanical dysfunction. Standard echocardiography techniques have demonstrated globally normal mechanical function in structurally normal hearts in LQTS. With the help of novel echocardiography or MRI techniques that allow to investigate regional tissue velocities and strain, however, more and more evidence is accumulating that diastolic relaxation may be impaired and contraction duration prolonged (Nador et al. 1991; Haugaa et al. 2010; Leren et al. 2015). Using *phase-contrast*-based MRI, a regionally heterogeneously reduced diastolic function (reduced peak diastolic relaxation velocities and prolonged time-to-diastolic peak as a marker for contraction duration) was demonstrated in transgenic LQT2 rabbits (Odening et al. 2013) (Fig. 15.2a–e) thus supplementing the clinical data with important insights into regional mechanical heterogeneity. In addition, a spatial correlation between the extent of electrical dysfunction (prolongation of APD) and the impaired diastolic function (Odening et al. 2013) and a correlation between mechanical dysfunction and arrhythmic risk were revealed (Lang et al. 2016a): LQT2 rabbits exhibiting arrhythmia featured a more prolonged time-to-diastolic peak and a more pronounced mechanical dispersion than those without arrhythmia (Fig. 15.2e) with time-to-diastolic peak proving to be a better predictor than QT/APD (Lang et al. 2016a). This strongly suggests a potential future role for the assessment of mechanical dysfunction for risk stratification. Transgenic LQT rabbit models may offer vast possibilities for further research investigating

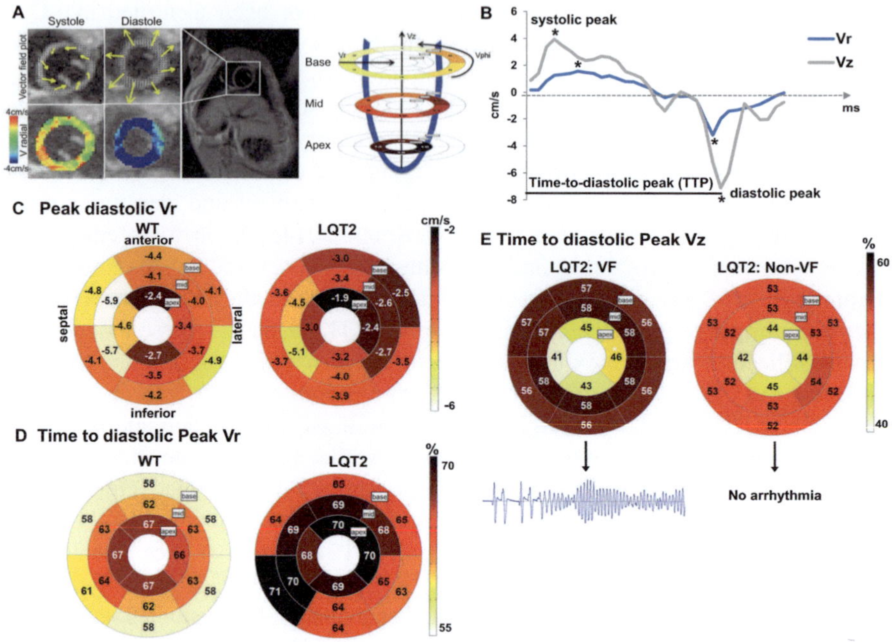

Fig. 15.2 Mechanical phenotype of transgenic LQT1 and LQT2 rabbits. (**a**) Color-coded radial velocities and summation vectors in the left ventricle in early systole and diastole are illustrated as recorded during tissue phase mapping (TPM) MRI. (**b**) Representative velocity graphs (velocities over time) in radial (Vr) and longitudinal (Vz) directions are displayed. Also indicated are systolic and diastolic peak radial velocities (asterisk) in the left ventricle and time to diastolic peak velocity (TTP) calculated from TPM. (**c**) Bull's-eye plots displaying averaged diastolic peak radial velocities (cm/s) in LV base, mid, and apical segments from TPM in wild-type (WT) ($n = 10$) and LQT2 rabbits ($n = 9$). LQT2 rabbits demonstrate a diastolic dysfunction with heterogeneously reduced peak radial velocities compared to wild type. (**d**) Bull's-eye plots displaying averaged heart rate—corrected time to diastolic radial peak velocities (TTP, %) from TPM. Regional TTP—a marker for contraction duration—prolonged in LQT2 ($n = 9$) compared to WT females ($n = 10$). (**e**) Bull's-eye plots displaying averaged heart rate—corrected time to diastolic longitudinal peak velocities (%). TTP are prolonged in LQT2 rabbits with ventricular fibrillation during Langendorff perfusion (VF, $n = 10$) compared to those without arrhythmia (Non-VF, $n = 33$). Modified and adapted from Odening et al. (2013) and Lang (2016a, b)

mechanisms underlying the observed mechanical dysfunction and its causative link to arrhythmogenesis.

15.2.2 Transgenic Animal Models for Sodium Channelopathies: Long QT Syndrome Type 3, Brugada Syndrome, Cardiac Conduction Disease, and Overlap Syndrome

The *SCN5A* gene encodes the alpha subunit of the cardiac voltage-gated sodium channel. While mutations in potassium channel genes alter cardiac repolarization (see Sect. 15.2.1), *SCN5A* sodium channel mutations may influence cardiac depolarization and repolarization, thus causing different channelopathies such as LQT3 syndrome (Schwartz et al. 2001), Brugada syndrome (BrS) (Priori et al. 2002a), cardiac conduction diseases (CCD), and overlap syndromes (Remme et al. 2008).

Mutations causing LQT3 are considered to disrupt fast inactivation of the sodium current, allowing for a persistent (late) sodium current ($I_{Na,L}$) during the plateau phase of the action potential (*gain-of-function* mutations), whereas mutations causing Brugada syndrome and cardiac conduction disease are considered to reduce the total amount of available sodium current (*loss-of-function* mutations). Additionally, an overlap phenotype may result from biophysical overlap with opposing alterations of the peak sodium current (I_{Na}) and the persistent sodium current component ($I_{Na,L}$) (Remme et al. 2006, 2008; Watanabe et al. 2011).

Notably, a single mutation in *SCN5A* may even present with different phenotypes (LQT3, BrS, or CCD) in different family members of one family (Remme et al. 2008). Hence, in addition to the concept that defined mutations of *SCN5A* result in a *gain-of-function* (LQT3) or *loss-of-function* (BrS or CCD) of the voltage-gated sodium channel, other modifiers—such as sex, age, alternative splicing, single nucleotide polymorphisms, or coinheritance of other genetic variations—may contribute to disease penetrance and phenotype (Remme et al. 2008; Derangeon et al. 2012).

Since the cardiac voltage-gated sodium channel constitutes the main depolarizing ion channel in various species including human and mice (Nerbonne and Kass 2005), genetically modified mouse models have been largely used to investigate the electrical consequences of different *SCN5A* mutations. Similarly as in human subjects, mouse models of sodium channelopathies may demonstrate prominent QT prolongation (LQT3 pattern), features of BrS, cardiac conduction disease, or even an overlap syndrome [reviewed in detail in Salama and London (2007), Charpentier et al. (2008), Remme et al. (2008) and Derangeon et al. (2012); Table 15.4]. Therefore, in the following, the different mouse (and pig) models are presented and discussed based on their "main" clinical and electrical features.

15.2.2.1 Genetically Modified Mouse Models with Features of LQT3 (QT Prolongation)

Imitation of Long QT Phenotype—APD, QT, and Arrhythmia

Genetically altered or transgenic LQT3 mouse models are either heterozygous for the knock-in +/Δ KPQ deletion in *Scn5a* (Nuyens et al. 2001; Head et al. 2005) or overexpress the human *SCN5A* point mutations N1325S or F1759A (Tian et al. 2004; Wan et al. 2016). The +/Δ KPQ mouse model demonstrated increased sodium current (I_{Na} and $I_{Na,L}$) and prolonged APD and QT intervals and also mimicked the arrhythmic phenotype with increased EAD formation and spontaneous and inducible polymorphic VT (pVT) (Nuyens et al. 2001; Head et al. 2005). Additionally, the +/Δ KPQ mutation was associated with AV block in both human subjects and mice (Zareba et al. 2001; Fabritz et al. 2010).

Similar to these knockout models, both transgenic mouse models (*SCN5A*-N1325S and *SCN5A*-F1759A) demonstrated prolonged APD and QT interval and imitated LQT3 with the occurrence of spontaneous premature ventricular contractions (PVC) and pVT (Tian et al. 2004; Wan et al. 2016). *SCN5A*-N1325S mice even died from premature cardiac death, and *SCN5A*-F1759A mice developed atrial fibrillation (the primary focus of the study by Wan et al.) (Tian et al. 2004; Wan et al. 2016). In cardiomyocytes of both transgenic models, only the persistent sodium current ($I_{Na,L}$) was increased, and peak sodium current (I_{Na}) was not altered (Wan et al. 2016).

LQT3 mouse models demonstrate prolonged APD and QT intervals; they develop spontaneous and/or inducible ventricular arrhythmia and thus mimic the human LQT3 phenotype considerably well. Additionally, the *SCN5A*-F1759A mice impressively demonstrated $I_{Na,L}$-induced damage to the heart, since beside electrical alterations, these mice developed pronounced macro- and microscopic alterations (atrial and ventricular enlargement, myofibril disarray, fibrosis, and mitochondrial injury) (Wan et al. 2016).

Mechanisms of Long QT-Related Arrhythmia

In human LQT3 patients, arrhythmia typically occurs during sleep and rest (Schwartz et al. 2001). Similarly, cholinergic stimulation with carbachol (imitating parasympathetic tone as during sleep and rest) induced arrhythmia in +/Δ KPQ mice (Fabritz et al. 2003). Increase in APD dispersion and EAD formation during bradycardia were detected as factors responsible for ventricular ectopy and pVT in these mice. Accordingly, ventricular pacing suppressed EADs and prevented pVT by reducing APD dispersion (Fabritz et al. 2003). However, +/Δ KPQ and *SCN5A*-N1325S mouse models also displayed APD prolongation with (sudden) acceleration of heart rate that was associated with EADs and triggered arrhythmia (Nuyens et al. 2001; Tian et al. 2004), indicating an additional pro-arrhythmic mechanism in these mouse models compared to human subjects.

Investigation of Anti-arrhythmic Therapeutic Drug Effects

Aiming at optimizing therapeutic strategies for LQT3 patients, beta-blockers and sodium channel blockers were extensively assessed in LQT3 mouse models (Table 15.4).

Conflicting data has been presented regarding the efficacy of beta-blockers in LQT3. Earlier studies in LQT3 mouse models found that propranolol and esmolol did not exert anti-arrhythmic effects (Head et al. 2005; Fabritz et al. 2010), while adrenergic agonists suppressed arrhythmia (Nuyens et al. 2001; Fabritz et al. 2010). However, a more recent study using (the same) +/Δ KPQ mice demonstrated that pretreatment with propranolol protected +/Δ KPQ mice from carbachol-induced VT/VF (Calvillo et al. 2014). Additionally, a recent study in LQT3 patients found a reduced risk for arrhythmia with beta-blocker therapy in female LQT3 patients, while the efficacy in males could not be determined conclusively due to the low number of arrhythmic events (Wilde et al. 2016). To date, beta-blockers are recommended as a cornerstone for the treatment of all human LQTS patients—including LQT3 (Priori et al. 2015).

Since an increased late $I_{Na,L}$ has been demonstrated to be causally linked to LQT3, the pharmacological blockade of this current should exert a mechanism-directed, genotype-specific anti-arrhythmic effect. Indeed, it was demonstrated in *Scn5a* +/Δ KPQ and *SCN5A*-N1325S mouse models that sodium blockers mexiletine and flecainide suppressed arrhythmia in LQT3 (Tian et al. 2004; Head et al. 2005; Fabritz et al. 2010). Similarly, recent clinical data demonstrate that LQT3 patients might benefit from sodium channel blockers (Moss et al. 2005, 2008; Priori et al. 2015). The ESC guidelines for the prevention of SCD (2015) thus mention sodium channel blockers (mexiletine, flecainide, or ranolazine) as a considerable add-on therapy in patients with LQT3 and a QTc greater than 500 ms with a class IIb recommendation (Priori et al. 2015). Since available agents do not only block late sodium current but to some extent also peak sodium current, this therapeutic approach should only be applied with caution in patients presenting with LQT3 and overlap syndrome (Moreno and Clancy 2012). However, more recently GS967 has been demonstrated to be a more selective inhibitor of $I_{Na,L}$ in *Scn5a*$^{1798insD/+}$ mice (Portero et al. 2017) and wild-type rabbit (Belardinelli et al. 2013).

15.2.2.2 Genetically Modified Mouse and Pig Models with Features of Brugada Syndrome, Cardiac Conduction Disease, and Overlap Phenotype

Imitation of Brugada Syndrome and Cardiac Conduction Disease Phenotype
Brugada syndrome (BrS) is characterized by typical right precordial coved-type ST segment elevations and an increased risk for VT and SCD (Priori et al. 2002a). Besides the characteristic ECG changes, patients with BrS frequently have conduction abnormalities, including prolonged PR and QRS intervals, particularly in case of *SCN5A* mutations (Antzelevitch et al. 2005). Importantly, arrhythmia can be triggered by increased body temperature (Antzelevitch and Brugada 2002).

In the heterozygous *Scn5a*$^{+/-}$ knockout mouse model and the *Scn5a*$^{\Delta SIV/\Delta SIV}$ knock-in mouse models, I_{Na} was decreased and significant cardiac conduction disease with PR and QRS prolongation as well as AV conduction block was present. QT intervals, however, were normal in both models (Papadatos et al. 2002; Shy et al. 2014) (Table 15.4). In *Scn5a*$^{+/-}$ mice ventricular refractoriness was prolonged, and VT was inducible with programmed ventricular stimulation and even occurred spontaneously upon aging (Papadatos et al. 2002), thus also imitating some "Brugada-like" features. *Scn5a*$^{\Delta SIV/\Delta SIV}$ mice, in contrast, did not develop arrhythmia, and decreased I_{Na} was attributed to defective cell surface expression of sodium channels (Shy et al. 2014).

Recently a transgenic pig model with features of BrS was generated using a human *SCN5A*$^{E558X/+}$ mutation resulting in decreased peak I_{Na} (Park et al. 2015) (Table 15.4). Typical BrS-type ST elevations were not present at baseline and could not be induced by sodium channel blocker flecainide, and no SCD was observed during the first 2 years of life in *SCN5A*$^{E558X/+}$ pigs. In Langendorff-perfused *SCN5A*$^{E558X/+}$ pig hearts, however, VT/VF was inducible by ventricular pacing and by short-coupled ventricular premature beats. Of note, the arrhythmia often initiated in the RV free wall (Park et al. 2015)—similarly as observed in human patients in whom the right ventricular outflow tract (RVOT) seems to be the source of PVCs and arrhythmia (Rudic et al. 2016). The development of arrhythmia during fever, a classical feature of BrS, was also imitated in $SCN5A^{E558X/+}$ pigs' hearts ex vivo: Spontaneous and inducible VF occurred at 39 °C, while hearts were stable and non-inducible at 35 °C (Park et al. 2015). Other non-specific features of BrS, e.g., conduction abnormalities such as a prolonged atrial-His and His-ventricular conduction intervals, were also present (Park et al. 2015). This model thus demonstrated some characteristic features of BrS, such as the sensitivity to increased temperature and increased VF inducibility (ex vivo), yet imitated the human phenotype incompletely. Due to pronounced similarities of the pig's cardiac physiology and anatomy to human with similar heart rate, heart size, ion channels/currents, action potential shape, and autonomic innervation (Park et al. 2015), the $SCN5A^{E558X/+}$ pig model may reveal more mechanistic insights into BrS and potential (pharmacological and interventional) therapeutic strategies in the future.

Imitation of Overlap Syndromes

In contrast to the above described sodium channel mutations leading to distinct features of LQT3 (predominant QT prolongation) or BrS (e.g., temperature sensitivity, VF inducibility), other mutations in *Scn5a* resulted in an "overlap" phenotype—clinically as well as biophysically (Table 15.4): In heterozygous $Scn5a^{1798insD/+}$ knock-in and transgenic *SCN5A*-D1275N mouse models, peak I_{Na} current (I_{Na}) was decreased, and significant cardiac conduction disease with PR and QRS prolongation as well as AV conduction block was present (Remme et al. 2006; Watanabe et al. 2011). In addition, preferential conduction slowing in the right ventricle was observed—similarly as in BrS (Remme et al. 2006). Persistent I_{Na} current ($I_{Na,L}$), however, was increased in both mouse models resulting in prolonged APD and QT intervals as in LQT3 (Remme et al. 2006; Watanabe et al. 2011)—similar as observed in the human overlap phenotype. These mouse models were thus able to demonstrate that one single *SCN5A* mutation may indeed be sufficient to cause an overlap syndrome of cardiac sodium channelopathy by differentially altering peak and late sodium current components (Remme et al. 2006; Watanabe et al. 2011). Moreover, further insights into BrS-associated arrhythmogenesis could be gathered with $Scn5a^{1798insD/+}$ models, in which the reduced peak I_{Na} current unmasked the maintenance of embryonic slow conduction in the adult RVOT that may account for the preferentially slowed conduction and arrhythmia initiation in RVOT observed in BrS (Boukens et al. 2013).

15.2.3 Genetically Modified Mouse Models for Catecholaminergic Polymorphic VT

The clinical characteristics of catecholaminergic polymorphic VT (CPVT) are physical or psychological stress-induced bidirectional (bVT) or polymorphic VT (pVT) and SCD in patients with normal baseline ECG and a structurally normal heart (Priori et al. 2002b). CPVT is caused by mutations in genes encoding the cardiac ryanodine receptor (*RyR2*) or calsequestrin 2 (*CASQ2*), leading to defective ("leaky") RyR2 channels with enhanced Ca^{2+} releases from the sarcoplasmic reticulum (SR) during adrenergic stimulation, delayed afterdepolarizations, and triggered activity (Priori et al. 2001b; Lahat et al. 2001).

Several mouse models of CPVT have been generated by genetic modification of *RyR2*, *CASQ2*, and the RyR2 stabilizer *FKBP12.6* (Wehrens 2003; Cerrone et al. 2005; Song et al. 2007; Katz et al. 2010) (Table 15.5). All these mouse models (except $RyR2^{+/Ex3\text{-}del}$ (Liu et al. 2014)) demonstrated bidirectional/polymorphic VT and ventricular fibrillation (VF) during physical activity and after adrenergic stimulation or caffeine challenge and thus mimicked the human CPVT phenotype. Consecutively, these models have led to a better understanding of the pathophysiology of CPVT, and several insights into pro- and anti-arrhythmic mechanisms have been successfully implemented into improved anti-arrhythmic therapies in the clinics (Watanabe et al. 2009; van der Werf et al. 2011).

15.2.3.1 Imitation of CPVT Phenotype and Insights into Arrhythmic Mechanisms

The first model of CPVT was generated in 2003 by knockout of *FKBP12.6*, which normally stabilizes RyR2. $FKBP12.6^{-/-}$ mice died from exercise-induced ventricular arrhythmia (Wehrens 2003). Although reduced binding of FKBP12.6 to RyR2 is acknowledged as part of the pathophysiology of CPVT, thus far to the best of our knowledge, no mutations in *FKBP12.6* have been identified as causative for CPVT in human patients (Song et al. 2007; Katz et al. 2010; Priori et al. 2013; Sumitomo 2016).

In 2005, the first *RyR2* knock-in mouse model was generated containing a heterozygous $RyR2^{R4496C}$ mutation, the murine equivalent of the human $RyR2^{R4497C}$ mutation (Cerrone et al. 2005). $RyR2^{R4496C/+}$ mice imitated the human CPVT phenotype and developed typical bVT and pVT during exercise stress testing and administration of catecholamine or caffeine. Optical mapping demonstrated concentric epicardial breakthrough patterns with focal origin in the His-Purkinje network as origin of the VT (Cerrone et al. 2007). Moreover, in these mice a mechanistic link between the gene defect and arrhythmia was established demonstrating abnormal Ca^{2+} release during diastole that was further increased by beta-adrenergic stimulation (Fernandez-Velasco 2009). Several other CPVT mouse models have been generated successfully based on other *RyR2* mutations ($RyR2^{R176Q/+}$ (Kannankeril 2006), $RyR2^{P2328S/+}$ and $RyR2^{P2328S/P2328S}$ (Goddard 2008; Zhang 2013), $RyR2^{R2474S/+}$ (Kobayashi et al. 2010), $RyR2^{S2246L10/+}$ (Suetomi et al. 2011), and $RyR2^{A4860G/+}$ (Zhao 2015).

Furthermore, two CPVT mouse models have been created involving mutations in *CASQ2* gene, which encodes a calcium-binding reservoir protein within the SR (Lahat et al. 2001). Both mouse models, $CASQ2^{\Delta E9/\Delta E9}$ and $CASQ2^{D307H/D307H}$, imitate the human CPVT phenotype with exercise- and catecholamine-induced bidirectional and polymorphic VT (Knollmann 2006; Song et al. 2007; Katz et al. 2010; Kurtzwald-Josefson et al. 2014) (Table 15.5). *RyR2* and *CASQ2* mouse models all mimicked the human CPVT phenotype and were consecutively used for pharmacological investigations of anti-arrhythmic treatment options (see below).

15.2.3.2 Investigation of Anti-arrhythmic Therapeutic Drug Effects

Beta-blockers are recommended in human CPVT patients as first-line therapy (Priori et al. 2015). However, clinical studies have shown that despite being treated with beta-blockers, 47% of CPVT patients still develop arrhythmia (Priori et al. 2002b). Accordingly, propranolol prevented epinephrine- and exercise-induced VT only unreliably in *RyR2*$^{\mathrm{R4496C/+}}$, *CASQ2*$^{\Delta\mathrm{E9}/\Delta\mathrm{E9}}$, and *CASQ2*$^{D\mathrm{307H/D307H}}$ mice (Cerrone et al. 2005, 2007; Katz et al. 2010). Due to only partial anti-arrhythmic efficacy of beta-blockers, potential anti-arrhythmic effects of various other agents were studied in detail in CPVT mouse models.

Recent data gathered in *CASQ2*$^{\Delta\mathrm{E9}/\Delta\mathrm{E9}}$ mice suggest that catecholamines may trigger arrhythmia not only via beta-receptor stimulation but also via alpha-receptor stimulation: The alpha-agonist phenylephrine provoked VT in *CASQ2*$^{\Delta\mathrm{E9}/\Delta\mathrm{E9}}$ mice, while the alpha-antagonist phentolamine or the alpha-/beta-antagonist labetalol abolished exercise- and epinephrine-induced arrhythmia, indicating that concomitant block of beta- and alpha-adrenergic receptors could become a therapeutic option for patients suffering from beta-blocker refractory arrhythmia (Kurtzwald-Josefson et al. 2014).

Among class I anti-arrhythmic drugs, flecainide and propafenone effectively suppressed exercise- and catecholamine-induced ventricular arrhythmia in CPVT mice (Watanabe et al. 2009; Hwang et al. 2011), whereas lidocaine and procainamide had no anti-arrhythmic effects (Watanabe et al. 2009; Katz et al. 2010; Hwang et al. 2011). Interestingly, direct inhibition of RyR2 by flecainide with suppression of spontaneous Ca^{2+} release from SR in addition to the suppression of triggered beats by sodium channel block was identified as its underlying anti-arrhythmic mechanisms (Watanabe et al. 2009). Based on these experimental and first clinical data (Watanabe et al. 2009; van der Werf et al. 2011) (as described in detail in the Chap. 10), the ESC guidelines for the prevention of SCD (2015) recommend the use of flecainide as add-on therapy in CPVT patients that remain symptomatic with syncope or VT while on beta-blocker (Priori et al. 2013). Of note, however, in another study flecainide surprisingly did not exert a relevant anti-arrhythmic effect in *CASQ2*$^{\Delta\mathrm{E9}/\Delta\mathrm{E9}}$ and *CASQ2*$^{D\mathrm{307H/D307H}}$ mice (Katz et al. 2010).

Calcium blocker verapamil significantly lowered VT prevalence in *CASQ2*$^{\Delta\mathrm{E9}/\Delta\mathrm{E9}}$ and *CASQ2*$^{D\mathrm{307H/D307H}}$ mice and similarly effectively abolished stress-induced ventricular arrhythmia in roughly half of CPVT patients still symptomatic despite propranolol therapy (Katz et al. 2010). Besides classical anti-arrhythmic agents, magnesium and dantrolene—both inhibiting RyR2 – decreased the incidence of catecholamine-induced VT in *CASQ2*$^{\Delta\mathrm{E9}/\Delta\mathrm{E9}}$, *CASQ2*$^{D\mathrm{307H/D307H}}$ (Song et al. 2007), *RyR2*$^{\mathrm{S2246L10/+}}$ (Suetomi et al. 2011), and *RyR2*$^{\mathrm{R2474S/+}}$ mice (Kobayashi et al. 2010).

15.3 Advantages and Limitations of Current Transgenic Animal Models for Channelopathies

Small animal models have the advantage of allowing conducting longitudinal studies in subjects with a defined genetic background and without confounding comorbidities for the assessment of factors that may alter the arrhythmic disease phenotype. This particularly applies to mouse models since they are genetically identical, resulting in a less heterogeneous phenotype (Davisson 1999). Moreover, different genetic backgrounds of different mouse strains can be used to study the modulatory effect of genetic modifiers on disease severity (Remme et al. 2009). As mice have relatively short generation times, their handling is rather cost-effective, and they can be more easily subjected to genetic manipulation than larger animals. Thanks to novel developments in animal transgenesis (Bősze et al. 2016), rabbits that have the added advantage of more closely mimicking human cardiac electrophysiology (Nerbonne 2000) have also entered the range of species in whom genetic manipulation can be more easily performed.

However, as described in detail in the introduction and the disease-specific subchapters, species differences exist in cardiac electrophysiological properties, channel composition, and Ca^{2+} handling (Nakata and Hearse 1990; Williams et al. 2000; Nerbonne et al. 2001; Salama and London 2007; Baczkó et al. 2016). Therefore, most animal models of channelopathies can only mimic certain aspects of the disease phenotype (see details in Sects. 15.2.1–15.2.3).

Apart from these species differences in cardiac ion channels and Ca^{2+} handling properties, some other aspects may limit the use and transferability of findings to human disease management:

1. A variety of different mutations in different genes can cause the abovementioned channelopathies (or overlap syndromes), which can have very different impacts on biophysical ion channel properties (e.g., reduction vs. complete loss of current, changes in voltage dependency, changes in gating properties) and on arrhythmic risk. When designing transgenic animal models for a given disease, one must choose a single disease-causing mutation—ideally one that produces a pronounced phenotype. Mutation-specific aspects on electrical dysfunction and arrhythmogenesis, however, cannot be assessed with the limited amount of different disease subtypes currently available or even one single animal model for each channelopathy subtype. Here, "high-throughput" approaches combined with computational modeling may add important information.
2. Moreover, species differences may exist also in electrical remodeling mechanisms occurring secondary to ion channel mutations. In mouse models of LQTS, compensatory upregulation of non-affected repolarizing ion currents—partially restoring the reduced repolarization reserve—has been described (Koren 2004; Salama and London 2007), while in rabbit models likewise, remodeling of repolarizing ion currents was observed that may aggravate the disease phenotype (Brunner et al. 2008; Major et al. 2016). Whether electrical remodeling mechanisms

occurring in human LQTS cardiomyocytes more closely resemble the changes in transgenic rabbit or mouse models still remains to be elucidated.

3. In addition, the current techniques used to generate genetic animal models, e.g., the complete knockout of cardiac ion channels or the dominant-negative approach with cardio-selective overexpression of (mouse or human) disease-causing genes rather than replacement of an endogenous gene with the human disease-causing gene thus replicating the heterozygous situation found in human patients, may cause additional differences in electrical remodeling. In the future, this limitation may be overcome with novel genetic techniques such as CRISPR/Cas9, which allows integration of the mutant human gene into the animal DNA [as reviewed in Bősze et al. (2016)], to generate future transgenic animal models for channelopathies that more closely resemble human patients genetically.

15.4 Clinical Implications/Translational Aspects

To improve medical anti-arrhythmic therapy in patients with inherited channelopathies—particularly in those who are refractory to standard therapy—novel (mechanistic) therapeutic concepts have to be evaluated preclinically. Since in vitro testing in cellular systems cannot assess all multidimensional factors of arrhythmogenesis, animal models of channelopathies are indispensable tools for preclinical identification of anti-arrhythmic mechanisms and drug testing to improve patient safety. Indeed, in all currently available genetic animal models for channelopathies, mechanisms underlying anti-arrhythmic effects of currently used standard therapies have been revealed, and several novel anti-arrhythmic treatment options have been identified; some of these even have already entered the guidelines:

1. The identification of anti-arrhythmic effects of progesterone in LQT2 models (Odening et al. 2012) has raised the awareness of the potential benefit of contraceptive drugs (minipills) in female LQTS patients (Moss 2012; Odening et al. 2016).
2. Studies identifying the anti-arrhythmic properties of late $I_{Na,L}$ blockers in LQT3 models (Fabritz et al. 2003, 2010; Tian et al. 2004; Head et al. 2005) have led to the development of more selective late $I_{Na,L}$ inhibitors [such as GS967 (Belardinelli et al. 2013) and GS6615 (Rajamani et al. 2016)], which have already entered phase 1 and 2 clinical trials (NCT01849003).
3. The identification of anti-arrhythmic effects of beta-blockers even in LQT3 models and the underlying mode of action (Calvillo et al. 2014) has led to a change in the clinical practice in the treatment of LQT3 patients.
4. The identification of anti-arrhythmic effects of flecainide in CPVT models (Hwang et al. 2011) has resulted in a class IIa recommendation to use flecainide in CPVT patients still symptomatic while on beta-blocker therapy (Priori et al. 2015).

15.5 Outlook

Despite their above discussed limitations, transgenic animal models for human channelopathies have already been instrumental for identifying arrhythmic mechanisms on the whole heart, cellular, and molecular levels (see details in disease-specific Sects. 15.2.1–15.2.3) and will certainly further improve our mechanistic insights into arrhythmogenesis in channelopathies. This knowledge will be crucial to develop novel, mechanism-directed, genotype-specific therapeutic strategies in the future. However, due to limitations intrinsic to the currently available transgenic animal models (see Sect. 15.3), several additional experimental and clinical steps have to be taken to be able to transfer these insights from bench to bedside.

1. Additional techniques such as the integration of experimental in vivo, whole-heart, cellular, and ion channel data into computational models are warranted—particularly to assess potential mutation-specific aspects since only limited animal models with different mutations are currently available for the different channelopathies.
2. Novel techniques such as CRISPR/Cas9 must be employed to generate future transgenic animal models for channelopathies that more closely resemble human patients genetically [as reviewed in Bősze et al. (2016)]. Here, the generation of additional animal models for the above presented channelopathies with different mutations (that will allow experimentally assessing and comparing mutation-specific disease patho-mechanisms and treatment options) as well as novel transgenic animal models for other channelopathies are clearly warranted. In this regard, transgenic rabbit models for short QT syndrome and LQT3 are already in the pipeline to be published.
3. As genetic techniques develop further (and get less time-consuming and less expensive), other species that more closely resemble (all) the different aspects of human cardiac pathophysiology may be used for future generation of transgenic animal models for channelopathies that may facilitate bench-to-bedside translation.
4. Most importantly, several of the molecular findings gathered in transgenic channelopathy animal models must still be verified in human diseased tissue and cells (such as iPS-CM from patients) prior to their clinical application. Although these iPS-CM also have some limitations in regard to ion current composition compared to mature ventricular cardiomyocytes, here, quite some work has already been done to generate and characterize disease- and patient-specific iPS-CM for LQTS, BrS, and CPVT [as reviewed in Hoekstra et al. (2012)], which will certainly supplement the insights gathered on the in vivo and whole heart levels using animal models.

Compliance with Ethical Standards

Sources of Funding None.

Conflict of Interest Katja E. Odening and David Ziupa declare that they have no conflict of interest.

Ethical Approval All animal studies summarized and reviewed in this article were conducted based on international, national, and/or institutional guidelines for the care and use of animals.

References

Antzelevitch C, Brugada R. Fever and Brugada syndrome. Pacing Clin Electrophysiol. 2002;25 (11):1537–9.

Antzelevitch C, Brugada P, Borggrefe M, Brugada J, Brugada R, Corrado D, Gussak I, LeMarec H, Nademanee K, Perez Riera AR, Shimizu W, Schulze-Bahr E, Tan H, Wilde A. Brugada syndrome: report of the second consensus conference: endorsed by the Heart Rhythm Society and the European Heart Rhythm Association. Circulation. 2005;111(5):659–70.

Babij P, Askew GR, Nieuwenhuijsen B, Su CM, Bridal TR, Jow B, Argentieri TM, Kulik J, DeGennaro LJ, Spinelli W, Colatsky TJ. Inhibition of cardiac delayed rectifier K+ current by overexpression of the long-QT syndrome HERG G628S mutation in transgenic mice. Circ Res. 1998;83(6):668–78.

Baczkó I, Jost N, Virág L, Bősze Z, Varró A. Rabbit models as tools for preclinical cardiac electrophysiological safety testing: importance of repolarization reserve. Prog Biophys Mol Biol. 2016;121(2):157–68.

Baker LC, London B, Choi BR, Koren G, Salama G. Enhanced dispersion of repolarization and refractoriness in transgenic mouse hearts promotes reentrant ventricular tachycardia. Circ Res. 2000;86(4):396–407.

Balasubramaniam R, Grace AA, Saumarez RC, Vandenberg JI, Huang CL. Electrogram prolongation and nifedipine-suppressible ventricular arrhythmias in mice following targeted disruption of KCNE1. J Physiol. 2003;552(Pt 2):535–46.

Barry DM, Xu H, Schuessler RB, Nerbonne JM. Functional knockout of the transient outward current, long-QT syndrome, and cardiac remodeling in mice expressing a dominant-negative Kv4 alpha subunit. Circ Res. 1998;83(5):560–7.

Belardinelli L, Liu G, Smith-Maxwell C, Wang WQ, El-Bizri N, Hirakawa R, Karpinski S, Li CH, Hu L, Li XJ, Crumb W, Wu L, Koltun D, Zablocki J, Yao L, Dhalla AK, Rajamani S, Shryock JC. A novel, potent, and selective inhibitor of cardiac late sodium current suppresses experimental arrhythmias. J Pharmacol Exp Ther. 2013;344(1):23–32.

Bősze Z, Major P, Baczkó I, Odening KE, Bodrogi L, Hiripi L, Varró A. The potential impact of new generation transgenic methods on creating rabbit models of cardiac diseases. Prog Biophys Mol Biol. 2016;121(2):123–30.

Boukens BJ, Sylva M, de Gier-de Vries C, Remme CA, Bezzina CR, Christoffels VM, Coronel R. Reduced sodium channel function unmasks residual embryonic slow conduction in the adult right ventricular outflow tract. Circ Res. 2013;113(2):137–41.

Brunner M, Guo W, Mitchell GF, Buckett PD, Nerbonne JM, Koren G. Characterization of mice with a combined suppression of I(to) and I(K,slow). Am J Physiol Heart Circ Physiol. 2001;281 (3):H1201–9.

Brunner M, Peng X, Liu GX, Ren XQ, Ziv O, Choi BR, Mathur R, Hajjiri M, Odening KE, Steinberg E, Folco EJ, Pringa E, Centracchio J, Macharzina RR, Donahay T, Schofield L, Rana N, Kirk M, Mitchell GF, Poppas A, Zehender M, Koren G. Mechanisms of cardiac arrhythmias and sudden death in transgenic rabbits with long QT syndrome. J Clin Invest. 2008;118(6):2246–59.

Calvillo L, Spazzolini C, Vullo E, Insolia R, Crotti L, Schwartz PJ. Propranolol prevents life-threatening arrhythmias in LQT3 transgenic mice: implications for the clinical management of LQT3 patients. Heart Rhythm. 2014;11(1):126–32.

Casimiro MC, Knollmann BC, Ebert SN, Vary JC Jr, Greene AE, Franz MR, Grinberg A, Huang SP, Pfeifer K. Targeted disruption of the Kcnq1 gene produces a mouse model of Jervell and Lange-Nielsen syndrome. Proc Natl Acad Sci U S A. 2001;98(5):2526–31.

Cerrone M, Colombi B, Santoro M, di Barletta MR, Scelsi M, Villani L, Napolitano C, Priori SG. Bidirectional ventricular tachycardia and fibrillation elicited in a knock-in mouse model carrier of a mutation in the cardiac ryanodine receptor. Circ Res. 2005;96(10):e77–82.

Cerrone M, Noujaim SF, Tolkacheva EG, Talkachou A, O'Connell R, Berenfeld O, Anumonwo J, Pandit SV, Vikstrom K, Napolitano C, Priori SG, Jalife J. Arrhythmogenic mechanisms in a mouse model of catecholaminergic polymorphic ventricular tachycardia. Circ Res. 2007;101 (10):1039–48.

Charpentier F, Merot J, Riochet D, Le Marec H, Escande D. Adult KCNE1-knockout mice exhibit a mild cardiac cellular phenotype. Biochem Biophys Res Commun. 1998;251(3):806–10.

Charpentier F, Bourgé A, Mérot J. Mouse models of SCN5A-related cardiac arrhythmias. Prog Biophys Mol Biol. 2008;98(2–3):230–7.

Choy L, Yeo JM, Tse V, Chan SP, Tse G. Cardiac disease and arrhythmogenesis: mechanistic insights from mouse models. Int J Cardiol Heart Vasc. 2016;12:1–10.

Davisson MT. Genetic and phenotypic definition of laboratory mice and rats. National Research Council (US) International Committee of the Institute for Laboratory Animal Research, 1999.

Demolombe S, Lande G, Charpentier F, van Roon MA, van den Hoff MJ, Toumaniantz G, Baro I, Guihard G, Le Berre N, Corbier A, de Bakker J, Opthof T, Wilde A, Moorman AF, Escande D. Transgenic mice overexpressing human KvLQT1 dominant-negative isoform. Part I: phenotypic characterisation. Cardiovasc Res. 2001;50(2):314–27.

Derangeon M, Montnach J, Baró I, Charpentier F. Mouse models of SCN5A-related cardiac arrhythmias. Front Physiol. 2012;3:210.

Drici MD, Arrighi I, Chouabe C, Mann JR, Lazdunski M, Romey G, Barhanin J. Involvement of IsK-associated K+ channel in heart rate control of repolarization in a murine engineered model of Jervell and Lange-Nielsen syndrome. Circ Res. 1998;83(1):95–102.

Fabritz L, Kirchhof P, Franz MR, Nuyens D, Rossenbacker T, Ottenhof A, Haverkamp W, Breithardt G, Carmeliet E, Carmeliet P. Effect of pacing and mexiletine on dispersion of repolarisation and arrhythmias in DeltaKPQ SCN5A (long QT3) mice. Cardiovasc Res. 2003;57(4):1085–93.

Fabritz L, Damke D, Emmerich M, Kaufmann SG, Theis K, Blana A, Fortmüller L, Laakmann S, Hermann S, Aleynichenko E, Steinfurt J, Volkery D, Riemann B, Kirchhefer U, Franz MR, Breithardt G, Carmeliet E, Schäfers M, Maier SK, Carmeliet P, Kirchhof P. Autonomic modulation and antiarrhythmic therapy in a model of long QT syndrome type 3. Cardiovasc Res. 2010;87(1):60–72.

Fernandez-Velasco M, Rueda A, Rizzi N, Benitah JP, Colombi B, Napolitano C, Priori SG, Richard S, Gomez AM. Increased Ca2+ sensitivity of the ryanodine receptor mutant RyR2R4496C underlies catecholaminergic polymorphic ventricular tachycardia. Circ Res. 2009;104:201–9.

Goddard CA, Ghais NS, Zhang Y, Williams AJ, Colledge WH, Grace AA, Huang CL. Physiological consequences of the P2328S mutation in the ryanodine receptor (RyR2) gene in genetically modified murine hearts. Acta Physiol (Oxf). 2008;194:123–40.

Guo W, Li H, London B, Nerbonne JM. Functional consequences of elimination of I(to,f) and i(to, s): early afterdepolarizations, atrioventricular block, and ventricular arrhythmias in mice lacking Kv1.4 and expressing a dominant-negative Kv4 alpha subunit. Circ Res. 2000;87(1):73–9.

Guo W, Jung WE, Marionneau C, Aimond F, Xu H, Yamada KA, Schwarz TL, Demolombe S, Nerbonne JM. Targeted deletion of Kv4.2 eliminates I(to,f) and results in electrical and molecular remodeling, with no evidence of ventricular hypertrophy or myocardial dysfunction. Circ Res. 2005;97(12):1342–50.

Haugaa KH, Smedsrud MK, Steen T, Kongsgaard E, Loennechen JP, Skjaerpe T, Voigt JU, Willems R, Smith G, Smiseth OA, Amlie JP, Edvardsen T. Mechanical dispersion assessed by myocardial strain in patients after myocardial infarction for risk prediction of ventricular arrhythmia. JACC Cardiovasc Imaging. 2010;3(3):247–56.

Head CE, Balasubramaniam R, Thomas G, Goddard CA, Lei M, Colledge WH, Grace AA, Huang CL. Paced electrogram fractionation analysis of arrhythmogenic tendency in DeltaKPQ Scn5a mice. J Cardiovasc Electrophysiol. 2005;16(12):1329–40.

Hoekstra M, Mummery CL, Wilde AA, Bezzina CR, Verkerk AO. Induced pluripotent stem cell derived cardiomyocytes as models for cardiac arrhythmias. Front Physiol. 2012;3:346.

Hondeghem LM. Disturbances of cardiac wavelength and repolarization precede torsade de pointes and ventricular fibrillation in Langendorff perfused rabbit hearts. Prog Biophys Mol Biol. 2016;121(1):3–10.

Hwang HS, Hasdemir C, Laver D, Mehra D, Turhan K, Faggioni M, Yin H, Knollmann BC. Inhibition of cardiac Ca2+ release channels (RyR2) determines efficacy of class I antiarrhythmic drugs in catecholaminergic polymorphic ventricular tachycardia. Circ Arrhythm Electrophysiol. 2011;4(2):128–35.

Jung B, Odening KE, Dall'Armellina E, Foll D, Menza M, Markl M, Schneider JE. A quantitative comparison of regional myocardial motion in mice, rabbits and humans using in-vivo phase contrast CMR. J Cardiovasc Magn Reson. 2012;14:87.

Kannankeril PJ, Mitchell BM, Goonasekera SA, Chelu MG, Zhang W, Sood S, Kearney DL, Danila CI, De Biasi M, Wehrens XH, Pautler RG, Roden DM, Taffet GE, Dirksen RT, Anderson ME, Hamilton SL. Mice with the R176Q cardiac ryanodine receptor mutation exhibit catecholamine-induced ventricular tachycardia and cardiomyopathy. Proc Natl Acad Sci USA. 2006;103:12179–84.

Katz G, Khoury A, Kurtzwald E, Hochhauser E, Porat E, Shainberg A, Seidman JG, Seidman CE, Lorber A, Eldar M, Arad M. Optimizing catecholaminergic polymorphic ventricular tachycardia therapy in calsequestrin-mutant mice. Heart Rhythm. 2010;7(11):1676–82.

Kim TY, Kunitomo Y, Pfeiffer Z, Patel D, Hwang J, Harrison K, Patel B, Jeng P, Ziv O, Lu Y, Peng X, Qu Z, Koren G, Choi B-R. Complex excitation dynamics underlie polymorphic ventricular tachycardia in a transgenic rabbit model of long QT syndrome type 1. Heart Rhythm. 2015;12(1):220–8.

Knollmann BC, Chopra N, Hlaing T, Akin B, Yang T, Ettensohn K, Knollmann BE, Horton KD, Weissman NJ, Holinstat I, Zhang W, Roden DM, Jones LR, Franzini-Armstrong C, Pfeifer K. Casq2 deletion causes sarcoplasmic reticulum volume increase, premature Ca2+ release, and catecholaminergic polymorphic ventricular tachycardia. J Clin Invest. 2006;116:2510–20.

Kobayashi S, Yano M, Uchinoumi H, Suetomi T, Susa T, Ono M, Xu X, Tateishi H, Oda T, Okuda S, Doi M, Yamamoto T, Matsuzaki M. Dantrolene, a therapeutic agent for malignant hyperthermia, inhibits catecholaminergic polymorphic ventricular tachycardia in a RyR2 (R2474S/+) knock-in mouse model. Circ J. 2010;74(12):2579–84.

Kodirov SA, Brunner M, Nerbonne JM, Buckett P, Mitchell GF, Koren G. Attenuation of I(K, slow1) and I(K,slow2) in Kv1/Kv2DN mice prolongs APD and QT intervals but does not suppress spontaneous or inducible arrhythmias. Am J Physiol Heart Circ Physiol. 2004;286 (1):368–74.

Koren G. Electrical remodeling and arrhythmias in long-QT syndrome: lessons from genetic models in mice. Ann Med. 2004;36(Suppl 1):22–7.

Kuo H-C, Cheng C-F, Clark RB, Lin JJ-C, Lin JL-C, Hoshijima M, Nguyêñ-Trân VTB, Yusu G, Ikeda Y, Chu P-H, Jr JR, Giles WR, Chien KR. A defect in the Kv channel-interacting protein 2 (KChIP2) gene leads to a complete loss of I(to) and confers susceptibility to ventricular tachycardia. Cell. 2001;107(6):801–13.

Kupershmidt S, Yang T, Anderson ME, Wessels A, Niswender KD, Magnuson MA, Roden DM. Replacement by homologous recombination of the minK gene with lacZ reveals restriction of minK expression to the mouse cardiac conduction system. Circ Res. 1999;84(2):146–52.

Kurtzwald-Josefson E, Hochhauser E, Bogachenko K, Harun-Khun S, Katz G, Aravot D, Seidman JG, Seidman CE, Eldar M, Shainberg A, Arad M. Alpha blockade potentiates CPVT therapy in calsequestrin-mutant mice. Heart Rhythm. 2014;11(8):1471–9.

Lahat H, Pras E, Olender T, Avidan N, Ben-Asher E, Man O, Levy-Nissenbaum E, Khoury A, Lorber A, Goldman B, Lancet D, Eldar M. A missense mutation in a highly conserved region of CASQ2 is associated with autosomal recessive catecholamine-induced polymorphic ventricular tachycardia in Bedouin families from Israel. Am J Hum Genet. 2001;69(6):1378–84.

Lande G, Demolombe S, Bammert A, Moorman A, Charpentier F, Escande D. Transgenic mice overexpressing human KvLQT1 dominant-negative isoform. Part II: Pharmacological profile. Cardiovasc Res. 2001;50(2):328–34.

Lang CN, Menza M, Jochem S, Franke G, Perez Feliz S, Brunner M, Koren G, Zehender M, Bugger H, Jung BA, Foell D, Bode C, Odening KE. Electro-mechanical dysfunction in long QT syndrome. Prog Biophys Mol Biol. 2016a;120(1–3):255–69.

Lang CN, Koren G, Odening KE. Transgenic rabbit models to investigate the cardiac ion channel disease long QT syndrome. Prog Biophys Mol Biol. 2016b;121(2):142–56.

Lau E, Kossidas K, Kim TY, Kunitomo Y, Ziv O, Zhen S, Taylor C, Schofield L, Yammine J, Liu G, Peng X, Qu Z, Koren G, Choi B-R. Spatially discordant alternans and arrhythmias in tachypacing-induced cardiac myopathy in transgenic LQT1 rabbits: the importance of IKs and Ca2+ cycling. PLoS One. 2015;10(5):e0122754.

Lee MP, Ravenel JD, Hu RJ, Lustig LR, Tomaselli G, Berger RD, Brandenburg SA, Litzi TJ, Bunton TE, Limb C, Francis H, Gorelikow M, Gu H, Washington K, Argani P, Goldenring JR, Coffey RJ, Feinberg AP. Targeted disruption of the Kvlqt1 gene causes deafness and gastric hyperplasia in mice. J Clin Invest. 2000;106(12):1447–55.

Lees-Miller JP, Guo J, Somers JR, Roach DE, Sheldon RS, Rancourt DE, Duff HJ. Selective knockout of mouse ERG1 B potassium channel eliminates I(Kr) in adult ventricular myocytes and elicits episodes of abrupt sinus bradycardia. Mol Cell Biol. 2003;23(6):1856–62.

Leren IS, Hasselberg NE, Saberniak J, Håland TF, Kongsgård E, Smiseth OA, Edvardsen T, Haugaa KH. Cardiac mechanical alterations and genotype specific differences in subjects with long QT syndrome. JACC Cardiovasc Imaging. 2015;8(5):501–10.

Li H, Guo W, Yamada KA, Nerbonne JM. Selective elimination of I(K,slow1) in mouse ventricular myocytes expressing a dominant negative Kv1.5alpha subunit. Am J Physiol Heart Circ Physiol. 2004;286(1):H319–28.

Liu GX, Choi BR, Ziv O, Li W, de Lange E, Qu Z, Koren G. Differential conditions for early after-depolarizations and triggered activity in cardiomyocytes derived from transgenic LQT1 and LQT2 rabbits. J Physiol. 2012;590(5):1171–80.

Liu Y, Wang R, Sun B, Mi T, Zhang J, Mu Y, Chen J, Bround MJ, Johnson JD, Gillis AM, Chen SR. Generation and characterization of a mouse model harboring the exon-3 deletion in the cardiac ryanodine receptor. PLoS One. 2014;9:e95615.

London B, Jeron A, Zhou J, Buckett P, Han X, Mitchell GF, Koren G. Long QT and ventricular arrhythmias in transgenic mice expressing the N terminus and first transmembrane segment of a voltage-gated potassium channel. Proc Natl Acad Sci U S A. 1998a;95(6):2926–31.

London B, Wang DW, Hill JA, Bennett PB. The transient outward current in mice lacking the potassium channel gene Kv1.4. J Physiol. 1998b;509(Pt 1):171–82.

London B, Guo W, Pan X, Lee JS, Shusterman V, Rocco CJ, Logothetis DA, Nerbonne JM, Hill JA. Targeted replacement of KV1.5 in the mouse leads to loss of the 4-aminopyridine-sensitive component of I(K,slow) and resistance to drug-induced qt prolongation. Circ Res. 2001;88 (9):940–6.

London B, Baker LC, Petkova-Kirova P, Nerbonne JM, Choi B-R, Salama G. Dispersion of repolarization and refractoriness are determinants of arrhythmia phenotype in transgenic mice with long QT. J Physiol. 2007;578(Pt 1):115–29.

Major P, Baczkó I, Hiripi L, Odening KE, Juhász V, Kohajda Z, Horváth A, Seprényi G, Kovács M, Virág L, Jost N, Prorok J, Ördög B, Doleschall Z, Nattel S, Varró A, Bősze Z. A novel transgenic rabbit model with reduced repolarization reserve: long QT syndrome caused by a dominant-negative mutation of the KCNE1 gene. Br J Pharmacol. 2016;173(12):2046–61.

McLerie M, Lopatin AN. Dominant-negative suppression of I(K1) in the mouse heart leads to altered cardiac excitability. J Mol Cell Cardiol. 2003;35(4):367–78.

Moreno JD, Clancy CE. Pathophysiology of the cardiac late Na current and its potential as a drug target. J Mol Cell Cardiol. 2012;52(3):608–19.

Morita H, Wu J, Zipes DP. The QT syndromes. Lancet. 2008;372(9640):750–63.

Moshal KS, Zhang Z, Roder K, Kim TY, Cooper L, Patedakis Litvinov B, Lu Y, Reddy V, Terentyev D, Choi BR, Koren G. Progesterone modulates SERCA2a expression and function in rabbit cardiomyocytes. Am J Physiol Cell Physiol. 2014;307(11):C1050–7.

Moss AJ. Sex hormones and ventricular tachyarrhythmias in LQTS: new insights regarding antiarrhythmic therapy. Heart Rhythm. 2012;9(5):833–4.

Moss AJ, Windle JR, Hall WJ, Zareba W, Robinson JL, McNitt S, Severski P, Rosero S, Daubert JP, Qi M, Cieciorka M, Manalan AS. Safety and efficacy of flecainide in subjects with long QT-3 syndrome (DeltaKPQ mutation): a randomized, double-blind, placebo-controlled clinical trial. Ann Noninvasive Electrocardiol. 2005;10(4 Suppl):59–66.

Moss AJ, Zareba W, Schwarz KQ, Rosero S, McNitt S, Robinson JL. Ranolazine shortens repolarization in patients with sustained inward sodium current due to type-3 long-QT syndrome. J Cardiovasc Electrophysiol. 2008;19(12):1289–93.

Nador F, Beria G, De Ferrari GM, Stramba-Badiale M, Locati EH, Lotto A, Schwartz PJ. Unsuspected echocardiographic abnormality in the long QT syndrome. Diagnostic, prognostic, and pathogenetic implications. Circulation. 1991;84(4):1530–42.

Nakata T, Hearse DJ. Species differences in vulnerability to injury by oxidant stress: a possible link with calcium handling? Cardiovasc Res. 1990;24(10):857–64.

Nattel S, Duker G, Carlsson L. Model systems for the discovery and development of antiarrhythmic drugs. Prog Biophys Mol Biol. 2008;98(2–3):328–39.

Nerbonne JM. Molecular basis of functional voltage-gated K+ channel diversity in the mammalian myocardium. J Physiol. 2000;525(Pt 2):285–98.

Nerbonne JM, Kass RS. Molecular physiology of cardiac repolarization. Physiol Rev. 2005;85 (4):1205–53.

Nerbonne JM, Nichols CG, Schwarz TL, Escande D. Genetic manipulation of cardiac K(+) channel function in mice: what have we learned, and where do we go from here? Circ Res. 2001;89 (11):944–56.

Nuyens D, Stengl M, Dugarmaa S, Rossenbacker T, Compernolle V, Rudy Y, Smits JF, Flameng W, Clancy CE, Moons L, Vos MA, Dewerchin M, Benndorf K, Collen D, Carmeliet E, Carmeliet P. Abrupt rate accelerations or premature beats cause life-threatening arrhythmias in mice with long-QT3 syndrome. Nat Med. 2001;7(9):1021–7.

Odening KE, Kohl P. Follow the white rabbit: experimental and computational models of the rabbit heart provide insights into cardiac (patho-) physiology. Prog Biophys Mol Biol. 2016;121 (2):75–6.

Odening KE, Koren G, Kirk M. Normalization of QT interval duration in a long QT syndrome patient during pregnancy and the postpartum period due to sex hormone effects on cardiac repolarization. Heart Rhythm Case Rep. 2016;2(3):223–7.

Odening KE, Koren G. How do sex hormones modify arrhythmogenesis in long QT syndrome? Sex hormone effects on arrhythmogenic substrate and triggered activity. Heart Rhythm. 2014;11 (11):2107–15.

Odening KE, Hyder O, Chaves L, Schofield L, Brunner M, Kirk M, Zehender M, Peng X, Koren G. Pharmacogenomics of anesthetic drugs in transgenic LQT1 and LQT2 rabbits reveal genotype-specific differential effects on cardiac repolarization. Am J Physiol Heart Circ Physiol. 2008;295(6):H2264–72.

Odening KE, Kirk M, Brunner M, Ziv O, Lorvidhaya P, Liu GX, Schofield L, Chaves L, Peng X, Zehender M, Choi BR, Koren G. Electrophysiological studies of transgenic long QT type 1 and type 2 rabbits reveal genotype-specific differences in ventricular refractoriness and his conduction. Am J Physiol Heart Circ Physiol. 2010;299(3):H643–55.

Odening KE, Choi BR, Liu GX, Hartmann K, Ziv O, Chaves L, Schofield L, Centracchio J, Zehender M, Peng X, Brunner M, Koren G. Estradiol promotes sudden cardiac death in transgenic long QT type 2 rabbits while progesterone is protective. Heart Rhythm. 2012;9 (5):823–32.

Odening KE, Jung BA, Lang CN, Cabrera Lozoya R, Ziupa D, Menza M, Relan J, Franke G, Perez Feliz S, Koren G, Zehender M, Bode C, Brunner M, Sermesant M, Föll D. Spatial correlation of action potential duration and diastolic dysfunction in transgenic and drug-induced LQT2 rabbits. Heart Rhythm. 2013;10(10):1533–41.

Opthof T, Remme CA, Jorge E, Noriega F, Wiegerinck RF, Tasiam A, Beekman L, Alvarez-Garcia J, Munoz-Guijosa C, Coronel R, Cinca J. Cardiac activation-repolarization patterns and ion channel expression mapping in intact isolated normal human hearts. Heart Rhythm. 2017;14 (2):265–72.

Organ-Darling LE, Vernon AN, Giovanniello JR, Lu Y, Moshal K, Roder K, Li W, Koren G. Interactions between hERG and KCNQ1 alpha-subunits are mediated by their COOH termini and modulated by cAMP. Am J Physiol Heart Circ Physiol. 2013;304(4):H589–99.

Papadatos GA, Wallerstein PM, Head CE, Ratcliff R, Brady PA, Benndorf K, Saumarez RC, Trezise AE, Huang CL, Vandenberg JI, Colledge WH, Grace AA. Slowed conduction and ventricular tachycardia after targeted disruption of the cardiac sodium channel gene Scn5a. Proc Natl Acad Sci U S A. 2002;99(9):6210–5.

Park DS, Cerrone M, Morley G, Vasquez C, Fowler S, Liu N, Bernstein SA, Liu FY, Zhang J, Rogers CS, Priori SG, Chinitz LA, Fishman GI. Genetically engineered SCN5A mutant pig hearts exhibit conduction defects and arrhythmias. J Clin Invest. 2015;125(1):403–12.

Portero V, Casini S, Hoekstra M, Verkerk AO, Mengarelli I, Belardinelli L, Rajamani S, Wilde AAM, Bezzina CR, Veldkamp MW, Remme CA. Anti-arrhythmic potential of the late sodium current inhibitor GS-458967 in murine Scn5a-1798insD+/– and human SCN5A-1795insD+/– iPSC-derived cardiomyocytes. Cardiovasc Res. 2017;113(7):829–38.

Priori SG, Bloise R, Crotti L. The long QT syndrome. Europace. 2001a;3:16–27.

Priori SG, Napolitano C, Tiso N, Memmi M, Vignati G, Bloise R, Sorrentino V, Danieli GA. Mutations in the cardiac ryanodine receptor gene (hRyR2) underlie catecholaminergic polymorphic ventricular tachycardia. Circulation. 2001b;103(2):196–200.

Priori SG, Napolitano C, Gasparini M, Pappone C, Della Bella P, Giordano U, Bloise R, Giustetto C, De Nardis R, Grillo M, Ronchetti E, Faggiano G, Nastoli J. Natural history of Brugada syndrome: insights for risk stratification and management. Circulation. 2002a;105 (11):1342–7.

Priori SG, Napolitano C, Memmi M, Colombi B, Drago F, Gasparini M, De Simone L, Coltorti F, Bloise R, Keegan R, Cruz Filho FE, Vignati G, Benatar A, De Logu A. Clinical and molecular characterization of patients with catecholaminergic polymorphic ventricular tachycardia. Circulation. 2002b;106(1):69–74.

Priori SG, Wilde AA, Horie M, Cho Y, Behr ER, Berul C, Blom N, Brugada J, Chiang CE, Huikuri H, Kannankeril P, Krahn A, Leenhardt A, Moss A, Schwartz PJ, Shimizu W, Tomaselli G, Tracy C. HRS/EHRA/APHRS expert consensus statement on the diagnosis and management of patients with inherited primary arrhythmia syndromes: document endorsed by HRS, EHRA, and APHRS in May 2013 and by ACCF, AHA, PACES, and AEPC in June 2013. Heart Rhythm. 2013;10(12):1932–63.

Priori SG, Blomström-Lundqvist C, Mazzanti A, Blom N, Borggrefe M, Camm J, Elliott PM, Fitzsimons D, Hatala R, Hindricks G, Kirchhof P, Kjeldsen K, Kuck KH, Hernandez-Madrid A, Nikolaou N, Norekvål TM, Spaulding C, Van Veldhuisen DJ. ESC guidelines for the management of patients with ventricular arrhythmias and the prevention of sudden cardiac death: the task force for the management of patients with ventricular arrhythmias and the prevention of sudden cardiac death of the European Society of Cardiology (ESC). Endorsed by: Association for European Paediatric and Congenital Cardiology (AEPC). Eur Heart J. 2015;36 (41):2793–867.

Quinn TA, Kohl P. Rabbit models of cardiac mechano-electric and mechano-mechanical coupling. Prog Biophys Mol Biol. 2016;121(2):110–22.

Rajamani S, Liu G, El-Bizri N, Guo D, Li C, Chen XL, Kahlig KM, Mollova N, Elzein E, Zablocki J, Belardinelli L. The novel late Na+ current inhibitor, GS-6615 (eleclazine) and its anti-arrhythmic effects in rabbit isolated heart preparations. Br J Pharmacol. 2016;173 (21):3088–98.

Remme CA, Verkerk AO, Nuyens D, van Ginneken AC, van Brunschot S, Belterman CN, Wilders R, van Roon MA, Tan HL, Wilde AA, Carmeliet P, de Bakker JM, Veldkamp MW, Bezzina CR. Overlap syndrome of cardiac sodium channel disease in mice carrying the equivalent mutation of human SCN5A-1795insD. Circulation. 2006;114(24):2584–94.

Remme CA, Wilde AA, Bezzina CR. Cardiac sodium channel overlap syndromes: different faces of SCN5A mutations. Trends Cardiovasc Med. 2008;18(3):78–87.

Remme CA, Scicluna BP, Verkerk AO, Amin AS, van Brunschot S, Beekman L, Deneer VH, Chevalier C, Oyama F, Miyazaki H, Nukina N, Wilders R, Escande D, Houlgatte R, Wilde AA, Tan HL, Veldkamp MW, de Bakker JM, Bezzina CR. Genetically determined differences in sodium current characteristics modulate conduction disease severity in mice with cardiac sodium channelopathy. Circ Res. 2009;104(11):1283–92.

Ren XQ, Liu GX, Organ-Darling LE, Zheng R, Roder K, Jindal HK, Centracchio J, McDonald TV, Koren G. Pore mutants of HERG and KvLQT1 downregulate the reciprocal currents in stable cell lines. Am J Physiol Heart Circ Physiol. 2010;299(5):H1525–34.

Rudic B, Chaykovskaya M, Tsyganov A, Kalinin V, Tülümen E, Papavassiliu T, Dösch C, Liebe V, Kuschyk J, Röger S, El-Battrawy I, Akin I, Yakovleva M, Zaklyazminskaya E, Shestak A, Kim S, Chmelevsky M, Borggrefe M. Simultaneous non-invasive epicardial and endocardial mapping in patients with Brugada syndrome: new insights into arrhythmia mechanisms. J Am Heart Assoc. 2016;5(11):pii: e004095.

Salama G, London B. Mouse models of long QT syndrome. J Physiol. 2007 Jan 1;578(Pt 1):43–53.

Salama G, Baker L, Wolk R, Barhanin J, London B. Arrhythmia phenotype in mouse models of human long QT. J Interv Card Electrophysiol. 2009;24(2):77–87.

Salata JJ, Jurkiewicz NK, Jow B, Folander K, Guinosso PJ Jr, Raynor B, Swanson R, Fermini B. IK of rabbit ventricle is composed of two currents: evidence for IKs. Am J Phys. 1996;271(6 Pt 2): H2477–89.

Sanbe A, James J, Tuzcu V, Nas S, Martin L, Gulick J, Osinska H, Sakthivel S, Klevitsky R, Ginsburg KS, Bers DM, Zinman B, Lakatta EG, Robbins J. Transgenic rabbit model for human troponin I-based hypertrophic cardiomyopathy. Circulation. 2005;111(18):2330–8.

Sanguinetti MC, Curran ME, Spector PS, Keating MT. Spectrum of HERG K+−channel dysfunction in an inherited cardiac arrhythmia. Proc Natl Acad Sci U S A. 1996 Mar 5;93(5):2208–12.

Sauer AJ, Moss AJ, McNitt S, Peterson DR, Zareba W, Robinson JL, Qi M, Goldenberg I, Hobbs JB, Ackerman MJ, Benhorin J, Hall WJ, Kaufman ES, Locati EH, Napolitano C, Priori SG, Schwartz PJ, Towbin JA, Vincent GM, Zhang L. Long QT syndrome in adults. J Am Coll Cardiol. 2007;49(3):329–37.

Schwartz PJ, Priori SG, Spazzolini C, Moss AJ, Vincent GM, Napolitano C, Denjoy I, Guicheney P, Breithardt G, Keating MT, Towbin JA, Beggs AH, Brink P, Wilde AA, Toivonen L, Zareba W, Robinson JL, Timothy KW, Corfield V, Wattanasirichaigoon D, Corbett C, Haverkamp W, Schulze-Bahr E, Lehmann MH, Schwartz K, Coumel P, Bloise R. Genotype-phenotype correlation in the long-QT syndrome: gene-specific triggers for life-threatening arrhythmias. Circulation. 2001;103(1):89–95.

Shy D, Gillet L, Ogrodnik J, Albesa M, Verkerk AO, Wolswinkel R, Rougier JS, Barc J, Essers MC, Syam N, Marsman RF, van Mil AM, Rotman S, Redon R, Bezzina CR, Remme CA, Abriel H. PDZ domain-binding motif regulates cardiomyocyte compartment-specific NaV1.5 channel expression and function. Circulation. 2014;130(2):147–60.

Song L, Alcalai R, Arad M, Wolf CM, Toka O, Conner DA, Berul CI, Eldar M, Seidman CE, Seidman JG. Calsequestrin 2 (CASQ2) mutations increase expression of calreticulin and ryanodine receptors, causing catecholaminergic polymorphic ventricular tachycardia. J Clin Invest. 2007 Jul;117(7):1814–23.

Suetomi T, Yano M, Uchinoumi H, Fukuda M, Hino A, Ono M, Xu X, Tateishi H, Okuda S, Doi M, Kobayashi S, Ikeda Y, Yamamoto T, Ikemoto N, Matsuzaki M. Mutation-linked defective interdomain interactions within ryanodine receptor cause aberrant ca(2)(+)release leading to catecholaminergic polymorphic ventricular tachycardia. Circulation. 2011;124(6):682–94.

Sumitomo N. Current topics in catecholaminergic polymorphic ventricular tachycardia. J Arrhythm. 2016;32:344–51.

Thomas G, Killeen MJ, Gurung IS, Hakim P, Balasubramaniam R, Goddard CA, Grace AA, Huang CL-H. Mechanisms of ventricular arrhythmogenesis in mice following targeted disruption of KCNE1 modelling long QT syndrome 5. J Physiol. 2007;578(Pt 1):99–114.

Tian X-L, Yong SL, Wan X, Wu L, Chung MK, Tchou PJ, Rosenbaum DS, van Wagoner DR, Kirsch GE, Wang Q. Mechanisms by which SCN5A mutation N1325S causes cardiac arrhythmias and sudden death in vivo. Cardiovasc Res. 2004;61(2):256–67.

Valentin JP, Hoffmann P, Clerck F, Hammond TG, Hondeghem L. Review of the predictive value of the Langendorff heart model (Screenit system) in assessing the proarrhythmic potential of drugs. J Pharmacol Toxicol Methods. 2004;49(3):171–81.

van der Werf C, Kannankeril PJ, Sacher F, Krahn AD, Viskin S, Leenhardt A, Shimizu W, Sumitomo N, Fish FA, Bhuiyan ZA, Willems AR, van der Veen MJ, Watanabe H, Laborderie J, Haissaguerre M, Knollmann BC, Wilde AA. Flecainide therapy reduces exercise-induced ventricular arrhythmias in patients with catecholaminergic polymorphic ventricular tachycardia. J Am Coll Cardiol. 2011;57(22):2244–54.

Wan E, Abrams J, Weinberg RL, Katchman AN, Bayne J, Zakharov SI, Yang L, Morrow JP, Garan H, Marx SO. Aberrant sodium influx causes cardiomyopathy and atrial fibrillation in mice. J Clin Invest. 2016;126(1):112–22.

Watanabe H, Chopra N, Laver D, Hwang HS, Davies SS, Roach DE, Duff HJ, Roden DM, Wilde AA, Knollmann BC. Flecainide prevents catecholaminergic polymorphic ventricular tachycardia in mice and humans. Nat Med. 2009;15(4):380–3.

Watanabe H, Yang T, Stroud DM, Lowe JS, Harris L, Atack TC, Wang DW, Hipkens SB, Leake B, Hall L, Kupershmidt S, Chopra N, Magnuson MA, Tanabe N, Knollmann BC, George AL Jr, Roden DM. Striking in vivo phenotype of a disease-associated human SCN5A mutation producing minimal changes in vitro. Circulation. 2011;124(9):1001–11.

Wehrens XH, Lehnart SE, Huang F, Vest JA, Reiken SR, Mohler PJ, Sun J, Guatimosim S, Song LS, Rosemblit N, D'Armiento JM, Napolitano C, Memmi M, Priori SG, Lederer WJ, Marks AR. FKBP12.6 deficiency and defective calcium release channel (ryanodine receptor) function linked to exercise-induced sudden cardiac death. Cell. 2003;113:829–40.

Wickenden AD, Lee P, Sah R, Huang Q, Fishman GI, Backx PH. Targeted expression of a dominant-negative K(v)4.2 K(+) channel subunit in the mouse heart. Circ Res. 1999;85(11):1067–76.

Wilde AA, Moss AJ, Kaufman ES, Shimizu W, Peterson DR, Benhorin J, Lopes C, Towbin JA, Spazzolini C, Crotti L, Zareba W, Goldenberg I, Kanters JK, Robinson JL, Qi M, Hofman N, Tester DJ, Bezzina CR, Alders M, Aiba T, Kamakura S, Miyamoto Y, Andrews ML, Mc Nitt S, Polonsky B, Schwartz PJ, Ackerman MJ. Clinical aspects of type 3 long-QT syndrome: an international multicenter study. Circulation. 2016;134(12):872–82.

Williams H, Kerr PM, Suleiman MS, Griffiths EJ. Differences in the calcium-handling response of isolated rat and Guinea-pig cardiomyocytes to metabolic inhibition: implications for cell damage. Exp Physiol. 2000;85(5):505–10.

Xu H, Barry DM, Li H, Brunet S, Guo W, Nerbonne JM. Attenuation of the slow component of delayed rectification, action potential prolongation, and triggered activity in mice expressing a dominant-negative Kv2 alpha subunit. Circ Res. 1999;85(7):623–33.

Zareba W, Sattari MN, Rosero S, Couderc JP, Moss AJ. Altered atrial, atrioventricular, and ventricular conduction in patients with the long QT syndrome caused by the DeltaKPQ SCN5A sodium channel gene mutation. Am J Cardiol. 2001;88(11):1311–4.

Zaritsky JJ, Eckman DM, Wellman GC, Nelson MT, Schwarz TL. Targeted disruption of Kir2.1 and Kir2.2 genes reveals the essential role of the inwardly rectifying K(+) current in K(+)-mediated vasodilation. Circ Res. 2000;87(2):160–6.

Zhang Y, Wu J, Jeevaratnam K, King JH, Guzadhur L, Ren X, Grace AA, Lei M, Huang CL, Fraser JA. Conduction slowing contributes to spontaneous ventricular arrhythmias in intrinsically active murine RyR2-P2328S hearts. J Cardiovasc Electrophysiol. 2013;24:210–8.

Zhao YT, Valdivia CR, Gurrola GB, Powers PP, Willis BC, Moss RL, Jalife J, Valdivia HH. Arrhythmogenesis in a catecholaminergic polymorphic ventricular tachycardia mutation that depresses ryanodine receptor function. Proc Natl Acad Sci USA. 2015;112:E1669–77.

Ziupa D, Beck J, Franke G, Perez Feliz S, Hartmann M, Koren G, Zehender M, Bode C, Brunner M, Odening KE. Pronounced effects of HERG-blockers E-4031 and erythromycin on APD, spatial APD dispersion and triangulation in transgenic long-QT type 1 rabbits. PLoS One. 2014;9(9): e107210.

Ziv O, Morales E, Song YK, Peng X, Odening KE, Buxton AE, Karma A, Koren G, Choi BR. Origin of complex behaviour of spatially discordant alternans in a transgenic rabbit model of type 2 long QT syndrome. J Physiol. 2009;587(Pt 19):4661–80.

Induced Pluripotent Stem Cell-Derived Cardiomyocytes: Towards Personalized Therapeutic Strategies?

16

Daniel Sinnecker and Alessandra Moretti

Abstract

Human induced pluripotent stem cells, which can be generated from somatic cells of healthy or diseased subjects, hold a great potential for research in the field of cardiac channelopathies. This review discusses how these cells can be used for disease modelling, as a safety pharmacology platform to assess the proarrhythmic potential of drug candidates, and for developing personalized therapeutic strategies by predicting individual drug responses as well as providing a system for patient-specific cardiac gene expression profiling and in vitro testing of patient-specific therapies.

16.1 Induced Pluripotent Stem Cells

Human pluripotent stem cells, such as embryonic stem cells (ES cells) derived from the inner cell mass of a blastocyst, provide intriguing possibilities for cardiovascular research (Fig. 16.1). On one hand, their capacity for indefinite self-renewal allows their unlimited propagation and amplification in the laboratory. On the other hand, their differentiation potential—pluripotency is defined as the ability to differentiate into any cell type of the adult organism—makes them especially appealing to researchers interested in myocardial biology, as primary human cardiomyocytes from human patients or healthy individuals can be obtained only in a limited number through invasive procedures and cannot be cultured in vitro for longer time periods.

D. Sinnecker (✉) · A. Moretti
Klinik und Poliklinik für Innere Medizin I, Klinikum rechts der Isar, Technical University of Munich, Munich, Germany

DZHK (German Centre for Cardiovascular Research)—Partner Site Munich Heart Alliance, Munich, Germany
e-mail: sinnecker@mytum.de

D. Thomas, C. A. Remme (eds.), *Channelopathies in Heart Disease*, Cardiac and Vascular Biology, https://doi.org/10.1007/978-3-319-77812-9_16

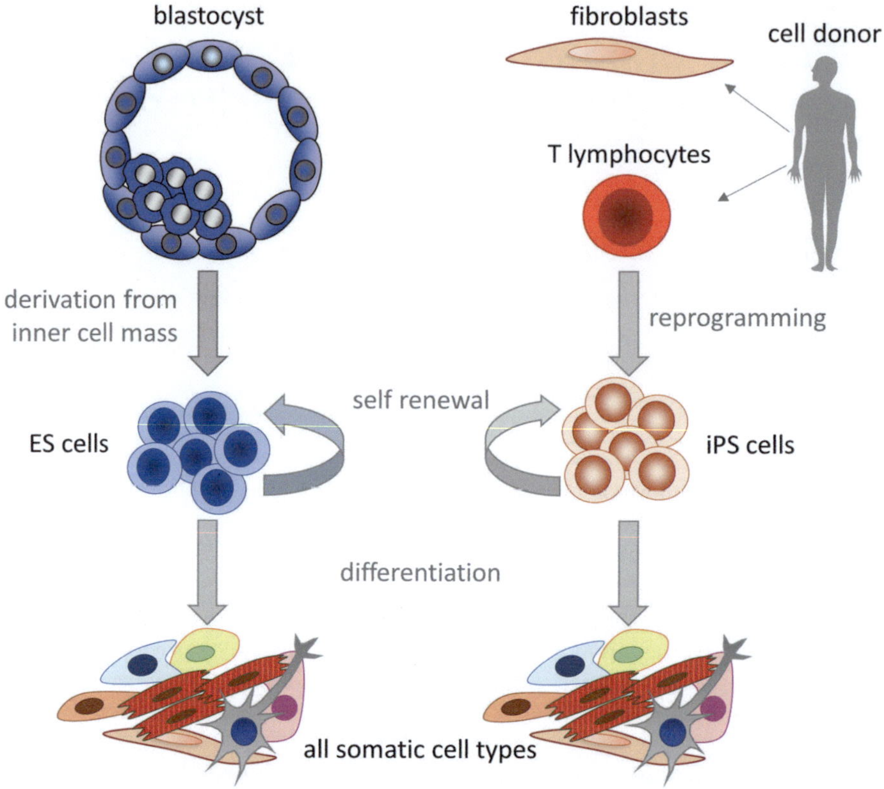

Fig. 16.1 Pluripotent stem cells as an inexhaustible source of cardiomyocytes and other cell types for cardiovascular research. ES cells, embryonic stem cells; iPS cells, induced pluripotent stem cells

However, ethical considerations and consequent regulatory as well as funding restrictions have interfered with a widespread use of these cells.

In 2006, Shinya Yamanaka and colleagues demonstrated that murine fibroblasts could be "reprogrammed" to a pluripotent, ES cell-like state by forced expression of a specific cocktail of transcription factors (Takahashi and Yamanaka 2006). These cells, which were termed "induced pluripotent stem cells" (iPSCs), were not only morphologically similar to murine ES cells but exhibited crucial properties of pluripotent stem cells including the capacity to self-renew and the differentiation potential to cell types representing all three germ layers (see Fig. 16.1), which was demonstrated by teratoma formation after injection into immunocompromised mice. Later, it was shown that iPSCs could also contribute to an embryo by chimera formation and even be transmitted to the next generation through the germ line (Okita et al. 2007).

Only one year after the first published report on murine iPSCs, the method was successfully transferred to human cells independently in two laboratories (Takahashi et al. 2007; Yu et al. 2007). This achievement has paved the way for disease modelling studies using patient-specific iPSC lines. The basic principle of such disease modelling studies is to generate iPSC lines from a patient affected by a genetically caused disease and to then differentiate these patient-specific iPSCs into the disease-relevant somatic cell types (e.g. cardiomyocytes). The somatic cells, which carry the patient's genetic background, are then investigated in vitro to learn more about the pathophysiology of the disease or to investigate disease-specific therapies.

16.2 Modelling Channelopathies with Patient-Specific iPSC Lines

The first cardiac diseases that have been successfully modelled using patient-specific iPSCs were channelopathies such as the long-QT syndrome (Moretti et al. 2010). The decision of several independent research groups to concentrate on this group of diseases for their first iPSC-based disease modelling endeavours may have been based on a number of reasons. First, channelopathies represent a relevant clinical problem affecting many patients. For example, the prevalence of congenital long-QT syndromes may be as high as 1 of 2000 newborn infants (Schwartz et al. 2009). Second, important aspects of the pathophysiology—such as the prolongation of the action potential in long-QT syndromes—develop in a cell-autonomous manner, making this group of diseases amenable to in vitro modelling using single iPSC-derived cardiomyocytes. Third, electrophysiological methods such as patch clamp recordings allow the investigation of the disease phenotype at the molecular level. Finally, due to the profound interspecies differences in cardiac electrophysiology, genetically defined animal models for channelopathies have been often inconclusive (London 2001; Nerbonne et al. 2001), making the human cell-based iPSC approach particularly valuable.

The first channelopathy successfully modelled with patient-specific iPSCs was the long-QT syndrome type 1 (LQT1), caused by a *KCNQ1* R190Q mutation (Moretti et al. 2010). This study, which will be briefly described in the following paragraphs, can be seen as exemplary for several disease modelling studies that were subsequently published, using similar methodology to investigate other channelopathies, e.g. LQT1 caused by other *KCNQ1* mutations (Egashira et al. 2012; Ma et al. 2015), Jervell and Lange-Nielsen syndrome (Zhang et al. 2014), LQT2 (Itzhaki et al. 2011; Matsa et al. 2011; Lahti et al. 2012; Bellin et al. 2013; Jouni et al. 2015), LQT3 (Fatima et al. 2013; Ma et al. 2013; Terrenoire et al. 2013; Malan et al. 2016), LQT3/Brugada overlap syndrome (Davis et al. 2012), Timothy syndrome (LQT8) (Yazawa et al. 2011), and catecholaminergic polymorphic ventricular tachycardia (CPVT1) (Fatima et al. 2011; Jung et al. 2012; Itzhaki et al. 2012; Kujala et al. 2012; Zhang et al. 2013; Di Pasquale et al. 2013) and CPVT2 (Novak et al. 2012, 2015).

To generate a patient-specific iPSC model of LQT1 (Moretti et al. 2010), skin biopsies were obtained from two family members affected by LQT1 caused by an R190Q mutation in the *KCNQ1* gene. This gene encodes the alpha subunit of the cardiac ion channel responsible for one important repolarizing potassium current, the slow component of the delayed rectifier current I_{Ks}. The electrocardiograms (ECGs) of both patients showed a prolonged QT interval. Skin fibroblasts were isolated from the biopsies, and, as a control, fibroblasts from two healthy subjects were obtained. Reprogramming to pluripotency was achieved by retroviral transduction of the fibroblasts with the classical Yamanaka factors OCT3/4, SOX2, KLF4, and c-MYC.

To induce differentiation, the iPSC lines were aggregated to form so-called embryoid bodies, in which differentiation into cell types representing all three germ layers occurs spontaneously. After around 20 days, cardiomyocyte-containing areas of the embryoid bodies started to exhibit spontaneous contractions. These spontaneously contracting areas were microdissected and cultured further to promote differentiation. For physiological characterization of the iPSC-derived cardiomyocytes, the cells were then enzymatically dissociated to single cells and investigated using whole-cell patch clamp electrophysiology.

Important aspects of the clinical phenotype of the LQT1 patients were recapitulated in the patient-specific iPSC-derived cardiomyocytes. Action potentials in patient-specific ventricular cardiomyocytes were significantly longer than in controls, as expected given the prolonged QT interval of the patients. Most importantly, the shortening of the cardiac ventricular action potentials in response to catecholamine stimulation or increased pacing rates—which is thought to be mediated by activation of the I_{Ks} current—was blunted in the patient cells (Fig. 16.2a). Current measurements showed that the I_{Ks} current was reduced by 75% in patient cardiomyocytes as compared to control cells (Fig. 16.2b). Additional experiments showed that this was due to a dominant-negative effect of the R190Q mutation: Trafficking of mutated KCNQ1 from the endoplasmic reticulum to the plasma membrane was impaired, resulting in a reduction of the number of functional ion channels available in the plasma membrane. Since the I_{Ks} channel contains a tetramer of four KCNQ1 subunits, the incorporation of mutated subunits into tetramers containing healthy subunits also interferes with the trafficking of these healthy channel subunits, resulting in a current reduction by more than 50% in the patients heterozygous for the R190Q mutation.

Besides action potential prolongation, another typical hallmark of the clinical phenotype of long-QT syndromes are arrhythmias, which are typically triggered by early afterdepolarizations (EADs) of the cell membrane, i.e. depolarizations that occur during phase 2 or 3 of the action potential. Such EADs were also observed in the patient-specific iPSC-derived cardiomyocytes, especially under conditions of adrenergic stimulation. This effect was ameliorated by application of a beta blocker, a drug clinically used to suppress arrhythmias in LQT1.

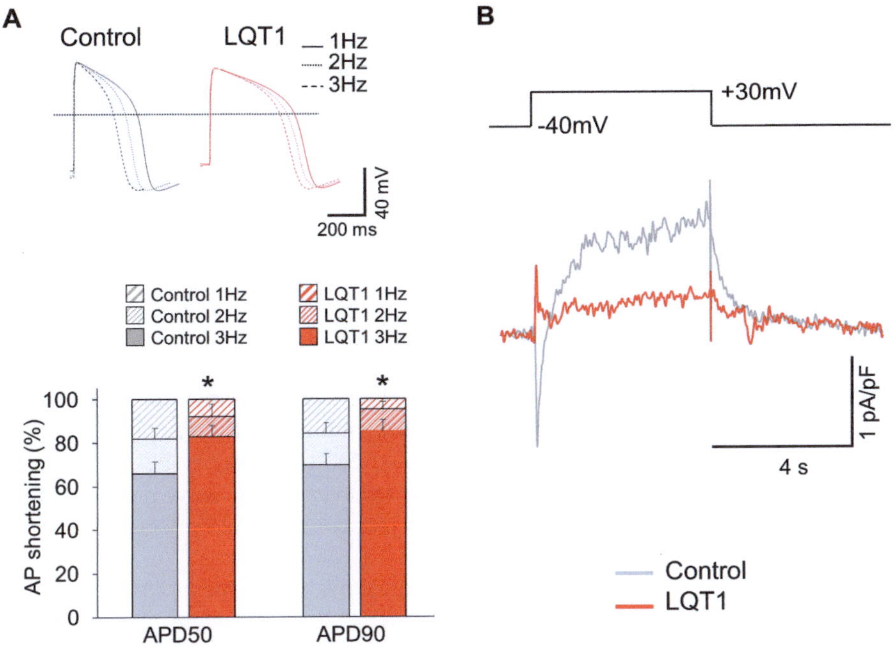

Fig. 16.2 Modelling long-QT syndrome type 1 with patient-specific induced pluripotent stem cells. Results of patch clamp experiments performed on ventricular cardiomyocytes generated from a patient-derived induced pluripotent stem cell line carrying an R190Q *KCNQ1* mutation (LQT1) or from a control line derived from a healthy individual (Control) are shown. (**a**) Action potentials recorded at three different pacing frequencies (1 Hz, 2 Hz, and 3 Hz) are shown. The bar graph shows the relative shortening of the action potential duration at 50% and at 90% of repolarization (APD50 and APD90, respectively) at higher pacing rates, compared to 1 Hz pacing. (**b**) Tracings of the I_{Ks} current (the slow component of the delayed rectifier potassium current), measured as the E4031-sensitive component of the current elicited by a voltage step from −40 mV to +30 mV. Based on data published in Moretti et al. (2010)

16.3 Technical Developments in Cardiac Disease Modelling with Patient-Specific iPSC Lines

Since the first studies, much progress has been made in different areas relevant for iPSC-based cardiac disease modelling. These areas comprise reprogramming methods, protocols for differentiation of iPSCs to cardiomyocytes, and novel phenotypic assays to characterize the iPSC-derived cardiomyocytes.

The conventional method of iPSC generation relies on retroviral gene transfer in order to achieve expression of the reprogramming factors in the cells to be reprogrammed (Takahashi and Yamanaka 2006). Since retroviruses stably integrate into the genome of the host cells, this bears the potential of disrupting endogenous genes or gene regulatory regions, which might result in unexpected behaviour of the iPSCs and their progeny. Therefore, many efforts have been made to develop “footprint-free” reprogramming techniques, which ideally leave the genetic background of the cells unaltered.

Footprint-free reprogramming has been achieved by using nonintegrating viral vectors, such as adenoviruses (Stadtfeld et al. 2008) or Sendai viruses (Fusaki et al. 2009), or virus-free DNA vectors such as plasmids (Okita et al. 2008), episomal vectors (Yu et al. 2009), or transposons that can be excised from the genome after successful reprogramming (Woltjen et al. 2009; Kaji et al. 2009). Alternative approaches such as the use of synthetic modified messenger RNAs (Warren et al. 2010), the application of the reprogramming factors as recombinant proteins (Kim et al. 2009), or even a completely chemically defined strategy using a cocktail of small molecules (Hou et al. 2013) have been reported.

The progress in reprogramming technology has additional benefits besides the ability to generate genetically unaltered iPSC lines. For example, the increasingly used Sendai virus-based approach allows reprogramming of T lymphocytes from peripheral blood (Seki et al. 2010; Orban et al. 2015), which makes iPSCs even more easily available, given that drawing a blood sample is a less invasive procedure than performing a skin biopsy, which also means that consent can be usually obtained more easily.

Another area in which much progress has been made over the last 10 years is the development of differentiation protocols to generate cardiomyocytes from iPSCs. A simple method used in many of the initial disease modelling studies is the above-described “embryoid body” technique, which relies on spontaneous differentiation. A better understanding of the regulation of cardiac differentiation has resulted in optimized protocols based on the sequential application of developmental cues that artificially direct the differentiation of iPSCs towards the cardiomyocytic lineage (Kattman et al. 2011; Burridge et al. 2012, 2014).

Cardiac differentiation strategies typically result in the generation of a mixture of all subtypes of cardiomyocytes, i.e. ventricular, atrial, and nodal cells. Thus, another area of intensive investigation is the development of subtype-specific differentiation protocols. Efforts in this area have led to the identification of specific culture conditions capable of directing differentiation more specifically towards the atrial, ventricular, or nodal lineage (Zhang et al. 2011; Karakikes et al. 2014; Devalla et al. 2015; Protze et al. 2017).

Another technical aspect worth considering is the choice of the phenotypic assays used to investigate the iPSC-derived cardiomyocytes. In the field of channelopathies, patch clamp electrophysiology can be still considered the gold standard, as it allows the assessment of different aspects of cellular electrophysiology, from single-channel currents to action potential morphology and arrhythmias, with a high precision. One important aspect of this method is that, by current clamp recordings of action potentials, it permits the subtype identification of each investigated cardiomyocyte based on the known subtype-specific action potential characteristics. This may compensate for the lack of subtype specificity of the current differentiation protocols. However, a major limitation of this method is its limited throughput.

In recent times, optical methods have been increasingly used as a higher-throughput alternative to image electrical activity in iPSC-derived cardiomyocytes.

These methods rely either on voltage-sensitive small-molecule dyes (Lee et al. 2012; Lopez-Izquierdo et al. 2014; Kim et al. 2015) or on genetically encoded voltage sensors (Chang Liao et al. 2015; Shinnawi et al. 2015; Song et al. 2015; Chen et al. 2017). While the information that can be obtained from such optical sensors is limited compared to patch clamp electrophysiology (e.g. it is not possible to measure actual transmembrane potentials, and the ability to track fast processes like the action potential upstroke is hampered by the intrinsic kinetics of the optical sensors), the approach provides an interesting complementation to standard methodology and has some additional advantages besides the increased throughput, like the noninvasiveness that allows to perform serial measurements in the same cell over extended time periods.

By expressing a voltage-sensitive fluorescent protein under the control of cardiomyocyte subtype-specific promoters, it is even possible to perform subtype-specific action potential recordings in ventricular, atrial, or nodal iPSC-derived cardiomyocytes (Chen et al. 2017).

16.4 Important Limitations of iPSC-Derived Cardiomyocytes

Despite these technical advances, there are limitations of iPSC-derived cardiomyocytes that should be considered in the planning and in the interpretation of experiments using them as a model system for channelopathies. Cardiomyocytes generated from iPSCs are more similar to foetal or neonatal than to adult cardiomyocytes. This immature phenotype is not limited to cellular and subcellular morphology [e.g. rounded rather than rod-shaped cell morphology, less organization and alignment of myofibrils and lower numbers of mitochondria as compared to adult cardiomyocytes, absence of T-tubules (Gherghiceanu et al. 2011)] but also pertains to physiological parameters.

For example, ventricular-like iPSC-derived cardiomyocytes frequently show automaticity (Zhang et al. 2009), while isolated adult ventricular cardiomyocytes are electrically silent unless stimulated. The maximum diastolic potential of hiPSC-derived cardiomyocyte is typically more positive than that of adult cardiomyocytes (Ma et al. 2011; Honda et al. 2011). One explanation for these peculiarities of iPSC-derived cardiomyocytes as compared to adult human cardiomyocytes is a reduced level or even absence of the inwardly rectifying potassium current I_{K1} (Ma et al. 2011; Doss et al. 2012), a potassium current that plays an important role in setting the resting membrane potential to negative values.

Several approaches to address this issue have been proposed. Selection of quiescent rather than spontaneously beating cardiomyocytes for further analysis may be a way to investigate more mature cells with less depolarized action potentials (Davis et al. 2012). Overexpression of the ion channel subunit responsible for I_{K1} (Vaidyanathan et al. 2016) or artificial addition of simulated I_{K1} in patch clamp studies using the dynamic clamp technique (Meijer van Putten et al. 2015; Bartolucci et al. 2015; Rocchetti et al. 2017) are alternative approaches to make the action potential of iPSC-derived cardiomyocytes more similar to that of adult cardiomyocytes.

16.5 Safety Pharmacology: A First "Real-World" Application of iPSC-Derived Cardiomyocytes?

Already in the first reports of the generation of human iPSCs, it was speculated that these cells bear a great potential for drug development and safety pharmacology (Takahashi et al. 2007; Zaehres and Schöler 2007). It is quite likely that this approach, which has been thoroughly pursued over the last decade, will be the first broad application of iPSC-derived cardiomyocytes outside of academic research laboratories.

Safety pharmacology in the field of cardiac arrhythmias suffers from the fact that the profound species differences in cardiac electrophysiology between laboratory animals such as mice and human patients result in a limited validity of animal models. Expectations were that human iPSC-derived cardiomyocytes could be used as an alternative model system that might more closely reflect the electrophysiology of the human heart. Indeed, by now a large body of evidence exists showing that cardiomyocytes generated from human iPSCs exhibit the expected responses to cardiovascular drugs, such as beta receptor agonists or beta blockers, or drugs with known cardiovascular side effects like QT interval prolongation (Bellin et al. 2012; Sinnecker et al. 2014).

Drug-induced QT interval prolongation is a major concern in cardiac safety pharmacology. A large number of drugs developed for cardiac and noncardiac conditions have been shown to prolong the QT interval in the ECG, which is associated with torsades de pointes, a potentially fatal form of ventricular tachycardia. After a number of withdrawals of already-approved medications from the market due to QT interval prolongation, a new preclinical and clinical screening paradigm has been installed as an integral component of the drug development process. This paradigm, which is formulated in the International Conference on Harmonization (ICH) guidelines S7B and E14 (International Conference on Harmonization 2005a, b), comprises preclinical assays in which drug actions on overexpressed ion channels (most importantly, the hERG channel) and on cardiomyocytes from different laboratory animal species are investigated, as well as clinical studies (thorough QT studies or TQT studies) in which the medication is administered to healthy individuals whose QT interval is then monitored.

Since the adoption of this screening paradigm, there have been no more drug withdrawals due to QT interval prolongation, indicating a sufficiently high sensitivity of these assays. However, there have been concerns that the specificity is not high enough, which would mean that it is possible that substances which might never cause torsades de pointes are identified as potentially dangerous and, thus, not developed further. Indeed, several now widely used drugs which do not cause torsades would have never reached the market if they would have undergone this safety assessment (Sager et al. 2014).

Human iPSC-derived cardiomyocytes, which can be used as a platform for safety pharmacology assays (Liang et al. 2013; Mehta et al. 2013), might more closely reflect the drug actions in patients compared to other preclinical models and, therefore, could represent one way to overcome the known limitation of the currently used screening standards. Indeed, the use of iPSC-derived cardiomyocytes is one integral part of a recently proposed new proarrhythmia screening paradigm termed the *Comprehensive In Vitro Proarrhythmia Assay (CiPA)* (Sager et al. 2014), which is now being validated in multicentre studies and more and more adopted by pharmaceutical companies (Colatsky et al. 2016).

16.6 Towards Personalized Therapeutic Strategies

It is increasingly recognized that the success of therapeutic interventions may critically depend on patient-specific factors, such as the genetic background, and that individualized therapies might be more effective than "one-size-fits-all" pharmacotherapy in reducing disease burden in various areas of medicine. This notion, combined with recent advances in next-generation sequencing and the ever-growing armamentarium of *-omics* methods, has culminated in the proclamation of the "age of personalized medicine" (Vaidyanathan 2012) and the launching of the *Precision Medicine Initiative* (Collins and Varmus 2015). So far, the most remarkable advances in personalized therapeutic strategies have been achieved in oncology, owing on one hand to the realization that different tumours may carry different genetic aberrations that may be targeted by different pathway-specific drugs and on the other hand to the practical consideration that tumour specimens that can be analysed by molecular methods are easily and routinely obtained from patients. Cardiomyocytes generated from patient-specific iPSCs bear the potential to be used in a similar way to elucidate specific molecular pathways involved in the patient's disease and to predict the efficacy of patient-specific therapies for cardiac diseases.

If patient-specific iPSC-derived cardiomyocytes are to be used to predict whether a specific patient will respond to a medication, it has to be established to which extent the drug response of the stem cell-derived cardiomyocytes indeed reflects the reaction of the patient's heart to the same medication. The most extensive study so far addressing this question (Stillitano et al. 2017) was based on a cohort of 92 healthy subjects who were challenged with a single 80 mg oral dose of sotalol, a drug known to prolong the QT interval in susceptible patients. The change of the rate-corrected QT interval (QTc) due to this pharmacological challenge was determined for every subject, and the 20 subjects with the most extreme reaction (i.e. the 10 subjects with the smallest and the 10 subjects with the largest QTc change) underwent skin biopsies from which iPSC lines were generated, which was successful for 17 individuals. The iPSCs were differentiated to cardiomyocytes, and microelectrode arrays were used to record the field potential duration (FPD), which is an in vitro measure for the duration of the cardiac action potential that is analogous to the QT interval in the ECG. The effect of different doses of sotalol on this parameter was assessed. Indeed, in the iPSC lines derived from the individual with the largest sotalol-induced QTc change, the sotalol-induced FPD change was significantly bigger than in the lines derived from the subjects with the smallest QTc change. Moreover, sotalol-induced arrhythmias were common in iPSC lines derived from the individuals with the largest QTc change but not in those derived from the subjects with the smallest QTc change. While these results are encouraging, it remains to be seen whether a similar correlation between human subjects and their corresponding iPSC lines will be also found for other drugs and physiological parameters other than the QT interval.

Thanks to advances in DNA sequencing technology, the generation of genome-wide expression data from tissue samples has become easier and more affordable. Concurrently, bioinformatic tools to analyse such data become more and more sophisticated and allow to infer which pathways are up- or downregulated in a sample based on gene expression data. Cardiomyocytes generated from patient-specific iPSCs might represent a platform to analyse cardiac gene expression without the need of obtaining heart tissue from patients. A recent study (Matsa et al. 2016) indicates that gene expression data from iPSC-derived cardiomyocytes indeed provides meaningful data and might be even used to predict toxicity of certain drugs. The authors of this study generated iPSC lines from fibroblasts of five healthy female subjects. These iPSC lines were differentiated into cardiomyocytes, RNA was isolated from these cardiomyocytes, and genome-wide gene expression analysis was performed. Global gene expression patterns obtained by principal component analysis and unsupervised hierarchical clustering clearly showed that gene expression signatures of cardiomyocytes generated from different iPSC clones from the same subject were more closely related to each other than to cardiomyocytes generated from different subjects. Notably, those genes that showed the highest amount of inter-subject variation showed similar expression levels in different clones from the same individuum. Then, iPSC lines from three patients with dilated cardiomyopathy, of whom heart tissue specimens had been obtained during heart transplantation or left ventricular assist device placement, were generated. Gene expression signatures obtained from iPSC-derived cardiomyocytes were compared with those obtained from native heart tissue samples from the same patients. This analysis revealed that, albeit the expression levels of several cardiac maturation markers were lower in iPSC-derived cardiomyocytes, gene expression patterns in hiPSC-derived cardiomyocytes to some extent mimicked those in heart specimens. Encouraged by these results, the authors performed toxicology analysis using a commercially available tool that predicts the overall likelihood of adverse cellular function based on the expression levels of interacting genes involved in certain pathways. This analysis predicted inter-subject differences in the susceptibility to oxidative stress mediated through the nuclear factor erythroid 2-like 2 (NEE2L2 or NRF2) pathway. By using the drugs tacrolimus and rosiglitazone, the authors of the study were able to recapitulate the predicted differences in susceptibility to cardiotoxicity in cardiomyocytes generated from the different iPSC lines.

Innovative therapies for genetically caused heart diseases may be patient- or mutation-specific. Human iPSC-derived cardiomyocytes might represent an ideal in vitro system to evaluate such mutation-specific therapies. For example, it has been demonstrated that an allele-specific small interfering RNA (siRNA) could rescue the action potential prolongation caused by a *KCNH2* mutation associated with long-QT syndrome 2 (Matsa et al. 2014). The *KCNH2* gene encodes the alpha subunit of the hERG potassium channel, and the mutated KCNH2 subunits are not glycosylated properly and are, thus, not properly transported to the plasma membrane. Moreover, the hERG channel contains four KCNH2 subunits, and the mutated subunits also interfere with intracellular trafficking of healthy subunits (the patient being a heterozygous mutation carrier), resulting in a dominant-negative effect. It was demonstrated using patient-specific iPSC-derived cardiomyocytes that treatment with a siRNA that specifically knocked down the mutated KCNH2 allele could overcome this dominant-negative mechanism, resulting in a significant reduction of the action potential duration.

16.7 Conclusion

Human induced pluripotent stem cells provide intriguing possibilities for cardiovascular research (Fig. 16.3). Patient-specific iPSC models have been widely used to investigate patho-mechanisms and possible treatments of cardiac channelopathies. Moreover, these cells hold great promise as a safety pharmacology platform to assess

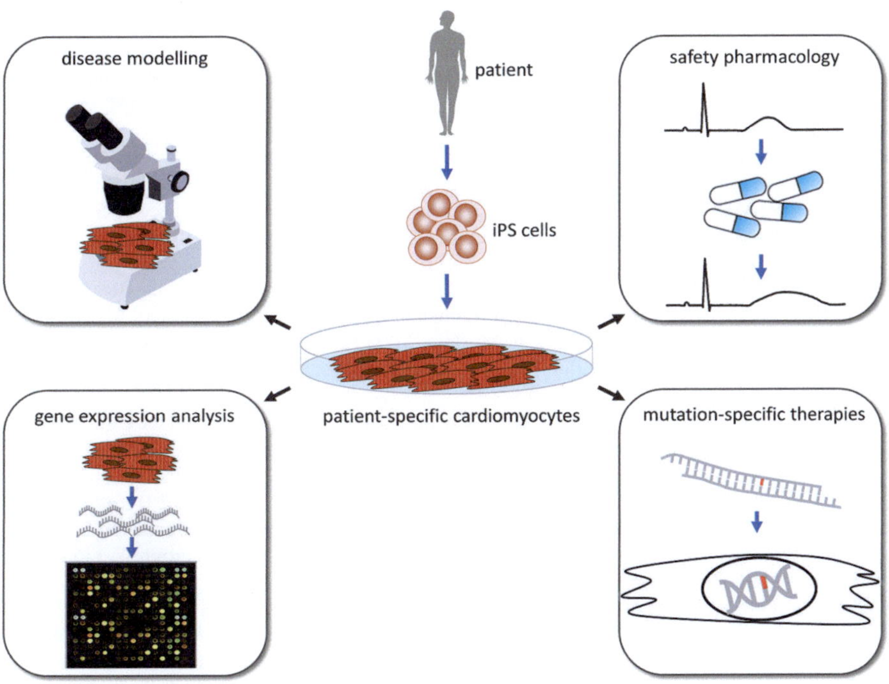

Fig. 16.3 Patient-specific induced pluripotent stem cell-derived cardiomyocytes as a basis for personalized diagnostic and therapeutic strategies

the proarrhythmic potential of drug candidates, which might become the first broad application of iPSC-derived cardiomyocytes outside of academic research laboratories. Finally, these cells might be instrumental in developing personalized therapeutic strategies for heart disease by predicting individual drug responses, by providing a platform for patient-specific cardiac gene expression profiling, and as an in vitro test system for patient-specific therapies.

Compliance with Ethical Standards

Sources of Funding The authors have been supported by the German Research Foundation (Mo 2217/1–1, Si 1747/1–1, Research Unit 923), the German Centre for Cardiovascular Research (DZHK), partner site Munich Heart Alliance, and the Else Kröner-Fresenius Stiftung.

Conflict of Interest The authors declared that they have no conflict of interest.

References

Bartolucci C, Altomare C, Bennati M, Furini S, Zaza A, Severi S. Combined action potential- and dynamic-clamp for accurate computational modelling of the cardiac IKr current. J Mol Cell Cardiol. 2015;79:187–94.

Bellin M, Marchetto MC, Gage FH, Mummery CL. Induced pluripotent stem cells: the new patient? Nat Rev Mol Cell Biol. 2012;13:713–26.

Bellin M, Casini S, Davis RP, D'Aniello C, Haas J, Ward-van Oostwaard D, Tertoolen LG, Jung CB, Elliott DA, Welling A, Laugwitz KL, Moretti A, Mummery CL. Isogenic human pluripotent stem cell pairs reveal the role of a KCNH2 mutation in long-QT syndrome. EMBO J. 2013;32:3161–75.

Burridge PW, Keller G, Gold JD, Wu JC. Production of de novo cardiomyocytes: human pluripotent stem cell differentiation and direct reprogramming. Cell Stem Cell. 2012;10:16–28.

Burridge PW, Matsa E, Shukla P, Lin ZC, Churko JM, Ebert AD, Lan F, Diecke S, Huber B, Mordwinkin NM, Plews JR, Abilez OJ, Cui B, Gold JD, Wu JC. Chemically defined generation of human cardiomyocytes. Nat Methods. 2014;11:855–60.

Chang Liao ML, de Boer TP, Mutoh H, Raad N, Richter C, Wagner E, Downie BR, Unsöld B, Arooj I, Streckfuss-Bömeke K, Döker S, Luther S, Guan K, Wagner S, Lehnart SE, Maier LS, Stühmer W, Wettwer E, van Veen T, Morlock MM, Knöpfel T, Zimmermann WH. Sensing cardiac electrical activity with a cardiac myocyte-targeted optogenetic voltage indicator. Circ Res. 2015;117:401–12.

Chen Z, Xian W, Bellin M, Dorn T, Tian Q, Goedel A, Dreizehnter L, Schneider CM, Ward-van Oostwaard D, Ng JK, Hinkel R, Pane LS, Mummery CL, Lipp P, Moretti A, Laugwitz KL, Sinnecker D. Subtype-specific promoter-driven action potential imaging for precise disease modelling and drug testing in hiPSC-derived cardiomyocytes. Eur Heart J. 2017;38:292–301.

Colatsky T, Fermini B, Gintant G, Pierson JB, Sager P, Sekino Y, Strauss DG, Stockbridge N. The comprehensive in vitro proarrhythmia assay (CiPA) initiative—update on progress. J Pharmacol Toxicol Methods. 2016;81:15–20.

Collins FS, Varmus H. A new initiative on precision medicine. N Engl J Med. 2015;372:793–5.

Davis RP, Casini S, van den Berg CW, Hoekstra M, Remme CA, Dambrot C, Salvatori D, Oostwaard DW, Wilde AA, Bezzina CR, Verkerk AO, Freund C, Mummery CL. Cardiomyocytes derived from pluripotent stem cells recapitulate electrophysiological characteristics of an overlap syndrome of cardiac sodium channel disease. Circulation. 2012;125: 3079–91.

Devalla HD, Schwach V, Ford JW, Milnes JT, El-Haou S, Jackson C, Gkatzis K, Elliott DA, Chuva de Sousa Lopes SM, Mummery CL, Verkerk AO, Passier R. Atrial-like cardiomyocytes from human pluripotent stem cells are a robust preclinical model for assessing atrial-selective pharmacology. EMBO Mol Med. 2015;7:394–410.

Di Pasquale E, Lodola F, Miragoli M, Denegri M, Avelino-Cruz JE, Buonocore M, Nakahama H, Portararo P, Bloise R, Napolitano C, Condorelli G, Priori SG. CaMKII inhibition rectifies arrhythmic phenotype in a patient-specific model of catecholaminergic polymorphic ventricular tachycardia. Cell Death Dis. 2013;4:e843.

Doss MX, Di Diego JM, Goodrow RJ, Wu Y, Cordeiro JM, Nesterenko VV, Barajas-Martínez H, Hu D, Urrutia J, Desai M, Treat JA, Sachinidis A, Antzelevitch C. Maximum diastolic potential of human induced pluripotent stem cell-derived cardiomyocytes depends critically on I(Kr). PLoS One. 2012;7:e40288.

Egashira T, Yuasa S, Suzuki T, Aizawa Y, Yamakawa H, Matsuhashi T, Ohno Y, Tohyama S, Okata S, Seki T, Kuroda Y, Yae K, Hashimoto H, Tanaka T, Hattori F, Sato T, Miyoshi S, Takatsuki S, Murata M, Kurokawa J, Furukawa T, Makita N, Aiba T, Shimizu W, Horie M, Kamiya K, Kodama I, Ogawa S, Fukuda K. Disease characterization using LQTS-specific induced pluripotent stem cells. Cardiovasc Res. 2012;95:419–29.

Fatima A, Xu G, Shao K, Papadopoulos S, Lehmann M, Arnáiz-Cot JJ, Rosa AO, Nguemo F, Matzkies M, Dittmann S, Stone SL, Linke M, Zechner U, Beyer V, Hennies HC, Rosenkranz S, Klauke B, Parwani AS, Haverkamp W, Pfitzer G, Farr M, Cleemann L, Morad M, Milting H, Hescheler J, Saric T. In vitro modeling of ryanodine receptor 2 dysfunction using human induced pluripotent stem cells. Cell Physiol Biochem Int J Exp Cell Physiol Biochem Pharmacol. 2011;28:579–92.

Fatima A, Kaifeng S, Dittmann S, Xu G, Gupta MK, Linke M, Zechner U, Nguemo F, Milting H, Farr M, Hescheler J, Šarić T. The disease-specific phenotype in cardiomyocytes derived from induced pluripotent stem cells of two long QT syndrome type 3 patients. PLoS One. 2013;8: e83005.

Fusaki N, Ban H, Nishiyama A, Saeki K, Hasegawa M. Efficient induction of transgene-free human pluripotent stem cells using a vector based on Sendai virus, an RNA virus that does not integrate into the host genome. Proc Jpn Acad Ser B Phys Biol Sci. 2009;85:348–62.

Gherghiceanu M, Barad L, Novak A, Reiter I, Itskovitz-Eldor J, Binah O, Popescu LM. Cardiomyocytes derived from human embryonic and induced pluripotent stem cells: comparative ultrastructure. J Cell Mol Med. 2011;15:2539–51.

Honda M, Kiyokawa J, Tabo M, Inoue T. Electrophysiological characterization of cardiomyocytes derived from human induced pluripotent stem cells. J Pharmacol Sci. 2011;117:149–59.

Hou P, Li Y, Zhang X, Liu C, Guan J, Li H, Zhao T, Ye J, Yang W, Liu K, Ge J, Xu J, Zhang Q, Zhao Y, Deng H. Pluripotent stem cells induced from mouse somatic cells by small-molecule compounds. Science. 2013;341:651–4.

International Conference on Harmonization of Technical Requirements for Registration of Pharmaceuticals for Human Use. The non-clinical evaluation of the potential for delayed ventricular repolarization (QT interval prolongation) by human pharmaceuticals (S7B). 2005a. http://www.ich.org/fileadmin/Public_Web_Site/ICH_Products/Guidelines/Safety/S7B/Step4/S7B_Guideline.pdf

International Conference on Harmonization of Technical Requirements for Registration of Pharmaceuticals for Human Use. The clinical evaluation of QT/QTc interval prolongation and proarrhythmic potential for non-antiarrhythmic drugs (E14). 2005b. http://www.ich.org/fileadmin/Public_Web_Site/ICH_Products/Guidelines/Efficacy/E14/E14_Guideline.pdf

Itzhaki I, Maizels L, Huber I, Zwi-Dantsis L, Caspi O, Winterstern A, Feldman O, Gepstein A, Arbel G, Hammerman H, Boulos M, Gepstein L. Modelling the long QT syndrome with induced pluripotent stem cells. Nature. 2011;471:225–9.

Itzhaki I, Maizels L, Huber I, Gepstein A, Arbel G, Caspi O, Miller L, Belhassen B, Nof E, Glikson M, Gepstein L. Modeling of catecholaminergic polymorphic ventricular tachycardia with patient-specific human-induced pluripotent stem cells. J Am Coll Cardiol. 2012;60:990–1000.

Jouni M, Si-Tayeb K, Es-Salah-Lamoureux Z, Latypova X, Champon B, Caillaud A, Rungoat A, Charpentier F, Loussouarn G, Baró I, Zibara K, Lemarchand P, Gaborit N. Toward personalized medicine: using cardiomyocytes differentiated from urine-derived pluripotent stem cells to recapitulate electrophysiological characteristics of type 2 long QT syndrome. J Am Heart Assoc. 2015;4:e002159.

Jung CB, Moretti A, Mederos y Schnitzler M, Iop L, Storch U, Bellin M, Dorn T, Ruppenthal S, Pfeiffer S, Goedel A, Dirschinger RJ, Seyfarth M, Lam JT, Sinnecker D, Gudermann T, Lipp P, Laugwitz KL. Dantrolene rescues arrhythmogenic RYR2 defect in a patient-specific stem cell model of catecholaminergic polymorphic ventricular tachycardia. EMBO Mol Med. 2012;4: 180–91.

Kaji K, Norrby K, Paca A, Mileikovsky M, Mohseni P, Woltjen K. Virus-free induction of pluripotency and subsequent excision of reprogramming factors. Nature. 2009;458:771–5.

Karakikes I, Senyei GD, Hansen J, Kong CW, Azeloglu EU, Stillitano F, Lieu DK, Wang J, Ren L, Hulot JS, Iyengar R, Li RA, Hajjar RJ. Small molecule-mediated directed differentiation of human embryonic stem cells toward ventricular cardiomyocytes. Stem Cells Transl Med. 2014;3:18–31.

Kattman SJ, Witty AD, Gagliardi M, Dubois NC, Niapour M, Hotta A, Ellis J, Keller G. Stage-specific optimization of activin/nodal and BMP signaling promotes cardiac differentiation of mouse and human pluripotent stem cell lines. Cell Stem Cell. 2011;8:228–40.

Kim D, Kim CH, Moon JI, Chung YG, Chang MY, Han BS, Ko S, Yang E, Cha KY, Lanza R, Kim KS. Generation of human induced pluripotent stem cells by direct delivery of reprogramming proteins. Cell Stem Cell. 2009;4(6):472.

Kim JJ, Yang L, Lin B, Zhu X, Sun B, Kaplan AD, Bett GCL, Rasmusson RL, London B, Salama G. Mechanism of automaticity in cardiomyocytes derived from human induced pluripotent stem cells. J Mol Cell Cardiol. 2015;81:81–93.

Kujala K, Paavola J, Lahti A, Larsson K, Pekkanen-Mattila M, Viitasalo M, Lahtinen AM, Toivonen L, Kontula K, Swan H, Laine M, Silvennoinen O, Aalto-Setälä K. Cell model of catecholaminergic polymorphic ventricular tachycardia reveals early and delayed after-depolarizations. PLoS One. 2012;7:e44660.

Lahti AL, Kujala VJ, Chapman H, Koivisto A-P, Pekkanen-Mattila M, Kerkelä E, Hyttinen J, Kontula K, Swan H, Conklin BR, Yamanaka S, Silvennoinen O, Aalto-Setälä K. Model for long QT syndrome type 2 using human iPS cells demonstrates arrhythmogenic characteristics in cell culture. Dis Model Mech. 2012;5:220–30.

Lee P, Klos M, Bollensdorff C, Hou L, Ewart P, Kamp TJ, Zhang J, Bizy A, Guerrero-Serna G, Kohl P, Jalife J, Herron TJ. Simultaneous voltage and calcium mapping of genetically purified human induced pluripotent stem cell-derived cardiac myocyte monolayers. Circ Res. 2012;110: 1556–63.

Liang P, Lan F, Lee AS, Gong T, Sanchez-Freire V, Wang Y, Diecke S, Sallam K, Knowles JW, Wang PJ, Nguyen PK, Bers DM, Robbins RC, Wu JC. Drug screening using a library of human induced pluripotent stem cell-derived cardiomyocytes reveals disease-specific patterns of cardiotoxicity. Circulation. 2013;127:1677–91.

London B. Cardiac arrhythmias: from (transgenic) mice to men. J Cardiovasc Electrophysiol. 2001;12:1089–91.

Lopez-Izquierdo A, Warren M, Riedel M, Cho S, Lai S, Lux RL, Spitzer KW, Benjamin IJ, Tristani-Firouzi M, Jou CJ. A near-infrared fluorescent voltage-sensitive dye allows for moderate-throughput electrophysiological analyses of human induced pluripotent stem cell-derived cardiomyocytes. Am J Physiol Heart Circ Physiol. 2014;307:H1370–7.

Ma J, Guo L, Fiene SJ, Anson BD, Thomson JA, Kamp TJ, Kolaja KL, Swanson BJ, January CT. High purity human-induced pluripotent stem cell-derived cardiomyocytes: electrophysiological properties of action potentials and ionic currents. Am J Physiol Heart Circ Physiol. 2011;301:H2006–17.

Ma D, Wei H, Zhao Y, Lu J, Li G, Sahib NB, Tan TH, Wong KY, Shim W, Wong P, Cook SA, Liew R. Modeling type 3 long QT syndrome with cardiomyocytes derived from patient-specific induced pluripotent stem cells. Int J Cardiol. 2013;168:5277–86.

Ma D, Wei H, Lu J, Huang D, Liu Z, Loh LJ, Islam O, Liew R, Shim W, Cook SA. Characterization of a novel KCNQ1 mutation for type 1 long QT syndrome and assessment of the therapeutic potential of a novel IKs activator using patient-specific induced pluripotent stem cell-derived cardiomyocytes. Stem Cell Res Ther. 2015;6:39.

Malan D, Zhang M, Stallmeyer B, Müller J, Fleischmann BK, Schulze-Bahr E, Sasse P, Greber B. Human iPS cell model of type 3 long QT syndrome recapitulates drug-based phenotype correction. Basic Res Cardiol. 2016;111:14.

Matsa E, Rajamohan D, Dick E, Young L, Mellor I, Staniforth A, Denning C. Drug evaluation in cardiomyocytes derived from human induced pluripotent stem cells carrying a long QT syndrome type 2 mutation. Eur Heart J. 2011;32:952–62.

Matsa E, Dixon J, Medway C, Georgiou O, Patel M, Morgan K, Kemp P, Staniforth A, Mellor I, Denning C. Allele-specific RNA interference rescues the long-QT syndrome phenotype in human induced pluripotent stem cell cardiomyocytes. Eur Heart J. 2014;35:1078–87.

Matsa E, Burridge PW, Yu KH, Ahrens JH, Termglinchan V, Wu H, Liu C, Shukla P, Sayed N, Churko JM, Shao N, Woo NA, Chao AS, Gold JD, Karakikes I, Snyder MP, Wu JC. Transcriptome profiling of patient-specific human iPSC-cardiomyocytes predicts individual drug safety and efficacy responses in vitro. Cell Stem Cell. 2016;19:311–25.

Mehta A, Chung Y, Sequiera GL, Wong P, Liew R, Shim W. Pharmacoelectrophysiology of viral-free induced pluripotent stem cell-derived human cardiomyocytes. Toxicol Sci Off J Soc Toxicol. 2013;131:458–69.

Meijer van Putten RM, Mengarelli I, Guan K, Zegers JG, Van Ginneken AC, Verkerk AO, Wilders R. Ion channelopathies in human induced pluripotent stem cell derived cardiomyocytes: a dynamic clamp study with virtual IK1. Front Physiol. 2015;6:7.

Moretti A, Bellin M, Welling A, Jung CB, Lam JT, Bott-Flügel L, Dorn T, Goedel A, Höhnke C, Hofmann F, Seyfarth M, Sinnecker D, Schömig A, Laugwitz K-L. Patient-specific induced pluripotent stem-cell models for long-QT syndrome. N Engl J Med. 2010;363:1397–409.

Nerbonne JM, Nichols CG, Schwarz TL, Escande D. Genetic manipulation of cardiac K(+) channel function in mice: what have we learned, and where do we go from here? Circ Res. 2001;89:944–56.

Novak A, Barad L, Zeevi-Levin N, Shick R, Shtrichman R, Lorber A, Itskovitz-Eldor J, Binah O. Cardiomyocytes generated from CPVTD307H patients are arrhythmogenic in response to β-adrenergic stimulation. J Cell Mol Med. 2012;16:468–82.

Novak A, Barad L, Lorber A, Gherghiceanu M, Reiter I, Eisen B, Eldor L, Itskovitz-Eldor J, Eldar M, Arad M, Binah O. Functional abnormalities in iPSC-derived cardiomyocytes generated from CPVT1 and CPVT2 patients carrying ryanodine or calsequestrin mutations. J Cell Mol Med. 2015;19:2006–18.

Okita K, Ichisaka T, Yamanaka S. Generation of germline-competent induced pluripotent stem cells. Nature. 2007;448:313–7.

Okita K, Nakagawa M, Hyenjong H, Ichisaka T, Yamanaka S. Generation of mouse induced pluripotent stem cells without viral vectors. Science. 2008;322:949–53.

Orban M, Goedel A, Haas J, Sandrock-Lang K, Gärtner F, Jung CB, Zieger B, Parrotta E, Kurnik K, Sinnecker D, Wanner G, Laugwitz KL, Massberg S, Moretti A. Functional comparison of induced pluripotent stem cell- and blood-derived GPIIbIIIa deficient platelets. PLoS One. 2015;10:e0115978.

Protze SI, Liu J, Nussinovitch U, Ohana L, Backx PH, Gepstein L, Keller GM. Sinoatrial node cardiomyocytes derived from human pluripotent cells function as a biological pacemaker. Nat Biotechnol. 2017;35:56–68.

Rocchetti M, Sala L, Dreizehnter L, Crotti L, Sinnecker D, Mura M, Pane LS, Altomare C, Torre E, Mostacciuolo G, Severi S, Porta A, De Ferrari GM, George AL Jr, Schwartz PJ, Gnecchi M, Moretti A, Zaza A. Elucidating arrhythmogenic mechanisms of long-QT syndrome CALM1-F142L mutation in patient-specific induced pluripotent stem cell-derived cardiomyocytes. Cardiovasc Res. 2017;113:531–41.

Sager PT, Gintant G, Turner JR, Pettit S, Stockbridge N. Rechanneling the cardiac proarrhythmia safety paradigm: a meeting report from the cardiac safety research consortium. Am Heart J. 2014;167:292–300.

Schwartz PJ, Stramba-Badiale M, Crotti L, Pedrazzini M, Besana A, Bosi G, Gabbarini F, Goulene K, Insolia R, Mannarino S, Mosca F, Nespoli L, Rimini A, Rosati E, Salice P, Spazzolini C. Prevalence of the congenital long-QT syndrome. Circulation. 2009;120:1761–7.

Seki T, Yuasa S, Oda M, Egashira T, Yae K, Kusumoto D, Nakata H, Tohyama S, Hashimoto H, Kodaira M, Okada Y, Seimiya H, Fusaki N, Hasegawa M, Fukuda K. Generation of induced pluripotent stem cells from human terminally differentiated circulating T cells. Cell Stem Cell. 2010;7:11–4.

Shinnawi R, Huber I, Maizels L, Shaheen N, Gepstein A, Arbel G, Tijsen AJ, Gepstein L. Monitoring human-induced pluripotent stem cell-derived cardiomyocytes with genetically encoded calcium and voltage fluorescent reporters. Stem Cell Rep. 2015;5:582–96.

Sinnecker D, Laugwitz K-L, Moretti A. Induced pluripotent stem cell-derived cardiomyocytes for drug development and toxicity testing. Pharmacol Ther. 2014;143:246–52.

Song L, Awari DW, Han EY, Uche-Anya E, Park SH, Yabe YA, Chung WK, Yazawa M. Dual optical recordings for action potentials and calcium handling in induced pluripotent stem cell models of cardiac arrhythmias using genetically encoded fluorescent indicators. Stem Cells Transl Med. 2015;4:468–75.

Stadtfeld M, Nagaya M, Utikal J, Weir G, Hochedlinger K. Induced pluripotent stem cells generated without viral integration. Science. 2008;322:945–9.

Stillitano F, Hansen J, Kong CW, Karakikes I, Funck-Brentano C, Geng L, Scott S, Reynier S, Wu M, Valogne Y, Desseaux C, Salem JE, Jeziorowska D, Zahr N, Li R, Iyengar R, Hajjar RJ, Hulot JS. Modeling susceptibility to drug-induced long QT with a panel of subject-specific induced pluripotent stem cells. elife. 2017;6:e19406.

Takahashi K, Yamanaka S. Induction of pluripotent stem cells from mouse embryonic and adult fibroblast cultures by defined factors. Cell. 2006;126:663–76.

Takahashi K, Tanabe K, Ohnuki M, Narita M, Ichisaka T, Tomoda K, Yamanaka S. Induction of pluripotent stem cells from adult human fibroblasts by defined factors. Cell. 2007;131:861–72.

Terrenoire C, Wang K, Tung KWC, Chung WK, Pass RH, Lu JT, Jean JC, Omari A, Sampson KJ, Kotton DN, Keller G, Kass RS. Induced pluripotent stem cells used to reveal drug actions in a long QT syndrome family with complex genetics. J Gen Physiol. 2013;141:61–72.

Vaidyanathan G. Redefining clinical trials: the age of personalized medicine. Cell. 2012;148: 1079–80.

Vaidyanathan R, Markandeya YS, Kamp TJ, Makielski JC, January CT, Eckhardt LL. IK1-enhanced human-induced pluripotent stem cell-derived cardiomyocytes: an improved cardiomyocyte model to investigate inherited arrhythmia syndromes. Am J Physiol Heart Circ Physiol. 2016;310:H1611–21.

Warren L, Manos PD, Ahfeldt T, Loh YH, Li H, Lau F, Ebina W, Mandal PK, Smith ZD, Meissner A, Daley GQ, Brack AS, Collins JJ, Cowan C, Schlaeger TM, Rossi DJ. Highly efficient reprogramming to pluripotency and directed differentiation of human cells with synthetic modified mRNA. Cell Stem Cell. 2010;7:618–30.

Woltjen K, Michael IP, Mohseni P, Desai R, Mileikovsky M, Hämäläinen R, Cowling R, Wang W, Liu P, Gertsenstein M, Kaji K, Sung HK, Nagy A. piggyBac transposition reprograms fibroblasts to induced pluripotent stem cells. Nature. 2009;458:766–70.

Yazawa M, Hsueh B, Jia X, Pasca AM, Bernstein JA, Hallmayer J, Dolmetsch RE. Using induced pluripotent stem cells to investigate cardiac phenotypes in Timothy syndrome. Nature. 2011; 471:230–4.

Yu J, Vodyanik MA, Smuga-Otto K, Antosiewicz-Bourget J, Frane JL, Tian S, Nie J, Jonsdottir GA, Ruotti V, Stewart R, Slukvin II, Thomson JA. Induced pluripotent stem cell linesderived from human somatic cells. Science. 2007;318:1917–20.

Yu J, Hu K, Smuga-Otto K, Tian S, Stewart R, Slukvin II, Thomson JA. Human induced pluripotent stem cells free of vector and transgene sequences. Science. 2009;324:797–801.

Zaehres H, Schöler HR. Induction of pluripotency: from mouse to human. Cell. 2007;131(5):834–5.

Zhang J, Wilson GF, Soerens AG, Koonce CH, Yu J, Palecek SP, Thomson JA, Kamp TJ. Functional cardiomyocytes derived from human induced pluripotent stem cells. Circ Res. 2009;104:e30–41.

Zhang Q, Jiang J, Han P, Yuan Q, Zhang J, Zhang X, Xu Y, Cao H, Meng Q, Chen L, Tian T, Wang X, Li P, Hescheler J, Ji G, Ma Y. Direct differentiation of atrial and ventricular myocytes from human embryonic stem cells by alternating retinoid signals. Cell Res. 2011;21:579–87.

Zhang X-H, Haviland S, Wei H, Sarić T, Fatima A, Hescheler J, Cleemann L, Morad M. Ca2+ signaling in human induced pluripotent stem cell-derived cardiomyocytes (iPS-CM) from normal and catecholaminergic polymorphic ventricular tachycardia (CPVT)-afflicted subjects. Cell Calcium. 2013;54:57–70.

Zhang M, D'Aniello C, Verkerk AO, Wrobel E, Frank S, Ward-van Oostwaard D, Piccini I, Freund C, Rao J, Seebohm G, Atsma DE, Schulze-Bahr E, Mummery CL, Greber B, Bellin M. Recessive cardiac phenotypes in induced pluripotent stem cell models of Jervell and Lange-Nielsen syndrome: disease mechanisms and pharmacological rescue. Proc Natl Acad Sci U S A. 2014;111:E5383–92.

MIX
Papier aus verantwortungsvollen Quellen
Paper from responsible sources
FSC® C105338

If you have any concerns about our products,
you can contact us on
ProductSafety@springernature.com

In case Publisher is established outside the EU,
the EU authorized representative is:
Springer Nature Customer Service Center GmbH
Europaplatz 3, 69115 Heidelberg, Germany

Printed by Libri Plureos GmbH
in Hamburg, Germany